Masonry Structural Design

About the Author
Richard E. Klingner is the L. P. Gilvin Professor in Civil Engineering at The University of Texas in Austin, where he teaches graduate and undergraduate courses, including a course in masonry engineering. His research interests include the behavior and design of masonry structures for earthquake loads. He is active in the technical committee work of The Masonry Society, the American Concrete Institute, the American Society of Civil Engineers, and the American Society for Testing and Materials. From 2002 to 2008, he was Chair of the Masonry Standards Joint Committee.

About the International Code Council
The **International Code Council (ICC)** is a nonprofit membership association dedicated to protecting the health, safety, and welfare of people by creating better buildings and safer communities. The mission of ICC is to provide the highest quality codes, standards, products, and services for all concerned with the safety and performance of the built environment. ICC is the publisher of the family of the International Codes® (I-Codes®), a single set of comprehensive and coordinated national model codes. This unified approach to building codes enhances safety, efficiency, and affordability in the construction of buildings. The Code Council is also dedicated to innovation, sustainability, and energy efficiency. Code Council subsidiary, ICC Evaluation Service, issues Evaluation Reports for innovative products and Reports of Sustainable Attributes Verification and Evaluation (SAVE). *Headquarters:* 500 New Jersey Avenue, NW, 6th Floor, Washington, D.C. 20001-2070; District Offices: Birmingham, AL; Chicago, IL; Los Angeles, CA, 1-888-422-7233, www.iccsafe.org.

Masonry Structural Design

Richard E. Klingner

New York Chicago San Francisco
Lisbon London Madrid Mexico City
Milan New Delhi San Juan
Seoul Singapore Sydney Toronto

Library of Congress Cataloging-in-Publication Data

Klingner, R. E.
    Masonry structural design / Richard E. Klingner.
       p.   cm.
    ISBN 978-0-07-163830-2 (alk. paper)
    1. Masonry. 2. Buildings. 3. Structural analysis (Engineering). 4. Brick walls.
I. Title.
TH1199.K575   2010
693'.1—dc22
                                                                                                                             2009050596

McGraw-Hill books are available at special quantity discounts to use as premiums and sales promotions, or for use in corporate training programs. To contact a representative please e-mail us at bulksales@mcgraw-hill.com.

Masonry Structural Design

Copyright © 2010 by The McGraw-Hill Companies, Inc. All rights reserved. Printed in the United States of America. Except as permitted under the United States Copyright Act of 1976, no part of this publication may be reproduced or distributed in any form or by any means, or stored in a database or retrieval system, without the prior written permission of the publisher.

2 3 4 5 6 7 8 9 0   DOH   15 14 13 12 11

ISBN 978-0-07-163830-2
MHID 0-07-163830-X

The pages within this book were printed on acid-free paper.

Sponsoring Editor Joy Bramble	**Proofreader** Shivani Arora, Glyph International
Acquisitions Coordinator Michael Mulcahy	**Indexer** Robert Swanson
Editorial Supervisor David E. Fogarty	**Production Supervisor** Pamela A. Pelton
Project Manager Smita Rajan, Glyph International	**Composition** Glyph International
Copy Editor Bhavna Gupta, Glyph International	**Art Director, Cover** Jeff Weeks

Information contained in this work has been obtained by The McGraw-Hill Companies, Inc. ("McGraw-Hill") from sources believed to be reliable. However, neither McGraw-Hill nor its authors guarantee the accuracy or completeness of any information published herein, and neither McGraw-Hill nor its authors shall be responsible for any errors, omissions, or damages arising out of use of this information. This work is published with the understanding that McGraw-Hill and its authors are supplying information but are not attempting to render engineering or other professional services. If such services are required, the assistance of an appropriate professional should be sought.

This book is dedicated to Ann.

Contents

	Illustrations	xi
	Tables	xxi
	Preface	xxv
	Notice of Use of Copyrighted Material	xxvii
1	**Basic Structural Behavior and Design of Low-Rise, Bearing Wall Buildings**	**1**
	1.1 Basic Structural Behavior of Low-Rise, Bearing Wall Buildings	3
	1.2 Basic Structural Design of Low-Rise, Masonry Buildings	4
2	**Materials Used in Masonry Construction**	**7**
	2.1 Basic Components of Masonry	9
	2.2 Masonry Mortar	15
	2.3 Masonry Grout	25
	2.4 General Information on ASTM Specifications for Masonry Units	27
	2.5 Clay Masonry Units	29
	2.6 Concrete Masonry Units	35
	2.7 Properties of Masonry Assemblages	37
	2.8 Masonry Accessory Materials	38
	2.9 Design of Masonry Structures Requiring Little Structural Calculation	47
	2.10 How to Increase Resistance of Masonry to Water Penetration	54
3	**Code Basis for Structural Design of Masonry Buildings**	**57**
	3.1 Introduction to Building Codes in the United States	59
	3.2 Introduction to the Calculation of Design Loading Using the 2009 IBC	64
	3.3 Gravity Loads according to the 2009 IBC	64
	3.4 Wind Loading according to the 2009 IBC	67
	3.5 Earthquake Loading	87
	3.6 Loading Combinations of the 2009 IBC	103
	3.7 Summary of Strength Design Provisions of 2008 MSJC *Code*	105
	3.8 Summary of Allowable-Stress Design Provisions of 2008 MSJC *Code*	109
	3.9 Additional Information on Code Basis for Structural Design of Masonry Buildings	113

4 Introduction to MSJC Treatment of Structural Design 117
4.1 Basic Mechanical Behavior of Masonry 119
4.2 Classification of Masonry Elements 120
4.3 Classification of Masonry Elements by Structural Function ... 120
4.4 Classification of Masonry Elements by Design Intent 121
4.5 Design Approaches for Masonry Elements 121
4.6 How Reinforcement Is Used in Masonry Elements 122
4.7 How This Book Classifies Masonry Elements 127

5 Strength Design of Unreinforced Masonry Elements 131
5.1 Strength Design of Unreinforced Panel Walls 133
5.2 Strength Design of Unreinforced Bearing Walls 147
5.3 Strength Design of Unreinforced Shear Walls 161
5.4 Strength Design of Anchor Bolts 168
5.5 Required Details for Unreinforced Bearing Walls and Shear Walls 177

6 Strength Design of Reinforced Masonry Elements 181
6.1 Strength Design of Reinforced Beams and Lintels 183
6.2 Strength Design of Reinforced Curtain Walls 191
6.3 Strength Design of Reinforced Bearing Walls 196
6.4 Strength Design of Reinforced Shear Walls 212
6.5 Required Details for Reinforced Bearing Walls and Shear Walls 223

7 Allowable-Stress Design of Unreinforced Masonry Elements 227
7.1 Allowable-Stress Design of Unreinforced Panel Walls 229
7.2 Allowable-Stress Design of Unreinforced Bearing Walls 243
7.3 Allowable-Stress Design of Unreinforced Shear Walls 259
7.4 Allowable-Stress Design of Anchor Bolts 265
7.5 Required Details for Unreinforced Bearing Walls and Shear Walls 274

8 Allowable-Stress Design of Reinforced Masonry Elements 279
8.1 Review: Behavior of Cracked, Transformed Sections 281
8.2 Allowable-Stress Design of Reinforced Beams and Lintels ... 295
8.3 Allowable-Stress Design of Curtain Walls 300
8.4 Allowable-Stress Design of Reinforced Bearing Walls 306
8.5 Allowable-Stress Design of Reinforced Shear Walls 319
8.6 Required Details for Reinforced Bearing Walls and Shear Walls 324

9 Comparison of Design by the Allowable-Stress Approach versus the Strength Approach 329
9.1 Comparison of Allowable-Stress and Strength Design of Unreinforced Panel Walls 331
9.2 Comparison of Allowable-Stress Design and Strength Design of Unreinforced Bearing Walls 332

	9.3	Comparison of Allowable-Stress Design and Strength Design of Unreinforced Shear Walls	332
	9.4	Comparison of Allowable-Stress and Strength Designs for Anchor Bolts	333
	9.5	Comparison of Allowable-Stress and Strength Designs for Reinforced Beams and Lintels	334
	9.6	Comparison of Allowable-Stress and Strength Designs for Reinforced Curtain Walls	335
	9.7	Comparison of Allowable-Stress and Strength Designs for Reinforced Bearing Walls	336
	9.8	Comparison of Allowable-Stress and Strength Designs for Reinforced Shear Walls	336
10	**Lateral Load Analysis of Shear-Wall Structures**	339
	10.1	Classification of Horizontal Diaphragms as Rigid or Flexible	342
	10.2	Lateral Load Analysis of Shear-Wall Structures with Rigid Floor Diaphragms	343
	10.3	Lateral Load Analysis and Design of Shear-Wall Structures with Flexible Floor Diaphragms	362
	10.4	The Simplest of All Possible Analytical Worlds	364
11	**Design and Detailing of Floor and Roof Diaphragms**	367
	11.1	Introduction to Design of Diaphragms	369
	11.2	Typical Connection Details for Roof and Floor Diaphragms	372
12	**Strength Design Example: One-Story Building with Reinforced Concrete Masonry**	375
	12.1	Introduction	377
	12.2	Design Steps for One-Story Building	378
	12.3	Step 1: Choose Design Criteria	378
	12.4	Step 2: Design Walls for Gravity plus Out-of-Plane Loads	389
	12.5	Step 3: Design Lintels	406
	12.6	Summary So Far	409
	12.7	Step 4: Conduct Lateral Force Analysis, Design Roof Diaphragm	410
	12.8	Step 5: Design Wall Segments	413
	12.9	Step 6: Design and Detail Connections	415
13	**Strength Design Example: Four-Story Building with Clay Masonry**	417
	13.1	Introduction	419
	13.2	Design Steps for Four-Story Example	419
	13.3	Step 1: Choose Design Criteria, Specify Materials	420
	13.4	Step 2: Design Transverse Shear Walls for Gravity plus Earthquake Loads	428

	13.5 Step 3: Design Exterior Walls for Gravity plus Out-of-Plane Wind	435
	13.6 Overall Comments on Four-Story Building Example	435
14	**Structural Design of AAC Masonry**	**437**
	14.1 Introduction to Autoclaved Aerated Concrete (AAC)	439
	14.2 Applications of AAC	444
	14.3 Structural Design of AAC Elements	444
	14.4 Design of Unreinforced Panel Walls of AAC Masonry	449
	14.5 Design of Unreinforced Bearing Walls of AAC Masonry	452
	14.6 Design of Unreinforced Shear Walls of AAC Masonry	462
	14.7 Design of Reinforced Beams and Lintels of AAC Masonry	467
	14.8 Design of Reinforced Curtain Walls of AAC Masonry	473
	14.9 Design of Reinforced Bearing Walls of AAC Masonry	473
	14.10 Design of Reinforced Shear Walls of AAC Masonry	481
	14.11 Seismic Design of AAC Structures	493
	14.12 Design Example: Three-Story AAC Shear-Wall Hotel	495
	14.13 References on AAC	524
	14.14 Additional References on AAC	525
	References	**529**
	General References	531
	ASTM Standards	532
	Index	**535**

Illustrations

Figure 1.1: Basic structural behavior of low-rise, bearing-wall buildings. (p. 4)
Figure 1.2: Basic structural configuration of reinforced masonry walls. (p. 5)
Figure 1.3: Starting point for reinforcement. (p. 5)
Figure 2.1: Orientations of masonry units in an element. (p. 12)
Figure 2.2: Typical bond patterns in a wall. (p. 13)
Figure 2.3: Wall types, classified by mode of water-penetration resistance. (p. 14)
Figure 2.4: Weathering indices in the United States. (p. 33)
Figure 2.5: Typical application of deformed reinforcement in grouted masonry wall. (p. 40)
Figure 2.6: Typical bed joint reinforcement. (p. 40)
Figure 2.7: Typical use of welded wire reinforcement. (p. 41)
Figure 2.8: Typical use of posttensioning tendons. (p. 42)
Figure 2.9: Typical veneer ties. (p. 43)
Figure 2.10: Typical adjustable pintle ties. (p. 43)
Figure 2.11: Typical connectors. (p. 44)
Figure 2.12: Sample calculation for stiffness of adjustable ties. (p. 44)
Figure 2.13: Placement of flashing at shelf angles in clay masonry veneer. (p. 45)
Figure 2.14: Horizontally oriented expansion joint under shelf angle. (p. 46)
Figure 2.15: Vertically oriented expansion joint. (p. 46)
Figure 2.16: Shrinkage control joint. (p. 47)
Figure 2.17: Example of control joints at openings in concrete masonry. (p. 47)
Figure 2.18: Overall starting point for reinforcement or structures requiring little structural calculation. (p. 50)
Figure 2.19: Example of overall modularity of a masonry structure in plan. (p. 51)
Figure 2.20: Foundation wall detail. (p. 51)
Figure 2.21: Detail of intersection between wall and precast concrete roof or floor slab. (p. 52)
Figure 2.22: Detail of wall and wooden roof truss. (p. 52)
Figure 2.23: Elevations showing locations of control joints in CMU wythe, and locations of expansion joints in clay masonry veneer wythe (fixed lintel and loose lintel, respectively). (p. 53)
Figure 2.24: Wall sections at lintels. (p. 54)
Figure 3.1: Schematic of process for development of design codes in the United States. (p. 60)

Illustrations

Figure 3.2: Graph showing permitted live-load reduction for roofs. (p. 67)

Figure 3.3: Basic Wind speed—Western Gulf of Mexico Hurricane Coastline. (p. 69)

Figure 3.4: External pressure coefficients, GC_{pf} for main wind force-resisting system ($h < 60$ ft). (p. 72)

Figure 3.5: Internal pressure coefficients for buildings, GC_{pi}. (p. 75)

Figure 3.6: External pressure coefficients, C_p for main wind force-resisting systems (all heights). (p. 76)

Figure 3.7: External pressure coefficients, GC_p for components and cladding. (p. 77)

Figure 3.8: Schematic view of building in Austin, Texas. (p. 80)

Figure 3.9: Idealized single-degree-of-freedom system. (p. 88)

Figure 3.10: Acceleration response spectrum, smoothed for use in design. (p. 89)

Figure 3.11: Figures 22-1 of ASCE 7-05. Maximum considered earthquake ground motion for the conterminous United States of 0.2 sec spectral response acceleration (5% of critical damping), site class B. (p. 92)

Figure 3.12: Figures 22-2 of ASCE 7-05. Maximum considered earthquake ground motion for the conterminous United States of 1.0 sec spectral response acceleration (5% of critical damping), site class B. (p. 94)

Figure 3.13: Figures 22-9 of ASCE 7-05. Maximum considered earthquake ground motion for region 4 of 0.2 and 1.0 sec spectral response acceleration (5% of critical damping), site class B. (p. 96)

Figure 3.14: Design acceleration response spectrum for example problem. (p. 101)

Figure 4.1: Examples of reinforcement in CMU lintels. (p. 123)

Figure 4.2: Examples of reinforcement in clay masonry lintels. (p. 123)

Figure 4.3: Example of placement of reinforcement in a masonry wall made of hollow units. (p. 125)

Figure 4.4: Examples of the placement of hollow units to form pilasters. (p. 126)

Figure 5.1: Example of an unreinforced panel wall. (p. 134)

Figure 5.2: Horizontal section showing connection of a panel wall to a column. (p. 134)

Figure 5.3: Schematic representation of an unreinforced, two-wythe panel wall as two sets of horizontal and vertical crossing strips. (p. 134)

Figure 5.4: Example panel wall to be designed. (p. 137)

Figure 5.5: Idealized cross-sectional dimensions of a nominal $8 \times 8 \times 16$ in. concrete masonry unit. (p. 138)

Figure 5.6: Idealized cross-sectional dimensions of a nominal $8 \times 8 \times 16$ in. concrete masonry unit with face-shell bedding. (p. 139)

Figure 5.7: Idealized cross-sectional dimensions of a nominal $8 \times 8 \times 16$ in. concrete masonry unit with face-shell bedding. (p. 141)

Figure 5.8: Idealization of a panel wall as an assemblage of crossing strips. (p. 144)

Figure 5.9: Idealization of a two-wythe panel wall as an assemblage of two sets of crossing strips. (p. 146)

Figure 5.10: Idealization of bearing walls as vertically spanning strips. (p. 148)

Figure 5.11: Effect of slenderness on the axial capacity of a column or wall. (p. 148)

Figure 5.12: Unreinforced masonry bearing wall with concentric axial load. (p. 150)

Figure 5.13: Assumed linear variation of bearing stresses under the bearing plate. (p. 153)

Figure 5.14: Unreinforced masonry bearing wall with eccentric axial load. (p. 153)

Figure 5.15: Unreinforced masonry bearing wall with eccentric axial load and wind load. (p. 156)
Figure 5.16: Unfactored moment diagrams due to eccentric axial load and wind. (p. 158)
Figure 5.17: Hypothetical unstable resistance mechanism in a wall with openings, involving vertically spanning strips only. (p. 160)
Figure 5.18: Stable resistance mechanism in a wall with openings, involving horizontally spanning strips in addition to vertically spanning strips. (p. 160)
Figure 5.19: Basic behavior of box-type buildings in resisting lateral loads. (p. 161)
Figure 5.20: Design actions for unreinforced shear walls. (p. 162)
Figure 5.21: Example problem for strength design of unreinforced shear wall. (p. 163)
Figure 5.22: Calculation of reaction on roof diaphragm, strength design of unreinforced shear wall. (p. 164)
Figure 5.23: Transmission of forces from roof diaphragm to shear walls. (p. 165)
Figure 5.24: Shear wall with openings. (p. 167)
Figure 5.25: Free body of one wall segment. (p. 168)
Figure 5.26: Common uses of anchor bolts in masonry construction. (p. 168)
Figure 5.27: Idealized conical breakout cones for anchor bolts loaded in tension. (p. 169)
Figure 5.28: Modification of projected breakout area, A_{pt}, by void areas or adjacent anchors. (p. 170)
Figure 5.29: Example involving a single tensile anchor, placed vertically in a grouted cell. (p. 171)
Figure 5.30: (a) Pryout failure and (b) shear breakout failure. (p. 173)
Figure 5.31: Design idealization associated with shear breakout failure. (p. 174)
Figure 5.32: Example of wall-to-foundation connection. (p. 178)
Figure 5.33: Example of wall-to-floor connection, planks perpendicular to wall. (p. 179)
Figure 5.34: Example of wall-to-floor connection, planks parallel to wall. (p. 179)
Figure 5.35: Example of wall-to-roof detail. (p. 180)
Figure 5.36: Examples of wall-to-wall connection details. (p. 180)
Figure 6.1: Assumptions used in strength design of reinforced masonry for flexure. (p. 184)
Figure 6.2: Equilibrium of internal stresses and external nominal moment for strength design of reinforced masonry for flexure. (p. 184)
Figure 6.3: Conditions corresponding to balanced reinforcement percentage for strength design. (p. 185)
Figure 6.4: Example of masonry lintel. (p. 186)
Figure 6.5: Example for strength design of a lintel. (p. 188)
Figure 6.6: Example showing placement of bottom reinforcement in lowest course of lintel. (p. 189)
Figure 6.7: Plan view of typical curtain wall construction. (p. 191)
Figure 6.8: Examples of use of curtain walls of clay masonry. (p. 193)
Figure 6.9: Examples of the use of curtain walls (freestanding or cantilevered fire wall with pilaster) with concrete masonry. (p. 194)
Figure 6.10: Ten anchors holding the ends of curtain wall strips to columns. (p. 196)
Figure 6.11: Effective width of a reinforced masonry bearing wall. (p. 197)

Figure 6.12: Idealized moment-axial force interaction diagram using strength design. (p. 197)
Figure 6.13: Location of neutral axis under balanced conditions, strength design. (p. 198)
Figure 6.14: Moment-axial force interaction (strength basis), calculated by hand. (p. 201)
Figure 6.15: Position of neutral axis at balanced conditions, strength calculation of moment-axial force interaction diagram by spreadsheet. (p. 202)
Figure 6.16: Position of the neutral axis for axial loads less than the balance-point axial load, strength design. (p. 203)
Figure 6.17: Position of the neutral axis for axial loads greater than the balance-point axial load, strength design. (p. 203)
Figure 6.18: Moment-axial force interaction diagram (strength approach), spreadsheet calculation. (p. 205)
Figure 6.19: Reinforced masonry wall loaded by eccentric gravity axial load plus out-of-plane wind load. (p. 207)
Figure 6.20: Unfactored moment diagrams due to eccentric axial load plus wind load. (p. 208)
Figure 6.21: Critical strain condition for a masonry wall loaded out of plane. (p. 211)
Figure 6.22: Design actions for reinforced masonry shear walls. (p. 213)
Figure 6.23: V_{nm} as a function of $(M_u/V_u d_v)$. (p. 213)
Figure 6.24: Idealized model used in evaluating the resistance due to shear reinforcement. (p. 214)
Figure 6.25: Maximum permitted nominal shear capacity as a function of $(M_u/V_u d_v)$. (p. 214)
Figure 6.26: Reinforced masonry shear wall to be designed. (p. 216)
Figure 6.27: Unfactored in-plane lateral loads, shear and moment diagrams for reinforced masonry shear wall. (p. 216)
Figure 6.28: Moment-axial force interaction (strength basis) for reinforced shear wall, neglecting slenderness effects. (p. 217)
Figure 6.29: Critical strain condition for strength design of masonry walls loaded in-plane, and for columns and beams. (p. 220)
Figure 6.30: Example of wall-to-foundation connection. (p. 223)
Figure 6.31: Example of wall-to-floor connection, planks perpendicular to wall. (p. 224)
Figure 6.32: Example of wall-to-floor connection, planks parallel to wall. (p. 224)
Figure 6.33: Example of wall-to-roof detail. (p. 225)
Figure 6.34: Examples of wall-to-wall unbonded and bonded intersections connection details. (p. 225)
Figure 7.1: Example of an unreinforced panel wall. (p. 230)
Figure 7.2: Horizontal section showing connection of a panel wall to a column. (p. 230)
Figure 7.3: Schematic representation of an unreinforced, two-wythe panel wall as two sets of horizontal and vertical crossing strips. (p. 230)
Figure 7.4: Example panel wall to be designed. (p. 232)
Figure 7.5: Idealized cross-sectional dimensions of a nominal 8 × 8 × 16 in. concrete masonry unit. (p. 234)

Figure 7.6: Idealized cross-sectional dimensions of a nominal 8 × 8 × 16 in. concrete masonry unit with face-shell bedding. (p. 235)

Figure 7.7: Idealized cross-sectional dimensions of a nominal 8 × 8 × 16 in. concrete masonry unit with face-shell bedding. (p. 237)

Figure 7.8: Idealization of a panel wall as an assemblage of crossing strips. (p. 240)

Figure 7.9: Idealization of a two-wythe panel wall as an assemblage of two sets of crossing strips. (p. 241)

Figure 7.10: Idealization of bearing walls as vertically spanning strips. (p. 243)

Figure 7.11: Effect of slenderness on the axial capacity of a column or wall. (p. 244)

Figure 7.12: Unreinforced masonry bearing wall with concentric axial load. (p. 246)

Figure 7.13: Assumed linear variation of bearing stresses under the bearing plate. (p. 249)

Figure 7.14: Unreinforced masonry bearing wall with eccentric axial load. (p. 249)

Figure 7.15: Unreinforced masonry bearing wall with eccentric axial load and wind load. (p. 252)

Figure 7.16: Unfactored moment diagrams due to eccentric axial load and wind. (p. 253)

Figure 7.17: Hypothetical unstable resistance mechanism in a wall with openings, involving vertically spanning strips only. (p. 257)

Figure 7.18: Stable resistance mechanism in a wall with openings, involving horizontally spanning strips in addition to vertically spanning strips. (p. 258)

Figure 7.19: Basic behavior of box-type buildings in resisting lateral loads. (p. 259)

Figure 7.20: Design actions for unreinforced shear walls. (p. 260)

Figure 7.21: Example problem for strength design of unreinforced shear wall. (p. 261)

Figure 7.22: Calculation of reaction on roof diaphragm, strength design of unreinforced shear wall. (p. 262)

Figure 7.23: Transmission of forces from roof diaphragm to shear walls. (p. 262)

Figure 7.24: Shear wall with openings. (p. 265)

Figure 7.25: Free body of one wall segment. (p. 265)

Figure 7.26: Common uses of anchor bolts in masonry construction. (p. 266)

Figure 7.27: Idealized conical breakout cones for anchor bolts loaded in tension. (p. 266)

Figure 7.28: Modification of projected breakout area, A_{pt}, by void areas or adjacent anchors. (p. 267)

Figure 7.29: Example involving a single tensile anchor, placed vertically in a grouted cell. (p. 269)

Figure 7.30: (a) Pryout failure and (b) shear breakout failure. (p. 270)

Figure 7.31: Design idealization associated with shear breakout failure. (p. 271)

Figure 7.32: Example of wall-to-foundation connection. (p. 274)

Figure 7.33: Example of wall-to-floor connection, planks perpendicular to wall. (p. 275)

Figure 7.34: Example of wall-to-floor connection, planks parallel to wall. (p. 276)

Figure 7.35: Example of wall-to-roof detail. (p. 276)

Figure 7.36: Examples of wall-to-wall connection details. (p. 277)

Figure 8.1: States of strain and stress in a cracked masonry section. (p. 282)

Figure 8.2: Location of the neutral axis for particular cases. (p. 283)

Figure 8.3: Slice of a cracked, transformed section showing triangular compressive stress blocks. (p. 285)

Figure 8.4: Slice of a cracked, transformed section showing equilibrium between shear forces and difference in shear forces. (p. 285)

Figure 8.5: Slice of a cracked, transformed section showing equilibrium of the compressive and tensile portions of the slice. (p. 286)

Figure 8.6: Slice of a cracked, transformed section, showing equilibrium of difference in compressive force and difference in tensile force. (p. 287)

Figure 8.7: Tensile portion of a slice, showing equilibrium between bond force and difference in tensile force in reinforcement. (p. 288)

Figure 8.8: Example of calculation of the position of the neutral axis. (p. 289)

Figure 8.9: Equilibrium of forces in the cross-section corresponding to allowable stress in the reinforcement. (p. 291)

Figure 8.10: Equilibrium of forces in the cross-section corresponding to allowable stress in the masonry. (p. 292)

Figure 8.11: Conditions of stress and strain corresponding to allowable-stress balanced reinforcement. (p. 293)

Figure 8.12: Example of masonry lintel. (p. 295)

Figure 8.13: Example for allowable-stress design of a lintel. (p. 296)

Figure 8.14: Example showing placement of bottom reinforcement in lowest course of lintel. (p. 297)

Figure 8.15: Equilibrium of forces on cross-section. (p. 298)

Figure 8.16: Plan view of typical curtain wall construction. (p. 300)

Figure 8.17: Examples of use of curtain walls of clay masonry. (p. 302)

Figure 8.18: Examples of the use of curtain walls with concrete masonry. (p. 303)

Figure 8.19: Anchors holding the ends of curtain wall strips to columns. (p. 305)

Figure 8.20: Effective width of a reinforced masonry bearing wall. (p. 306)

Figure 8.21: Location of neutral axis under allowable-stress balanced conditions. (p. 308)

Figure 8.22: Plot of allowable-stress moment-axial force interaction diagram calculated by hand. (p. 311)

Figure 8.23: Conditions of strain and stress at allowable-stress balanced conditions. (p. 312)

Figure 8.24: Conditions of strain and stress for values of k less than the allowable-stress balanced value. (p. 313)

Figure 8.25: Conditions of strain and stress for values of k greater than the allowable-stress balanced value. (p. 314)

Figure 8.26: Plot of allowable-stress interaction calculated by spreadsheet. (p. 315)

Figure 8.27: Example of reinforced bearing wall loaded out-of-plane. (p. 317)

Figure 8.28: Unfactored moment diagrams due to eccentric axial dead load and wind. (p. 318)

Figure 8.29: Design actions for reinforced masonry shear walls. (p. 319)

Figure 8.30: Reinforced masonry shear wall to be designed. (p. 321)

Figure 8.31: Unfactored in-plane lateral loads, shear and moment diagrams for reinforced masonry shear wall. (p. 321)

Figure 8.32: Plot of allowable-stress moment-axial force interaction diagram calculated by spreadsheet. (p. 323)

Figure 8.33: Example of wall-to-foundation connection. (p. 325)

Figure 8.34: Example of wall-to-floor connection, planks perpendicular to wall. (p. 325)

Figure 8.35: Example of wall-to-floor connection, planks parallel to wall. (p. 326)
Figure 8.36: Example of wall-to-roof detail. (p. 326)
Figure 8.37: Examples of wall-to-wall connection details (unbonded and bonded intersections). (p. 327)
Figure 10.1: Example of building with perforated walls. (p. 342)
Figure 10.2: Solution to example problem using Method 1 (finite element method). (p. 345)
Figure 10.3: Shearing deformation of a wall segment. (p. 346)
Figure 10.4: Plan lengths of wall segments for example problem using simplest hand method (Method 2a). (p. 347)
Figure 10.5: Shears in wall 2 and wall 4 of example using the simplest hand method (Method 2a). (p. 348)
Figure 10.6: Shears in wall segments of wall 4 using the simplest hand method (Method 2a). (p. 348)
Figure 10.7: Examples of location of center of rigidity for symmetrical buildings. (p. 349)
Figure 10.8: Examples of location of center of rigidity for unsymmetrical buildings. (p. 349)
Figure 10.9: Decomposition of lateral load into a lateral load applied through the center of rigidity plus pure torsion about the center of rigidity. (p. 350)
Figure 10.10: Location of the center of rigidity in one direction. (p. 350)
Figure 10.11: Free-body diagram of diaphragm showing applied loads and reactions from shear walls. (p. 351)
Figure 10.12: Decomposition of lateral load into lateral load through center of rigidity plus torsion about the center of rigidity. (p. 351)
Figure 10.13: Lateral load applied through the center of rigidity. (p. 352)
Figure 10.14: Pure rotation in plan of a structure with lateral load applied through the center of rigidity. (p. 353)
Figure 10.15: Structure loaded by a combination of load through the center of rigidity plus plan torsion about the center of rigidity. (p. 355)
Figure 10.16: Application of rigid-diaphragm analysis to the structure considered in this section (Method 2b). (p. 355)
Figure 10.17: Location of center of rigidity for the example of this section (Method 2b). (p. 356)
Figure 10.18: Shear forces acting on walls due to direct shear and due to torsion (Method 2b). (p. 358)
Figure 10.19: Combined shear forces acting on walls of example structure (Method 2b). (p. 358)
Figure 10.20: Distribution of shears to segments of the east wall of example structure (Method 2b). (p. 358)
Figure 10.21: Example of perforated wall with segments of unequal height. (p. 359)
Figure 10.22: Plan view of example building with flexible roof diaphragm. (p. 363)
Figure 10.23: Results of example problem, assuming flexible diaphragm. (p. 363)
Figure 10.24: Example of a flexible horizontal diaphragm with more than two points of lateral support. (p. 364)
Figure 11.1: Example of a flexible diaphragm. (p. 370)
Figure 11.2: Example of design of a flexible diaphragm for shear and moment. (p. 371)

Figure 11.3: Example of computation of diaphragm chord forces. (p. 371)

Figure 11.4: Example of a connection detail between a CMU wall and steel joists. (p. 372)

Figure 11.5: Example of a connection detail between a CMU wall and wooden joists. (p. 373)

Figure 12.1: Plan of example one-story building. (p. 379)

Figure 12.2: Elevation of example one-story building. (p. 379)

Figure 12.3: Locations of control joints on north and south facades. (p. 379)

Figure 12.4: Locations of control joints on west facade. (p. 379)

Figure 12.5: Spacing of control joints on east facade. (p. 379)

Figure 12.6: Three-dimensional view of one-story building. (p. 380)

Figure 12.7: Assumed variation of bearing stresses under bearing plate. (p. 389)

Figure 12.8: Tributary area of typical bar joist on west wall. (p. 390)

Figure 12.9: West bearing wall of example one-story building. (p. 390)

Figure 12.10: Unfactored moment diagrams due to eccentric dead load and wind. (p. 392)

Figure 12.11: Design moment-axial force interaction diagram for west wall of example one-story building. (p. 394)

Figure 12.12: East wall of example one-story building. (p. 396)

Figure 12.13: Strength moment-axial force interaction diagram for solidly grouted 8-in. CMU wall loaded out-of-plane (#5 bars @ 48 in.). (p. 396)

Figure 12.14: Trial design Wall Segment B of east wall as governed by out-of-plane wind load. (p. 397)

Figure 12.15: Design moment-axial force interaction diagram for Wall Segment B of one-story example building. (p. 397)

Figure 12.16: Design of lintel on east wall for out-of-plane loads. (p. 399)

Figure 12.17: Placement of bar joists adjacent to north and south walls. (p. 400)

Figure 12.18: Cross section of typical pilaster in north and south walls of example one-story building. (p. 401)

Figure 12.19: Tributary area supported by typical pilaster. (p. 402)

Figure 12.20: Distribution of bearing stresses under bearing plates of pilasters. (p. 402)

Figure 12.21: Factored moment diagrams due to eccentric dead load and wind on pilasters. (p. 403)

Figure 12.22: Effective depth, d, of pilasters. (p. 404)

Figure 12.23: Strength moment-axial force interaction diagram for typical pilaster. (p. 404)

Figure 12.24: Bearing plate under long-span joists. (p. 406)

Figure 12.25: East facade of one-story building, showing critical 20-ft lintel. (p. 406)

Figure 12.26: Tributary area supported by bar joists bearing on lintel of east wall. (p. 407)

Figure 12.27: Section through 20-ft lintel of east wall. (p. 408)

Figure 12.28: Reinforcement in east wall of one-story building. (p. 409)

Figure 12.29: Wind load transmitted to roof diaphragm. (p. 410)

Figure 12.30: Plan view of one-story building showing wind loads transferred to roof diaphragm. (p. 411)

Figure 12.31: Assumed variation of shear and moment in each segment of east wall. (p. 414)

Figure 13.1: Plan view of typical floor of four-story example building. (p. 420)
Figure 13.2: Plan view of typical floor facade of four-story example building. (p. 421)
Figure 13.3: Design response spectrum for Charleston, South Carolina. (p. 424)
Figure 13.4: Factored design shears and moments for four-story example building. (p. 427)
Figure 13.5: Effective flange width used for each transverse shear wall. (p. 429)
Figure 13.6: Strength moment-axial force interaction diagram for transverse masonry shear wall. (p. 430)
Figure 14.1: Close-up view of AAC. (p. 440)
Figure 14.2: Examples of AAC elements. (p. 441)
Figure 14.3: Overall steps in manufacture of AAC. (p. 442)
Figure 14.4: AAC residence in Monterrey, Mexico. (p. 445)
Figure 14.5: AAC hotel in Tampico, Mexico. (p. 445)
Figure 14.6: AAC cladding, Monterrey, Mexico. (p. 446)
Figure 14.7: Integrated U.S. design background for AAC elements and structures. (p. 446)
Figure 14.8: Example panel wall to be designed using AAC masonry. (p. 450)
Figure 14.9: Unreinforced AAC masonry bearing wall with concentric axial load. (p. 453)
Figure 14.10: Assumed linear variation of bearing stresses under bearing plate of AAC masonry wall. (p. 456)
Figure 14.11: Unreinforced AAC masonry bearing wall with eccentric axial load. (p. 456)
Figure 14.12: Unreinforced masonry bearing wall with eccentric axial load and wind load. (p. 459)
Figure 14.13: Unfactored moment diagrams due to eccentric axial load and wind. (p. 460)
Figure 14.14: Design actions for unreinforced shear walls. (p. 463)
Figure 14.15: Example problem for strength design of unreinforced shear wall. (p. 463)
Figure 14.16: Calculation of reaction on roof diaphragm, strength design of unreinforced AAC masonry shear wall. (p. 464)
Figure 14.17: Transmission of forces from roof diaphragm to shear walls. (p. 465)
Figure 14.18: Example of masonry lintel. (p. 468)
Figure 14.19: Example for design of an AAC masonry lintel. (p. 469)
Figure 14.20: Example showing placement of bottom reinforcement in lowest course of lintel. (p. 470)
Figure 14.21: Moment-axial force interaction diagram (strength approach), spreadsheet calculation. (p. 473)
Figure 14.22: Reinforced masonry wall loaded by eccentric gravity axial load plus out-of-plane wind load. (p. 475)
Figure 14.23: Unfactored moment diagrams due to eccentric axial load plus wind load. (p. 476)
Figure 14.24: Critical strain condition for an AAC masonry wall loaded out-of-plane. (p. 479)
Figure 14.25: Design actions for reinforced AAC masonry shear walls. (p. 481)
Figure 14.26: Idealized model used in evaluating the resistance due to shear reinforcement. (p. 482)

Figure 14.27: Maximum permitted nominal shear capacity of AAC masonry as a function of $(M_u/V_u d_v)$. (p. 483)

Figure 14.28: Reinforced AAC masonry shear wall to be designed. (p. 484)

Figure 14.29: Unfactored in-plane lateral loads, shear, and moment diagrams for reinforced AAC masonry shear wall. (p. 484)

Figure 14.30: Moment-axial force interaction for reinforced AAC shear wall, neglecting slenderness effects. (p. 487)

Figure 14.31: Critical strain condition for design of AAC masonry walls loaded in-plane, and for columns and beams. (p. 490)

Figure 14.32: Plan of 3-story hotel example using AAC masonry. (p. 496)

Figure 14.33: Elevation of 3-story hotel, typical facade example using AAC masonry. (p. 496)

Figure 14.34: Figures 22-1 of ASCE 7-05. Maximum considered earthquake ground motion for the conterminous United States of 0.2 sec spectral response acceleration (5% of critical damping), site class B. (p. 500)

Figure 14.35: Figures 22-2 of ASCE 7-05. Maximum considered earthquake ground motion for the conterminous United States of 1.0 sec spectral response acceleration (5% of critical damping), site class B. (p. 502)

Figure 14.36: Design response spectrum for Asheville, North Carolina. (p. 509)

Figure 14.37: Graphs of factored design shears and moments for 3-story hotel example using AAC masonry. (p. 513)

Figure 14.38: Typical transverse shear wall of 3-story hotel example with AAC masonry. (p. 514)

Figure 14.39: Strength interaction diagram by spreadsheet, AAC transverse shear wall. (p. 516)

Figure 14.40: Plan view of section of exterior wall, 3-story example with AAC masonry. (p. 518)

Figure 14.41: Plan view of AAC floor diaphragm, 3-story hotel example with AAC masonry. (p. 519)

Figure 14.42: Sectional view of AAC floor diaphragm, 3-story hotel example with AAC masonry. (p. 520)

Figure 14.43: Section of panel-to-panel joint, AAC floor diaphragm. (p. 521)

Figure 14.44: Section of panel-to-bond beam joint, AAC floor diaphragm. (p. 521)

Figure 14.45: Truss model for design of AAC diaphragm. (p. 522)

Figure 14.46: Loaded nodes for design of AAC diaphragm. (p. 522)

Figure 14.47: Unloaded notes for design of AAC diaphragm. (p. 523)

Tables

Table 2.1: Classification of Masonry Units (p. 10)
Table 2.2: Approximate Proportion Requirements for Cement-Lime Mortar (p. 19)
Table 2.3: Property Requirements for Cement-Lime Mortar (p. 20)
Table 2.4: Approximate Proportion Requirements for Masonry-Cement Mortar (p. 21)
Table 2.5: Property Requirements for Masonry-Cement Mortar (p. 21)
Table 2.6 : Approximate Proportion Requirements for Mortar-Cement Mortar (p. 22)
Table 2.7: Property Requirements for Mortar-Cement Mortar (p. 22)
Table 2.8: Proportion Requirements for Grout for Masonry (p. 26)
Table 2.9: Summary of ASTM Requirements for Clay Masonry Units (p. 34)
Table 3.1: Minimum Live Loads (L) for Floors (p. 65)
Table 3.2: Table 1607.9.1 Live Load Element Factor, K_{LL} (p. 66)
Table 3.3: Conversion of Wind Speeds from 3-s Gust to Fastest Mile (p. 70)
Table 3.4: Wind Directionality Factor, K_d (p. 70)
Table 3.5: Velocity Pressure Exposure Coefficients, K_h and K_z (p. 71)
Table 3.6: Velocity Pressure Coefficients for Building of Sec 3.4.2 (p. 80)
Table 3.7: Spreadsheet for Wind Forces, Sec. 3.4.2 (p. 84)
Table 3.8: Velocity Pressure Exposure Coefficients for Building of Sec. 3.4.3 (p. 85)
Table 3.9: Spreadsheet for Components and Cladding Pressures, Windward Side of Sec. 3.4.2 (p. 86)
Table 3.10: Spreadsheet for Components and Cladding Pressures, Leeward Side of Sec. 3.4.2 (p. 87)
Table 3-11: Table 20.3-1 Site Classification (p. 97)
Table 3.12: Table 11.4-1 Site Classification (p. 97)
Table 3.13: Table 11.4-2 Site Classification, F_v (p. 98)
Table 3.14: Table 11.5-1 Importance Factors (p. 101)
Table 3.15: Table 11.6-1 Seismic Design Category Based on Short Period Response Acceleration Parameter (p. 101)
Table 3.16: Table 11.6-2 Seismic Design Category Based on 1-s Period Response Acceleration Parameter (p. 102)
Table 3.17: Strength-Reduction Factors (p. 106)
Table 3.18: Summary of Steps for Strength Design of Unreinforced Panel Walls (p. 106)
Table 3.19: Summary of Steps for Strength Design of Unreinforced Bearing Walls (p. 107)

Table 3.20: Summary of Steps for Strength Design of Unreinforced Shear Walls (p. 107)
Table 3.21: Summary of Steps for Strength Design of Reinforced Beams and Lintels (p. 108)
Table 3.22: Summary of Steps for Strength Design of Reinforced Curtain Walls (p. 108)
Table 3.23: Summary of Steps for Strength Design of Reinforced Bearing Walls (p. 109)
Table 3.24: Summary of Steps for Strength Design of Reinforced Shear Walls (p. 109)
Table 3.25: Summary of Steps for Allowable-Stress Design of Unreinforced Panel Walls (p. 110)
Table 3.26: Summary of Steps for Allowable-Stress Design of Unreinforced Bearing Walls (p. 111)
Table 3.27: Summary of Steps for Allowable-Stress Design of Unreinforced Shear Walls (p. 111)
Table 3.28: Summary of Steps for Allowable-Stress Design of Reinforced Beams and Lintels (p. 111)
Table 3.29: Summary of Steps for Allowable-Stress Design of Reinforced Curtain Walls (p. 112)
Table 3.30: Summary of Steps for Allowable-Stress Design of Reinforced Bearing Walls (p. 112)
Table 3.31: Summary of Steps for Strength Design of Reinforced Shear Walls (p. 113)
Table 5.1: Modulus of Rupture (p. 136)
Table 5.2: Section Properties for Masonry Walls (p. 143)
Table 5.3: Section Properties for Masonry Units (p. 150)
Table 5.4: Weights of Hollow CMU Walls (p. 150)
Table 6.1: Physical Properties of Steel Reinforcing Wire and Bars (p. 187)
Table 6.2: Spreadsheet for Computing Moment-Axial Force Interaction Diagram (Strength Approach) (p. 206)
Table 7.1: Allowable Flexural Tension for Clay and Concrete Masonry, psi (p. 231)
Table 7.2: Section Properties for Masonry Walls (p. 239)
Table 7.3: Section Properties for Masonry Units (p. 246)
Table 7.4: Weights of Hollow CMU Walls (p. 247)
Table 8.1: Physical Properties of Steel Reinforcing Wire and Bars (p. 289)
Table 8.2: Spreadsheet for Calculating Allowable-Stress Interaction Diagram for Wall Loaded Out-of-Plane (p. 316)
Table 9.1: Comparison of Allowable-Stress and Strength Design for Unreinforced Panel Walls (p. 332)
Table 9.2: Comparison of Allowable-Stress and Strength Design for Unreinforced Shear Walls (p. 333)
Table 9.3: Comparison of Allowable-Stress and Strength Design for Anchor Bolts, Masonry Controls (p. 334)
Table 9.4: Comparison of Allowable-Stress and Strength Design for Anchor Bolts, Steel Controls (p. 334)
Table 9.5: Comparison of Allowable-Stress and Strength Design for Reinforced Beams and Lintels (p. 335)
Table 9.6: Comparison of Allowable-Stress and Strength Design for Reinforced Beams and Lintels (p. 335)
Table 9.7: Comparison of Allowable-Stress and Strength Design for Reinforced Bearing Walls (p. 336)

Table 9.8: Comparison of Allowable-Stress and Strength Design for Reinforced Bearing Walls (p. 337)

Table 9.9: Comparison of Allowable-Stress and Strength Design for Reinforced Shear Walls (p. 337)

Table 10.1: Comparison of Results Obtained for Calculating Shear-Wall Forces by Each Method (p. 360)

Table 10.2: Comparison of Results Obtained for Calculating Shear-Wall Forces by Each Method (p. 360)

Table 12.1: Velocity Pressure Exposure Coefficients for One-Story Example Building (p. 381)

Table 12.2: Spreadsheet for Computation of Base Shear and Long-Span Joist Reactions for Example One-Story Building (MWFRS) (p. 383)

Table 12.3: Velocity Pressure Exposure Coefficients for One-Story Example Building (p. 384)

Table 12.4: Spreadsheet for Calculation of Wind Pressure on Windward Side of One-Story Example Building (Components and Cladding) (p. 386)

Table 12.5: Spreadsheet for Calculation of Wind Pressure on Leeward Side of One-Story Example Building (Components and Cladding) (p. 387)

Table 12.6: Spreadsheet for Calculation of Wind Pressure on Roof of One-Story Example Building (Components and Cladding) (p. 388)

Table 12.7: Spreadsheet for Calculating Strength Moment-Axial Force Interaction for West Wall of One-Story Building (p. 395)

Table 12.8: Spreadsheet for Calculating Moment-Axial Force Interaction Diagram for Wall Segment B of One-Story Example Building (p. 398)

Table 12.9: Spreadsheet for Calculating Strength Moment-Axial Force Interaction Diagram for Typical Pilaster (p. 405)

Table 12.10: Factored Gravity Loads Acting on 20-ft Lintel of East Wall (p. 407)

Table 12.11: Design Shear in Each Segment of East Wall due to Design Wind Load (p. 413)

Table 13.1: Importance Factors (p. 424)

Table 13.2: Seismic Design Category Based on 1-s Period Response Acceleration Parameter (p. 424)

Table 13.3: Factored Design Lateral Forces for Four-Story Example Building (p. 427)

Table 13.4: Spreadsheet for Calculating Strength Moment-Axial Force Interaction Diagram for Transverse Shear Wall of Four-Story Building Example (p. 432)

Table 14.1: Typical Material Characteristics of AAC in Different Strength Classes (p. 443)

Table 14.2: Dimensions of Plain AAC Wall Units (p. 444)

Table 14.3: Dimensions of Reinforced AAC Wall Units (p. 444)

Table 14.4: Physical Properties of Steel Reinforcing Bars (p. 469)

Table 14.5: Spreadsheet for Computing Moment-Axial Force Interaction Diagram for AAC Bearing Wall (p. 474)

Table 14.6: Seismic Design Factors for Ordinary Reinforced AAC Masonry Shear Walls (p. 494)

Table 14.7: Site Classification (p. 505)

Table 14.8: Site Coefficient, F_a (p. 506)

Table 14.9: Site Coefficient, F_v (p. 506)

Table 14.10: Importance Factors (p. 509)
Table 14.11: Seismic Design Category based on Short Period Response Acceleration Parameter (p. 510)
Table 14.12: Seismic Design Category based on 1-s Period Response Acceleration Parameter (p. 510)
Table 14.13: Factored Design Shears and Moments for 3-Story Hotel Example Using AAC Masonry (p. 513)

Preface

This book was developed from a set of masonry course notes used for many years in a semester-long, undergraduate and graduate course in masonry engineering at The University of Texas at Austin. It covers the design of masonry structures using the 2009 *International Building Code*, the 2008 Masonry Standards Joint Committee *Code* and *Specification*, and other documents referenced by those standards.

The book is intended for an undergraduate or a graduate course in masonry as part of a civil engineering or architectural engineering curriculum. It can also be used for self-study and continuing education by practicing engineers. It emphasizes the strength design of masonry, and also includes allowable-stress design.

The first part of this book (Chaps. 1 and 2) begins, not with design calculations, but rather with a basic discussion of how box-type buildings behave, and how these buildings can be detailed and specified using masonry. The reason for this is that until the reader understands how the elements of a masonry building work together structurally, the design of those individual elements will not have a clear purpose. Many classes of masonry buildings require only the most rudimentary structural design, and the first part of this book is intended to show how to specify and detail such buildings correctly.

The next part of this book (Chaps. 3 and 4) shows where our structural design provisions for masonry come from—the relationship between the masonry design provisions developed by the Masonry Standards Joint Committee, loading and overall design documents such as ASCE 7, material specifications such as those of ASTM, and model codes such as the *International Building Code*. In particular, it discusses how different types and configurations of masonry elements are addressed by that code framework. It also gives detailed examples of the derivation of design wind and seismic loads according to the 2009 *International Building Code*, and provides summaries of the steps involved in the design of masonry elements by the strength approach and the allowable-stress approach.

The next part of this book (Chaps. 5 through 9) is a discussion of the design of masonry elements, first using the strength design provisions of the 2008

MSJC *Code* (in the context of ASCE 7-05 and the 2009 *International Building Code*), and then using the allowable-stress provisions. Designs begin with unreinforced masonry elements and continue with reinforced masonry elements.

The next part of this book (Chaps. 10 and 11) addresses the analysis of low-rise, wall-type buildings for lateral loads, and in particular the calculation of design shears and moments in the shear walls of such buildings. It also discusses the role of horizontal diaphragms, and their design for shear and bending moment.

The next part of this book (Chaps. 12 and 13) consists of two overall building design examples, carried out using the strength design provisions of the 2008 MSJC *Code* (in the context of ASCE 7-05 and the 2009 *International Building Code*). The first building is a low-rise commercial building, designed for gravity and wind loads; the second is a four-story hotel, designed for gravity and earthquake loads.

The last part of this book (Chap. 14) addresses autoclaved aerated concrete masonry (AAC), an innovative construction material recently introduced into the MSJC *Code* and *Specification*, and into the *International Building Code*. Background on AAC masonry is reviewed; design examples are presented; and a complete building design example is presented. The design example is a three-story hotel, designed for gravity and earthquake loads.

Notice of Use of Copyrighted Material

Portions of this publication reproduce sections from ASCE 7-05 (Supplement), published by the American Society of Civil Engineers, Reston, VA. Reproduced with permission. All rights reserved.

Portions of this publication reproduce sections from ASTM standards, published by the American Society for Testing and Materials, West Conshohocken, PA. Reproduced with permission. All rights reserved.

Portions of this publication reproduce material supplied by the Brick Industry Association, Reston, Virginia. Reproduced with permission. All rights reserved.

Portions of this publication reproduce tables/illustrations from the 2009 *International Building Code*, published by the International Code Council, Inc., Washington, D.C. Reproduced with permission. All rights reserved.

Portions of this publication reproduce material supplied by the National Concrete Masonry Association, Herndon, Virginia. Reproduced with permission. All rights reserved.

Portions of this publication reproduce material supplied by Xella Mexicana, Monterrey, Mexico. Reproduced with permission. All rights reserved.

CHAPTER 1
Basic Structural Behavior and Design of Low-Rise, Bearing Wall Buildings

1.1 Basic Structural Behavior of Low-Rise, Bearing Wall Buildings

This book does not start with the design of masonry elements. Rather, it starts with the behavior of low-rise, bearing wall buildings. The reason for this is that the behavior of masonry structural elements, and the design requirements for those elements, depends on the behavior of the structures comprising those elements.

Low-rise, bearing wall buildings resist lateral loads as shown in Fig. 1.1. This resistance mechanism involves three steps:

- Walls oriented perpendicular to the direction of lateral load transfer those loads to the level of the foundation and the levels of the horizontal diaphragms. The walls are idealized and designed as vertically oriented strips.
- The roof and floors act as horizontal diaphragms, transferring their forces to walls oriented parallel to the direction of lateral load.
- Walls oriented parallel to the direction of applied load must transfer loads from the horizontal diaphragms to the foundation. In other words, they act as shear walls.

This overall mechanism demands that the horizontal roof diaphragm have sufficient strength and stiffness to transfer the required loads.

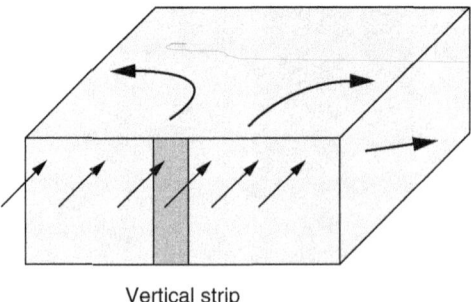

FIGURE 1.1 Basic structural behavior of low-rise, bearing-wall buildings.

The addition of vertical loads from sources other than self-weight places the vertical strips in compression, and makes the walls bearing walls.

1.2 Basic Structural Design of Low-Rise, Masonry Buildings

The fundamental design premise of low-rise, masonry buildings is that they are composed of masonry walls only.

There are no embedded steel or concrete frame elements. That's right. None.

Low-rise, bearing wall masonry buildings are designed for gravity loads and lateral loads. Lateral loads from wind are presumed to act separately in each principal plan direction. Depending on the direction in which they act, walls can be bearing walls or shear walls.

1.2.1 Basic Structural Configuration

Masonry walls are generally composed of hollow masonry units, held together by mortar. Vertical reinforcement is placed in continuous vertical cells, and horizontal reinforcement is placed in horizontal courses (bond beams). Cells with reinforcement, and bond beams, and possibly other cells as well, are filled with grout (a fluid concrete mixture). A typical arrangement for the case of hollow concrete masonry units is shown in Fig. 1.2. Construction with hollow clay masonry units would be quite similar to construction with hollow concrete masonry units.

Because the walls are reinforced, design is straightforward and reasonably familiar even to those with little or no experience in masonry design.

- Vertical strips resisting combinations of gravity loads and out-of-plane loads act as reinforced beam-columns. Behavior of reinforced masonry is quite similar to that of reinforced concrete, and can be

Basic Structural Behavior and Design of Low-Rise, Bearing Wall Buildings

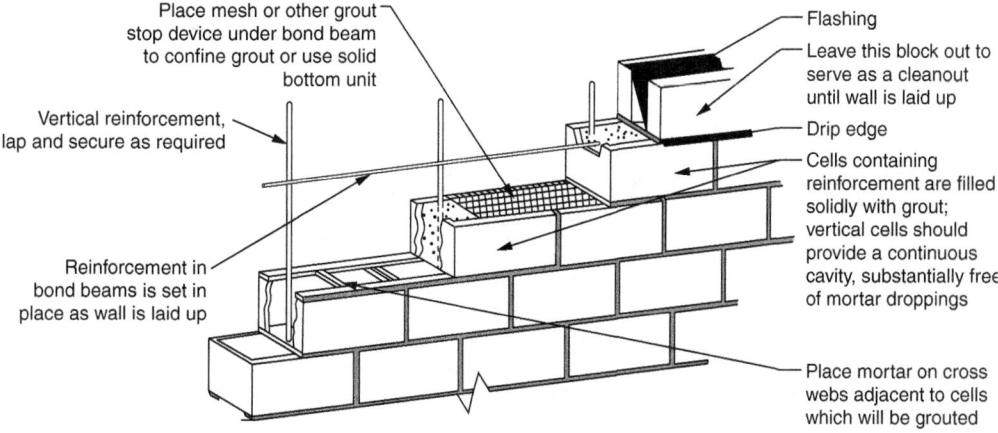

FIGURE 1.2 Basic structural configuration of reinforced masonry walls. (*Source*: Figure 1 of NCMA TEK 3-2A.)

described by a moment-axial force interaction diagram for out-of-plane bending.
- Shear walls act in flexure as cantilever beam-columns. Behavior of reinforced masonry is quite similar to that of reinforced concrete, and can be described by a moment-axial force interaction diagram for in-plane bending.

1.2.2 Overall Starting Point for Reinforcement

The overall starting point for reinforcement is shown in Fig. 1.3.

Structural design is carried out using the strength provisions of the Masonry Standards Joint Committee Code and Specification (MSJC, 2008a,b), because this document is referenced by most model codes, and its strength provisions can be easily learned by designers familiar with strength provisions for reinforced concrete.

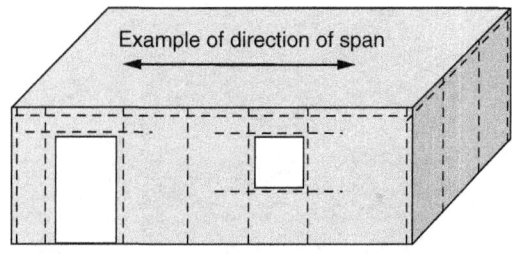

Vertical reinforcement consisting of #4 bars at corners, jambs, and intervals of about 4 ft

Horizontal reinforcement consisting of 2 #4 bars in bond beams, and above and below openings

Increase horizontal reinforcement to 2 #5 bars over openings with spans > 6 ft

FIGURE 1.3 Starting point for reinforcement.

… CHAPTER 2

Materials Used in Masonry Construction

2.1 Basic Components of Masonry

Masonry can be used in a wide variety of architectural applications, including

- Walls (bearing, shear, structural, decorative, bas-relief, mosaic)
- Arches, domes, and vaults
- Beams and columns

Masonry, while often simple and elegant in form, can be complex in behavior. Also, unlike concrete, it cannot be ordered by the cubic yard. To understand its behavior, and to be able to specify masonry correctly, we must examine each of its basic components: units; mortar; grout; and accessory materials.

Immediately below, each component (and related concepts) is discussed briefly. In later sections, more details are provided.

2.1.1 Preliminary Discussion of Masonry Units

Masonry units, as noted in Table 2.1, can be classified in a wide variety of ways.

In this book, we shall emphasize the behavior and use of structural masonry units, of fired clay or of concrete.

Unfired clay masonry units	Adobe
Fired clay masonry units	Roofing tile
	Drain tile
	Refractory brick
	Wall tile
	Glazed facing tile (terra-cotta, ceramic veneer)
	Structural clay products
	Structural tile
	Facing tile
	Glazed
	Textured
	Floor tile
	Brick (solid, frogged, cored, hollow)
	Facing and building brick
	Glazed brick
	Floor and paving brick
	Industrial
	Paving
	Patio
	Chemical resistant brick
	Sewer brick
	Chimney lining brick
Concrete masonry units	Concrete block (solid, hollow)
Other masonry units	Glass
	Stone (artificial shape)
	Rock (natural shape)

TABLE 2.1 Classification of Masonry Units

2.1.2 Preliminary Discussion of Masonry Mortar

In the United States, three basic cementitious systems are used for mortar: cement-lime mortar; masonry-cement mortar; and mortar-cement mortar. The first two are widely used, the third has been recently introduced.

Cement-lime mortar is made from different proportions of portland cement or other cements, hydrated masons' lime, and masonry sand, mixed with water. It can be batched by hand on-site using material from bags, or batched automatically on-site using material from silos.

Masonry-cement mortar is made from different proportions of masonry cement and sand, mixed with water. It may also contain additional portland cement or other cements. Masonry-cement formulations and manufacturing processes are manufacturer-specific. Ingredients are not required to be identified, and usually are not. Masonry cement generally consists of portland cement, pozzolanic cement, or slag cement, plasticizing additives, air-entraining additives, water-retention additives, and finely ground limestone (added primarily as a filler, but with some plasticizing and cementitious effect).

Mortar-cement mortar is made from different proportions of mortar cement and sand, mixed with water. It may also contain additional portland cement or other cements. Mortar cement formulations and manufacturing processes are manufacturer-specific. Ingredients are not required to be identified, and usually are not. Mortar cement generally consists of portland cement, pozzolanic cement, or slag cement, plasticizing additives, air-entraining additives, water-retention additives, and finely ground limestone (added primarily as a filler, but with some plasticizing and cementitious effect). It differs from masonry cement in that it is formulated specifically for tensile bond strength comparable to that of cement-lime mortar.

2.1.3 Preliminary Discussion of Masonry Grout

Grout is fluid concrete, usually with pea-gravel aggregate. It can be used to fill some or all cells in hollow units, or between wythes.

2.1.4 Preliminary Discussion of Masonry Accessory Materials

Masonry accessory materials include reinforcement, connectors, sealants, flashing, coatings, and vapor barriers.

- Connectors (of galvanized or stainless steel) connect a masonry wall to another wall, or a masonry wall to a frame, or a masonry wall to something else.
- Sealants are used in expansion joints (clay masonry), control joints (concrete masonry), and construction joints.
- Flashing is a flexible waterproof membrane used for drainage walls.
- Coatings include paints and clear water-repellent coatings.
- Moisture barriers and vapor barriers are used as parts of wall systems to retard the passage of water in liquid form and vapor form, respectively.

2.1.5 Preliminary Discussion of Masonry Dimensions

Masonry unit dimensions are typically described in terms of thickness × height × length. Typically, the length is the largest dimension, the thickness is next, and the height is the smallest dimension.

For example, a typical clay masonry unit has dimensions of 4 × 2.67 × 8 in. These are *nominal dimensions*, that is, the distances occupied by the unit plus one-half a joint width on each side. Joints are normally 3/8-in. thick. The *specified dimensions* of the unit themselves are smaller; in this case, 3-5/8 × 2-1/4 × 7-5/8 in. The *actual dimensions* are the measured size, and should fall within the specified dimensions, plus or minus the tolerance permitted by the governing material specification.

The sides of a masonry unit are often designated in literature by special names

- The bed is the side formed by thickness × length
- The face is the side formed by height × length
- The head is the side formed by thickness × height

2.1.6 Orientation of Masonry Units in an Element

Masonry units can be placed in a wall or other element in many orientations, as shown in Fig. 2.1 (looking perpendicular to the plane of the element). The stretcher orientation is the most common, and the soldier orientation is often used above or below wall openings.

2.1.7 Bond Patterns

Masonry units can be placed in a wall or other element in many bond patterns (arrangements), as shown in Fig. 2.2 (again, looking perpendicular to the plane of the element). In this figure, the horizontal joints are referred to as bed joints, and the vertical joints are referred to head joints.

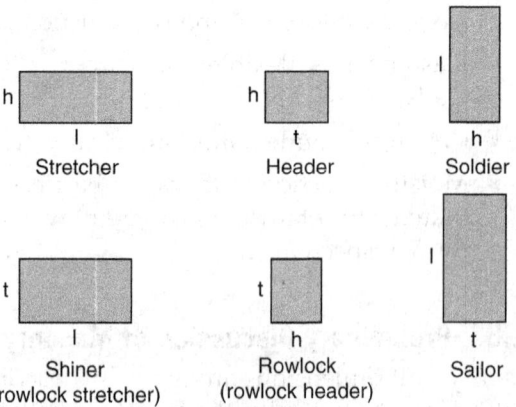

FIGURE 2.1 Orientations of masonry units in an element.

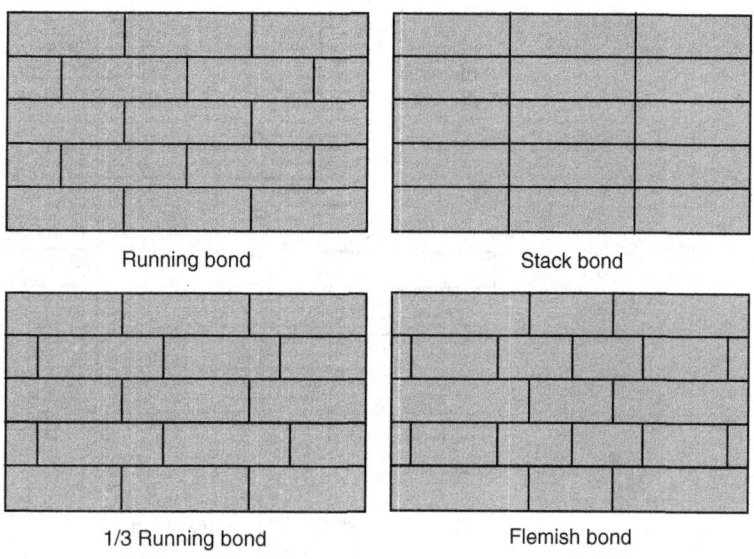

FIGURE 2.2 Typical bond patterns in a wall.

In all the bond patterns of this figure, the bed joints are continuous along every course (level) of masonry.

- In running bond, the head joints align in alternate courses, and are aligned with the middle of the units in adjacent courses.
- In stack bond, the head joints align in adjacent courses.
- In 1/3 running bond, the head joints align in alternate courses, and are aligned one-third of the way along the units in adjacent courses.
- In Flemish bond, units of two different lengths are used.

Many other bond patterns are possible.

2.1.8 Types of Walls

Masonry is most commonly used in walls. Masonry walls can be classified in many different ways. For now, we shall classify masonry walls according to how they resist water penetration. Using this criterion, masonry walls can be classified as barrier walls or drainage walls. Examples of each type are shown in Fig. 2.3.

- Barrier walls resist water penetration primarily by virtue of their thickness. They may have coatings. They may be single-wythe (one thickness of masonry), or multiwythe. Multiwythe barrier

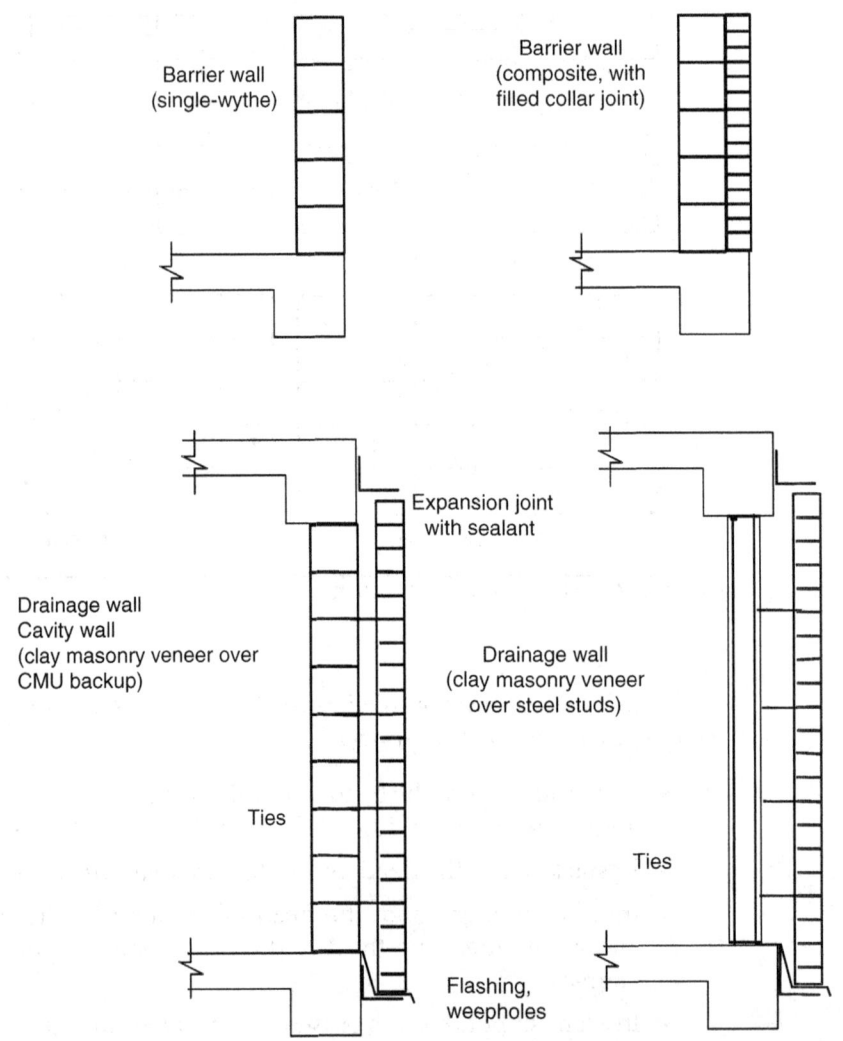

FIGURE 2.3 Wall types, classified by mode of water-penetration resistance.

walls can have the wythes connected by bonded headers [masonry units placed in header orientation, or by a filled collar joint (space between wythes)].

- Drainage walls resist water penetration by a combination of thickness and drainage details. Drainage walls may also have coatings. Drainage details include an airspace (at least 2 in. wide), flashing, and weepholes. Drainage walls can be single-wythe (veneer over steel studs) or multiwythe (veneer over masonry backup). The latter are often also called "cavity walls."

2.1.9 Overview of How Masonry Is Specified

To further discuss how masonry is specified, it is necessary to recognize the following:

1. Unlike the steel or concrete industries, individual segments of the masonry industry rarely produce a finished masonry assembly.
2. It is therefore necessary to specify precisely each component of masonry (units, mortar, grout, accessory materials). In the United States, this is done through standard specifications, methods of sampling and testing, test methods, and practices. Most applicable standards are developed by the American Society for Testing and Materials (ASTM). A few are produced by model code organizations (e.g., the International Code Council). We will focus on ASTM standards, using them as a frame of reference for the specification of masonry elements.

2.2 Masonry Mortar

Masonry mortar holds masonry units together, and also holds them apart (compensates for their dimensional tolerances). Mortar for unit masonry is addressed by ASTM C270, which in turn cites other ASTM specifications.

Specifying masonry mortar under ASTM C270 requires three choices:

- The designer must choose a cementitious system. Three options are possible: cement-lime mortar, masonry-cement mortar, or mortar-cement mortar.
- The designer must choose a mortar type, basically related to the proportion of cement in the mortar.
- The designer must choose whether ASTM C270 will be enforced by proportions of the different ingredients, or by the properties of the final mortar. The proportion specification is the default, and is assumed to govern if the designer does not state otherwise.

Each choice is discussed in more detail in Sec. 2.2.5. For now, to help explain the background and significance of these choices, it is useful to discuss the chemistry of masonry mortar.

2.2.1 Introduction to the Chemistry of Masonry Mortar

Masonry mortars can be broadly classified as sand-lime mortars and hydraulic mortars. The former harden (set) only in the presence of air. The latter can harden under water.

Chemistry of Sand-Lime Mortar

Since the time of the Romans, masonry mortar has been made from a mixture of lime and sand. Limestone is first calcined (heated) to produce quicklime (calcium oxide). The chemical formula for this reaction, and its corresponding verbal explanation, is shown below:

Limestone + heat = calcium oxide + carbon dioxide
(quicklime)

$$CaCO_3 = CaO + CO_2$$

To form mortar, the quicklime is mixed with water to produce hydrated lime, plus large amounts of heat:

Calcium oxide + water = calcium hydroxide + heat
(quicklime) (hydrated lime)

$$CaO + H_2O = Ca(OH)_2 + heat$$

Finally, exposure to the atmosphere converts the calcium hydroxide to calcium carbonate. This reaction takes place over several years:

Calcium hydroxide + air = calcium carbonate + water
(hydrated lime) (carbon dioxide) (limestone)

$$Ca(OH)_2 + CO_2 = CaCO_3 + H_2O$$

Sand-lime mortar is found in many historic buildings. It hardens very slowly, but also has the ability to deform slowly over time without cracking. Sand-lime mortar is not a hydraulic-cement mortar because the last step in its hardening process (conversion of calcium hydroxide to calcium carbonate) occurs only in the presence of air.

Chemistry of Hydraulic-Cement Mortars

Hydraulic cements harden as a result of a chemical reaction of minerals with water. Hydraulic cements have been used since prehistoric times. Their cementitious ingredients include pozzolanic cements, gypsum cements, portland cements, and other cements.

Chemistry of Pozzolanic Cements These were discovered by the Greeks. The word "pozzolan" comes from a site in Italy (Pozzuoli, near the volcano Vesuvius) where these minerals were found and used by the Romans. A pozzolan possesses little or no cementitious properties on its own, but reacts with calcium hydroxide and water to form cementitious compounds. An example of a natural pozzolan is quartz, whose chemical formula ($SiO_2 \cdot XH_2O$) denotes silica (silicon dioxide) combined chemically with water. When finely ground quartz is mixed with hydrated lime [calcium hydroxide, or $Ca(OH)_2$], the following reaction occurs:

$$SiO_2 \cdot XH_2O + Ca(OH)_2 = Ca_{1\text{-}3}SiO_3 \cdot H_2O$$
$$\text{(calcium silicate,}$$
$$\text{a natural cement)}$$

Chemistry of Gypsum Cement "Plaster of Paris" Gypsum reacts much faster with water than lime or pozzolans do. Pure gypsum sets in about 5 min. Commercial gypsum (such as Hydrostone®) sets in about 45 min because it has a retarder with it.

Gypsum rock is calcined (heated) like limestone, but requires less energy:

$$CaSO_4 \cdot 2H_2O + \text{heat} = CaSOH_4 \cdot 1/2H_2O + 3/2H_2O$$
$$\text{(gypsum rock)} \qquad \text{(plaster of Paris)}$$

When water is added to the calcined gypsum, it reverts to its original state:

$$CaSO_4 \cdot 1/2H_2O + 3/2H_2O = CaSOH_4 \cdot 2H_2O + \text{heat}$$
$$\text{(plaster of paris)} \qquad \text{(gypsum rock)}$$

The resulting cement is very strong and stiff as concrete. Its main disadvantage is that it expands slowly over time as it absorbs water from the outside air. This produces large splitting forces if the gypsum is restrained.

In the calcining operation, if the gypsum rock is heated too much, the following undesirable reaction results, producing a powder that is not useful for building:

$$CaSO_4 \cdot 2H_2O + \text{heat} = CaSOH_4 + 2H_2O$$
$$\text{(gypsum rock)} \qquad \text{(anhydrite)}$$

Chemistry of Portland Cement
Portland cement is a particular class of hydraulic cement. It was first manufactured in England in the early 1800s, and was so named because its color was thought to resemble that of a natural limestone from the Isle of Portland.

Hardened portland cement is the result of the hydration of four principal chemical constituents:

Name	Chemical formula	Abbreviated name
Tricalcium silicate	$3CaO \cdot SiO_2$	C_3S
Dicalcium silicate	$2CaO \cdot SiO_2$	C_2S
Tricalcium aluminate	$3CaO \cdot Al_2O_3$	C_3A
Tetracalcium aluminoferrite	$4CaO \cdot Al_2O_3 \cdot Fe_2O_3$	C_4AF

Dry (unhydrated) cement consists of these compounds in powdered form. When water is added, the compounds combine with water in an exothermic (heat-producing) reaction, to form calcium hydroxide (about 25% by weight) and calcium silicate hydrate (about 50% by weight).

Chemistry of Other Hydraulic Cements In recent years, portland cement has increasingly been used in combination with other hydraulic cements, particularly pozzolanic cements and slag cements. Each has its own ASTM specification. Pozzolanic cements combine with calcium hydroxide to produce calcium silicate hydrate. Slag cements (usually produced from ground granulated blast-furnace slag or GGBF slag) are combinations of silicates and aluminosilicates. Slag cements, when hydrated, produce primarily calcium silicate hydrates as well.

2.2.2 Cementitious Systems Used in Modern Masonry Mortar

Modern masonry mortar is composed of cementitious agents (portland cement or other hydraulic cements and hydrated lime, or masonry cement, or mortar cement), sand, and water. Each of these can be referred to as a cementitious system. Three cementitious systems are defined by ASTM C270:

- Cement and lime
- Masonry cement
- Mortar cement

The first of these (cement and lime) is self-explanatory. The second and third (masonry cement and mortar cement) are generally mixtures of portland or blended cement and plasticizing materials (such as hydrated lime or finely ground limestone), together with other materials introduced to enhance performance. These other materials generally include air-entraining and water-retention additives, intended to improve freeze-thaw durability, workability, and water retention. These are discussed further in this chapter.

2.2.3 Types of Masonry Mortar

ASTM C270 defines, for all cementitious systems, different mortar types. In general, these are distinguished by the proportion of cement in the mortar. As discussed in Sec. 2.2.4, types of masonry mortar are designated by ASTM C270 using the letters M, S, N, O, and K, representing every second letter of the phrase, "mason work" (**M** a **S** o **N** w **O** r **K**).

Toward the "M" end of the spectrum, mortars have a higher volume proportion of cement; toward the "N" end, a lower proportion. This designation was selected intentionally (rather than, for example, "A, B, C, D"), to avoid the implication that a "Type A" mortar would always be the best.

2.2.4 Characteristics of Different Types of Masonry Mortar

- *Type M*: High compressive and tensile bond strength
- *Type S*: Moderate compressive and tensile bond strength
- *Type N*: Low compressive and tensile bond strength
- *Type O*: Very low compressive and tensile bond strength
- *Type K*: No longer used

Type S mortar is a good all-purpose mortar.

Now let's look at how mortar is specified using each cementitious system. Specification is either by proportion or by property. Specification by proportion is the default. If the specifier does not say "by property," the specification is assumed to be by proportion.

Cement-Lime Mortar

Approximate proportion requirements for cement-lime mortar are shown in Table 2.2. The proportions given in this table are near the midpoints of

Mortar type	Mortar proportions by volume		
	Portland cement or other cements	Hydrated lime	Mason's sand (2-1/4–3 times volume of cementitious materials)
M	1	≤1/4	3
S	1	1/2	4-1/2
N	1	1	6
O	1	2	9

Source: ASTM C270.

TABLE 2.2 Approximate Proportion Requirements for Cement-Lime Mortar

the ranges of proportions required by ASTM C270. When specifying a mortar by the proportion specifications of ASTM C270, it is not necessary to specify the proportions, only the mortar type.

ASTM C270 refers in turn to other ASTM specifications. Even if sand doesn't meet grading limits, it can still pass "by use" if mortar made with it can pass the property specifications of ASTM C270.

ASTM C207: Hydrated Lime for Masonry Purposes
Type N: No oxide limits (Type NA is air-entrained)
Type S: Oxide limits (Type SA is air-entrained)
ASTM C144: Aggregate for Masonry Mortar (sand gradations)

Property requirements for cement-lime mortar from ASTM C270 are repeated in Table 2.3.

Masonry-Cement Mortar

Approximate proportion requirements for masonry-cement mortar are given in Table 2.4. The proportions given in this table are near the midpoints of the ranges of proportions required by ASTM C270. When specifying a mortar by the proportion specifications of ASTM C270, it is not necessary to specify the proportions, only the mortar type.

The most common types are single-bag mixes (the first four lines of the table). However, Types M and S masonry-cement mortar can also be made by adding Portland cement to Type N masonry cement.

	Property requirements for cement-lime mortar		
Mortar type	Compressive strength, psi	Water retention	Maximum air content
M	2500	75%	12%
S	1800	75%	12%
N	750	75%	14% (12% if reinforced)
O	350	75%	14% (12% if reinforced)

Source: ASTM C270.
Note: These property requirements apply only to laboratory-prepared mortar, with a flow of about 110. They are not requirements for field mortar. See the end of this section for an explanation of flow.

TABLE 2.3 Property Requirements for Cement-Lime Mortar

Mortar type	Mortar proportions by volume				
	Portland cement or blended cement	Masonry cement type			Mason's sand (2-1/4–3 times volume of cementitious materials)
		M	S	N	
M		1			3
S			1		3
N				1	3
O				1	3
M	1			1	6
S	1/2			1	4-1/2

Source: ASTM C270.

TABLE 2.4 Approximate Proportion Requirements for Masonry-Cement Mortar

Property requirements for masonry-cement mortar from ASTM C270 are repeated in Table 2.5.

Mortar-Cement Mortar

Approximate proportion requirements for mortar-cement mortar are given in Table 2.6. The proportions given in that table are near the midpoints of the ranges of proportions required by ASTM C270. When

Masonry cement mortar type	Property requirements for masonry-cement mortar		
	Compressive strength, psi	Water retention	Maximum air content
M	2500	75%	18%
S	1800	75%	18%
N	750	75%	20% (18% if reinforced)
O	350	75%	20% (18% if reinforced)

Source: ASTM C270.
Note: These property requirements apply only to laboratory-prepared mortar, with a flow of about 110. They are not requirements for field mortar.

TABLE 2.5 Property Requirements for Masonry-Cement Mortar

	Mortar proportions by volume				
Mortar type	Portland cement	Mortar cement type			Mason's sand (2-1/4–3 times volume of cementitious materials)
		M	S	N	
M		1			3
S			1		3
N				1	3
O			1		3
M	1			1	6
S	1/2			1	4-1/2

Source: ASTM C270.

TABLE 2.6 Approximate Proportion Requirements for Mortar-Cement Mortar

specifying a mortar by the proportion specifications of ASTM C270, it is not necessary to specify the proportions, only the mortar type.

By far the most common types are single-bag mixes (the first four lines of the table). Types M and S mortar-cement mortar can also be made, however, by adding portland cement to Type N mortar cement.

Property requirements for mortar-cement mortar from ASTM C270 are repeated in Table 2.7.

Mortar cement mortar type	Property requirements for mortar-cement mortar		
	Compressive strength, psi	Water retention	Maximum air content
M	2500	75%	12%
S	1800	75%	12%
N	750	75%	14% (12% if reinforced)
O	350	75%	14% (12% if reinforced)

Source: ASTM C270.
Note: These property requirements apply only to laboratory-prepared mortar, with a flow of about 110. They are not requirements for field mortar.

TABLE 2.7 Property Requirements for Mortar-Cement Mortar

2.2.5 Characteristics of Fresh Mortar

The most important characteristic of fresh mortar is workability, generally defined as the ability to be easily spread on masonry units using a trowel. In the context of ASTM C270, workability is defined and measured very simply, in terms of flow. A standard-shaped, circular sample of mortar 4 in. in diameter is placed on a flow table, which is then dropped 25 times. The flow of that mortar is defined as the increase in diameter of the sample, divided by the original diameter and multiplied by 100. Thus, if the final diameter is 8 in., the flow is $(8-4)/4$, or 100. Laboratory mixed mortars have a flow of about 110 ± 5; field mortars, about 130 to 150. Field mortars should be retempered (water added) as necessary to maintain workability, but should not be used beyond 2-1/2 h after mixing. Workability can also be measured with a cone penetrometer.

According to ASTM C270, mortar can be specified by proportion (the default) or by property. If mortar is specified by proportion, the following characteristics of fresh mortar are controlled indirectly as a result of complying with the required proportions. If mortar is specified by property, they are controlled directly:

1. *Retentivity*: This is the ratio of the flow after suction to the initial flow. Flow after suction is measured using mortar from which some of the water has been removed using a standard vacuum apparatus. In one other specification, the mortar is spread on a masonry unit and allowed to sit for 1 min. According to ASTM C270, mortar is required to have a retentivity of at least 75 percent.

2. *Air content*: Percent air by volume (ASTM C91). Cement-lime mortar and mortar-cement mortar usually have a maximum permissible air content of 12 percent. Masonry-cement mortar usually has a maximum permitted air content of 18 percent if used in reinforced masonry.

2.2.6 Characteristics of Hardened Mortar

Characteristics of hardened mortar include compressive strength and tensile bond strength. Only the first is controlled by ASTM C270.

If mortar is specified by the property specification of ASTM C270, compressive strength is controlled directly. It is measured using 2 in. mortar cubes, made with laboratory flow mortar, cured for 28 days at 100 percent relative humidity and 70°F. It typically ranges from 500 to 3000 psi. It does not significantly affect the compressive strength of masonry assemblages. ASTM C270 requires minimum compressive strengths of 2500, 1800i, 750, and 350 psi for Types M, S, N, and O mortar, respectively.

If mortar is specified by the proportion specification of ASTM C270, compressive strength is controlled indirectly. Masonry-cement mortar meeting the proportion specification usually has a compressive strength slightly greater than the minimum value specified in the property specification. Cement-lime mortar meeting the proportion specification usually has a compressive strength considerably greater than the minimum value specified in the property specification.

2.2.7 Other Characteristics of Hardened Mortar

One other characteristic of hardened mortar is tensile bond strength (the tensile strength of the bond between mortar and units). Cement-lime mortar has traditionally satisfied practical requirements for tensile bond strength. Tensile bond strength is not specified directly for cement-lime mortar, or for masonry-cement mortar. It is addressed directly in the specification for mortar-cement mortar. Strictly speaking, it can be measured only in conjunction with units, and is therefore not a property of the hardened mortar alone. Nevertheless, certain characteristics of the mortar itself contribute to good tensile bond strength regardless of the type of unit used. High tensile bond strength can be obtained using cement-lime mortar or mortar-cement mortar. It is also enhanced by the use of mortars with air content below 12 percent.

2.2.8 Note on Cement-Lime Mortars versus Masonry-Cement Mortars

At times in the past, and to some extent even to this day, controversy has existed within the masonry technical community over the comparative performance of cement-lime mortar and masonry-cement mortar. Each cementitious system has advantages and disadvantages. Each has demonstrated general suitability for use and also general cost-effectiveness for suppliers and users.

Masonry cement complies with the physical property requirements of ASTM C91. Because the standard specifications are based on properties rather than ingredients, specific formulations of masonry cement vary from manufacturer to manufacturer. Ingredients and formulations are not required to be disclosed, and generally are not. Masonry cement is generally delivered to the jobsite in prepackaged form. It consists of a mixture of portland or blended cement and plasticizing materials (such as hydrated lime or finely ground limestone) together with other materials introduced to enhance performance. These other materials generally include air-entraining and water-retention additives, intended to improve freeze-thaw durability, workability, and water retention.

The primary advantage of cement-lime mortar is its high tensile bond strength. Its disadvantages are the additional complexity of mixing three ingredients, and some lack of workability (stickiness) if not retempered. The first disadvantage can be overcome by single-bag or silo mixes. The second can sometimes be overcome by retempering.

The advantages of masonry-cement mortar are its relative simplicity of batching and its good workability. It has a "fluffy" consistency (because of its entrained air), which leads to good productivity. Its lower tensile bond strength is accounted for by lower allowable stresses in design codes. In part because of these lower bond strengths, and in part because of tradition, masonry cement is prohibited in structural masonry zones of high seismic risk in the United States.

Considerable anecdotal evidence, and some controlled experimental evidence indicates that other things being equal, walls laid with cement-lime mortar leak less than walls with masonry-cement mortar. In the author's judgment, this is true. It is also true, however, that acceptably water-resistant walls can be constructed using either cementitious system, however the cementitious system is not the most important choice to make when specifying masonry. The proper type of wall (drainage vs. barrier) and proper drainage details, if applicable, are more important.

From the viewpoint of cement producers, masonry cement is probably a profitable "niche" product. A 70-lb bag of masonry cement typically contains about 40 percent or less (28 lb or less) of Portland cement or other cements, and about 40 lb of ground limestone. The rest is air-entraining additives, and possibly additives for water-retention and plasticity. A 70-lb bag of masonry cement (28 lb cement, 40 lb limestone, and additives) commonly sells for the same price as a 94-lb bag of Portland cement.

Mortar cement was introduced in the 1990s to preserve the construction advantages and potential profitability of masonry cement, while at the same time increasing the tensile bond strength of the resulting mortar to values comparable to those of cement-lime mortar. Mortar cement is regarded by building codes as the equivalent of cement-lime mortar, and is permitted in all seismic zones of the United States.

2.3 Masonry Grout

Masonry grout is essentially fluid concrete. It is used to fill spaces in masonry, and to surround reinforcement and anchors. It is specified using ASTM C476 (Grout for Masonry).

Grout for masonry is composed of portland cement, sand, and (for coarse grout) pea gravel. It is permitted to contain a small amount of hydrated masons' lime, but usually does not. It is permitted to be specified by proportion or by compressive strength. Neither of these is the default.

	Grout proportions by volume			
Grout type	Portland cement	Hydrated lime	Mason's sand	Pea gravel
Fine	1	≤1/10	2-1/4 to 3	—
Coarse	1	≤1/10	2-1/4 to 3	1 to 2

Source: ASTM C476.

TABLE 2.8 Proportion Requirements for Grout for Masonry

2.3.1 Proportion Specifications for Grout for Masonry

The proportion requirements of ASTM C476 for grout for masonry are repeated in Table 2.8. Note that hydrated lime is permitted but not required.

2.3.2 Properties of Fresh Grout

The most important property of fresh grout is its ability to flow. Masonry grout should be placed with a slump of 8 to 11 in., so that it will flow freely into the cells of the masonry. Because of its high water-cement ratio at time of grouting, masonry grout undergoes considerable plastic shrinkage as the excess water is absorbed by the surrounding units. To prevent the formation of voids due to this process, the grout is consolidated during placement, and reconsolidated after initial plastic shrinkage. Grouting admixtures containing plasticizers and water-retention agents can also be useful in the grouting process.

2.3.3 Properties of Hardened Grout

The most important property of hardened grout is its compressive strength. If grout is specified by compressive strength, that strength must be at least 2000 psi. If it is specified by proportion, its compressive strength is controlled indirectly to at least that value, by the ingredients used.

Because of its high water-cement ratio at the time of grouting, masonry grout cast into impermeable molds has a very low compressive strength, which is not representative of its strength under field conditions, when the surrounding units absorb water from it. For this reason, ASTM C1019 (Sampling and Testing Grout) requires that the compression specimen be cast using permeable molds. The most common way of preparing such a mold is to arrange masonry units so that they enclose a rectangular solid whose base is 2 in.^{2} and whose height is equal to the height of the units. The rectangular solid is surrounded by paper towels or filter paper, so that the compressive specimen's water-cement ratio is similar to that of grout in the actual wall.

2.3.4 Self-Consolidating Grout

Starting with the 2008 MSJC Code, self-consolidating grout is permitted to be used in masonry. Self-consolidating grout is a highly fluid and stable grout, typically with admixtures, that remains homogeneous when placed and does not require puddling or vibration for consolidation. This type of grout has the ability to flow easily into even small voids in the masonry, and to surround reinforcement without the need for mechanical consolidation. This ability is imparted by a combination of super-plasticizing admixtures and aggregate characteristics. Test methods associated with self-consolidating grout are provided in ASTM C1611.

2.4 General Information on ASTM Specifications for Masonry Units

Definitions of terms are given in ASTM C1180 (Standard Terminology of Mortar and Grout for Unit Masonry) and in ASTM C1232 (Standard Terminology of Masonry).

2.4.1 General Information on ASTM Specifications for Clay or Shale Masonry Units

ASTM specifications for clay or shale masonry units are summarized below:

- ASTM C62: Building Brick (Solid Masonry Units Made from Clay or Shale)
- ASTM C216: Facing Brick (Solid Masonry Units Made from Clay or Shale)
- ASTM C410: Industrial Floor Brick
- ASTM C652: Hollow Brick (Hollow Masonry Units Made from Clay or Shale)
- ASTM C902: Pedestrian and Light Traffic Paving Brick
- ASTM C1272: Heavy Vehicular Paving Brick

ASTM specifications for methods of sampling and testing clay or shale masonry units are given in ASTM C67 (Sampling and Testing Brick and Structural Clay Tile) and in ASTM C1006 (Splitting Tensile Strength of Masonry Units).

2.4.2 General Information on ASTM Specifications for Concrete Masonry Units

ASTM specifications for concrete masonry units are summarized below:

- ASTM C55: Concrete Building Brick
- ASTM C90: Hollow Load-Bearing Concrete Masonry Units
- ASTM C129: Hollow Non-Load-Bearing Concrete Masonry Units
- ASTM C139: Concrete Masonry Units for Construction of Catch Basins and Manholes
- ASTM C744: Prefaced Concrete and Calcium Silicate Masonry Units
- ASTM C936: Solid Concrete Interlocking Paving Units
- ASTM C1319: Concrete Grid Paving Units
- ASTM C1372: Dry-Cast Segmental Retaining Wall Units

ASTM specifications for methods of sampling and testing concrete masonry units are given in ASTM C140, which references ASTM C426 (Drying Shrinkage).

2.4.3 General Information on ASTM Specifications for Masonry Assemblages

ASTM specifications for standard methods of test for masonry assemblages are summarized below:

- ASTM E72: Strength Tests of Panels for Building Construction (lateral load by air bag)
- ASTM E514: Water Permeance of Masonry
- ASTM E518: Flexural Bond Strength of Masonry (modulus of rupture)
- ASTM E519: Diagonal Tension (Shear) in Masonry Assemblages
- ASTM C1072: Measurement of Masonry Flexural Bond Strength (bond wrench)
- ASTM C1314: Measurement of Compressive Strength of Masonry Prisms to Determine Compliance with f'_m
- ASTM C1357: Evaluating Masonry Bond Strength
- ASTM C1717: Conducting Strength Tests of Masonry Wall Panels

2.4.4 Concluding Remarks on ASTM Specifications for Masonry

The above apparently bewildering list of applicable ASTM specifications covers almost every possible aspect of masonry mortar, units, and assemblages. While the ASTM specifications can help organize the field, they may have indirect rather than direct relationships with the performance of the finished masonry. To shed more light on this point, we must investigate the desired performance characteristics of masonry materials, and study the relation between those characteristics and the ASTM specifications.

2.5 Clay Masonry Units

2.5.1 Geology Associated with Clay Masonry Units

Clay masonry units are formed of clay, a sedimentary mineral. Clay is found in the form of surface clay, shale (naturally compressed and hardened clay), or fire clay (deeper clays). In the United States, clay is found primarily in central Texas and the east coast, although small amounts are found in sedimentary deposits throughout the country.

2.5.2 Chemistry Associated with Clay Masonry Units

Clays and shales are about 65 percent silicon oxide and 20 percent aluminum oxide. They may also contain varying amounts of other metallic oxides (manganese, phosphorus, calcium, magnesium, sodium, potassium, and vanadium). These metallic oxides give a fired clay units a distinctive color, decrease the unit's vitrification temperature, and also affect its appearance and durability. For example, small amounts of chromite, added to light-colored (buff) clay, give it a gray color; small amounts of manganese, added to buff clay, give it a brown color.

2.5.3 Manufacturing of Clay Masonry Units

Three processes are in use today for manufacturing clay masonry units:

1. *Soft mud process*: Clay containing 20 to 30 percent water by weight is molded. This process is used occasionally in the United States, but more often in Europe.
2. *Stiff mud process*: Clay containing 12 to 15 percent water by weight is mixed, forced through a die, and cut with wire. This is the most common process in the United States.
3. *Dry press process*: Mixed clay containing 7 to 10 percent water by weight, form in hydraulic press. This process is rare. It is used, for example, to make fire brick.

After forming, various surface textures can be imparted to the unit: wire-cut, rug (heavy scratches), matte (light scratches), or sand finished.

The clay units are then placed on specially insulated railway cars, and subjected to the firing process. This involves six basic steps.

The units then move into a tunnel kiln, which is kept relatively cool at the entrance, hot in the middle, and cooler again at the exit. The heat comes from burning fuel within the kiln itself. Over a period of 12 h to as long as 3 days, the units pass from the entrance to the hottest section, and then to the cooler exit. Temperatures in the different sections are regulated to produce different results. The units pass through the following steps:

1. *Preheating*: The "green" units are dried at about 350°F, in drying ovens heated by exhaust gases from the kiln. During this process, the units shrink.
2. *Dehydration*: The units continue to dry at temperatures from 300 to about 800°F.
3. *Oxidation*: At temperatures from about 800 to 1800°F, organic material burns.
4. *Vitrification (or incipient vitrification)*: At temperatures of 1600 to 2400°F, the clay begins to vitrify. Silicates in the clay begin to fuse, binding the unvitrified clay particles together. This point is termed "incipient fusion." The temperature used depends on the type of clay. Most clays will undergo incipient fusion at about 2000°F. The purest clays, which are used for refractory brick, are fired at temperatures up to 2400°F.
5. *Control of Oxygen*: The color of metallic oxides can be changed by feeding additional air into the kiln at this point to promote an oxidizing environment, or by intentionally withholding air to produce a reducing environment. The latter is termed "reduction firing," or "flashing."
6. *Cooling*: The units are then slowly cooled.

2.5.4 Visual and Serviceability Characteristics of Clay Masonry Units

The following characteristics are covered by ASTM C62 (Standard Specification for Building Brick) or by ASTM C216 (Standard Specification for Facing Brick):

- Dimensional tolerances
- Durability
- Freeze-thaw resistance
- Appearance

Different ASTM requirements for clay masonry units are summarized in a table at the end of this section, and are described in more detail below:

1. Dimensional tolerances for building brick (ASTM C62) vary with nominal dimensions, but are typically ± 1/4 in. For facing brick (ASTM C216), corresponding typical required tolerances are ± 8/32 in. for Type FBS (face brick, standard) and ± 5/32 in. for Type FBX (face' brick, extra). Tolerances are also specified for distortion.

2. Durability is controlled indirectly in terms of boiling-water absorption. Units are dried at 230 to 239°F, placed in boiling water for 5 h, then reweighed. Boiling water absorption equals weight gain divided by original dry weight.

   Boiling-water absorption is taken as a general index of durability. Building brick (ASTM C62) must have a boiling-water absorption (average of 5 units) of at most 17 percent for Grade SW (severe weathering), and at most 22 percent for Grade MW (moderate weathering). No limit is imposed for Grade NW (negligible weathering). Facing brick (ASTM C216) must have corresponding boiling-water absorptions of 17 percent (Grade SW) and 22 percent (Grade MW). Grade NW does not exist under ASTM C216.

3. Freeze-thaw resistance is controlled indirectly in terms of a "saturation coefficient," defined as follows:

   a. *Cold-water absorption (24 h)*: Units are dried at 230 to 239°F, placed in cold water for 24 h, then reweighed. Cold-water absorption equals weight gain divided by original dry weight.

   b. *Boiling-water absorption (5 h)*: After the cold-water absorption test described above, units are placed in boiling water for an additional 5 h, and again weighed. Boiling-water absorption equals weight gain (cold plus boiling-water absorption) divided by original dry weight.

   c. *Saturation coefficient (c/b ratio)*: Cold-water absorption (24 h) divided by boiling water absorption (5 h).

   The saturation coefficient is a measure of the additional void space available in the units after saturation by cold water. A saturation coefficient of 1.0 indicates no additional void space. The lower the saturation coefficient, the more additional void space is available. This is taken as a rough index of resistance to freeze-thaw degradation (additional void space is available for the freezing water).

Building brick (ASTM C62) must have a saturation coefficient (average of 5 units) of at most 0.78 for Grade SW (severe weathering), 0.88 for Grade MW (moderate weathering), and at most 1.0 (no limit) for Grade NW (negligible weathering). Facing brick (ASTM C216) must have corresponding saturation coefficients of 0.78 (Grade SW) and 0.88 (Grade MW). Grade NW does not exist under ASTM C216.

4. Appearance is not addressed by ASTM C62 (Building brick). ASTM C216 (Facing brick) addresses chippage and efflorescence potential.

   a. *Chippage*: Under ASTM C216, up to 10 percent of units complying with Grade FBS can have chips up to 1 in. in size; for Grade FBX, the corresponding percentage is 5 percent.

   b. *Efflorescence*: This is a white or colored chemical residue on the surface of the masonry. It is produced by water-soluble compounds within or in contact with the masonry. If water gains access into the masonry in sufficient amounts, and comes in contact with the water-soluble compounds for a sufficient time, it dissolves those compounds into positive and negative ions. The water containing the dissolved compounds (in ionic form) moves through the masonry and evaporates; the dissolved ions combine to form deposits. If these deposits form on the surface of the masonry, they are called "efflorescence;" if they form within the pores of the masonry near the surface, they are called "cryptoflorescence." The positive ions are usually potassium, sodium, or calcium. The negative ions are usually sulfates, chlorides, or hydroxides. In general, because the positive and negative ions are present in all masonry, efflorescence is reduced by limiting the amount of water in contact with the masonry, and by limiting the passage of water to the surface of the masonry.

When required to be tested, Facing Brick (ASTM C216) are required to show "no efflorescence." Efflorescence Testing (in accordance with ASTM C67), uses distilled water and is not a complete check for efflorescence.

2.5.5 Mechanical Characteristics of Clay Masonry Units

The compressive strength varies from about 1200 to 30,000 psi. It is typically 8000 to 15,000 psi. Building brick (ASTM C62) must have a minimum compressive strength (average of 5 units, tested flatwise) of 1500 psi for Grade NW, 2500 psi for Grade MW, and 3000 psi for Grade SW. Building brick (ASTM C216) must have corresponding compressive strengths of 2500 psi for Grade MW, and 3000 psi for Grade SW. Grade NW does not exist under ASTM C216.

2.5.6 Specifications of Clay Masonry Units

Clay masonry units are specified in accordance with the required appearance and durability (refer to the summary table at the end of this section).

- The required appearance determines whether the units need to conform to C62 or to C216. Under C216, the required dimensional tolerances determine the Type (FBS or FBX).

- The required durability and freeze-thaw resistance determine whether the units need to conform to Grade NW, MW, or SW. The required durability and freeze-thaw resistance depend on the "weathering index" (product of the average annual number of freezing cycle days and the average annual winter rainfall in inches), defined in detail in ASTM C62 and depicted in Fig. 2.4. "Negligible" weathering regions have weathering indices of less than 50; "moderate" weathering regions have weathering indices between 50 and 500; and "severe" weathering regions have weathering indices in excess of 500.

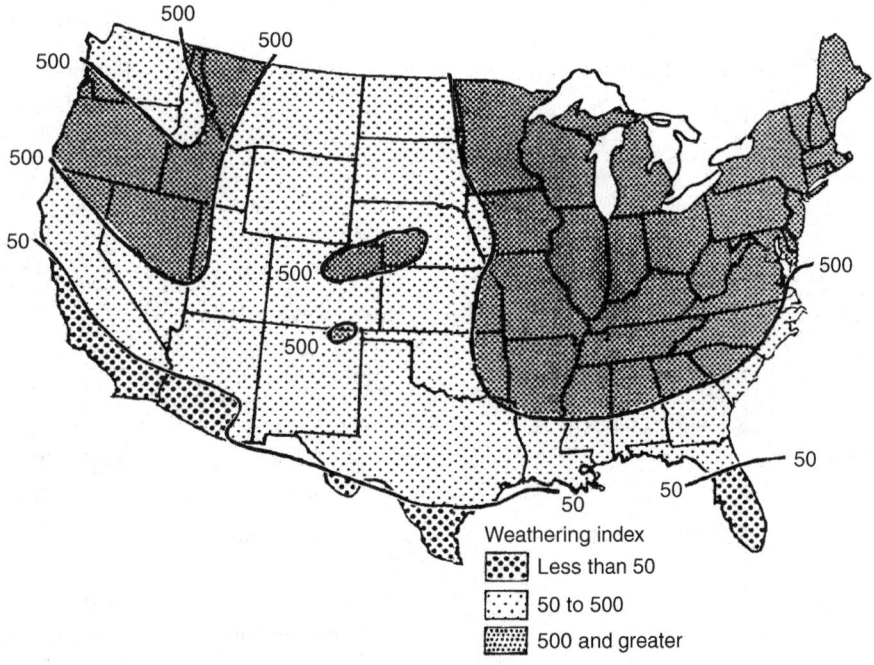

FIGURE 2.4 Weathering indices in the United States. (*Source*: Figure 1 of ASTM C62.)

Under ASTM C62 (Building Brick), Grade NW brick are recommended for interior use only. Grade MW brick are permitted for use in "severe" weathering regions. Grade SW brick are recommended for use in "severe" weathering regions, and whenever brick are in contact with the ground, or laid in horizontal surfaces, or likely to be permeated with water.

Under ASTM C216 (Facing Brick), there are no Grade NW brick. Grade MW brick are permitted for use in "severe" weathering regions. Grade SW brick are required for use whenever the weathering index is greater than or equal to 500, and whenever brick in other than vertical surfaces are in contact with soil.

Requirements of ASTM C62 and C216 for clay units are summarized in Table 2.9.

	Requirement according to	
Characteristic	**ASTM C62**	**ASTM C216**
Dimensional Tolerance	±1/4 in.	Type FBS ±1/4 in. Type FBX ±5/32 in.
Chippage	No requirements	Type FBS 10% Type FBX 5%
Efflorescence	No requirements	Required to show "not effloresced" by ASTM C67 (distilled water, units only)
Compressive strength	Grade NW: 1500 psi Grade MW: 2500 psi Grade SW: 3000 psi	Grade MW: 2500 psi Grade SW: 3000 psi
Durability (boiling water absorption)	Grade NW no requirement Grade MW ≤ 22% Grade SW ≤ 17%	Grade MW ≤ 22% Grade SW ≤ 17%
Saturation coefficient (c/b ratio)	Grade NW ≤ 1.0 Grade MW ≤ 0.88 Grade SW ≤ 0.78	Grade MW ≤ 0.88 Grade SW ≤ 0.78
Design criteria	Grade NW interior use only Grade MW permitted for WI ≤ 500 Grade SW recommended for WI > 500	Grade MW permitted for WI ≤ 500 Grade SW required for WI > 500

TABLE 2.9 Summary of ASTM Requirements for Clay Masonry Units

Materials Used in Masonry Construction

2.5.7 Other Characteristics of Clay Masonry Units

The following other characteristics of clay masonry units are not addressed by ASTM specifications:

1. *Color*: Metallic oxides give different units their characteristic color. Colors are commonly judged by eye from sample panels. Systems for color tolerance exist, but are not widely used.

2. *Tensile strength*: Parallel to the grain (in the direction of extrusion), this is typically 20 to 30 percent of the corresponding compressive strength. Perpendicular to the grain, it is typically 10 to 20 percent of the corresponding compressive strength.

3. *Initial rate of absorption (IRA)*: This is defined as the number of grams of water absorbed in 1 min per 30 in.^{2} of bed area. An ideal range is 10 to 30. Many clay masonry units used in Texas have IRAs exceeding 30. Units with an IRA greater than 30 should be wetted briefly before laying. A simple field test for IRA is as follows: Place 20 drops of water in a quarter-sized area. If it takes longer than 1.5 min for the water to be absorbed, the units do not need to be wetted before laying.

4. *Tensile bond strength (strength between mortar and units)*: This is typically about 100 psi when cement-lime mortar or mortar-cement mortar is used, and about 50 psi or less when masonry-cement mortar is used. Tensile bond strength is increased by compatibility between mortar and units: units with high IRA should be used with mortar having high water retention (high lime content); low IRA units should be used with low-retentivity mortar.

5. *Modulus of elasticity*: $1.4 - 5 \times 10^6$ psi.

6. *Freeze-thaw expansion*: Clay units exposed to cycles of freezing and thawing undergo permanent expansion (mean, standard deviation, and 97-percentile value of 118, 96, and 300 µε, respectively).

7. *Moisture expansion*: Clay units exposed to moisture undergo permanent expansion caused by adsorption of water into unvitrified clay molecules (mean, standard deviation, and 97-percentile values of 200, 190, and 540 µε, respectively).

8. *Coefficient of thermal expansion*: $3 - 4$ µε/°F.

2.6 Concrete Masonry Units

2.6.1 Materials and Manufacturing of Concrete Masonry Units

Concrete masonry units are formed from zero-slump concrete, sometimes using lightweight aggregate. The concrete mixture is usually vibrated under pressure in multiple-block molds. After stripping the molds, the units are usually cured under atmospheric conditions in a chamber that is maintained at warm and humid conditions by the presence of the curing units. Atmospheric steam or high-pressure steam (autoclaving) can also be used for curing. Concrete units normally have a much higher void ratio than clay units, making determination of (c/b) ratios unnecessary.

2.6.2 Visual and Serviceability Characteristics of Concrete Masonry Units

The following visual and serviceability characteristics are addressed by ASTM C90 (Standard Specification for Hollow Load-Bearing Concrete Masonry Units):

1. *Dimensional tolerances*: ASTM C90 prescribes maximum dimensional tolerances of ± 1/8 in. Thicknesses of face shells and webs are specified.

2. *Chippage*: According to ASTM C90, up to 5 percent of a shipment may contain units with chips up to 1 in. in size.

Other visual and serviceability characteristics, such as color, are not addressed by ASTM C90. Color is gray or white, unless metallic oxide pigments are used.

2.6.3 Mechanical Characteristics of Concrete Masonry Units

The following mechanical characteristics are covered by ASTM C90, ASTM C140, and ASTM C426:

1. Compressive strength is typically 1500 to 3000 psi on the net area (actual area of concrete). ASTM C90 requires a minimum compressive strength (average of 3 units) of 1900 psi, measured on the net area.

2. Absorption (used to measure void volume) is evaluated in the following manner. The unit is immersed in cold water for 24 h. It is weighed immersed (weight F), and weighed in air while still wet (weight E). It is then dried for at least 24 h at a temperature of 212 to 239°F, and again weighed (weight C). Absorption in lb/ft³ is calculated as $[(E - C)/(E - F)] \times 62.4$. Maximum permissible absorption is 18 lb/ft³ for light-weight units (less than 105 lb/ft³ oven-dried weight), 15 lb/ft³ for medium-weight units (105 – 125 lb/ft³), and 13 lb/ft³ for normal-weight units (more than 125 lb/ft³).

3. Shrinkage of concrete masonry units due to drying and carbonation is 300–600 με. In general, shrinkage is controlled by controlling the concrete mix used to make the units, and by limiting the moisture content of the units between the time of production and when they are placed in the wall.

The concrete masonry industry formerly produced Type I (moisture-controlled) units, which due to a combination of inherent characteristics and packaging were designed to shrink less, and Type II units (nonmoisture-controlled). This distinction was not as successful as originally hoped, because it was difficult to control the condition of Type I units in the field. As a result, ASTM C90 now does not refer to Type I and Type II units. All C90 units must demonstrate a potential drying shrinkage of less than 0.065 percent (650 με), which was the shrinkage requirement that formerly applied to Type II units.

2.6.4 Other Characteristics of Concrete Masonry Units

The following characteristic are not covered by ASTM specifications:

1. Surface texture can be smooth, slump block, split-face block, ribbed block, various patterns, or polished face.
2. Tensile strength is about 10 percent of compressive strength.
3. Tensile bond strength (strength between mortar and CMU) is typically about 40 to 75 psi when portland cement-lime mortar is used, and about 35 psi or less when masonry-cement mortar is used.
4. Initial rate of absorption (IRA) is typically 40 to 160 g/min per 30 in.^{2} of bed area. It is much less than this in units with integral water-repellent admixtures. In contrast to clay units, the tensile bond strength of concrete masonry units is not sensitive to initial rate of absorption. For this reason, specifications for concrete masonry units do not require determination of IRA.
5. Modulus of elasticity is typically $1 - 3 \times 10^6$ psi.
6. Coefficient of thermal expansion is typically: $4 - 5$ με/°F.

2.7 Properties of Masonry Assemblages

The following characteristics of masonry assemblages are covered by ASTM Specifications E72, C1072, C1388, C1389, C1390, C1391, C1357, and C1314:

1. *Compressive strength*: This is often denoted by f_m. Using ASTM C1314, it is measured using stack-bonded prisms whose maximum ratio of height divided by least lateral dimension is between 1.3 and 5. For example:

    a. Hollow concrete masonry units measuring 8 × 8 × 16 in., tested as a 2-high prism, would have a height of about 16 in. and a minimum lateral dimension of about 8 in., for a ratio of height to least lateral dimension of about 2.

    b. Modular clay units measuring 4 × 2-2/3 × 8 in., tested as a 6-high prism, would have a height of about 16 in. and a minimum base dimension of about 4 in., for a ratio of height to least lateral dimension of about 4.

    The compressive strength of a clay masonry prism is less than that of the mortar or the unit tested alone. This is because clay masonry prisms typically fail due to transverse splitting. The mortar is usually more flexible than the units. Under compression perpendicular to the bed joints, it expands laterally, placing the units in transverse biaxial tension. The prism cracks perpendicular to the bed joints (parallel to the direction of the applied load).

    Because concrete masonry prisms typically have mortar and units of similar strengths and elasticity, these tend to fail like a concrete cylinder.

2. Tensile bond strength can be measured by tests on wall specimens (E72), by modulus of rupture tests on masonry beams (E518), by bond wrench tests (C1072), or by crossed-brick couplet tests. Results from these tests are not equal. Strict protocols for bond-wrench testing are specified in C1357.

3. Shear strength can be measured by diagonal compression tests (E519).

4. Water permeability is measured in terms the amount of water passing through a wall under a standard pressure gradient, simulating the effects of wind-driven rain (E514).

2.8 Masonry Accessory Materials

Masonry accessory materials include reinforcement, connectors, sealants, flashing, coatings, and vapor barriers and moisture barriers. Each of these is described further below.

Reinforcement	
Connectors (galvanized or stainless steel)	Ties (connect a masonry wall to another wall)
	Anchors (connect a masonry wall to a frame)
	Fasteners (connect something else to a masonry wall)
Sealants	Expansion joints (clay masonry)
	Control joints (concrete masonry)
	Construction joints
Flashing	
Coatings	Paints
	Water-repellent coatings
Vapor barriers and moisture barriers	

2.8.1 Reinforcement

Reinforcement consists of the following:

- Steel deformed reinforcing bars meeting the requirements of ASTM A615 (billet steel) or A996 (rail and axle steel), or ASTM A706 (low-alloy weldable steel)
- Joint reinforcement (ASTM A951)
- Deformed reinforcing wire (ASTM A496)
- Steel welded wire reinforcement for concrete (ASTM A497)
- Steel prestressing strand (ASTM A416)

Typical uses of each type of reinforcement are shown in Figs. 2.5 through 2.8. Figure 2.5 shows deformed reinforcing bars in a grouted masonry wall. Figure 2.6 shows joint reinforcement. Figure 2.7 shows welded wire reinforcement in the topping of a floor slab connected to a masonry wall. Figure 2.8 shows posttensioning tendons in a masonry wall.

2.8.2 Connectors

Connectors are addressed by the following ASTM specifications:

- ASTM F1554 (plate, headed and bent bar anchors)
- ASTM A325 (high-strength bolt anchors)
- ASTM A1008 (sheet steel anchors and ties)
- ASTM A185 (steel wire mesh ties)
- ASTM A82 (steel wire ties and anchors)
- ASTM A167 (stainless steel sheet anchors and ties)

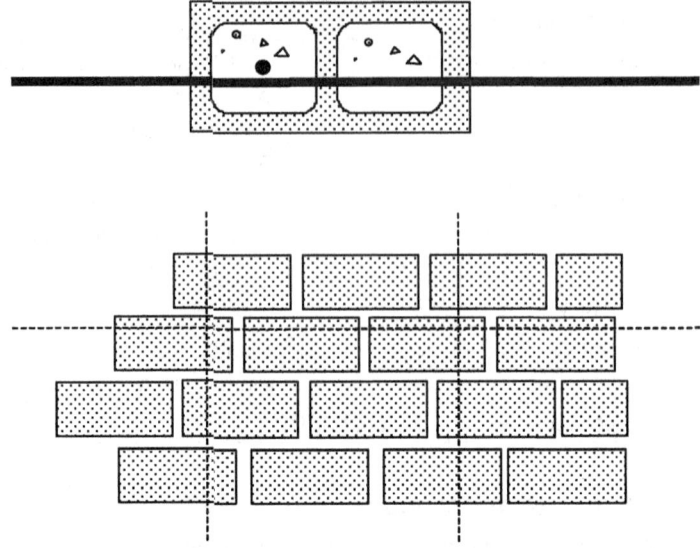

FIGURE 2.5 Typical application of deformed reinforcement in grouted masonry wall.

- ASTM A193-B7 (high-strength threaded rod anchors)
- ASTM A641, A153, or A 653 (galvanized steel connectors)

Typical uses of each type of connector are shown in Figs. 2.9 through 2.11. Figure 2.9 shows typical veneer ties. Figure 2.10 shows typical adjustable pintle ties. Figure 2.11 shows typical connectors.

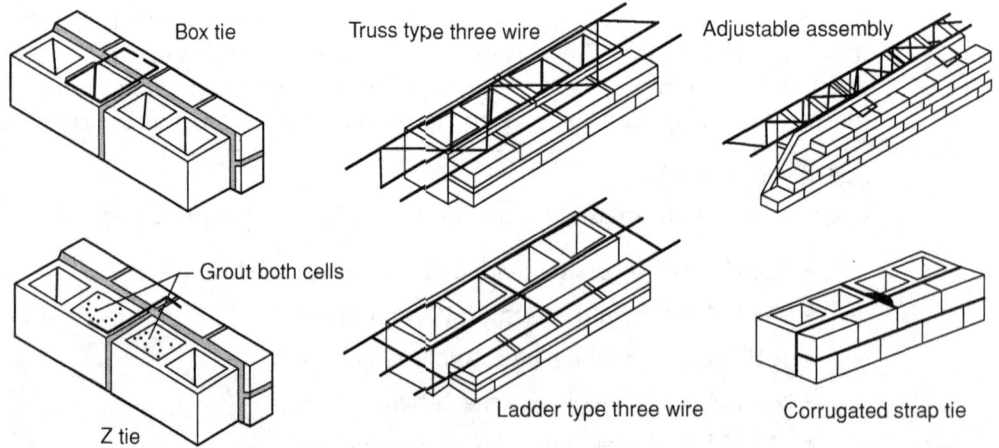

FIGURE 2.6 Typical bed joint reinforcement. (*Source*: Figure 1 of National Concrete Masonry Association TEK 12-01A. Typical wall ties.)

Materials Used in Masonry Construction 41

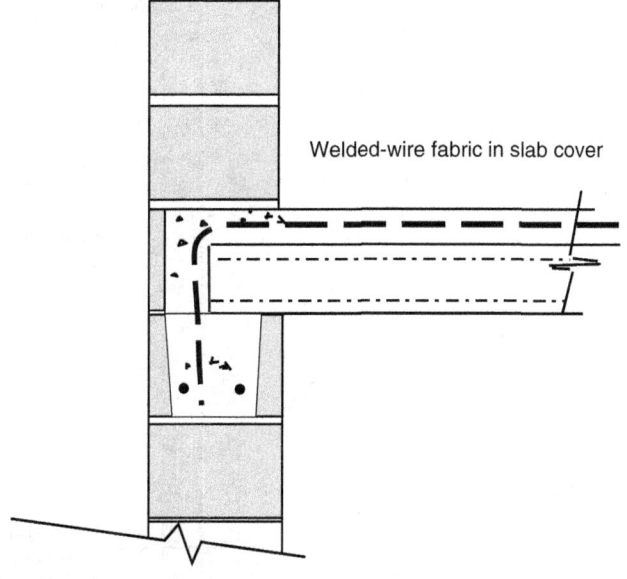

FIGURE 2.7 Typical use of welded wire reinforcement.

Adjustable ties are useful for accommodating differences in elevation of bed joints. As shown in Fig. 2.12, adjustable ties can be very flexible if they are used at large eccentricities (difference in elevation of bed joints).

2.8.3 Sealants

Sealants are used to prevent the passage of water at places where gaps are intentionally left in masonry walls. Three basic kinds of gaps (joints) are used

- Expansion joints are used in clay masonry to accommodate expansion
- Control joints are used in concrete masonry to conceal cracking due to shrinkage
- Construction joints are placed between different sections of a structure

Sealants are most commonly formulated using synthetic polymers such as silicone, neoprene, latex, or butyl rubber. Their elastic properties include compressibility, expressed as the ratio of minimum thickness to original thickness. Because the polymers comprising them deteriorate under exposure to ultraviolet light and ozone, the normal life of sealants in exterior exposures is about 7 years. Sealants should be replaced at intervals approximately equal to their expected life.

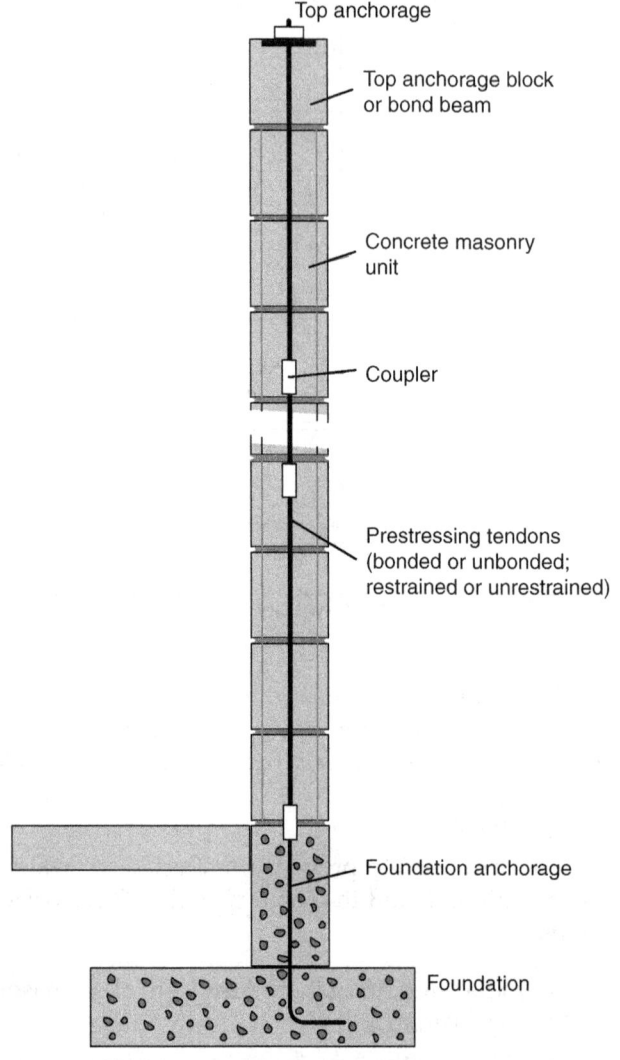

FIGURE 2.8 Typical use of posttensioning tendons.

2.8.4 Flashing

Flashing is a flexible waterproof barrier, intended to permit water that has penetrated the outer wythe to re-exit the wall. It is placed at every interruption of the vertical drainage cavity, including the following locations: at the bottom of each story level (on shelf angles or foundations, as shown in Fig. 2.13), over window and door lintels, and under window and door sills.

Materials Used in Masonry Construction

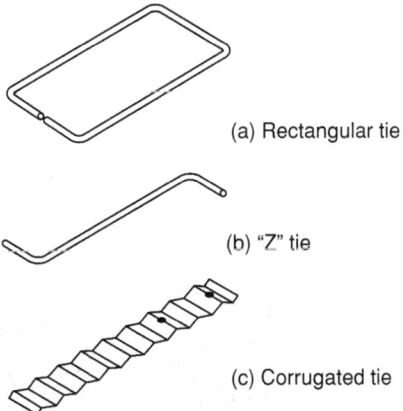

FIGURE 2.9 Typical veneer ties. (*Source*: Technical Note 44B. © Brick Industry Association.)

Flashing is made of stainless steel, copper, plastic-coated aluminum, plastic, rubberized asphalt, EPDM (ethylene-propylene-diene monomer), or PVC (polyvinyl chloride). Metallic flashing lasts much longer than plastic flashing. Nonmetallic flashings are subject to tearing. Modern self-adhering flashing of rubberized asphalt is a good compromise between durability and ease of installation.

Flashing should be applied above shelf angles, above door and window openings, and below door and window openings. Flashing should be

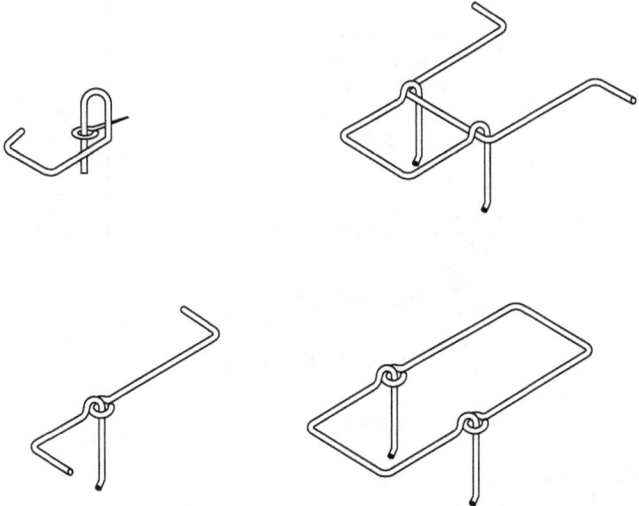

FIGURE 2.10 Typical adjustable pintle ties. (*Source*: Technical Note 44B. © Brick Industry Association.)

44 Chapter Two

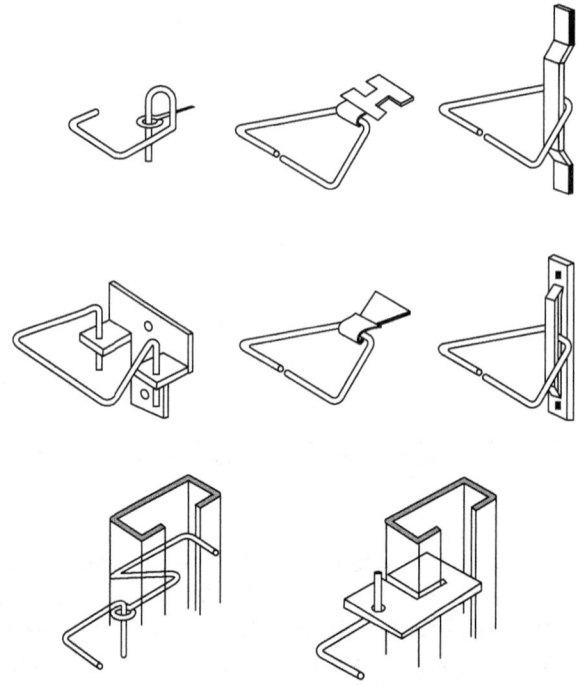

Figure 2.11 Typical connectors. (*Source*: Technical Note 44B. © Brick Industry Association.)

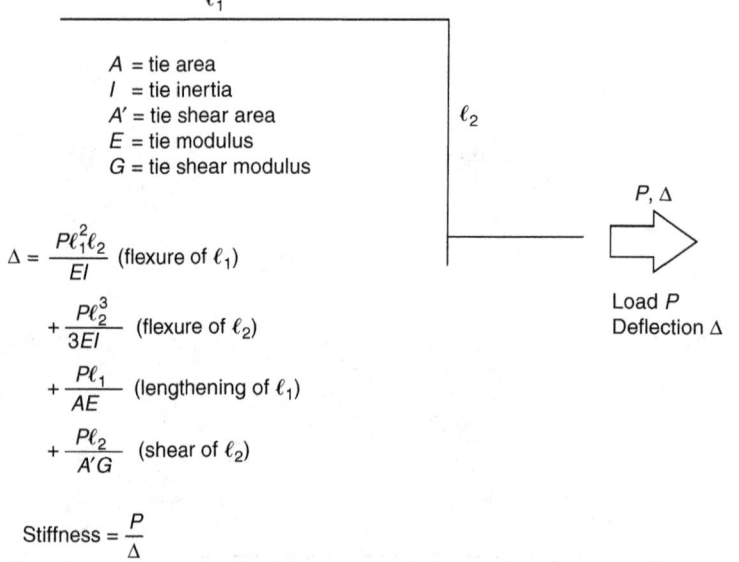

A = tie area
I = tie inertia
A' = tie shear area
E = tie modulus
G = tie shear modulus

$$\Delta = \frac{P\ell_1^2 \ell_2}{EI} \text{ (flexure of } \ell_1\text{)}$$

$$+ \frac{P\ell_2^3}{3EI} \text{ (flexure of } \ell_2\text{)}$$

$$+ \frac{P\ell_1}{AE} \text{ (lengthening of } \ell_1\text{)}$$

$$+ \frac{P\ell_2}{A'G} \text{ (shear of } \ell_2\text{)}$$

Load P
Deflection Δ

$$\text{Stiffness} = \frac{P}{\Delta}$$

Figure 2.12 Sample calculation for stiffness of adjustable ties.

Materials Used in Masonry Construction 45

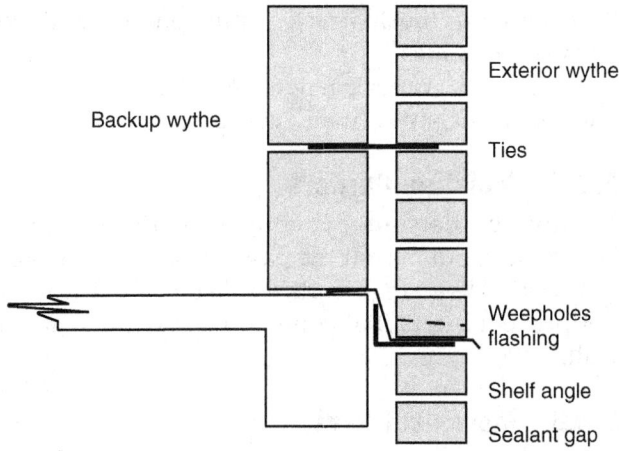

FIGURE 2.13 Placement of flashing at shelf angles in clay masonry veneer.

lapped, and ends of flashing should be defined by end dams (flashing turned up at ends). Directly above the level of the flashing, weepholes should be provided at 24-in. spacing.

2.8.5 Coatings

Coatings include paints and water-repellent coatings.

- Paint is less durable than the masonry it covers.
- Water-repellent coatings cannot bridge wide cracks. They tend to trap water behind them, causing freeze-thaw damage behind the coating, and also crypto-florescence. They are generally unnecessary for clay masonry, and generally less effective than integral water-repellent admixtures for concrete masonry.

2.8.6 Vapor Barriers

Vapor barriers are waterproof membranes (usually polyethylene or PVC). They are intended to prevent the passage of water in vapor or liquid form, and thereby prevent interstitial condensation within the air space of a drainage wall.

In warm climates, warm air from the outside of the building can pass through the outer wythe of a cavity wall and condense within the interior wythe. The vapor barrier is therefore placed against the exterior face of the interior wythe.

In cold climates, warm air from the inside of the building will pass through the inner wythe of a cavity wall and condense within the cavity. The vapor barrier should therefore be placed against the interior face of the exterior wythe. In practice, this conflicts with the above requirements for warm climates.

As a result, vapor barriers are often placed on the exterior face of the interior wythe anyway, as a compromise.

2.8.7 Moisture Barriers

Moisture barriers are membranes that prevent the passage of water in liquid form, but permit the passage of water in vapor form. One example is Tyvek®. They are intended to keep liquid water out of walls. They do not prevent interstitial condensation within the air space of a drainage wall.

2.8.8 Movement Joints

Three basic kinds of movement joints are used in masonry construction: expansion joints, control joints, and construction joints.

1. Expansion joints, shown in Figs. 2.14 and 2.15, are used in clay masonry to accommodate expansion. In these figures the backer rod is shown separated from the sealant, so that they can be distinguished. In reality, they are in contact.

2. Control joints, shown in Fig. 2.16, are used in concrete masonry to conceal cracking due to shrinkage.

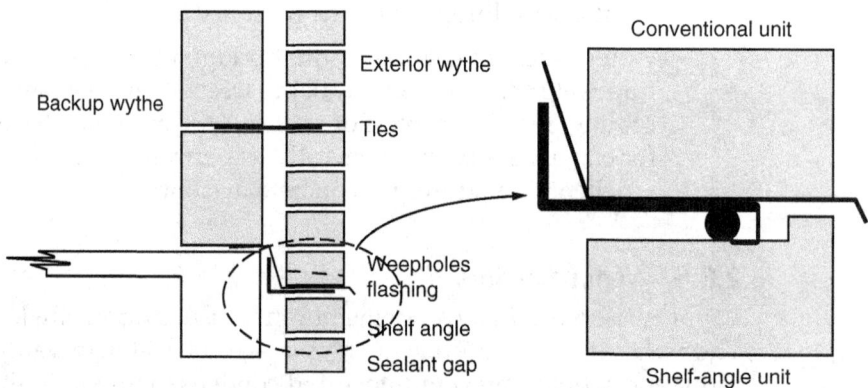

FIGURE 2.14 Horizontally oriented expansion joint under shelf angle.

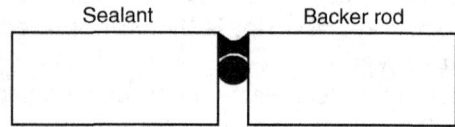

FIGURE 2.15 Vertically oriented expansion joint.

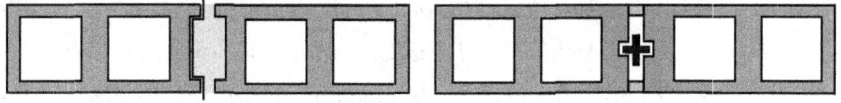

FIGURE 2.16 Shrinkage control joint.

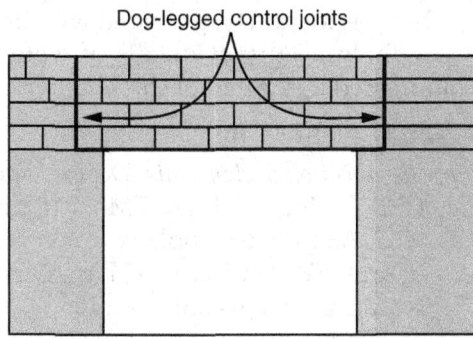

FIGURE 2.17 Example of control joints at openings in concrete masonry.

Control joints are placed at openings in concrete masonry. As shown in Fig. 2.17, the joints are "dog-legged" so that the lintel can be supported by the masonry on both sides of the opening, and also so that it can be restrained against uplift by vertical reinforcement at the edges of the opening.

3. Construction joints are placed between different sections of a structure.

2.9 Design of Masonry Structures Requiring Little Structural Calculation

2.9.1 Design Steps for Structures Requiring Little Structural Calculation

Many masonry structures require little structural calculation. Their primary design steps are layout, design involving primarily structural layout, detailing, and material specification. In this section these steps are outlined. This section can be viewed as a summary of material previously presented in Chaps. 1 and 2.

1. Layout of overall structural configuration:

    a. Modularity: Adjust the plan dimensions to the nominal dimensions of the units to be used.

b. *Selection of the overall structural system*: Locate walls in plan.

  c. *Architectural details*: Locate windows and doors.

2. Specify type of wall system according to desired level of water-penetration resistance. In areas of severe driving rain, specify a drainage wall, or a fully grouted barrier wall with a thickness of at least 8 in. If a drainage wall is specified, it should have at least a 2-in. cavity, and be provided with drainage details (flashing and weepholes). Further details of water-penetration resistance are addressed at the end of this chapter.

3. Specify masonry units:

   a. *Specification of clay units*: Decide whether building brick (ASTM C62), facing brick (ASTM C216), or some other kind of unit (e.g., ASTM C652, hollow brick) is required. Specify the grade of unit (SW, MW, or NW) based on the weathering index at the building's geographic location.

   b. *Specification of concrete units*: Decide whether conventional hollow units (ASTM C90) or some other type of unit is required.

4. Specify mortar: Specify an ASTM C270 mortar, Type S or Type N (normally by proportion), and cementitious system (cement-lime, masonry-cement, or mortar-cement). The proportion specification is normally preferable to the property specification, because it avoids the additional cost of testing, and the additional difficulty of having to decide what to do if the test results do not comply with the required values. The property specification permits some savings in material costs, in return for increases in costs due to testing. Although it is theoretically not required, it is useful to insert the words "by proportion" in the specification as additional protection against inadvertent and possibly improper mortar testing. If tensile bond strength is important, use either cement-lime mortar or mortar-cement mortar. While water-penetration resistance can be enhanced by using cement-lime mortar, this choice is probably not as important as choosing a properly specified drainage wall.

5. Specify grout: Normally, specify a coarse grout, conforming to the proportion specifications of ASTM C476 (one part Portland cement or other cements, three parts sand, and two parts pea-gravel). The amount of water should be sufficient to obtain a slump of about 11 in.

6. Specify accessories: For simple structures with single-wythe walls, the only accessories will be deformed reinforcement, conforming to ASTM A615 (new steel). This is simply prescribed

(e.g., #4 bars @ 48 in. on centers) based on seismic design category or other considerations. Joint reinforcement could also be specified.

   a. For hollow units of concrete or clay, vertical reinforcement is placed in the continuous vertical cells, and horizontal reinforcement is placed in bond beam units (units with depressed webs).
   b. For solid units of concrete or clay, deformed reinforcement (vertical or horizontal) can be placed only in grouted spaces between wythes. Bed-joint reinforcement can be placed in the bed joints of a single-wythe wall.
   c. Specify connectors, flashing, and sealants.
7. Specify construction details:
   a. Foundation dowels.
   b. Splices between foundation dowels and vertical reinforcement (required lap length depends on bar diameter).
   c. Foundation details: See Sec. 2.9.3.
   d. Roof connection details: See Sec. 2.9.3.
8. Construction process:
   a. Decide whether or not to wet clay units: Check if the IRA exceeds 30, or use a field test (see if 20 drops of water, placed in a quarter-sized circle, are absorbed in 90 s or less). If the IRA exceeds 30, or if the drops are absorbed in the field test, wet the units briefly before laying.
   b. Place units in running or stack bond. Tool joints using concave tooling.
   c. Place reinforcement.
   d. Pour grout (clean cells, mist the cells with water, grout by high-lift or low-lift procedures).
   e. If possible, cure the masonry by keeping it damp.

2.9.2 Overall Starting Point for Reinforcement

An overall starting point for reinforcement for structures requiring little structural calculation is shown in Fig. 2.18. Structural design is discussed extensively in later chapters of this book.

2.9.3 Examples of Construction Details for Masonry Structures Requiring Little Structural Calculation

Examples of construction details for masonry structures requiring little calculation are given in the sections and figures below. These details are

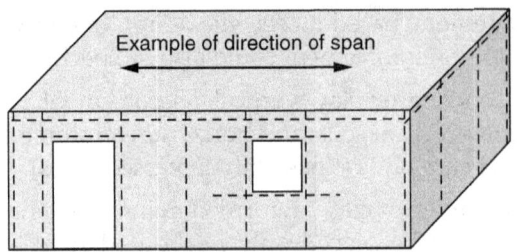

FIGURE 2.18 Overall starting point for reinforcement or structures requiring little structural calculation.

generic in nature. These can be supplemented by the details provided in NCMA and BIA technical notes.

1. Overall modularity:
    a. Overall modularity of the CMU wythe will be satisfied provided that the interior nominal dimensions of the CMU wythe are an even number of feet, and the units have a nominal thickness of 8 in. The exterior dimensions of each CMU wythe will be an even number of feet, plus 2 times 8 in. (two nominal CMU wythes). Any such exterior dimension can be laid out without cutting CMU, because the interior dimension is an even number of feet (some number of 16 in. units plus perhaps one 8 in. half-unit), and the exterior dimension is obtained by adding another nominal dimension of 16 in.
    b. Overall modularity of the clay wythe will be satisfied given the above, plus a 2-in. nominal air space and nominal 4 in. clay units. The exterior nominal dimensions of each clay wythe will be an even number of feet (see above), plus 2 times 8 in. (two nominal 8-in. CMU wythes), plus 2 times 2 in. (two nominal 2-in. air spaces), plus 2 times 4 in. (two nominal 4-in. clay wythes). Any such exterior dimension is an even number of feet, plus 16 in. plus 4 in. plus 8 in., or an even number of feet plus 28 in. This exterior dimension can always be made with nominal 8-in. units plus a nominal 4-in. half-unit.

    Overall modularity in this example is shown in Fig. 2.19.
2. Connections between floor slab and walls. These are exemplified in Fig. 2.20.
3. Connections between walls and roof. Examples of connections between walls and roof are exemplified by Figs. 2.21 and 2.22.
4. Locations of control joints in CMU wythe, and locations of expansion joints in clay wythe. An example of these is shown in Fig. 2.23.

Materials Used in Masonry Construction 51

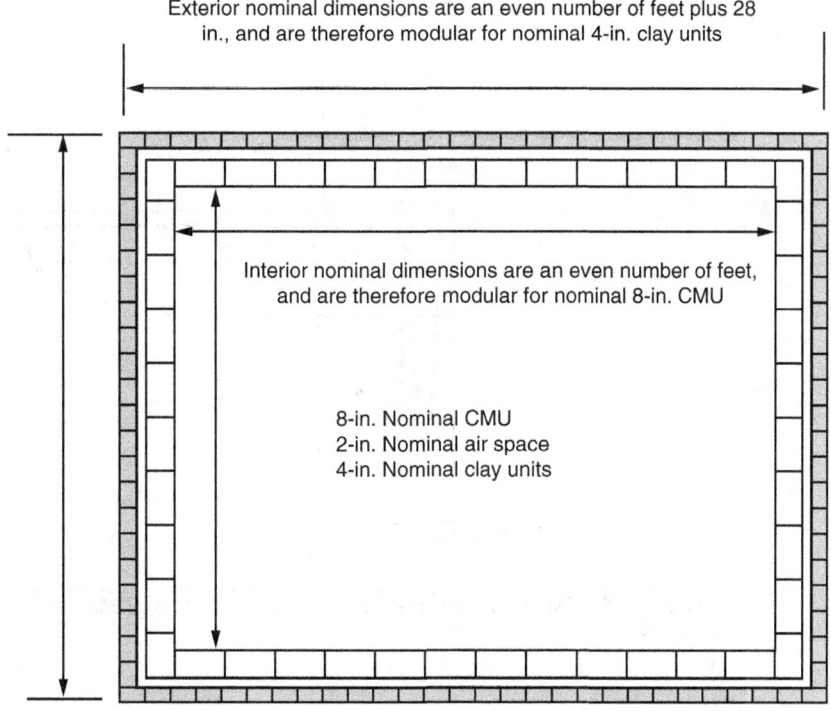

FIGURE 2.19 Example of overall modularity of a masonry structure in plan.

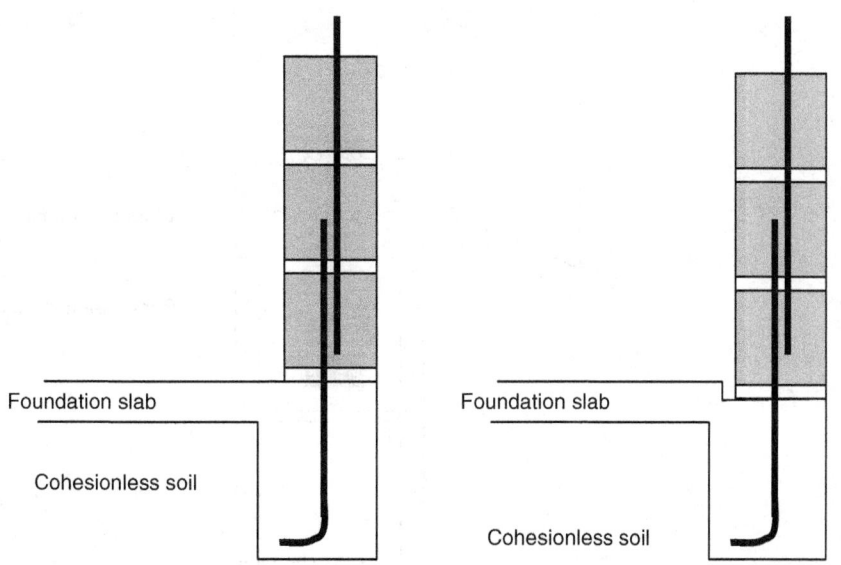

FIGURE 2.20 Foundation wall detail.

52   Chapter Two

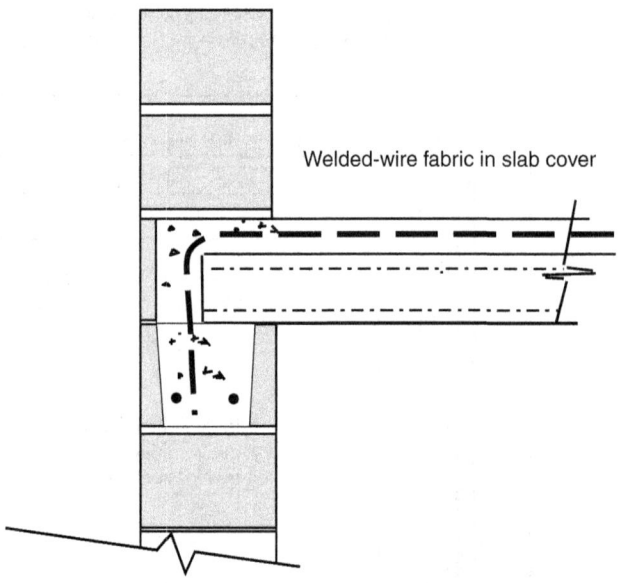

Figure 2.21 Detail of intersection between wall and precast concrete roof or floor slab.

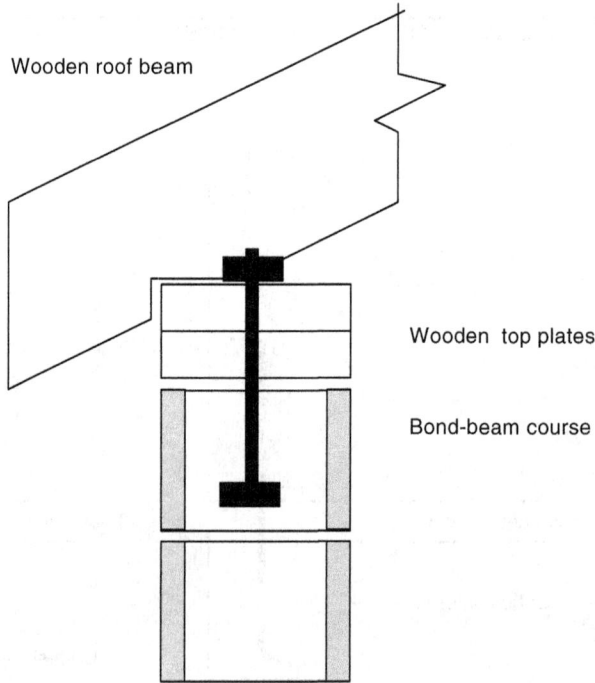

Figure 2.22 Detail of wall and wooden roof truss.

Materials Used in Masonry Construction 53

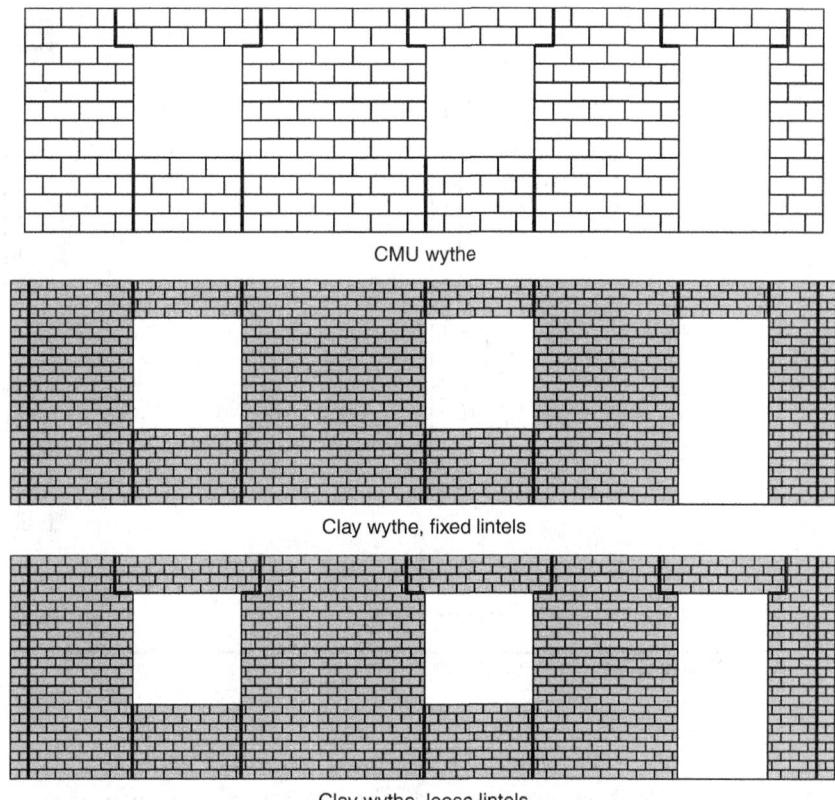

FIGURE 2.23 Elevations showing locations of control joints in CMU wythe, and locations of expansion joints in clay masonry veneer wythe (fixed lintel and loose lintel, respectively).

The CMU wythe has control joints above and below windows, and above doors. The control joints above the windows and doors are normally offset from the jams so that the lintels produced by these joints can have 8 in. of bearing at each end. The control joints below the windows are normally even with the window jams because there is no need for an offset. The clay wythe has expansion joints within 18 in. of the corners, and at window and door openings. If the lintels in the clay masonry wythe are supported on the CMU wythe (fixed lintels), the expansion joints at openings can be even with the jambs. If the lintels in the clay masonry wythe are supported only by the clay masonry wythe (loose lintels), the expansion joints above openings must be offset from the window and door jambs so that the lintels produced by these joints can have 8 in. of bearing at each end, just like the lintels in the CMU wythe.

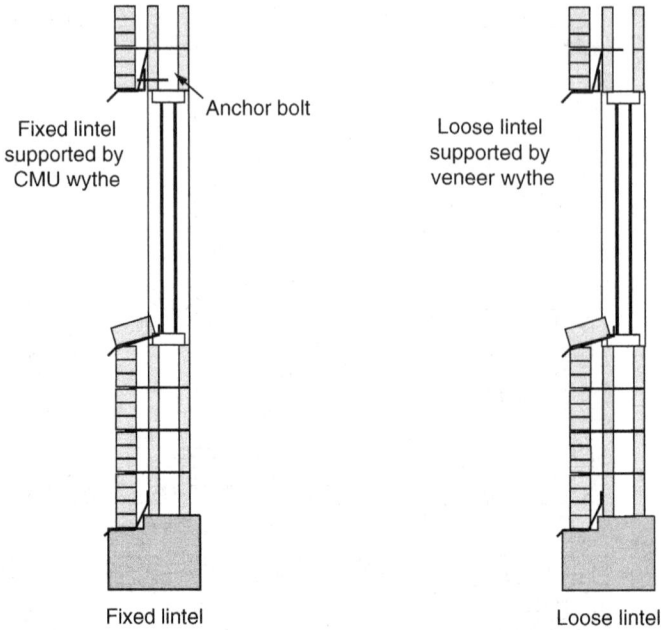

FIGURE 2.24 Wall sections at lintels.

5. Wall sections at windows. Details of wall sections at windows are shown in Fig. 2.24. The left and right figures correspond to fixed lintels and loose lintels, respectively.

The details are identical except that in the fixed-lintel detail, the lintel is supported by the CMU wythe (note the anchor bolt), whereas for the loose-lintel detail, the lintel is supported by the veneer wythe. In each case, the window head has weepholes in the veneer, and flashing with end dams. The window foot has an inclined masonry or precast concrete sill, with flashing and weepholes underneath. The wall has weepholes and flashing at the top of the foundation. In these details, the roof connection or parapet are not shown.

Sections at doors are the same, except that the door sill is at foundation level.

2.10 How to Increase Resistance of Masonry to Water Penetration

Water penetration resistance of masonry depends on wall type, workmanship, and materials. In this section, additional information is presented on each of these.

1. Specification and design:
    a. Specify and design wall types appropriate for the severity of driving rain expected in the geographic location of the building. In areas of severe driving rain, specify a drainage wall or a fully grouted barrier wall with a thickness of at least 8 in.
    b. If a drainage wall is specified, specify and design to reduce water penetration through the outer wythe.
        (1) Don't let the outer wythe crack under service loads. If masonry cement is used, consider the effect of its lower tensile bond strength in determining whether the veneer will crack. If steel studs are used, consider the effect of their flexibility in determining whether the veneer will crack. Follow industry recommendations that the out-of-plane deflection of the studs not exceed their span divided by 600.
        (2) Specify proper sealant joints in outer wythe.
    c. If a cavity wall is specified, specify and design to keep the cavity open and properly drained.
        (1) Specify proper weepholes and flashing to keep water out of cavity and direct it outward if it gets in.
        (2) Specify hot-dip galvanized or stainless-steel ties.
        (3) Specify a cavity at least 2 in. wide. Keep the cavity clean by beveling the back of the joint, using a board to catch mortar droppings, or both.
        (4) Use a vapor barrier on the exterior face of the interior wythe, to prevent interstitial condensation within the inner wythe.
    d. Design and detail to accommodate differential movement.
        (1) Provide horizontally oriented expansion joints in clay masonry walls under shelf angles. Over time, clay masonry typically expands about 300 µe, due to permanent moisture expansion and freeze-thaw expansion. In contrast, concrete and concrete masonry shrink is typically about 600 µe. These opposite tendencies combine to produce differential strain of about 1000 µe over a 12-ft (144 in.) story height, that corresponds to a differential deformation of 0.144 in. In other words, a gap of 0.144 in. under a shelf angle would be expected to close completely during the life of a building. Because the gap must be filled with sealant that can compress to about half of its original thickness, the gap must be twice the 0.144 in., or 0.29 in. A gap of 3/8 in. is typically used. If a sufficient gap is not used, the veneer will be loaded vertically, and can spall, crack, or even buckle under the load.

(2) Provide vertically oriented expansion joints in clay masonry walls near corners. Expansion of clay masonry, restrained at building corners, causes moments about vertical axes near the corners. Vertically oriented expansion joints prevent such moments and resulting cracking.

(3) Use bond breaker between clay masonry walls and concrete foundations, slabs, and roofs. Expansion of clay masonry (about 300 µe) and shrinkage of concrete (about 600 µe) combine to give differential strain of 1000 µe. If restrained, this differential strain can crack the concrete foundation.

(4) Use control joints at door and window openings of concrete masonry walls. These will prevent tensile stresses from restrained shrinkage.

(5) If composite brick-block masonry walls are used, consider the effect of restrained differential movement on interfacial stresses and wall deformations.

2. Construction:

a. Use compatible combinations of mortar and clay units. Clay units with high IRA should be wetted and used with high-retentivity mortar.

b. Mix and batch mortar properly: Measure ingredients accurately by volume. To distribute them thoroughly in the mix, combine cementitious ingredients with part of the sand and part of the water. If all the water is added at the beginning, the initial mix will be too fluid, and the cementitious ingredients will again not combine well. The following mixing sequence is taken from ASTM C780 (field mortar):

Add all cement and lime + 1/2 sand + 3/4 water
Mix 2 min
Add remaining sand and water
Mix 3 to 8 min

c. Clean and roughen the foundation before laying the first course.

d. Lay units within 1 min of spreading the mortar bed.

e. Lay units without excessive tapping.

f. Use full bed and head joints.

g. Use concave-tooled joints. In particular, use raked joints for interior masonry only.

h. Retemper mortar as required to maintain workability. Do not use mortar more than 2-1/2 h after initial mixing.

i. After laying, keep masonry walls damp to help the mortar cure properly.

CHAPTER 3
Code Basis for Structural Design of Masonry Buildings

3.1 Introduction to Building Codes in the United States

The United States has no national design code, primarily because the United States Constitution has been interpreted as delegating building code authority to the states, some of which in turn delegate it to municipalities and other local governmental agencies. Design codes used in the United States are developed by a complex process involving technical experts, industry representatives, code users, and building officials. As it applies to the development of design provisions for masonry, this process is shown in Fig. 3.1, and is then described.

1. Consensus design provisions and specifications for materials or methods of testing are first drafted in mandatory language by technical specialty organizations, operating under consensus rules approved by the American National Standards Institute (ANSI), or (in the case of ASTM) rules that are similar in substance. Those consensus rules vary from organization to organization, but include requirements for the following:

    a. Balance of interests (producer, user, and general interest).

    b. Written balloting of proposed provisions, with prescribed requirements for a successful ballot.

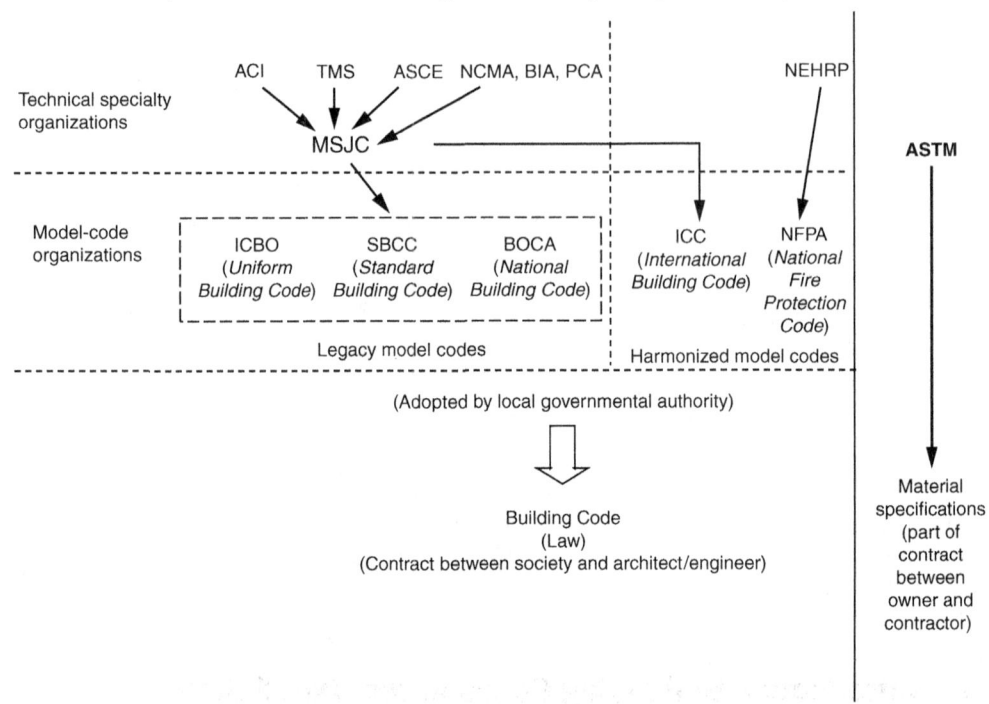

Figure 3.1 Schematic of process for development of design codes in the United States.

  c. Resolution of negative vote: Negative votes must be discussed and found nonpersuasive before a ballot item can pass. A single negative vote, if found persuasive, can prevent an item from passing.
  d. Public comment: After being approved within the technical specialty organization, the mandatory-language provisions must be published for review and comment by the general public. All comments are responded to, but do not necessarily result in further modification.
2. These consensus design and construction provisions are adopted, usually by reference and sometimes in modified form, by model code organizations, and take the form of model codes.
3. These model codes are adopted, sometimes in modified form, by local governmental agencies (such as states, cities, or counties). Upon adoption, but not before, they acquire legal standing as building codes.

3.1.1 Technical Specialty Organizations

Technical specialty organizations are open to designers, contractors, product suppliers, code developers, and end users. Their income (except for FEMA, a U.S. government agency) is derived from member dues and the sale of publications. Technical specialty organizations active in the general area of masonry include the following:

1. *American Society for Testing and Materials (ASTM):* Through its many technical committees, ASTM develops consensus specifications for materials and methods of test. Although some model code organizations use their own such specifications, most refer to ASTM specifications.

2. *American Concrete Institute (ACI):* Through its many technical committees, this group publishes a variety of design recommendations dealing with different aspects of concrete design. ACI Committee 318 develops design provisions for concrete structures. ACI is also involved with masonry, as one of the three sponsors of the Masonry Standards Joint Committee (MSJC). This committee was formed in 1978 to combine the masonry design provisions then being developed by ACI, ASCE, TMS, and industry organizations. ACI is the second of the three sponsoring societies of the Masonry Standards Joint Committee (MSJC), responsible for the development and updating of the MSJC *Code* and *Specification* (MSJC 2008a, b).

3. *American Society of Civil Engineers (ASCE):* ASCE is a joint sponsor of many ACI technical committees dealing with concrete or masonry. ASCE is the third of the three sponsoring societies of the Masonry Standards Joint Committee (MSJC). ASCE publishes *ASCE 7-05* (2005), which prescribes design loadings and load factors for all structures, independent of material type.

4. *The Masonry Society (TMS):* Through its technical committees, this group influences different aspects of masonry design. TMS is the lead sponsor of the Masonry Standards Joint Committee (MSJC). TMS also publishes a *Masonry Designers' Guide* to accompany the MSJC design provisions.

3.1.2 Industry Organizations

1. *Portland Cement Association (PCA):* This marketing and technical support organization is composed of cement producers. Its technical staff participates in technical committee work.

2. *National Concrete Masonry Association (NCMA):* This marketing and technical support organization is composed of producers of

concrete masonry units. Its technical staff participates in technical committee work and also produces technical bulletins which can influence consensus design provisions.

3. *Brick Industry Association (BIA):* This marketing, distributing, and technical support organization is composed of clay brick and tile producers and distributors. Its technical staff participates in technical committee work and also produces technical bulletins which can influence consensus design provisions.

4. *National Lime Association (NLA):* This marketing and technical support organization is composed of hydrated lime producers. Its technical staff participates in technical committee work.

5. *Expanded Shale Clay and Slate Institute (ESCSI):* This marketing and technical support organization is composed of producers. Its technical staff participates in technical committee meetings.

6. *International Masonry Institute (IMI):* This is a union contractor-craftworker collaborative supported by dues from union masons. Its technical staff participates in technical committee meetings.

7. *Mason Contractors' Association of America (MCAA):* This organization is composed of union and nonunion mason contractors. Its technical staff participates in technical committee meetings.

8. *Autoclaved Aerated Concrete Products Association (AACPA):* This organization is composed of producers of autoclaved aerated concrete units. Its technical staff participates in technical committee meetings.

3.1.3 Governmental Organizations

Federal Emergency Management Agency (FEMA): FEMA has jurisdiction over the National Earthquake Hazard Reduction Program, and develops and periodically updates the NEHRP provisions (NEHRP, 2003), a set of recommendations for earthquake-resistant design. That document includes provisions for masonry design. The document is published by the Building Seismic Safety Council (BSSC), which operates under a contract with the National Institute of Building Sciences (NIBS). BSSC is not a consensus organization. Its recommended design provisions are intended for consideration and possible adoption by consensus organizations. The *Recommended Provisions* (NEHRP, 2003) are the latest of a series of such documents, now issued at 6-year intervals, and pioneered by *ATC 3-06*, which was issued by the Applied Technology Council in 1978 under contract to the National Bureau of Standards. The 2003 NEHRP *Recommended Provisions* addresses the broad issue of seismic regulations for buildings. They contain chapters dealing with the determination of seismic loadings on structures, and with the design of masonry structures for those loadings. The role of the NEHRP *Provisions* is evolving, with the next edition anticipated to reference existing standards (ASCE 7 for Loads, MSJC for Masonry Design, etc.) much more, and to emphasize the development of innovative design and construction suggestions for consensus design and construction standards.

3.1.4 Model-Code Organizations

Model-code organizations are composed primarily of building officials, although designers, contractors, product suppliers, code developers, and end users can also be members. Their income is derived from dues and the sale of publications. Historically, the United States had three legacy model-code organizations:

1. *International Conference of Building Officials (ICBO)*: In the past, this group developed and published the Uniform Building Code (UBC).
2. *Southern Building Code Congress International (SBCCI)*: In the past, this group developed and published the Standard Building Code (SBC).
3. *Building Officials and Code Administrators International (BOCA)*: In the past, this group developed and published the National Building Code (NBC).

In the past, certain model codes were used more in certain areas of the country. The *Uniform Building Code* was used throughout the western United States and in the state of Indiana. It was used in California until January of 2008, in the slightly modified form of the *California Building Code*. The *Standard Building Code* was used in the southern part of the United States. The *National Building Code* was used in the eastern and northeastern United States.

In 1996, intensive efforts began in the United States to harmonize the three model building codes. The primary harmonized model building code is called the *International Building Code* (IBC). It has been developed by the International Code Council (ICC), composed of primarily of building code officials of the three legacy model-code organizations. The first edition of the *International Building Code* (IBC, 2000) was published in May 2000. In most cases, it references consensus design provisions and specifications. It is intended to take effect when adopted by local jurisdictions, and to replace the three legacy model building codes. Its latest edition was published in 2009. The IBC has been adopted or is scheduled for adoption in most governmental jurisdictions of the United States.

Another model code, adopted in only a few jurisdictions, is published by the National Fire Protection Association (NFPA 5000).

1. *International Code Council (ICC):* This group develops and publishes the *International Building Code* (IBC).
2. *National Fire Protection Association (NFPA):* This group develops and publishes *NFPA 5000*.

3.2 Introduction to the Calculation of Design Loading Using the 2009 IBC

Design loadings for buildings in general, including masonry buildings, are prescribed by the legally adopted building code. In most parts of the United States, the legally adopted building code is based on the *International Building Code* (IBC). In the following sections of this chapter, background information and sample calculations are presented for each of the principal code-mandated design loadings:

- Gravity loads (dead load and live load)
- Wind loads
- Earthquake loads

These loads are used in many design examples in subsequent chapters of this book.

3.3 Gravity Loads according to the 2009 IBC

3.3.1 Dead Load according to the 2009 IBC

Dead load is due to the weight of the structure itself, plus permanently attached components. Calculation of dead loads is not discussed further at this point. It is discussed in specific examples, as are IBC loading combinations including dead load.

3.3.2 Floor Live Load according to the 2009 IBC

Live load is prescribed by the 2009 IBC. The loading provisions of the 2009 IBC are discussed here using the section numbers taken from that document. The 2009 IBC itself uses loads that are almost identical to those prescribed by *ASCE 7-05 (Supplement)*. In the future, the IBC will tend more and more to reference ASCE 7 loads directly. Minimum live loads (L) for floors (from Table 1607.1 of the 2009 IBC) are given in Table 3.1.

Occupancy or use	Uniform (psf)	Concentrated (lb)
4. Assembly areas w/ moveable seats	100	—
5. Balconies (exterior) and decks	Same as occupancy served	
9. Corridors, except as otherwise indicated	100	
26. Offices	50	2000
27. Residential	40	
27. Residential, corridors, and public areas of hotels	100	
29. Ordinary flat roofs	20	300
30. School classrooms	40	1000
30. School corridors above first floor	80	1000
30. School corridors, first floor	100	1000
35. Stairs and exits	100	
35. Stairs and exits, 1- and 2-family dwellings	40	
37. Stores, retail, first floor	100	1000
37. Stores, retail, upper floors	75	1000

Source: Table 1607.1 of the 2009 IBC.

TABLE 3.1 Minimum Live Loads (L) for Floors

By Sec. 1607.9.1 of the 2009 IBC, live loads are permitted to be reduced based on the tributary area over which those live loads act. Live loads in public assembly areas (balconies, corridors, and stairs) are not permitted to be reduced. The live-load reduction factor, shown below, applies to elements for which the product $K_{LL} A_T$ equals or exceeds 400 ft² as shown here

$$L = L_o \left(0.25 + \frac{15}{\sqrt{K_{LL} A_T}} \right)$$

where L = reduced design live load per square foot of area supported by the member
L_o = unreduced design live load per square foot of area supported by the member (see Table 1607.1, IBC, 2009)
K_L = live element factor (see Table 1607.9.1, IBC, 2009)
A_T = tributary area, in square feet

Element	K_{LL}
Interior columns	4
Exterior columns without cantilever slabs	4
Edge columns with cantilever slabs	3
Corner columns with cantilever slabs	2
Edge beams without cantilever slabs	2
Interior beams	2

All other members not identified above including
    Edge beams with cantilever slabs
    Cantilever beams
    One-way slabs
    Two-way slabs
    Members without provisions for continuous shear transfer normal to their span
	1

TABLE 3.2 Table 1607.9.1 Live Load Element Factor, K_{LL}

L shall not be less than $0.50\,L_o$ for members supporting one floor and L shall not be less than $0.40\,L_o$ for members supporting more floors.

Live-load reduction factors are given in Table 1607.9.1 of the 2009 IBC, reproduced in this chapter as Table 3.2.

3.3.3 Example of Floor Live-Load Reduction according to the 2009 IBC

Consider an interior beam of an office floor with a tributary area of 400 ft² ($K_{LL} = 2$).

$$L = L_o\left(0.25 + \frac{15}{\sqrt{K_{LL}\,A_T}}\right) = L_o\left(0.25 + \frac{15}{\sqrt{2 \times 400}}\right) = 0.78\,L_o$$

The lower limit of 0.50 does not govern, and

$$L = 0.78\,L_o$$

3.3.4 Example of a Wall Live-Load Reduction according to the 2009 IBC

Consider an interior wall supporting 10 floors, each with tributary area 400 ft² (assume $K_{LL} = 4$).

$$L = L_o\left(0.25 + \frac{15}{\sqrt{K_{LL}\,A_T}}\right) = L_o\left(0.25 + \frac{15}{\sqrt{4 \times (10 \times 400)}}\right) = 0.37\,L_o$$

Code Basis for Structural Design of Masonry Buildings

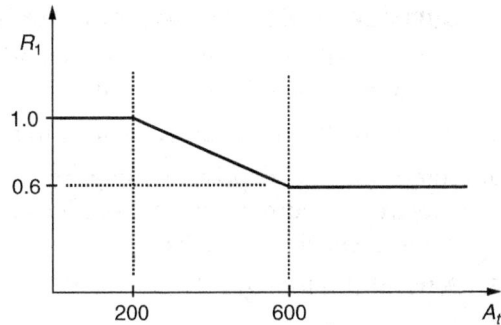

FIGURE 3.2 Graph showing permitted live-load reduction for roofs. (*Source:* Section 1607.11.2 of the 2009 IBC.)

The 0.40 limit governs, and

$$L = 0.40 L_o$$

3.3.5 Roof Live Load according to the 2009 IBC

In accordance with Sec. 1607.11.2 of the 2009 IBC, the minimum roof live load for most roofs is 20 psf. Roof live loads are permitted to be reduced in accordance with the following:

$$L_r = L_o R_1 R_2$$

where L_r = reduced roof live load per square foot
L_o = unreduced roof live load per square foot
R_1 = see Fig. 3.2
R_2 = 1.0 for flat roofs

The minimum reduced roof live load is 12 psf. The reduction is shown graphically in Fig. 3.2.

3.4 Wind Loading according to the 2009 IBC

According to Sec. 1609.1.1 of the 2009 IBC, wind loading is to be calculated using the provisions of *Minimum Design Loads for Buildings and Other Structures* (ASCE 7-05), or the simplified alternate all-heights method in Sec. 1609.6 of the 2009 IBC. ASCE 7 gives three procedures: a simplified procedure (ASCE 7-05, Sec. 6.4); an analytical procedure (ASCE 7-05, Sec. 6.5); and a wind tunnel procedure. We shall discuss the analytical procedure, because it is the most general. In the following discussion, section numbers refer to ASCE 7-05.

3.4.1 Summary of Design Procedure for Wind Loading

1. Determine the basic wind speed V and wind directionality factor K_d in accordance with Sec. 6.5.4.
2. Determine the importance factor I in accordance with Sec. 6.5.5.
3. Determine the exposure category or exposure categories and velocity pressure exposure coefficient K_z or K_h, as applicable, in accordance with Sec. 6.5.6.
4. Determine a topographic factor K_{zt} in accordance with Sec. 6.5.7.
5. Determine a gust effect factor G or G_f, as applicable, in accordance with Sec. 6.5.8.
6. Determine an enclosure classification in accordance with Sec. 6.5.9.
7. Determine an internal pressure coefficient GC_{pi} in accordance with Sec. 6.5.11.1.
8. Determine the external pressure coefficients C_p or GC_{pf}, or force coefficients C_f, as applicable, in accordance with Sec. 6.5.11.2 or 6.5.11.3, respectively.
9. Determine the velocity pressure q_z or q_h, as applicable, in accordance with Sec. 6.5.10.
10. Determine the design wind load P or F in accordance with Secs. 6.5.12, 6.5.13, 6.5.14, and 6.5.15.

Now let's discuss each step in more detail.

Step 1: Determine the basic wind speed V and wind directionality factor K_d in accordance with Sec. 6.5.4.

Refer to Fig. 3.3.

Basic wind speeds are described in terms of a 3-s gust speed (average speed over a 3-s window), with a 2 percent annual probability of exceedance (50-year wind).

In earlier codes, wind speeds were often described in terms of "fastest-mile wind speed" (the speed with which a group of hypothetical air particles would travel a distance of 1 mi). For a wind speed of 60 mi/h (1 mi/min), this would be equivalent to a 60-s gust speed. For a wind speed of 120 mi/h, it would be equivalent to a 30-s gust speed. For all practical cases, the 3-s gust speed is greater than the "fastest mile" speed. Therefore, design wind speeds in modern codes appear to be greater than they were 10 years ago. This is addressed by reductions in coefficients, so that actual wind loads are about the same in many cases. To convert equivalent basic wind speeds, use Table 3.3.

The wind directionality factor K_d is determined using Table 3.4.

Step 2: Determine the importance factor, I, in accordance with Sec. 6.5.5.

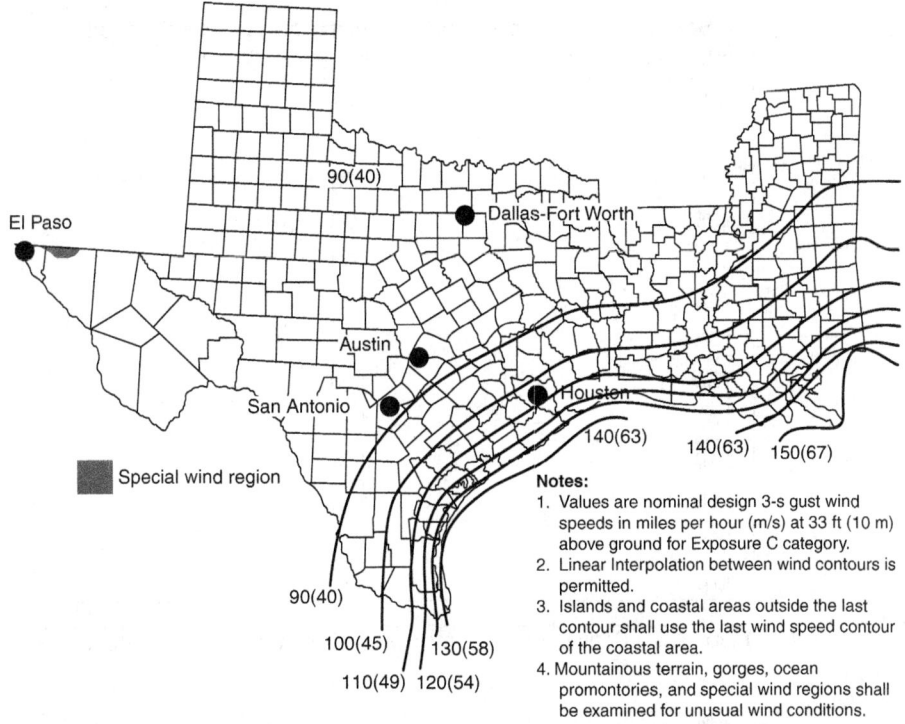

Figure 3.3 Basic Wind speed—Western Gulf of Mexico Hurricane Coastline. (*Source*: Information is adapted from ASCE 7-05.) (Adapted from Figure 6-1a of ASCE 7-05.)

The importance factor depends on the "classification" of a building, which is a function of its occupancy. Most buildings are classified in category II, which corresponds to an importance factor of 1.0. Refer to Table 6-1 and IBC Table 1604.5.

Step 3: Determine the exposure category or exposure categories and velocity pressure exposure coefficient, K_z or K_h, as applicable, in accordance with Sec. 6.5.6.

Exposure categories depend on Wind Direction and Sectors (Sec. 6.5.6.1) and Surface Roughness Categories (Sec. 6.5.6.2):

Exposure B: Urban and suburban areas

Exposure C: Open terrain with scattered obstructions

Exposure D: Flat, unobstructed areas exposed to wind flowing over open water

The velocity pressure exposure coefficients are defined in Table 3.5.

Essentially, Case 2 permits lower coefficients in return for a more complex calculation.

Figure 3.4 prescribes wind loading combinations.

3-Second gust wind speed, miles per hour	Fastest mile wind speed, miles per hour
85	71
90	76
100	85
105	90
110	95
120	104
125	109
130	114
140	123
145	128
150	133
160	142
170	152

TABLE 3.3 Conversion of Wind Speeds from 3-s Gust to Fastest Mile

Structure type	Directionality factor, K_d*
Buildings	
Main wind force resisting system	0.85
Components and cladding	0.85
Arched roofs	0.85
Chimneys, tanks, and similar structures	
Square	0.90
Hexagonal	0.95
Round	0.95
Solid signs	0.85
Open signs and lattice framework	0.85
Trussed towers	
Triangular, square, rectangular	0.85
All other cross sections	0.95

*Directionality factor, K_d, has been calibrated with combinations of loads specified in Sec. 2. This factor shall only be applied when used in conjuction with load combinations specified in Secs. 2.3 and 2.4 of IBC (2009).
Source: Table 6-4 of ASCE 7-05.

TABLE 3.4 Wind Directionality Factor, K_d

Height above ground level, z		Exposure (Note 1)			
		B		C	D
ft	(m)	Case 1	Case 2	Cases 1 & 2	Cases 1 & 2
0–15	(0–46)	0.70	0.57	0.85	1.03
20	(6.1)	0.70	0.62	0.90	1.08
25	(7.6)	0.70	0.66	0.94	1.12
30	(9.1)	0.70	0.70	0.98	1.16
40	(12.2)	0.76	0.76	1.04	1.22
50	(15.2)	0.81	0.81	1.09	1.27
60	(18)	0.85	0.85	1.13	1.31
70	(21.3)	0.89	0.89	1.17	1.34
80	(24.4)	0.93	0.93	1.21	1.38
90	(27.4)	0.96	0.96	1.24	1.40
100	(30.5)	0.99	0.99	1.26	1.43
120	(36.6)	1.04	1.04	1.31	1.48
140	(42.7)	1.09	1.09	1.36	1.52
160	(48.8)	1.13	1.13	1.39	1.55
180	(54.9)	1.17	1.17	1.43	1.58
200	(61.0)	1.20	1.20	1.46	1.61
250	(76.2)	1.28	1.28	1.53	1.68
300	(91.4)	1.35	1.35	1.59	1.73
350	(106.7)	1.41	1.41	1.64	1.78
400	(121.9)	1.47	1.47	1.69	1.82
450	(137.2)	1.52	1.52	1.73	1.86
500	(152.4)	1.56	1.56	1.77	1.89

Note:
1. Case 1: a. All components and cladding.
      b. Main wind force resisting system in low-rise building designed using Fig. 6-10.
   Case 2: a. All main wind force resisting systems in buildings except those in low-rise buildings designed using Fig. 6-10.
      b. All main wind force resisting systems in other structures.
2. The velocity pressure exposure coefficient, K_z, may be determined from the following formula:

   For $15 \text{ ft} \leq z < z_g$ For $z \leq 15 \text{ ft}$
   $K_z = 2.01(z/z_g)^{2/\alpha}$ $K_z = 2.01(15/z_g)^{2/\alpha}$

   Note: z shall not be taken less than 30 ft for Case 1 in exposure B.
3. α and z_g are tabulated in Table 6-2.
4. Linear interpolation for intermediate values of height z is acceptable.
5. Exposure categories are defined in Sec. 6.5.6.

Source: Table 6-3 of ASCE 7-05.

TABLE 3.5 Velocity Pressure Exposure Coefficients, K_h and K_z

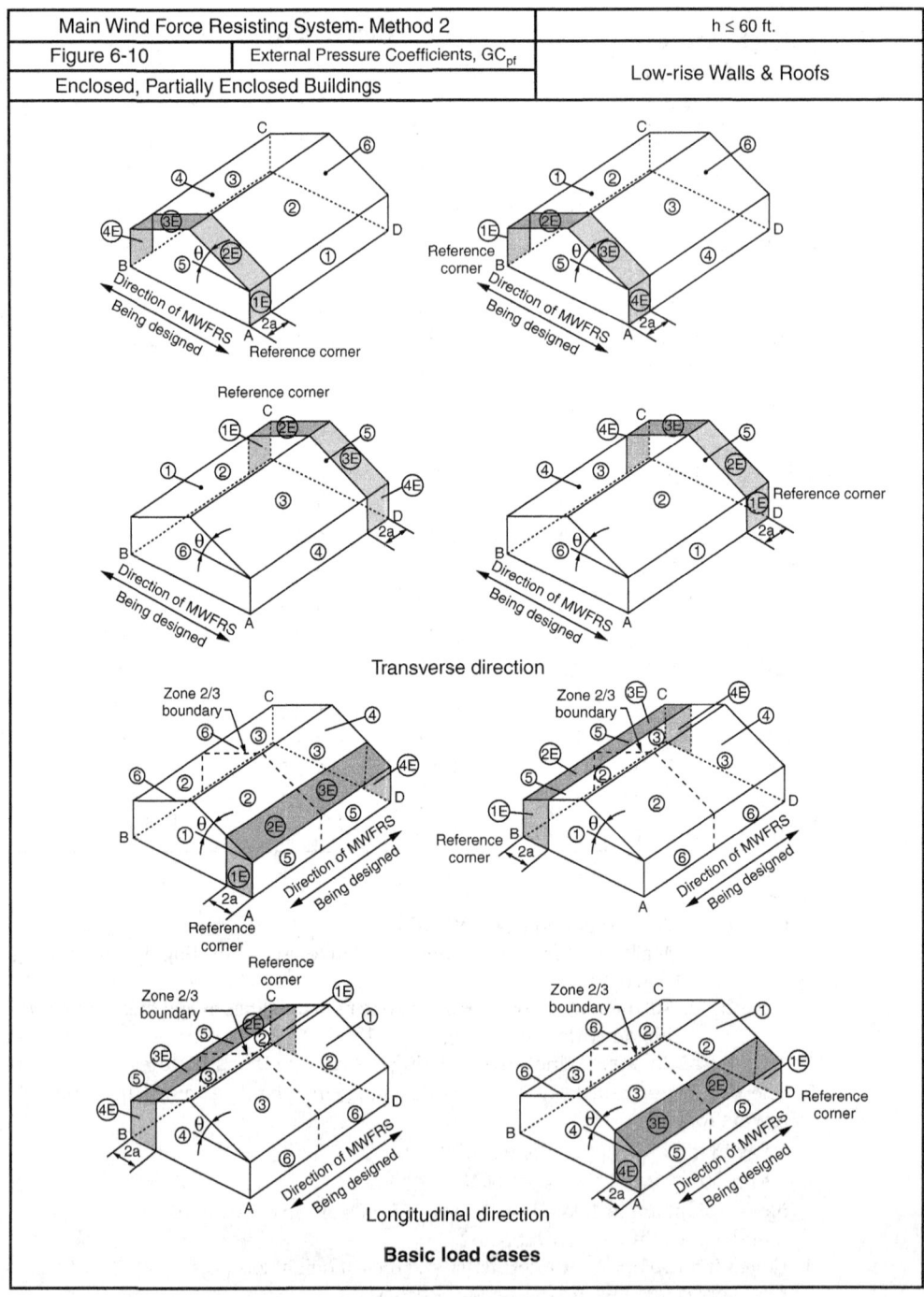

FIGURE 3.4 External pressure coefficients, GC_{pf} for main wind force-resisting system ($h < 60$ ft). (*Source:* Figure 6-10 of ASCE 7-05.)

Main Wind Force Resisting System- Method 2	$h \leq 60$ ft.
Figure 6-10 (cont'd) External Pressure Coefficients, GC_{pf} Enclosed, Partially Enclosed Buildings	Low-rise Walls & Roofs

Roof angle θ (degrees)	Building Surface									
	1	2	3	4	5	6	1E	2E	3E	4E
0–5	0.40	−0.69	−0.37	−0.29	−0.45	−0.45	0.61	−1.07	−0.53	−0.43
20	0.53	−0.69	−0.48	−0.43	−0.45	−0.45	0.80	−1.07	−0.69	−0.64
30–45	0.56	0.21	−0.43	−0.37	−0.45	−0.45	0.69	0.27	−0.53	−0.48
90	0.56	0.56	−0.37	−0.37	−0.45	−0.45	0.69	0.69	−0.48	−0.48

Notes:
1. Plus and minus signs signify pressures acting toward and away from the surfaces, respectively.
2. For values of θ other than those shown, linear interpolation is permitted.
3. The building must be designed for all wind directions using the 8 loading patterns shown. The load patterns are applied to each building corner in turn as the Reference Corner.
4. Combinations of external and internal pressures (see Figure 6–5) shall be evaluated as required to obtain the most severe loadings.
5. For the torsional load cases shown below, the pressures in zones designated with a "T" (1T, 2T, 3T, 4T) shall be 25% of the full design wind pressures (zones 1, 2, 3, 4).
    Exception: One story buildings with h less than or equal to 30 ft (9.1 m), buildings two stories or less framed with light frame construction, and buildings two stories or less designed with flexible diaphragms need not be designed for the torsional load cases.
   Torsional loading shall apply to all eight basic load patterns using the figures below applied at each reference corner.
6. Except for moment-resisting frames, the total horizontal shear shall not be less than that determined by neglecting wind forces on roof surfaces.
7. For the design of the MWFRS providing lateral resistance in a direction parallel to a ridge line or for flat roofs, use θ = 0° and locate the zone 2/3 boundary at the mid-length of the building.
8. The roof pressure coefficient GC_{pf}, when negative in Zone 2 or 2E, shall be applied in Zone 2/2E for a distance from the edge of roof equal to 0.5 times the horizontal dimension of the building parallel to the direction of the MWFRS being designed or 2.5 times the eave height, h_e, at the windward wall, whichever is less, the remainder of Zone 2/2E extending to the ridge line shall use the pressure coefficient GC_{pf} for Zone 3/3E.
9. Notation:
   a: 10 percent of least horizontal dimension or 0.4 h, whichever is smaller, but not less than either 4% of least horizontal dimension or 3 ft (0.9 m).
   h: Mean roof height, in feet (meters), except that eave height shall be used for θ ≤ 10°.
   θ: Angle of plane of roof from horizontal, in degrees.

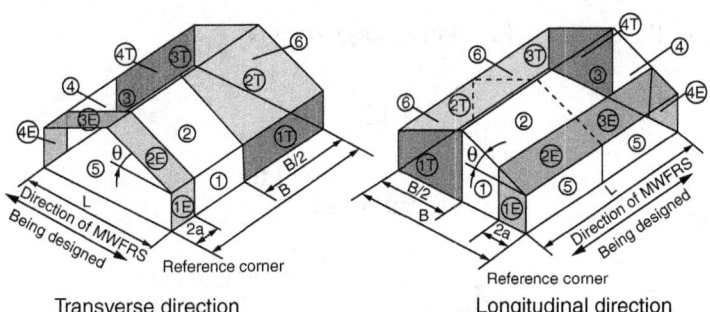

Transverse direction   Longitudinal direction
Torsional load cases

FIGURE 3.4 (*Continued*)

Step 4: Determine a topographic factor, K_{zt} in accordance with Sec. 6.5.7.

The topographic factor applies to structures located on a hill (higher than the surrounding terrain in all directions), ridge (higher than the surrounding terrain in two opposite directions), or escarpment (higher than the surrounding terrain in one direction only).

$$K_{zt} = (1 + K_1 K_2 K_3)^2$$

Values of K_1, K_2, and K_3 are given in Fig. 6-4 of ASCE 7-05. The default condition is $K_{zt} = 1.0$.

Step 5: Determine a gust effect factor, G or G_f, as applicable, in accordance with Sec. 6.5.8.

For rigid structures, the gust effect factor G is taken as 0.85 or calculated by an equation. For flexible structures, the gust effect factor G_f is calculated by an equation.

Step 6: Determine an enclosure classification in accordance with Sec. 6.5.9.

Classify the building as enclosed, partially enclosed, or open as defined in Sec 6.2. In these definitions,

A_o = total area of openings in a wall that receives positive external pressure
A_{oi} = sum of areas of openings in the building envelope not including A_o
A_{og} = total area of openings in building envelope
A_g = gross area of that wall in which A_o is identified
A_{gi} = gross area of building envelope not including A_g

- Open buildings have each wall at least 80 percent open ($A_o \geq 0.80\, A_g$).
- Partially enclosed buildings satisfy

$$A_o \geq 1.10 A_{oi}$$

$$A_o > \text{smaller of } \begin{cases} 4 \text{ ft}^2 \\ 0.01 A_g \end{cases}$$

$$\frac{A_{oi}}{A_{gi}} \leq 0.20$$

- Enclosed buildings are everything else

Enclosure classification	GC_{pi}
Open buildings	0.00
Partially enclosed buildings	+0.55 −0.55
Enclosed buildings	+0.18 −0.18

Notes:

1. Plus and minus signs signify pressures acting toward and away from the internal surfaces, respectively.
2. Values of GC_{pi} shall be used with q_z or q_h as specified in 6.5.12.
3. Two cases shall be considered to determine the critical load requirements for the appropriate condition:

    a. a positive value of GC_{pi} applied to all internal surfaces
    b. a negative value of GC_{pi} applied to all internal surfaces

FIGURE 3.5 Internal pressure coefficients for buildings, GC_{pi}. (*Source:* Figure 6-5 of ASCE 7-05.)

Step 7: Determine an internal pressure coefficient GC_{pi}, in accordance with Sec. 6.5.11.1.

Internal pressure coefficients are determined by Fig. 3.5.

Step 8: Determine the external pressure coefficients, C_p or GC_{pf}, or force coefficients C_f, as applicable, in accordance with Sec. 6.5.11.2 or 6.5.11.3, respectively.

Main Wind Force Resisting Systems

External pressure coefficients for main wind force resisting systems C_p are given in Fig. 3.6. Note that in the figure, the title is black on a white background, to emphasize the difference between lateral force-resisting systems and components and cladding.

Components and Cladding

External pressure coefficients for components and cladding GC_p are given (e.g., for buildings with flat roofs) in Fig. 3.7. Note that in the figure, the title is white on a black background, to emphasize the difference between lateral force-resisting systems and components and cladding.

76 Chapter Three

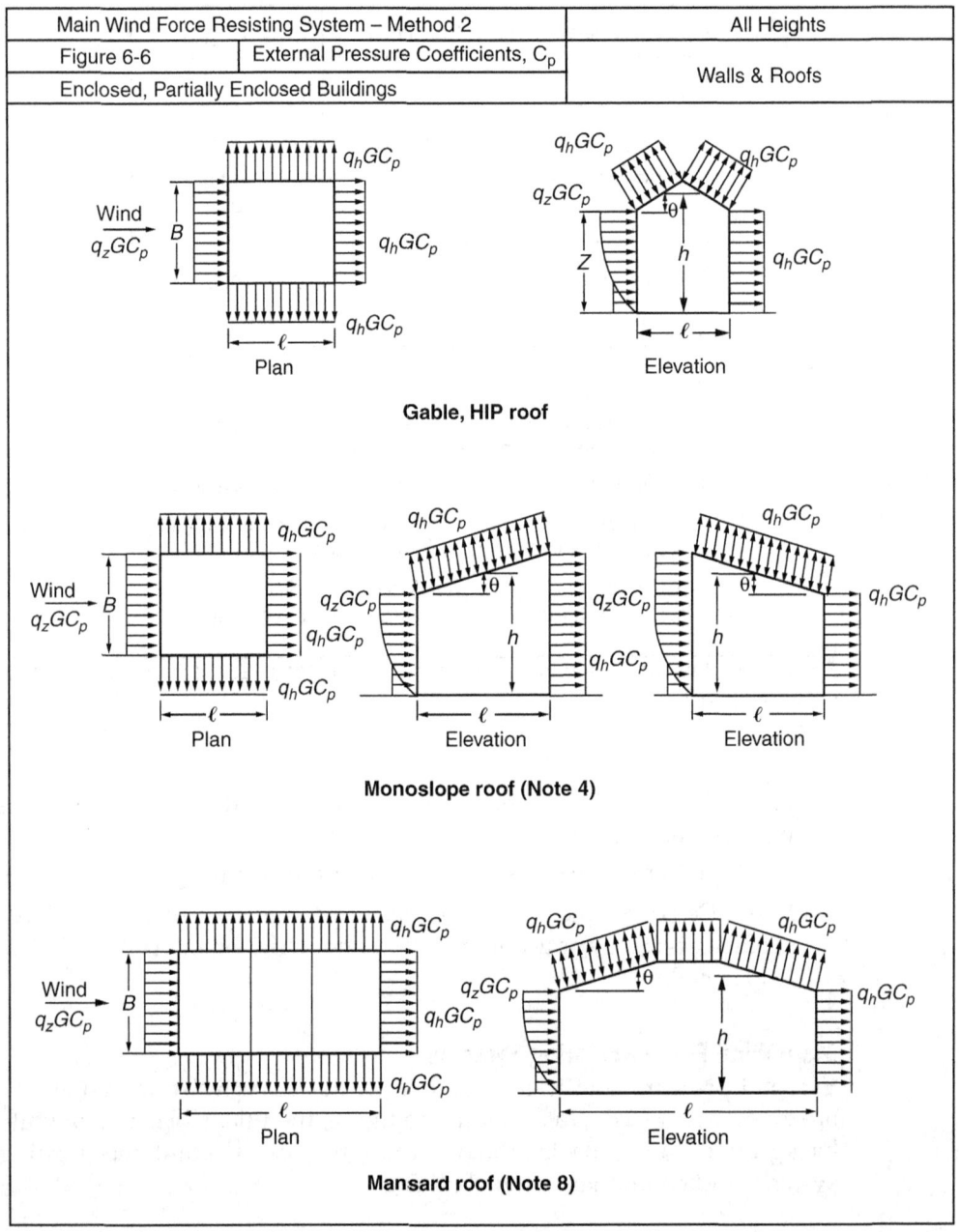

Figure 3.6 External pressure coefficients, C_p for main wind force-resisting systems (all heights). (*Source*: Figure 6-6 of ASCE 7-05.)

Code Basis for Structural Design of Masonry Buildings 77

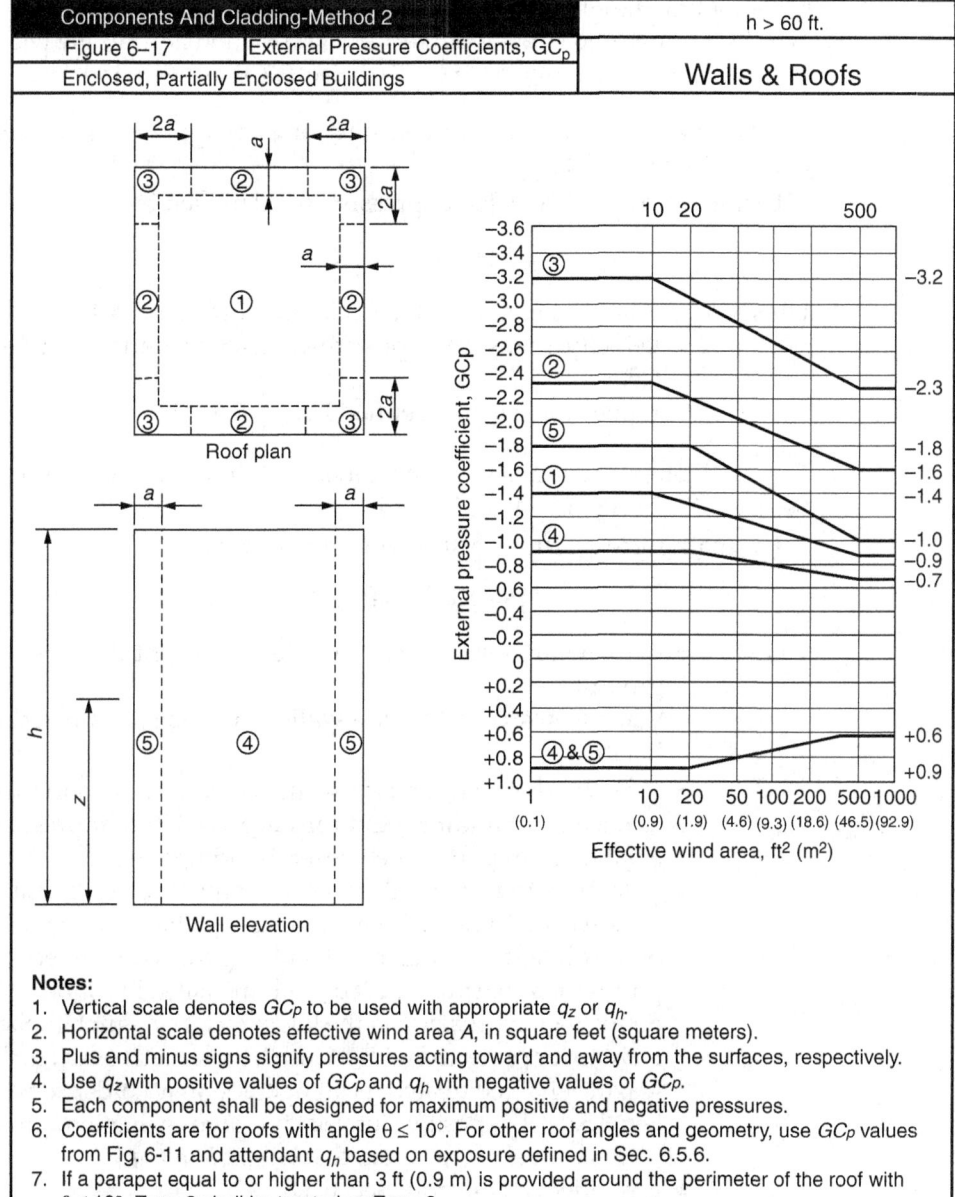

FIGURE 3.7 External pressure coefficients, GC_p for components and cladding. (*Source*: Figure 6-17 of ASCE 7-05.)

In computing the effective area of the cladding element, it is permitted to use an effective area equal to the product of the span and an effective width not less than one-third the span (ASCE 7-05, Sec. 6.2, Effective Wind Area).

Step 9: Determine the velocity pressure q_z or q_h, as applicable, in accordance with Sec. 6.5.10.

Using Sec. 6.5.10, the velocity pressure is calculated by

$$q_z = 0.00256\, K_z K_{zt} K_d V^2 I$$

where K_d = wind directionality factor defined in Sec. 6.5.4.4
K_z = velocity pressure exposure coefficient defined in Sec. 6.5.6.4
K_{zt} = topographic factor defined in Sec. 6.5.7.2

Step 10: Determine the design wind load P or F in accordance with Secs. 6.5.12, 6.5.13, 6.5.14, and 6.5.15.

For main force-resisting systems of rigid systems,

$$p = qGC_p - q_i(GC_{pi})$$

where $q = q_z$ for windward walls evaluated at height z above the ground
$= q_h$ for leeward walls, side walls, and roofs, evaluated at height h
$q_i = q_h$ for windward walls, side walls, leeward walls, and roofs of enclosed buildings and for negative internal pressure evaluation in partially enclosed buildings
$= q_z$ for positive internal pressure evaluation in partially enclosed buildings where height z is defined as the level of the highest opening in the building that could affect the positive internal pressure. For buildings sited in wind-borne debris regions, glazing in the lower 60 ft that is not impact-resistant or protected with an impact-resistant covering, the glazing shall be treated as an opening in accordance with Sec. 6.5.9.3. For positive internal pressure evaluation, q_i may conservatively be evaluated at height h ($q_i = q_h$)
G = gust effect factor from Sec. 6.5.8
C_p = external pressure coefficient from Figs. 6-6 through 6-10
C_{pi} = internal pressure coefficient from Fig. 6-5

For components and cladding of low-rise buildings and buildings with $h \leq 60$ ft.

$$p = q(GC_p) - q_i(GC_{pi})$$

where q_h = velocity pressure evaluated at mean roof height h using exposure defined in Sec. 6.5.6.3.1
(GC_p) = external pressure coefficients from Figs. 6-11 through 6-17
(GC_{pi}) = internal pressure coefficients from Fig. 6-5

For components and cladding of buildings with $h > 60$ ft

$$p = q(GC_p) - q_i(GC_{pi})$$

where $q = q_z$ for windward walls, evaluated at height z above the ground
  $= q_h$ for leeward walls, side walls, and roofs, evaluated at height h
  $q_i = q_h$ for windward walls, side walls, leeward walls, and roofs of enclosed buildings and for negative internal pressure evaluation in partially enclosed buildings
  $= q_z$ for positive internal pressure evaluation in partially enclosed buildings where height z is defined as the level of the highest opening in the building that could affect the positive internal pressure. For buildings sited in wind-borne debris regions, glazing in the lower 60 ft that is not impact-resistant or protected with an impact-resistant covering, the glazing shall be treated as an opening in accordance with Sec. 6.5.9.3. For positive internal pressure evaluation, q_i may conservatively be evaluated at height h ($q_i = q_h$)
  GC_p = external pressure coefficient from Fig. 6-17
  GC_{pi} = internal pressure coefficient from Fig. 6-5

3.4.2 Example of Wind Loading on Main Wind Force-Resisting System

Using the procedures of ASCE 7-05, compute the design base shear due to wind for the building of Fig. 3.8, located in the suburbs of Austin, Texas.

The critical direction will be NS, because the walls on the north and south sides have greater area, and the shear walls in the north and south directions have less area.

1. Determine the basic wind speed V and wind directionality factor K_d in accordance with Sec. 6.5.4.

80 Chapter Three

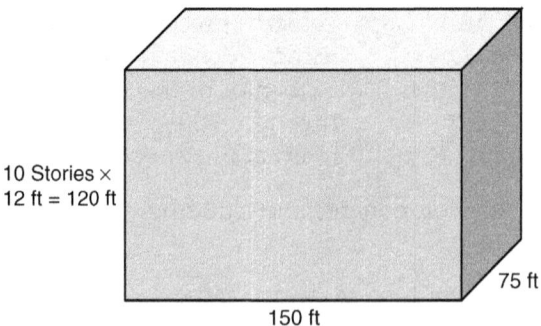

FIGURE 3.8 Schematic view of building in Austin, Texas.

The basic wind speed for Austin is 90 mi/h (ASCE 7-05, Fig. 6-1a). The wind directionality factor K_d is 0.85 (ASCE 7-05, Table 6-4, buildings).

2. Determine the importance factor I in accordance with Sec. 6.5.5.

   Assume that the importance factor is 1.0.

3. Determine the exposure category or exposure categories and velocity pressure exposure coefficient K_z or K_h, as applicable, in accordance with Sec. 6.5.6.

Height above ground level, z	K_h, K_z
1–15	0.57
20	0.62
25	0.66
30	0.70
40	0.76
50	0.81
60	0.85
70	0.89
80	0.93
90	0.96
100	0.99
120	1.04

Source: Table 6-3 of ASCE 7-05.

TABLE 3.6 Velocity Pressure Coefficients for Building of Sec 3.4.2

Code Basis for Structural Design of Masonry Buildings

Assume Exposure B (urban and suburban areas). The velocity pressure exposure coefficients K_h and K_z are determined from ASCE 7-05, Table 6-3, for Exposure B and Case 2 (all main wind force-resisting systems in other structures).

4. Determine a topographic factor K_{zt} in accordance with Sec. 6.5.7

   Because the structure is not located on a hill, ridge or escarpment, $K_{zt} = 1.0$.

5. Determine a gust effect factor G or G_f, as applicable, in accordance with Sec. 6.5.8.

   Assume a rigid structure; the gust effect factor, G, is 0.85.

6. Determine an enclosure classification in accordance with Sec. 6.5.9.

   Assume that the building is enclosed.

7. Determine an internal pressure coefficient GC_{pi} in accordance with Sec. 6.5.11.1

   The internal pressure coefficient GC_{pi} is ±0.18.

8. Determine the external pressure coefficients C_p or GC_{pf}, or force coefficients C_f, as applicable, in accordance with Sec. 6.5.11.2 or 6.5.11.3, respectively.

   The external pressure coefficients for main wind force resisting systems GC_p are given in Fig. 6-6 of ASCE 7-05.

   From the plan views in Fig. 6-6 of ASCE 7-05, the windward pressure is $q_z GC_p$. The leeward pressure is $q_h GC_p$. The difference between the q_z and the q_h is that the former varies as a function of the height above ground level, while the latter is uniform over the building height, and is evaluated using the height of the building.

   For wind blowing in the NS direction, $L/B = 0.5$. From Fig. 6-6 (cont'd), on the windward side of the building the external pressure coefficient C_p is 0.8. On the leeward side of the building, it is −0.5.

9. Determine the velocity pressure q_z or q_h, as applicable, in accordance with Sec. 6.5.10.

   The velocity pressure is

$$q_z = 0.00256\, K_z K_{zt} K_d V^2 I$$

$$I = 1.0$$

$$K_d = 0.85$$

$$V = 90 \text{ miles/hr}$$

$$k_{zt} = 1.0$$

$$q_z = 17.63 \, K_z \text{ lb/ft}^2$$

Note that the above expression for q_z has K_z embedded in it.

10. Determine the design wind load P or F in accordance with Secs. 6.5.12 and 6.5.13, as applicable.

For main force-resisting systems,

$$p = qGC_p - q_i(GC_{pi})$$

where $q = q_z$ for windward walls evaluated at height z above the ground
 $= q_h$ for leeward walls, side walls, and roofs, evaluated at height h
 $q_i = q_h$ for windward walls, side walls, leeward walls, and roofs of enclosed buildings and for negative internal pressure evaluation in partially enclosed buildings
 $= q_z$ for positive internal pressure evaluation in partially enclosed buildings where height z is defined as the level of the highest opening in the building that could affect the positive internal pressure. For buildings sited in wind-borne debris regions, glazing in the lower 60 ft that is not impact-resistant or protected with an impact-resistant covering, the glazing shall be treated as an opening in accordance with Sec. 6.5.9.3. For positive internal pressure evaluation, q_i may conservatively be evaluated at height h ($q_i = q_h$)
 G = gust effect factor from Sec. 6.5.8
 C_p = external pressure coefficient from Fig. 6-6 or other analogous figures
 C_{pi} = internal pressure coefficient from Table 6-5

Because the building is enclosed, the internal pressures on the windward and leeward sides are of equal magnitude and opposite direction, will produce zero net base shear, and therefore need not be considered.

On the windward side of the building,

$$p = q_z GC_p$$

$$p = (17.63 \, K_z)GC_p$$

Because C_p is positive in sign, this pressure is positive in sign, indicating that the pressure acts inward against the windward wall. If the wind

Code Basis for Structural Design of Masonry Buildings 83

comes from the south, for example, the force on the windward wall acts toward the north. These values are shown in the "windward side" columns of the spreadsheet in Table 3.7.

On the leeward side of the building,

$$p = q_h G C_p$$

$$p = (17.63 \, K_h) G C_p$$

Because C_p is negative in sign, this pressure is negative in sign, indicating that the pressure acts outward against the leeward wall. If the wind comes from the south, for example, the force on the leeward wall acts toward the north. These values are shown in the "leeward side" columns of the spreadsheet in Table 3.7.

The design base shear due to wind load is the summation of 177.92 kips acting inward on the upwind wall, and 140.26 kips acting outward on the downwind wall, for a total of 318.2 kips.

3.4.3 Example of Wind Loading on Components and Cladding

Using the procedures of ASCE 7-05, compute the design wind pressure on a cladding element near the corner of the top floor of the building of Sec. 3.4.1.

1. Determine the basic wind speed V and wind directionality factor K_d in accordance with Sec. 6.5.4.

   The basic wind speed for Austin is 90 miles per hour (ASCE 7-05, Fig. 6-1a). The wind directionality factor K_d is 0.85 (ASCE 7-05, Table 6-4, buildings).

2. Determine the importance factor I in accordance with Sec. 6.5.5.

   Assume that the importance factor is 1.0.

3. Determine the exposure category or exposure categories and velocity pressure exposure coefficient K_z or K_h, as applicable, in accordance with Sec. 6.5.6.

   Assume Exposure B (urban and suburban areas). The velocity pressure exposure coefficients K_h and K_z are as shown in Table 3.8.

4. Determine a topographic factor K_{zt} in accordance with Sec. 6.5.7

   Because the structure is not located on a hill, ridge, or escarpment, $K_{zt} = 1.0$.

5. Determine a gust effect factor G or G_f, as applicable, in accordance with Sec. 6.5.8.

   Assume a rigid structure; the gust effect factor, G, is 0.85.

Building floor	Height above ground	Tributary area	Windward side							Leeward side			
			K_z	q_z	G	C_p	p	Force	K_h	q_h	G	C_p	p	Force
Roof	120	900	1.04	18.34	0.85	0.8	12.47	11.22	1.04	18.34	0.85	−0.5	7.79	−7.01
10	108	1800	1.01	17.81	0.85	0.8	12.11	21.79	1.04	18.34	0.85	−0.5	7.79	−14.03
9	96	1800	0.98	17.28	0.85	0.8	11.75	21.15	1.04	18.34	0.85	−0.5	7.79	−14.03
8	84	1800	0.94	16.57	0.85	0.8	11.27	20.28	1.04	18.34	0.85	−0.5	7.79	−14.03
7	72	1800	0.9	15.87	0.85	0.8	10.79	19.42	1.04	18.34	0.85	−0.5	7.79	−14.03
6	60	1800	0.85	14.99	0.85	0.8	10.19	18.34	1.04	18.34	0.85	−0.5	7.79	−14.03
5	48	1800	0.8	14.10	0.85	0.8	9.59	17.26	1.04	18.34	0.85	−0.5	7.79	−14.03
4	36	1800	0.74	13.05	0.85	0.8	8.87	15.97	1.04	18.34	0.85	−0.5	7.79	−14.03
3	24	1800	0.65	11.46	0.85	0.8	7.79	14.03	1.04	18.34	0.85	−0.5	7.79	−14.03
2	12	1800	0.57	10.05	0.85	0.8	6.83	12.30	1.04	18.34	0.85	−0.5	7.79	−14.03
Ground	0	900	0.57	10.05	0.85	0.8	6.83	6.15	1.04	18.34	0.85	−0.5	7.79	−7.01
Total force								**177.92**						**−140.26**

TABLE 3.7 Spreadsheet for Wind Forces, Sec. 3.4.2

Height above ground level, z	K_h, K_z
100	0.99
120	1.04

Source: ASCE 7-05, Table 6-3, for Exposure B and Case 1 (Components and Cladding).

TABLE 3.8 Velocity Pressure Exposure Coefficients for Building of Sec. 3.4.3

6. Determine an enclosure classification in accordance with Sec. 6.5.9.

   Assume that the building is enclosed.

7. Determine an internal pressure coefficient GC_{pi} in accordance with Sec. 6.5.11.1.

   The internal pressure coefficient GC_{pi} is ± 0.18.

8. Determine the external pressure coefficients C_p or GC_{pf}, or force coefficients C_f, as applicable, in accordance with Sec. 6.5.11.2 or 6.5.11.3, respectively.

   The external pressure coefficients for components and cladding GC_p are given in Fig. 6-17 of ASCE 7-05.

   In computing the effective area of the cladding element, it is permitted to use an effective area equal to the product of the span and an effective width not less than one-third the span (ASCE 7-05, Sec. 6.2, Effective Wind Area).

   Assume a panel with a span equal to the story height of 12 ft minus a spandrel depth of 2 ft, or 10 ft. Assume an effective width of one-third of that span, or 3.33 ft. The resulting effective area is 33.3 ft². From Fig. 6-17, a panel in zone 5 has a positive pressure coefficient of 0.85, and a negative pressure coefficient of -1.7.

9. Determine the velocity pressure q_z or q_h, as applicable, in accordance with Sec. 6.5.10.

   The velocity pressure is

   $$q_z = 0.00256 K_z K_{zt} K_d V^2 I$$

   $$I = 1.0$$

   $$K_d = 0.85$$

$$V = 90 \text{ miles/hr}$$
$$k_{zt} = 1.0$$
$$q_z = 17.63\, K_z \text{ lb/ft}^2$$

Note that the above expression for q_z has K_z embedded in it.

10. Determine the design wind load P or F in accordance with Secs. 6.5.12 and 6.5.13, as applicable.

Since this is a building with $h > 60$ ft

$$p = q(GC_p) - q_i(GC_{pi})$$

Windward Side of Building

On the windward side of the building, the maximum inward pressure will be produced on the cladding, due to the combination of GC_p acting inward (positive sign) and GC_{pi} also acting inward (negative sign).

$q = q_z$ evaluated at the height of the element, or 120 ft

$q_i = q_h$ evaluated at the height of the building, or 120 ft

$(GC_p) = 0.85$ (Fig. 6-17)

$(GC_{pi}) = \pm 0.18$ (Fig. 6-5).

$$p = q\,(GC_p) - q_i(GC_{pi})$$
$$p = q_z(GC_p) - q_h(GC_{pi})$$
$$p = (17.63\, K_z)GC_p - (17.63\, K_h)GC_{pi}$$

These values are shown in the spreadsheet of Table 3.9. The maximum inward pressure is the sum of 15.58 psf on the outside plus 3.30 psf on the inside, for a total of 18.89 psf acting inward.

Building height, h	Height above ground, z	Maximum inward pressure (windward wall)								Total
		External pressure				Internal pressure				
		K_z	q_z	GC_p	$P_{outside}$	K_h	q_h	GC_{pi}	P_{inside}	P_{total}
120	120	1.04	18.34	0.85	15.58	1.04	18.34	−0.18	−3.30	18.89

TABLE 3.9 Spreadsheet for Components and Cladding Pressures, Windward Side of Sec. 3.4.2

| Building height, h | Height above ground, z | Maximum outward pressure (leeward wall) |||||||| Total |
| | | External pressure |||| Internal pressure |||| |
		K_h	q_h	GC_p	$P_{outside}$	K_h	q_h	GC_{pi}	P_{inside}	P_{total}
120	120	1.04	18.34	−1.7	−31.17	1.04	18.34	0.18	3.30	34.47

TABLE 3.10 Spreadsheet for Components and Cladding Pressures, Leeward Side of Sec. 3.4.2

Leeward Side of Building

On the leeward side of the building, the maximum outward pressure will be produced on the cladding, due to the combination of GC_p acting outward (negative sign) and GC_{pi} also acting outward (positive sign).

$q = q_h$ or 120 ft
$q_i = q_h$ or 120 ft
$(GC_p) = -1.7$ (Fig. 6-17)
$(GC_{pi}) = \pm 0.18$ (Fig. 6-5).

$$p = q(GC_p) - q_i(GC_{pi})$$
$$p = q_z(GC_p) - q_h(GC_{pi})$$
$$p = (17.63\ K_z)GC_p - (17.63\ K_h)GC_{pi}$$

These values are shown in the spreadsheet of Table 3.10. The maximum outward pressure is the sum of −31.17 psf on the outside plus 3.30 psf on the inside, for a total of 34.47 psf acting outward.

The cladding must therefore be designed for a pressure of 18.9 lb/ft² acting inward, and 34.5 lb/ft² acting outward.

3.5 Earthquake Loading

Design earthquake loads are calculated according to Sec. 1613 of the 2009 IBC. That section essentially references *ASCE 7-05 (Supplement)*. Seismic design criteria are given in Chap. 11. The seismic design provisions of *ASCE 7-05 (Supplement)* begin in Chap. 12, which prescribes basic requirements (including the requirement for continuous load paths) (Sec. 12.1); selection of structural systems (Sec. 12.2); diaphragm characteristics and other possible irregularities (Sec. 12.3); seismic load effects and combinations (Sec. 12.4); direction of loading (Sec. 12.5); analysis procedures (Sec. 12.6); modeling procedures (Sec. 12.7); and specific design approaches. Four procedures are prescribed: an equivalent lateral force procedure (Sec. 12.8); a modal response-spectrum analysis (Sec. 12.9); a simplified alternative procedure (Sec. 12.14); and a seismic response history procedure (Chap. 16). The equivalent lateral-force procedure is described here, because it is relatively simple, and is permitted in most situations. The simplified alternative procedure is permitted in only a few situations. The other procedures are permitted in all situations, and are required in only a few situations.

3.5.1 Background on Earthquake Loading

The basic approach to earthquake design is to idealize a building as a single-degree-of-freedom system—that is, a system whose configuration in space can be defined using a single variable (Fig. 3.9).

The equation of equilibrium for this system is

$$M\ddot{u} + 2\xi\omega M\dot{u} + \omega^2 Mu = -M\ddot{u}_g(t)$$

where: $\ddot{u}$ = relative acceleration
$\dot{u}$ = relative velocity
u = relative displacement
$\ddot{u}_g$ = ground acceleration
M = mass
K = stiffness
$\omega = \sqrt{K/M}$
ξ = equivalent viscous damping coefficient, whose value is chosen so that the energy dissipation of the system in the elastic range will be similar to that of the original structure

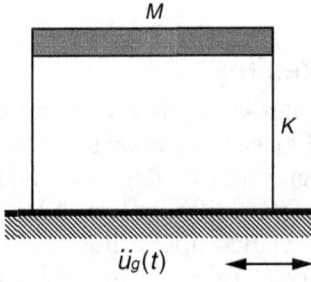

FIGURE 3.9 Idealized single-degree-of-freedom system.

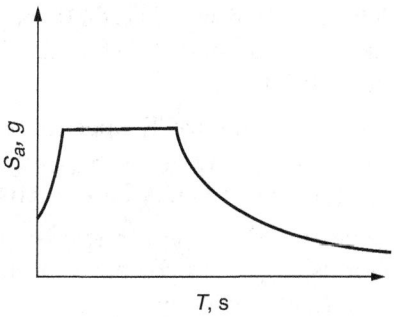

Figure 3.10 Acceleration response spectrum, smoothed for use in design.

For a given ground motion, the solution to the above equation can be calculated step by step using computer programs. The response of a structure depends on the strength of the ground motion, and also on the relationship between the characteristic frequencies of ground motion, and the frequency of the structure.

Of particular interest are the maximum values of the seismic response, which can be graphed in the form of a response spectrum, whose ordinates indicate the maximum response as a function of the period of vibration of the structure. For example, the acceleration response spectrum gives the values of absolute acceleration (which can be multiplied by mass to give the maximum inertial forces that act on the structure) in terms of period. An example of an acceleration response spectrum smoothed for use in design is given in Fig. 3.10.

Using a response spectrum, the maximum response of a structure can be calculated for a particular earthquake, with little effort. Such response spectra, smoothed as shown Fig. 3.10, can be used to calculate design forces as part of the process of seismic design.

In modern design codes, these design spectra are modified to address the effects of inelastic response, structural overstrength, and multimodal response.

3.5.2 Determine Seismic Ground Motion Values

1. Determine S_S, the mapped MCE (maximum considered earthquake), 5 percent damped, spectral response acceleration parameter at short periods as defined in Sec. 11.4.1 of ASCE 7-05.

2. Determine S_1, the mapped MCE, 5 percent damped, spectral response acceleration parameter at a period of 1 s as defined in Sec. 11.4.1.

3. Determine the site class (*A* through *F*, a measure of soil response characteristics and soil stability) in accordance with Sec. 20.3 and Table 20.3-1.
4. Determine the MCE spectral response acceleration for short periods (S_{MS}) and at 1 s (S_{M1}), adjusted for site-class effects, using Eqs. (11.4-1) and (11.4-2) respectively.
5. Determine the design response acceleration parameter for short periods, S_{DS}, and for a 1-s period, S_{D1}, using Eqs. (11.4-3) and (11.4-4), respectively.
6. If required, determine the design response spectrum curve as prescribed by Sec. 11.4.5.

3.5.3 Determine Seismic Base Shear Using the Equivalent Lateral Force Procedure

1. Determine the structure's importance factor, *I*, and occupancy category using Sec. 11.5.
2. Determine the structure's seismic design category using Sec. 11.6.
3. Calculate the structure's seismic base shear using Secs. 12.8.1 and 12.8.2.

3.5.4 Distribute Seismic Base Shear Vertically and Horizontally

1. Distribute seismic base shear vertically using Sec. 12.8.3.
2. Distribute seismic base shear horizontally using Sec. 12.8.4.

Now let's discuss each step in more detail, combining with an example for Charleston, South Carolina.

Step 1: Determine S_S, the mapped MCE (maximum considered earthquake), 5 percent damped, spectral response acceleration parameter at short periods as defined in Sec. 11.4.1.

Step 2: Determine S_1, the mapped MCE, 5 percent damped, spectral response acceleration parameter at a period of 1 s as defined in Sec. 11.4.1.

Determine the parameters S_s and S_1 from the 0.2- and 1-s spectral response maps shown in Figs. 22-1 through 22-7 of ASCE 7-05.

With the exception of some parts of the western United States (where design earthquakes have a deterministic basis), those maps generally

Code Basis for Structural Design of Masonry Buildings

correspond to accelerations with a 2 percent probability of exceedance within a 50-year period. The earthquake associated with such accelerations is sometimes described as a "2500-year earthquake." To see why, let p be the unknown annual probability of exceedance of that level of acceleration:

The probability of exceedance in a particular year is	p
The probability of non-exceedance in a particular year is	$(1-p)$
The probability of non-exceedance in 50 consecutive years is	$(1-p)^{50}$
The probability of exceedance within a 50-year period is	$[1-(1-p)^{50}]$
Solve for p, the annual probability of exceedance. Set the probability of exceedance within the 50-year period equal to the given 2%	$[1-(1-p)^{50}]=0.02$ $(1-p)^{50}=0.98$ $p=1-0.98^{\left(\frac{1}{50}\right)}$ $p=4.04\times10^{-4}$
The return period is the reciprocal of the annual probability of exceedance	$\frac{1}{p}=2475$
The approximate return period is	2500 years

For Charleston, South Carolina, for example, $S_S = 2.00\,g$, and $S_1 = 0.50\,g$.

Step 3: Determine the site class (A through F, a measure of soil response characteristics and soil stability) in accordance with Sec. 20.3 and Table 20.3-1.

In accordance with Table 3.11 (Table 20.3-1 of ASCE 7-05), site classes are assigned as follows:

Assume Site Class D (stiff soil).

Step 4: Determine the MCE spectral response acceleration for short periods (S_{MS}) and at 1 s (S_{M1}), adjusted for site class effects, using Eqs. (11.4-1) and (11.4-2), respectively and reproduced as Tables 3.12 and 3.13 in this chapter.

$$S_{MS} = F_a \cdot S_s \quad (11.4\text{-}1)$$

$$S_{M1} = F_v \cdot S_1 \quad (11.4\text{-}2)$$

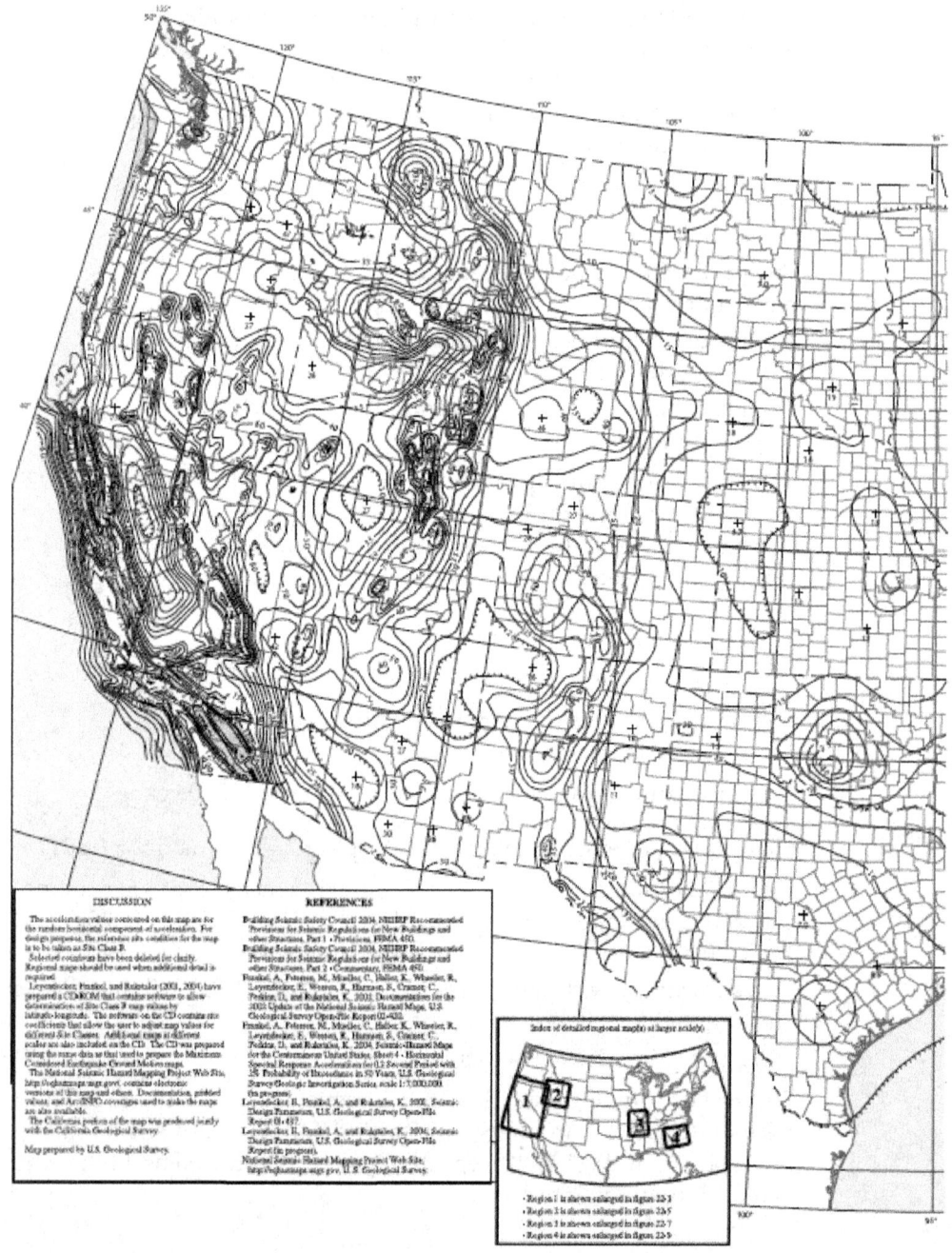

FIGURE 3.11 Figures 22-1 of ASCE 7-05. Maximum considered earthquake ground motion for the conterminous United States of 0.2 sec spectral response acceleration (5% of critical damping), site class B.

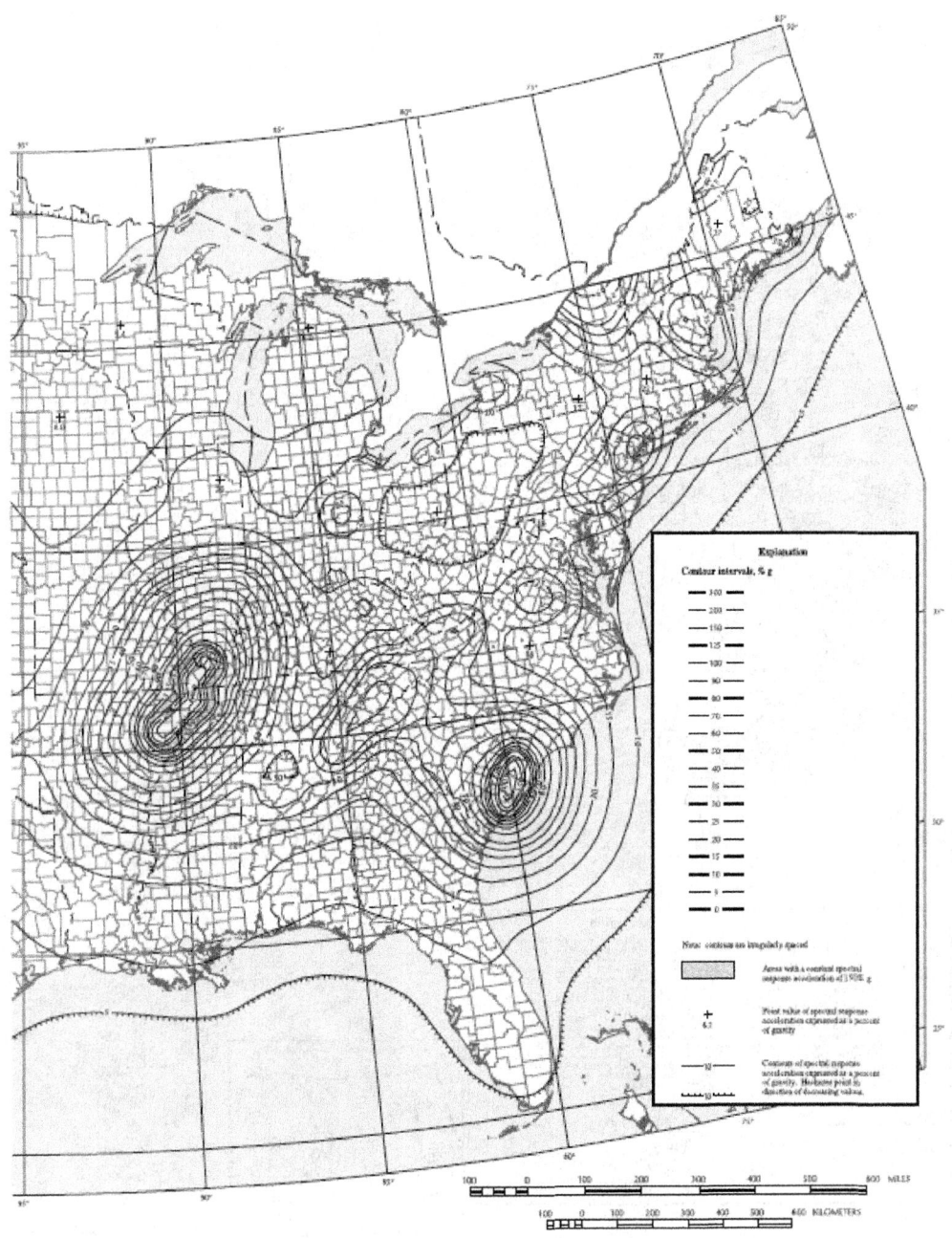

94 Chapter Three

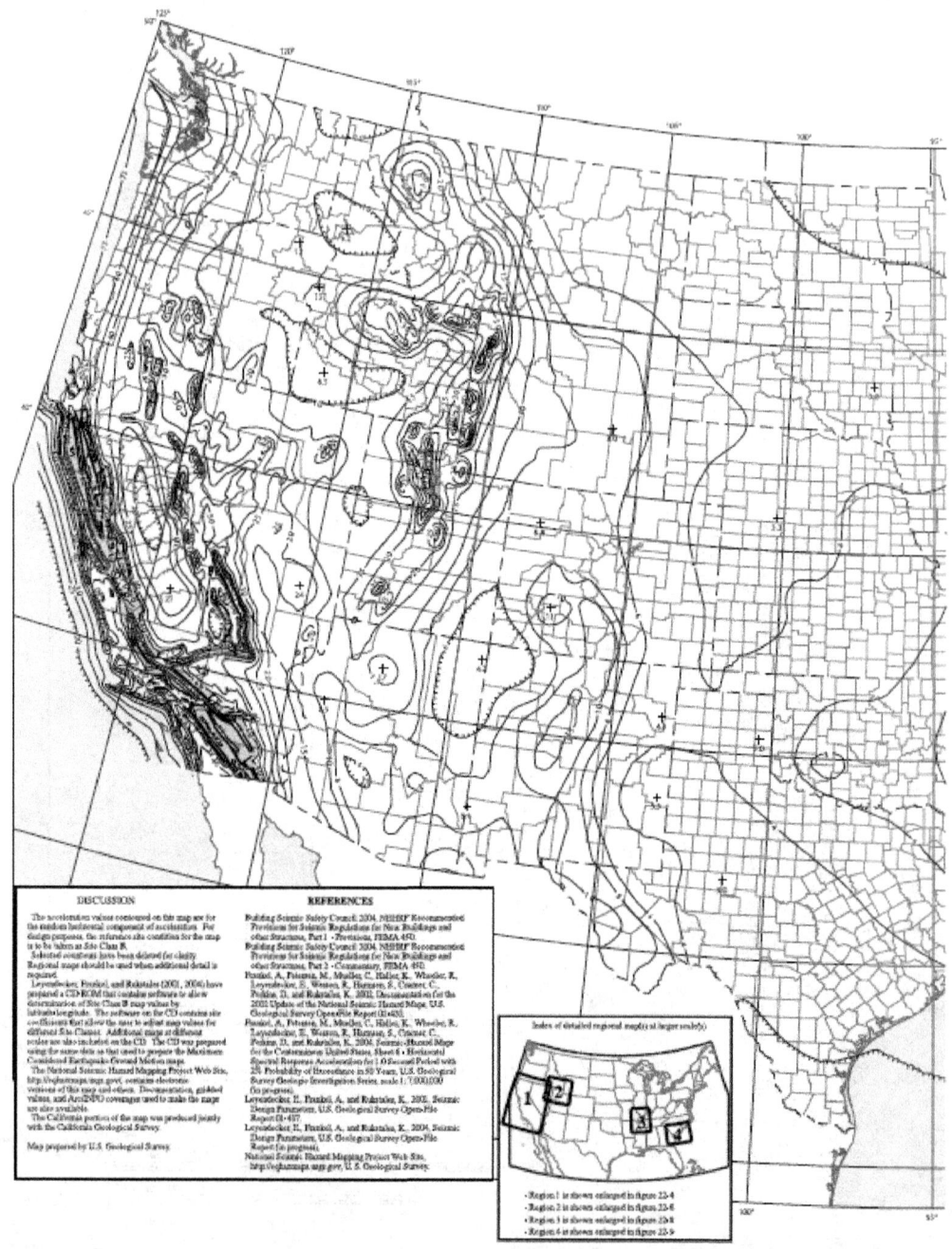

FIGURE 3.12 Figures 22-2 of ASCE 7-05. Maximum considered earthquake ground motion for the conterminous United States of 1.0 sec spectral response acceleration (5% of critical damping), site class B.

Code Basis for Structural Design of Masonry Buildings

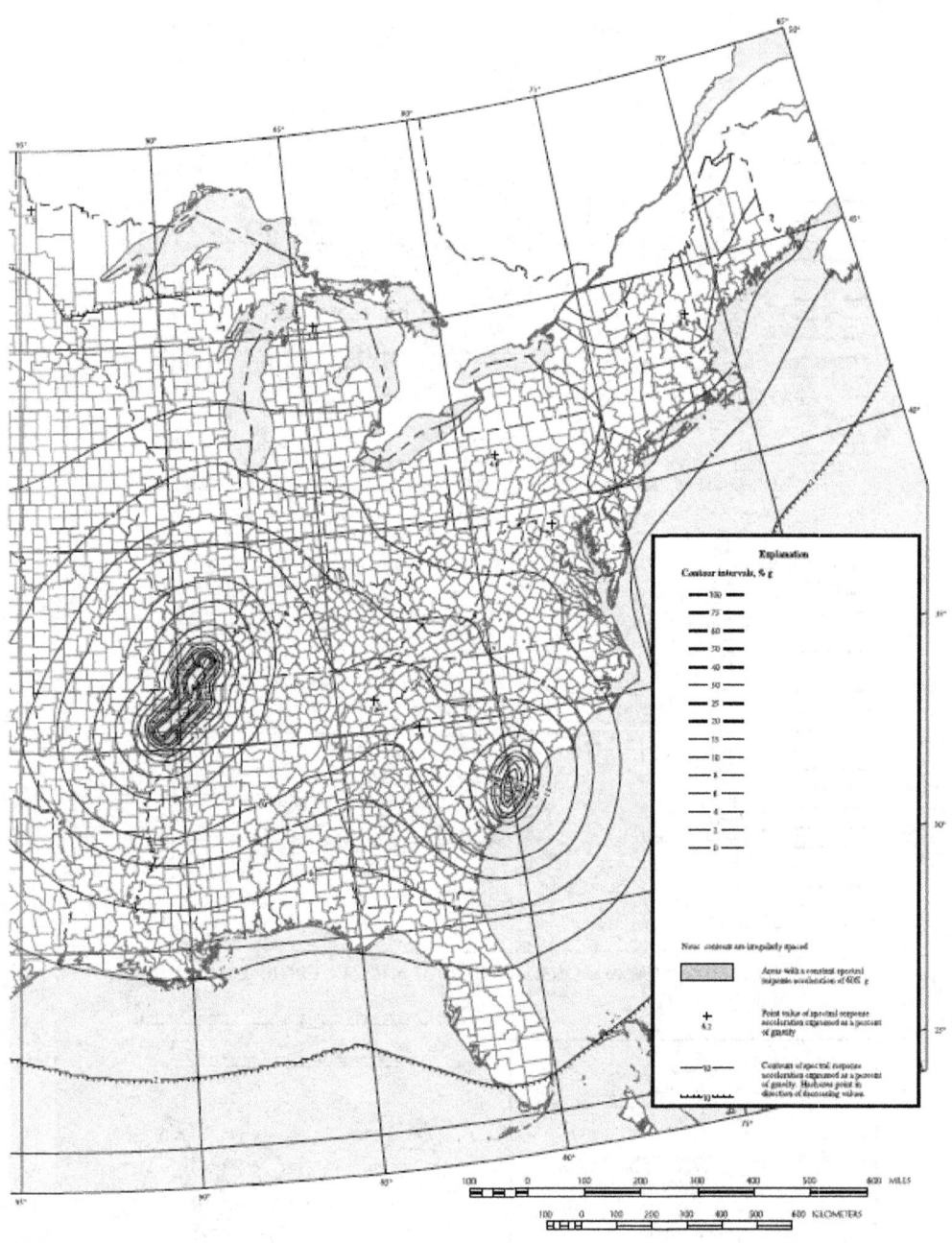

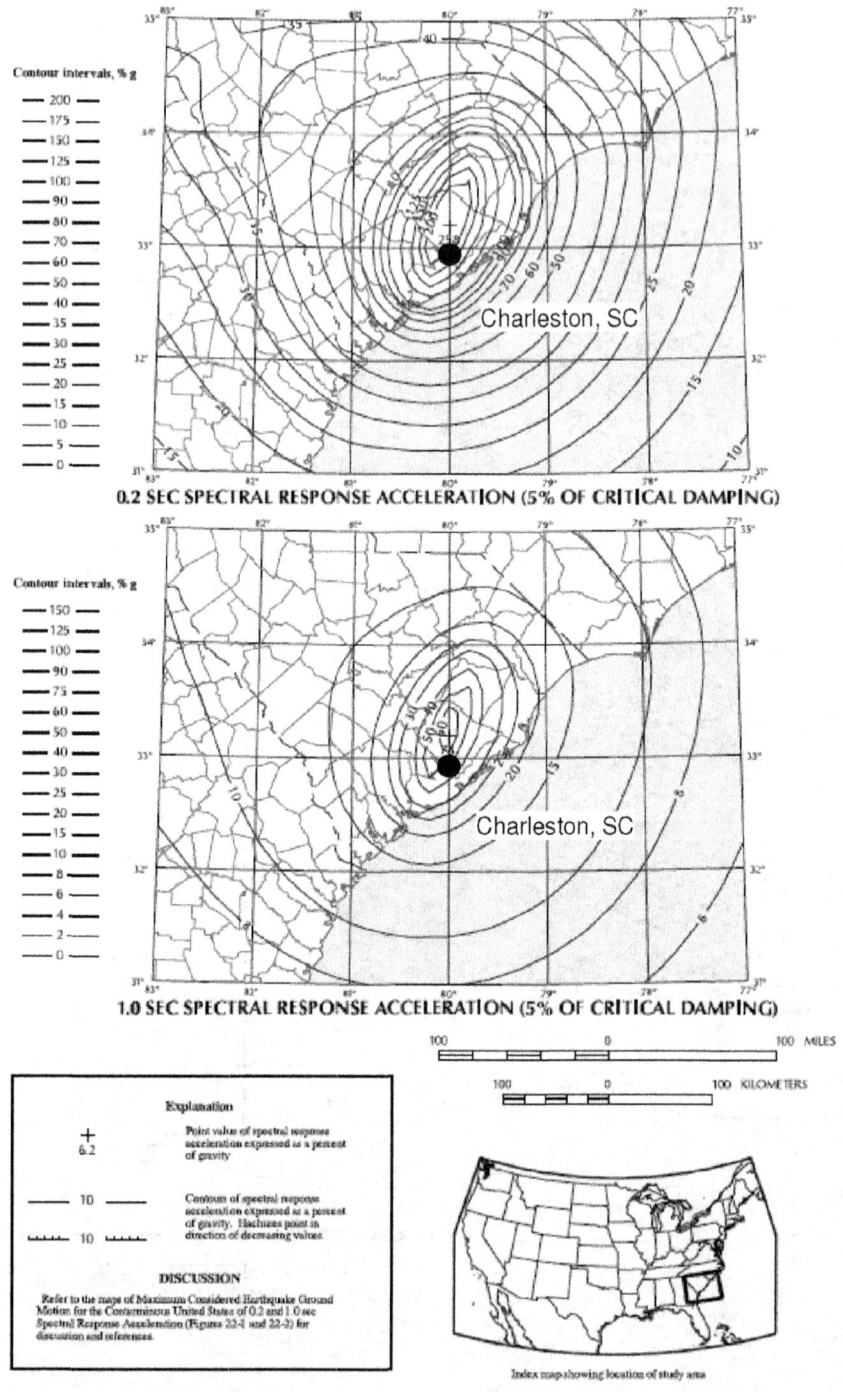

FIGURE 3.13 Figures 22-9 of ASCE 7-05. Maximum considered earthquake ground motion for region 4 of 0.2 and 1.0 sec spectral response acceleration (5% of critical damping), site class B.

Code Basis for Structural Design of Masonry Buildings

Site class	v_s	N or N_{ch}	$\bar{s}_u$
A. Hard rock	>5,000 ft/s	NA	NA
B. Rock	2,500–5,000 ft/s	NA	NA
C. Very dense soil and soft rock	1,200–2,500 ft/s	>50	>2,000 psf
D. Stiff soil	600–1,200 ft/s	15–50	1,000–2,000 psf
E. Soft clay soil	<600 ft/s	<15	<1,000 psf
	Any profile with more than 10 ft of soil having the following characteristics: — Plasticity index PI > 20 — Moisture content w ≥ 40%, and — Undrained shear strength $\bar{s}_u$ < 500 psf		
F. Soils requiring site response analysis in accordance with Sec. 21.1	See Sec. 20.3.1		

For SI: 1 ft/s = 0.3048 m/s; 1 lb/ft² = 0.0479 kN/m²
Source: Table 20.3-1 of ASCE 7-05.

TABLE 3-11 Table 20.3-1 Site Classification

Site class	Mapped maximum considered earthquake spectral response acceleration parameter at short period				
	$S_s \leq 0.25$	$S_s = 0.5$	$S_s = 0.75$	$S_s = 1.0$	$S_s \geq 1.25$
A	0.8	0.8	0.8	0.8	0.8
B	1.0	1.0	1.0	1.0	1.0
C	1.2	1.2	1.1	1.0	1.0
D	1.6	1.4	1.2	1.1	1.0
E	2.5	1.7	1.2	0.9	0.9
F	See Sec. 11.4.7				

Note: Use straight-line interpolation for intermediate values of S_s.
Source: Table 11.4-1 of ASCE 7-05.

TABLE 3.12 Table 11.4-1 Site Classification

The acceleration-dependent site coefficient, F_a, is 1.0. Table 3.12 (Table 11.4-1 of of ASCE 7-05).

The velocity-dependent site coefficient, F_v, is 1.5. Table 3.13 (Table 11.4-2 of of ASCE 7-05).

	Mapped maximum considered earthquake spectral response acceleration parameter at 1 s period				
Site class	$S_1 \leq 0.1$	$S_1 = 0.2$	$S_1 = 0.3$	$S_1 = 0.4$	$S_1 \geq 0.5$
A	0.8	0.8	0.8	0.8	0.8
B	1.0	1.0	1.0	1.0	1.0
C	1.7	1.6	1.5	1.4	1.3
D	2.4	2.0	1.8	1.6	1.5
E	3.5	3.2	2.8	2.4	2.4
F	See Sec. 11.4.7				

Note: Use straight-line interpolation for intermediate values of S_1.
Source: Table 11.4-2 of ASCE 7-05.

TABLE 3.13 Table 11.4-2 Site Classification, F_v

Then the maximum considered short-period response acceleration is

$$S_{MS} = F_a \cdot S_s = 1.0 \cdot 2.00g = 2.00g$$

and the maximum considered 1 s response acceleration is

$$S_{M1} = F_v \cdot S_1 = 1.5 \cdot 0.50g = 0.75g$$

Step 5: Determine the design response acceleration parameter for short periods, S_{DS}, and for a 1-s period, S_{D1}, using Eqs. (11.4-3) and (11.4-4), respectively.

The design response acceleration is two-thirds of the maximum considered acceleration:

$$S_{DS} = \frac{2}{3} \cdot S_{MS} \qquad (11.4\text{-}3)$$

$$S_{D1} = \frac{2}{3} \cdot S_{M1} \qquad (11.4\text{-}4)$$

With the exception of some parts of the western United States (where design earthquakes have a deterministic basis), these design spectral ordinates correspond to an earthquake with a 10 percent probability of exceedance within a 50-year period. Such an earthquake is sometimes described as a "500-year earthquake." To see why, let p be the unknown annual probability of exceedance of that level of acceleration.

The probability of exceedance in a particular year is	p
The probability of non-exceedance in a particular year is	$(1-p)$
The probability of non-exceedance in 50 consecutive years is	$(1-p)^{50}$
The probability of exceedance within a 50-year period is	$[1-(1-p)^{50}]$
Solve for p, the annual probability of exceedance. Set the probability of exceedance within the 50-year period equal to the given 10%	$[1-(1-p)^{50}] = 0.10$
	$(1-p)^{50} = 0.90$
	$p = 1 - 0.90^{\left(\frac{1}{50}\right)}$
	$p = 2.10 \times 10^{-3}$
The return period is the reciprocal of the annual probability of exceedance	$\dfrac{1}{p} = 475$
The approximate return period is	500 years

Continuing with our example for Charleston, South Carolina, the design response acceleration for short periods is

$$S_{DS} = \frac{2}{3} \cdot S_{MS} = \frac{2}{3} \cdot 2.00g = 1.33g$$

and the design response acceleration for a 1-s period is

$$S_{D1} = \frac{2}{3} \cdot S_{M1} = \frac{2}{3} \cdot 0.75g = 0.50g$$

Step 6: If required, determine the design response spectrum curve as prescribed by Sec. 11.4.5.

Because the equivalent lateral force procedure is being used, the response spectrum curve is not required. Nevertheless, for pedagogical completeness, it is developed here.

First, define

$$T_0 \equiv 0.2 \frac{S_{D1}}{S_{DS}}$$

and

$$T_S \equiv \frac{S_{D1}}{S_{DS}}$$

Then for our case,

$$T_0 \equiv 0.2 \frac{S_{D1}}{S_{DS}} = 0.2 \left(\frac{0.50g}{1.33g}\right) = 0.08 \text{ s}$$

$$T_S \equiv \frac{S_{D1}}{S_{DS}} = \left(\frac{0.50g}{1.33g}\right) = 0.38 \text{ s}$$

- For periods less than or equal to T_0, the design spectral response acceleration, S_a, is given by Eq. (11.4-5).

$$S_a = S_{DS}\left(0.4 + 0.6\frac{T}{T_0}\right) \qquad (11.4\text{-}5)$$

- For periods greater than T_0 and less than or equal to T_S, the design spectral response acceleration, S_a, is equal to S_{DS}.
- For periods greater than T_S and less than or equal to T_L (from Figs. 22-15 through 22-20), the design spectral response acceleration, S_a, is given by Eq. (11.4-6). In our case, $T_L = 8$ s.

$$S_a = \frac{S_{D1}}{T} \qquad (11.4\text{-}6)$$

- For periods greater than T_L, the design spectral response acceleration, S_a, is given by Eq. (11.4-7).

$$S_a = \frac{S_{D1} T_L}{T^2} \qquad (11.4\text{-}7)$$

The resulting design acceleration response spectrum is given in Fig. 3.14.

Step 7: Determine the structure's importance factor, I, and occupancy category using Sec. 11.5.

Assume that the structure is assigned an occupancy category II. This corresponds to an importance factor of 1.0.

Step 8: Determine the structure's seismic design category using Sec. 11.6.

Tables 3.15 and 3.16 (Tables 11.6-1 and 11.6-2 of ASCE 7-05) must be checked, and the higher seismic design category from those two tables applies.

In our case, S_{DS} is 1.33, and S_{D1} is 0.50. Because S_{DS} exceeds 0.50 (Table 11.6-1 of ASCE 7-05), and S_{D1} exceeds 0.20 (Table 11.6-2 of ASCE 7-05), the structure is assigned to Seismic Design Category D.

Code Basis for Structural Design of Masonry Buildings 101

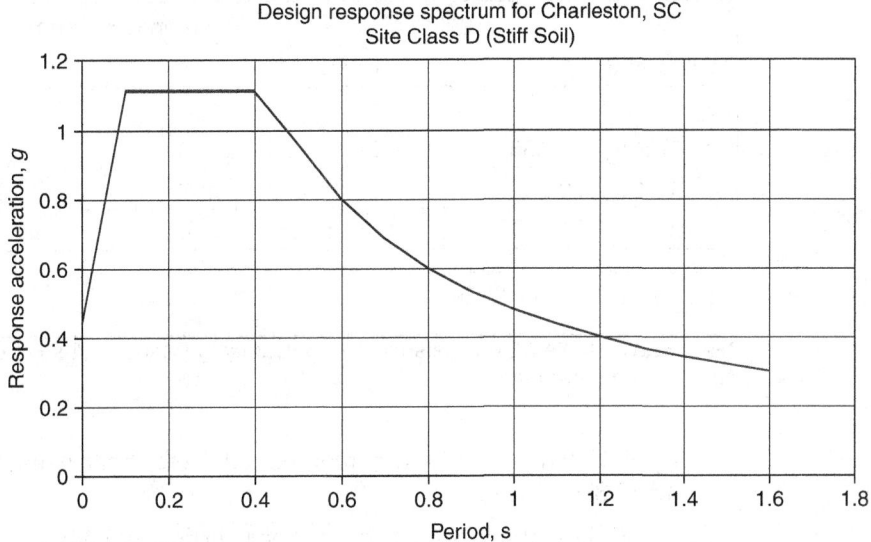

FIGURE 3.14 Design acceleration response spectrum for example problem.

Occupancy category	I
I or II	1.0
III	1.25
IV	1.5

Source: Table 11.5-1 of ASCE 7-05.

TABLE 3.14 Table 11.5-1 Importance Factors

	Occupancy category		
Value of S_{DS}	I or II	III	IV
$S_{DS} < 0.167$	A	A	A
$0.167 \leq S_{DS} < 0.33$	B	B	C
$0.33 \leq S_{DS} < 0.50$	C	C	D
$0.50 \leq S_{DS}$	D	D	D

Source: Table 11.6-1 of ASCE 7-05.

TABLE 3.15 Table 11.6-1 Seismic Design Category Based on Short Period Response Acceleration Parameter

	Occupancy category		
Value of S_{D1}	I or II	III	IV
$S_{D1} < 0.067$	A	A	A
$0.067 \leq S_{D1} < 0.133$	B	B	C
$0.133 \leq S_{D1} < 0.20$	C	C	D
$0.20 \leq S_{D1}$	D	D	D

Source: Table 11.6-2 of ASCE 7-05.

TABLE 3.16 Table 11.6-2 Seismic Design Category Based on 1-s Period Response Acceleration Parameter

Step 9: Calculate the structure's seismic base shear using Secs. 12.8.1 and 12.8.2.

Step 10: Distribute seismic base shear vertically using Sec. 12.8.3.

Step 11: Distribute seismic base shear horizontally using Sec. 12.8.4.

These last three steps are structure-dependent. They depend on the seismic response modification coefficient assigned to the structural system, on the structure's plan structural irregularities, on the structure's vertical structural irregularities, and on the structure's redundancy.

Plan structural irregularities include

- Plan eccentricities between the center of mass and the center of stiffness
- Re-entrant corners
- Out-of-plane offsets
- Nonparallel systems

These can increase seismic response.

Vertical structural irregularities include

- Stiffness irregularity
- Mass irregularity
- Vertical geometric irregularity
- In-plane discontinuity in vertical lateral-force-resisting elements
- Discontinuity in capacity—weak story

These can also increase seismic response.

Structures with low redundancy have a higher probability of failure, which is compensated for by increasing design seismic forces.

The above characteristics depend on the particular building, and are not addressed further here. These are addressed in an example problem at the end of this book.

3.6 Loading Combinations of the 2009 IBC

3.6.1 Strength Loading Combinations of the 2009 IBC (Sec. 1605.2.1)

Strength loading combinations from Sec. 1605.2.1 of the 2009 IBC are given below:

1. $1.4\,(D + F)$
2. $1.2\,(D + F + T) + 1.6\,(L + H) + 0.5\,(L_r \text{ or } S \text{ or } R)$
3. $1.2D + 1.6\,(L_r \text{ or } S \text{ or } R) + (L \text{ or } 0.8W)$
4. $1.2D + 1.6W + f_1 L + 0.5\,(L_r \text{ or } S \text{ or } R)$
5. $1.2D + 1.0E + f_1 L + f_2 S$
6. $0.9D + 1.6W + 1.6H$
7. $0.9D + 1.0E + 1.6H$

where f_1 = 1.0 for floors in places of public assembly, for live loads in excess of 100 lb/ft², and for parking garage live load
     = 0.5 for other live loads
  f_2 = 0.7 for roof configurations (such as saw tooth) that do not shed snow off the structure
     = 0.2 for other roof configurations
  D = dead load
  E = combined effect of horizontal and vertical earthquake induced forces as defined in Sec. 12.4.2 of ASCE 7
  F = load due to fluids with well-defined pressures and maximum heights
  H = load due to lateral earth pressure, ground water pressure, or pressure of bulk materials
  L = live load, except roof live load, including any permitted live load reduction
  L_r = roof live load including any permitted live load reduction
  R = rain load
  S = snow load
  T = self-straining force arising from contraction or expansion resulting from temperature change, shrinkage, moisture change, creep in component materials, movement due to differential settlement or combinations thereof
  W = wind load due to wind pressure

3.6.2 Basic Allowable-Stress Loading Combinations of the 2009 IBC (Sec. 1605.3.1)

Basic allowable-stress loading combinations from Sec. 1605.3.1 of the 2009 IBC are given below:

1. $D + F$
2. $D + H + F + L + T$
3. $D + H + F + (L_r$ or S or $R)$
4. $D + H + F + 0.75 (L + T) + 0.75 (L_r$ or S or $R)$
5. $D + H + F + (W$ or $0.7E)$
6. $D + H + F + 0.75 (W$ or $0.7E) + 0.75L + 0.75 (L_r$ or S or $R)$
7. $0.6D + W + H$
8. $0.6D + 0.7E + H$

where D = dead load
 E = combined effect of horizontal and vertical earthquake induced forces as defined in Sec. 12.4.2 of ASCE 7
 F = load due to fluids with well-defined pressures and maximum heights
 H = load due to lateral earth pressure, ground water pressure, or pressure of bulk materials
 L = live load, except roof live load, including any permitted live load reduction
 L_r = roof live load including any permitted live load reduction
 R = rain load
 S = snow load
 T = self-straining force arising from contraction or expansion resulting from temperature change, shrinkage, moisture change, creep in component materials, movement due to differential settlement or combinations thereof
 W = wind load due to wind pressure

Note on 1/3 Increase Permitted by the 2008 MSJC *Code*

The allowable-stress provisions of the 2008 MSJC *Code* permit allowable stresses to be increased by 1/3 for loading combinations involving wind or earthquake, for load standards that do not specifically prohibit such increase.

This 1/3 increase is based on experience only. No data exist to justify it based on rate effects, and because dead load always acts, there is no statistical justification for a reduction in wind or earthquake loading in combination with dead load. The 1/3 stress increase is in effect simply an extra increase in allowable stress for wind or earthquake loads. It might be justified in some cases by historically low allowable stresses for flexural reinforcement, but it does not seem justified in general. It is currently being reviewed by the MSJC.

The 2009 IBC specifically prohibits use of the 1/3 increase in conjunction with basic allowable-stress loading combinations (IBC 2009, Sec. 1605.3.1.1), while permitting use of the 1/3 stress increase in conjunction with the alternative allowable-stress loading combinations (IBC 2009, Sec. 1605.3.2). ASCE 7-05 specifically prohibits the use of the 1/3 increase. Because it is permitted in only limited circumstances, the 1/3 increase is not used in this book.

3.7 Summary of Strength Design Provisions of 2008 MSJC *Code*

In Chaps. 5 and 6 of this book, the strength design of masonry elements is discussed in detail. In this section, strength design provisions are summarized.

3.7.1 Strength Loading Combinations from Sec. 1605.2.1 of the 2009 IBC

Strength loading combinations from Sec. 1605.2.1 of the 2009 IBC are repeated below:

1. $1.4(D + F)$
2. $1.2(D + F + T) + 1.6(L + H) + 0.5(L_r \text{ or } S \text{ or } R)$
3. $1.2D + 1.6(L_r \text{ or } S \text{ or } R) + (L \text{ or } 0.8W)$
4. $1.2D + 1.6W + f_1 L + 0.5(L_r \text{ or } S \text{ or } R)$
5. $1.2D + 1.0E + f_1 L + f_2 S$
6. $0.9D + 1.6W + 1.6H$
7. $0.9D + 1.0E + 1.6H$

where f_1 = 1.0 for floors in places of public assembly, for live loads in excess of 100 pounds per square foot, and for parking garage live load

= 0.5 for other live loads

f_2 = 0.7 for roof configurations (such as saw tooth) that do not shed snow off the structure

= 0.2 for other roof configurations

Combination of actions	Strength-reduction factor
Combinations of flexure and axial load in reinforced masonry	0.90
Combinations of flexure and axial load in unreinforced masonry	0.60
Shear	0.80
Anchor bolts, strength controlled by steel	0.90
Anchor bolts, strength controlled by masonry breakout, crushing, or pryout	0.50
Anchor bolts, strength controlled by pullout	0.65
Bearing	0.60

Source: Section 3.1.4 of the 2008 MSJC *Code*.

TABLE 3.17 Strength-Reduction Factors

3.7.2 Strength-Reduction Factors from 2008 MSJC *Code*, Sec. 3.1.4

Strength-reduction factors, taken from Sec. 3.1.4 of the 2008 MSJC *Code*, are summarized in Table 3.17.

3.7.3 Summary of Steps for Strength Design of Unreinforced Panel Walls

Using the 2008 MSJC *Code*, steps for strength design of panel walls are summarized in Table 3.18.

Design step	2008 MSJC reference
For most boundary conditions, assume all load will be taken by the vertical strip in the interior wythe. Check that strip for maximum stresses. Because axial stresses are zero, maximum compressive stress will not govern, nor will axial capacity reduced by slenderness effects. Because masonry is unreinforced, maximum tensile stresses will govern. So only tensile stresses need to be checked.	*Code* 3.1.4 *Code* 3.2.2 *Code* Table 3.1.8.2.1 *Specification* Tables 1 and 2
Check one-way shear (usually will not govern).	*Code* 3.1.4 *Code* 3.2.4

TABLE 3.18 Summary of Steps for Strength Design of Unreinforced Panel Walls

Design step	2008 MSJC reference
Usually, all load is taken by vertical strips. Check typical vertical strip for slenderness-dependent axial capacity, maximum compressive stresses, and maximum tensile stresses.	*Code* 3.1.4 *Code* 3.2.2 *Specification* Tables 1 and 2
Check one-way shear (usually will not govern).	*Code* 3.1.4 *Code* 3.2.4

TABLE 3.19 Summary of Steps for Strength Design of Unreinforced Bearing Walls

3.7.4 Summary of Steps for Strength Design of Unreinforced Bearing Walls

Using the 2008 MSJC *Code*, steps for strength design of unreinforced bearing walls are summarized in Table 3.19.

3.7.5 Summary of Steps for Strength Design of Unreinforced Shear Walls

Using the 2008 MSJC *Code*, steps for strength design of unreinforced bearing walls are summarized in Table 3.20.

3.7.6 Summary of Steps for Strength Design of Reinforced Beams and Lintels

Using the 2008 MSJC *Code*, steps for strength design of reinforced beams and lintels are summarized in Table 3.21.

3.7.7 Summary of Steps for Strength Design of Reinforced Curtain Walls

Using the 2008 MSJC *Code*, steps for strength design of reinforced curtain walls are summarized in Table 3.22.

Design step	2008 MSJC reference
Check in-plane flexural capacity.	*Code* 3.1.4 *Code* 3.2.2 *Specification* Tables 1 and 2
Check in-plane shear capacity. Verify ability of roof diaphragm to transfer horizontal reactions to shear walls.	*Code* 3.1.4 *Code* 3.2.4

TABLE 3.20 Summary of Steps for Strength Design of Unreinforced Shear Walls

Design step	2008 MSJC reference
Check that depth is sufficient to ensure that shear can be resisted by masonry alone, without shear reinforcement.	Code 3.1.4 Code 3.3.4.1.2 Specification Tables 1 and 2
Compute required flexural reinforcement, approximating internal lever arm as 0.9d. Revise if necessary.	Code 3.1.4
$M_n \approx A_s f_y\, 0.9\,d$ Check nominal moment versus cracking capacity. Check maximum reinforcement.	Code 3.3.2 Code 3.3.4.2.2.2 Code 3.3.3.5

TABLE 3.21 Summary of Steps for Strength Design of Reinforced Beams and Lintels

Design step	2008 MSJC reference
All load must be taken by horizontal strips. Check that strip for stresses. Usually, reinforcement will be needed. Estimate required reinforcement using $jd = d - d'$, then recalculate if necessary: $M_n = A_s f_y (d - d')$.	Code 3.1.4 Code 3.3.2 Specification Tables 1 and 2
Check one-way shear (usually will not govern).	Code 3.1.4 Code 3.3.4.1.2.1

TABLE 3.22 Summary of Steps for Strength Design of Reinforced Curtain Walls

3.7.8 Summary of Steps for Strength Design of Reinforced Bearing Walls

Using the 2008 MSJC *Code*, steps for strength design of reinforced bearing walls are summarized in Table 3.23.

3.7.9 Summary of Steps for Strength Design of Reinforced Shear Walls

Using the 2008 MSJC *Code*, steps for strength design of reinforced shear walls are summarized in Table 3.24.

Design step	2008 MSJC reference
Usually, all load is taken by vertical strips. Verify ability of roof diaphragm to transfer horizontal reactions from those strips. Check typical vertical strip for stresses using column interaction diagram.	*Code* 3.1.4 *Code* 3.3.2 *Specification* Tables 1 and 2
Check one-way shear out-of-plane (usually will not govern).	*Code* 3.1.4 *Code* 3.3.4.1.2

TABLE 3.23 Summary of Steps for Strength Design of Reinforced Bearing Walls

Design step	2008 MSJC reference
Check for in-plane flexure plus axial loads. Check maximum reinforcement.	*Code* 3.1.4 *Code* 3.2.2 *Code* 3.3.3.5 *Specification* Tables 1 and 2
Check in-plane shear capacity. Verify ability of roof diaphragm to transfer horizontal reactions to shear walls.	*Code* 3.1.4 *Code* 3.3.4.1.2

TABLE 3.24 Summary of Steps for Strength Design of Reinforced Shear Walls

3.8 Summary of Allowable-Stress Design Provisions of 2008 MSJC *Code*

In Chaps. 7 and 8 of this book, the allowable-stress design of masonry elements is discussed in detail. In this section, allowable-stress design provisions are summarized.

3.8.1 Allowable-Stress Loading Combinations from Sec. 1605.3.1 of the 2009 IBC

Allowable-stress loading combinations from Sec. 1605.3.1 of the 2009 IBC are repeated below:

1. $D + F$
2. $D + H + F + L + T$
3. $D + H + F + (L_r$ or S or $R)$

4. $D + H + F + 0.75 (L + T) + 0.75 (L_r \text{ or } S \text{ or } R)$
5. $D + H + F + (W \text{ or } 0.7E)$
6. $D + H + F + 0.75 (W \text{ or } 0.7E) + 0.75L + 0.75 (L_r \text{ or } S \text{ or } R)$
7. $0.6D + W + H$
8. $0.6D + 0.7E + H$

3.8.2 Summary of Steps for Allowable-Stress Design of Unreinforced Panel Walls

Using the 2008 MSJC *Code*, steps for allowable-stress design of unreinforced panel walls are summarized in Table 3.25.

3.8.3 Summary of Steps for Allowable-Stress Design of Unreinforced Bearing Walls

Using the 2008 MSJC *Code*, steps for allowable-stress design of unreinforced bearing walls are summarized in Table 3.26.

3.8.4 Summary of Steps for Allowable-Stress Design of Unreinforced Shear Walls

Using the 2008 MSJC *Code*, steps for allowable-stress design of unreinforced bearing walls are summarized in Table 3.27.

3.8.5 Summary of Steps for Allowable-Stress Design of Reinforced Beams and Lintels

Using the 2008 MSJC *Code*, steps for allowable-stress design of reinforced beams and lintels are summarized in Table 3.28.

Design step	2008 MSJC reference
For most boundary conditions, assume all load will be taken by the vertical strip in the interior wythe. Check that strip for stresses. Because axial stresses are zero, (f_a/F_a) will always be zero, and buckling will never govern. Because masonry is unreinforced, tensile stresses will govern. So only tensile stresses need to be checked.	*Code* 2.2.3.1(c) *Code* Table 2.2.3.2
Check one-way shear (usually will not govern) $f_v = \dfrac{VQ}{I_n b} = \dfrac{3}{2} \dfrac{V}{A_n}$	*Code* Eq. (2-19) *Code* 2.2.5

TABLE 3.25 Summary of Steps for Allowable-Stress Design of Unreinforced Panel Walls

Code Basis for Structural Design of Masonry Buildings

Design step	2008 MSJC reference
Usually, all load is taken by vertical strips. Check typical vertical strip for compressive stresses. Use unity equation: $$\frac{f_a}{F_a} + \frac{f_b}{F_b} \leq 1$$ F_a depends on slenderness and f'_m. F_b is $(1/3)\, f'_m$.	Code Eq. (2-13) Code Eqs. (2-15), (2-16) Specification Tables 1 and 2 Code Eq. (2-17)
Check that strip for tensile stresses.	Code Table 2.2.3.2
Check that strip for buckling.	Code Eq. (2-18)
Check one-way shear out of plane (usually will not govern) $f_v = \dfrac{VQ}{I_n b} = \dfrac{3}{2}\dfrac{V}{A_n}$	Code Eq. (2-19) Code 2.2.5

TABLE 3.26 Summary of Steps for Allowable-Stress Design of Unreinforced Bearing Walls

Design step	2008 MSJC reference
Check for in-plane shear $f_v = \dfrac{VQ}{I_n b} = \dfrac{3}{2}\dfrac{V}{A_n}$ Verify ability of roof diaphragm to transfer horizontal reactions to shear walls.	Code Eq. 2-19 Code 2.2.5 Specification Tables 1 and 2
Check for in-plane flexure plus axial loads $f_t = \dfrac{Mc}{I} - \dfrac{P}{A} \leq F_t$	Code 2.2.3.2

TABLE 3.27 Summary of Steps for Allowable-Stress Design of Unreinforced Shear Walls

Design step	2008 MSJC reference
Check that depth is sufficient to ensure that shear can be resisted by masonry alone, without shear reinforcement.	Code 2.3.5 Specification Tables 1 and 2
Compute required flexural reinforcement, approximating internal lever arm as $0.9d$. Revise if necessary. $M \approx A_s F_s \left(\dfrac{7}{8}\right) d$	Code 2.3.2

TABLE 3.28 Summary of Steps for Allowable-Stress Design of Reinforced Beams and Lintels

Design step	2008 MSJC reference
All load must be taken by horizontal strips. Check that strip for stresses. Usually, reinforcement will be needed. Estimate required reinforcement using $k = 3/8$, $j = 7/8$, then calculate k and j and check: $$f_s = \frac{M_o}{A_s jd} \qquad f_m = \frac{2M_o}{jkbd^2}$$ $f_s \leq F_s \qquad f_m \leq F_m$ Allowable stress in masonry is $(1/3)\, f'_m$.	Code 2.3.3.2.2 Code 2.3.2 Specification Tables 1 and 2
Check one-way shear (usually will not govern) $$f_v = \frac{V}{bd}$$	Code Eq. (2-23) Code 2.3.5.2.2

TABLE 3.29 Summary of Steps for Allowable-Stress Design of Reinforced Curtain Walls

3.8.6 Summary of Steps for Allowable-Stress Design of Reinforced Curtain Walls

Using the 2008 MSJC *Code*, steps for allowable-stress design of reinforced curtain walls are summarized in Table 3.29.

3.8.7 Summary of Steps for Allowable-Stress Design of Reinforced Bearing Walls

Using the 2008 MSJC *Code*, steps for strength design of reinforced bearing walls are summarized in Table 3.30.

Design step	2008 MSJC reference
Usually, all load is taken by vertical strips. Verify ability of roof diaphragm to transfer horizontal reactions from those strips. Check typical vertical strip for stresses using column interaction diagram. Allowable stress in masonry is $(1/3)\, f'_m$.	Code 2.3.2.1 Code 2.3.3 Code 2.3.3.2.2 Specification Tables 1 and 2
Check one-way shear out of plane (usually will not govern) $f_v = \dfrac{V}{bd}$	Code Eq. (2-19) Code 2.3.5

TABLE 3.30 Summary of Steps for Allowable-Stress Design of Reinforced Bearing Walls

Design step	2008 MSJC reference
Check for in-plane shear $f_v = \dfrac{V}{bd}$ Verify ability of roof diaphragm to transfer horizontal reactions to shear walls.	*Code* Eq. (2-23) *Code* 2.3.5.2.2 (default case) or *Code* 2.3.5.2.3 (steel resists all shear) *Specification* Tables 1 and 2
Check for in-plane flexure plus axial loads. See "Reinforced Bearing Walls" above.	*Code* 2.3.2.1 *Code* 2.3.3 *Code* 2.3.3.2.2

TABLE 3.31 Summary of Steps for Strength Design of Reinforced Shear Walls

3.8.8 Summary of Steps for Allowable-Stress Design of Reinforced Shear Walls

Using the 2008 MSJC *Code*, steps for strength design of reinforced shear walls are summarized in Table 3.31.

3.9 Additional Information on Code Basis for Structural Design of Masonry Buildings

In this section, complete names and addresses are given for United States technical specialty organizations, industry organizations, governmental organizations, and model-code organizations.

3.9.1 Technical Specialty Organizations Related to Masonry

American Concrete Institute
38800 Country Club Drive
Farmington Hills, MI 48331
www.aci-int.org

American Institute of Steel
 Construction
400 North Michigan Avenue
Chicago, IL 60611-4185
http://www.aisc.org/

American Society of Civil
 Engineers
1801 Alexander Bell Drive
Reston, VA 20191
www.asce.org

TMS (The Masonry Society)
3970 Broadway, Suite 201-D
Boulder, CO 80304-1135
http://www.masonrysociety.org/

3.9.2 Industry Organizations Related to Masonry

Autoclaved Aerated Concrete
Products Association
info@aacpa.org
www.aacpa.org

Brick Industry Association
11490 Commerce Park Drive
Reston, VA 22091
www.bia.org

International Masonry Institute
The James Brice House
42 East Street
Annapolis, MD 21401
www.imiweb.org

Mason Contractors Association of
 America
33 South Roselle Road
Schaumburg, IL 60193
http://www.masoncontractors.org/

National Concrete Masonry
 Association
13750 Sunrise Valley Drive
Herndon, VA 22071-3406
www.ncma.org

Portland Cement Association
5420 Old Orchard Road
Skokie, IL 60077-1083
www.cement.org

Prestressed Concrete Institute
175 West Jackson Boulevard
Chicago, IL 60604
www.pci.org

3.9.3 Model-Code Development Organizations

International Code Council
 Headquarters
500 New Jersey Avenue, NW,
 6th Floor Washington,
 DC 20001-2070
www.iccsafe.org

ICC Chicago District Office
900 Montclair Road
Birmingham, AL 35213-1206

ICC Chicago District Office
4051 West Flossmoor Road
Country Club Hills, IL 60478-5795

ICC Los Angeles District Office
5360 Workman Mill Road
Whittier, CA 90601-2298

National Fire Protection Association
1 Batterymarch Park
Quincy, MA 02169-7471
www.nfpa.org

3.9.4 Specification Development Organizations

American Society for Testing
 and Materials, Inc.
100 Barr Harbor Drive
West Conshohocken, PA 19428-2959

www.astm.org

3.9.5 Governmental Organizations

Building Seismic Safety Council
1015 15th Street. N.W., Suite 700
Washington, D.C. 20005
http://www.nibs.org/index.php/bssc/

3.9.6 Other Organizations

American National Standards Institute, Inc.
25 West 43rd Street,
(between 5th and
6th Avenues), 4th Floor
New York, NY 10036
www.ansi.org

CHAPTER 4
Introduction to MSJC Treatment of Structural Design

4.1 Basic Mechanical Behavior of Masonry

Masonry is a composite material, comprising units, mortar, grout, and accessory materials. Because of this, its mechanical behavior is complex. Using nonlinear finite-element analysis, addressing the behavior of constituent materials and of the interface relationships between them, it is possible to describe the force-deformation behavior of masonry elements.

For design, however, this approach is neither practical nor necessary. For design purposes, masonry is normally idealized as an isotropic material, with nonlinear stress-strain behavior in compression (much like concrete) and linear stress-strain behavior in tension. Compressive capacity is governed by crushing (often characterized by complex local behavior) and tensile capacity, by the bond strength between units and mortar.

The crushing strength of masonry can be evaluated by compression tests on masonry prisms. Design of masonry elements is based on a specified compressive strength of masonry, f_m', whose role is analogous to that of the specified compressive strength of concrete, f_c', in concrete design. The specified compressive strength of masonry is the basis for design and forms part of the contract documents. These contract documents require verification that the masonry comply with the specified compressive strength, either by compression tests of prisms or by conservative relationships involving the compressive strengths of the units and the type of mortar. These are addressed in the MSJC Code and Specification (MSJC 2008a,b).

One advantage of using the conservative relationships involving the compressive strengths of the units and the type of mortar (the so-called "unit strength method" from Tables 1 and 2 of the MSJC Specification) is that it is possible to verify compliance with the specified compressive strength f_m' with no project-specific material testing whatsoever. The compressive strength of the units is verified by the manufacturer as part of quality control and compliance with the unit specification. The mortar can be specified by proportion, and compliance with that specification is verified by verifying proportions (no mortar testing). The grout can be specified by proportion and compliance with that specification is verified by verifying proportions (no grout testing). The minimum probable strength of the masonry is then obtained from Table 1 or 2 of the MSJC Specification (MSJC 2008b). Material tests can be performed for quality control or to verify compliance with a specified strength, but they are not necessary if the designer chooses specification criteria that do not require testing.

Masonry elements requiring structural calculation are designed using the specified compressive strength, verified as noted above, and prescribed tensile bond strengths based on extensive experimental investigation.

4.2 Classification of Masonry Elements

Masonry elements can be designed in at least two ways:

- According to their structural function
- According to the approach used to design them

4.3 Classification of Masonry Elements by Structural Function

- Nonload-bearing masonry: Supports vertical loads from self-weight only, plus possibly loads from out-of-plane wind or earthquake.
- Load-bearing masonry: Supports vertical loads from roof or overlying floors, plus possibly loads from in-plane shear, plus possibly loads from out-of-plane wind or earthquake.

Although these two types of masonry can be approached using exactly the same tools of engineering mechanics, they have been distinguished historically.

4.4 Classification of Masonry Elements by Design Intent

- Unreinforced masonry: Is designed assuming that flexural tensile stresses are resisted by masonry, and that the presence of any reinforcement is neglected in design. "Unreinforced masonry" can therefore actually have reinforcement in it, for structural integrity or by prescription. That reinforcement, however, is neglected in design calculations. Design is carried out in the linear elastic range.

- Reinforced masonry: Is designed assuming that flexural tensile stresses cannot be resisted by masonry, and are resisted by reinforcement only. Shear stresses can be resisted by masonry or by reinforcement, singly or in combination. Design can be carried out by allowable stress design, or by strength design. Using this definition, unreinforced masonry can actually have reinforcement (for integrity or to meet prescriptive requirements).

4.5 Design Approaches for Masonry Elements

The two design approaches (allowable-stress design and strength design) can each have the same result, and also the same level of safety (measured in terms of probability of failure under service loads). In allowable-stress design, the probability of failure is controlled directly by the factor of safety. In strength design the probability of failure depends on the quotient of the load factor and the capacity reduction factor (ϕ factor).

Most modern codes are based on strength design, because it gives a more uniform factor of safety against collapse.

- Strength design: In strength design, design actions (axial forces, shears, and moments) are computed using service loads, and are then increased by load factors. The factored design actions are then compared with nominal member strengths, decreased by strength-reduction factors.

$$\text{Service actions} \times \text{LF} \leq \phi \times \text{nominal capacity}$$

- Allowable-stress design: In allowable-stress design, stresses corresponding to service loads are compared with allowable stresses. The allowable stresses are material strengths, reduced by a factor of safety. Factors of safety for masonry typically range from 2.5 to 4.

$$\text{Stresses from service loads} \leq \frac{\text{failure stresses}}{\text{safety factor}}$$

- Empirical design: Empirical design (design carried out based on aspect ratios, dimensional limits, and approximate gravity-load stresses on gross areas) is addressed by Chap. 5 of the MSJC Code (2008a), and is permitted for masonry structures and elements in very limited circumstances. It is not discussed further in this book.
- Veneer design: According to Chap. 6 of the MSJC Code (2008a), masonry veneer is designed prescriptively through control of connector type and spacing. Design of veneer is not discussed further in this book.
- Glass-block masonry design: Design of glass unit masonry is addressed by Chap. 7 of the MSJC Code (2008a). It is not discussed further in this book.
- AAC masonry design: Design of autoclaved aerated concrete (AAC) masonry is addressed by App. A of the MSJC Code (2008a). It is covered in detail in Chap. 14 of this book.

4.6 How Reinforcement Is Used in Masonry Elements

4.6.1 Masonry Beams and Lintels

These require horizontal reinforcement placed in hollow bond-beam units, or in fully grouted cavities between wythes of solid clay masonry units. Examples of these are shown in Figs. 4.1 and 4.2, respectively. As noted in later sections dealing with the design of masonry beams and lintels, it is not necessary to use so-called "trough units" in concrete masonry; ordinary stretcher units can be used if supported by a shoring board.

4.6.2 How Reinforcement Is Used in Masonry Curtain Walls

Masonry curtain walls are normally single-wythe, made of clay masonry units. Reinforcement is oriented horizontally and is placed in bed joints. The reinforcement can be bed-joint reinforcement or smooth (No. 2) bars. Deformed bars cannot be used because the outer diameter of their deformations normally exceeds the specified width of a bed joint (3/8 in.). Also, the MSJC Code (2008a) requires that deformed reinforcement be surrounded by grout.

4.6.3 How Reinforcement Is Used in Masonry Walls

When solid units are used, masonry walls are reinforced horizontally with bed joint reinforcement. Alternatively, the wall can be constructed in two wythes, and a curtain of reinforcement is placed between the wythes, and grout is then poured between the wythes.

Introduction to MSJC Treatment of Structural Design

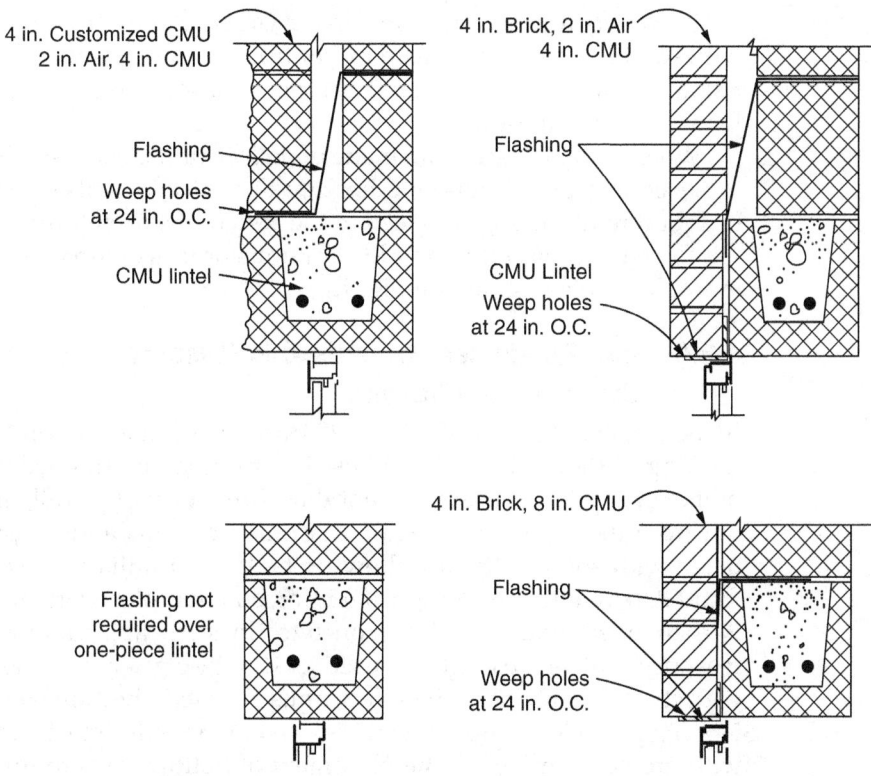

FIGURE 4.1 Examples of reinforcement in CMU lintels.

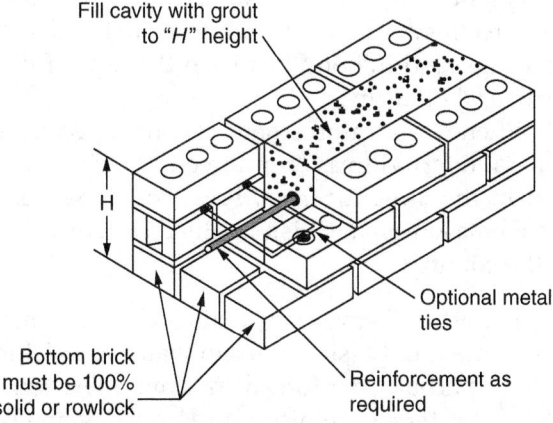

FIGURE 4.2 Examples of reinforcement in clay masonry lintels. (*Source*: Figure 7 of Technical Notes on Brick Construction 7B. © Brick Industry Association.)

In other countries (but rarely in the United States), masonry walls laid with solid units are reinforced by continuous horizontal and vertical elements of reinforced concrete. This type of masonry is sometimes referred to as "confined masonry."

When masonry walls are made of hollow units, vertical reinforcement is placed in grouted cells, and horizontal reinforcement consists either of bed-joint reinforcement, placed in the bed joints, or deformed horizontal reinforcement, placed in bond-beam units or units with cutout webs. An example of this is shown in Fig. 4.3.

4.6.4 How Reinforcement Is Used in Masonry Columns and Pilasters

In the context of the MSJC Code (2008a), a column is an isolated element, meeting certain dimensional restrictions, that carries axial load and moment. A pilaster is an element that forms part of a wall, and projects out from the plane of the wall. Masonry columns and pilasters can be made with solid units or hollow units. If solid units are used, they are formed to make a box. A cage of reinforcement is placed in the box, which is then filled with grout or concrete. In such applications, the solid masonry units are essentially used as stay-in-place cover and formwork with structural function. If hollow units are used, they are laid in an overlapping pattern. Reinforcement is placed in the cells, which are then filled with grout. Examples of the placement of hollow units to form pilasters are shown in Fig. 4.4

4.6.5 Nomenclature Associated with Reinforced Masonry

As we have discussed, "unreinforced masonry" may actually have reinforcement in it, but is designed ignoring the structural action of that reinforcement. Nominal reinforcement is placed at corners, around openings, and in bond beams at the tops of walls, primarily for general structural integrity.

Whenever reinforcement is considered in design, the masonry is in general referred to as "reinforced." Because masonry is a regional tradition in the United States, however, some terms associated with reinforced masonry have historically meant different things in different parts of the country.

- East of Denver (approximately), masonry with bed-joint reinforcement only (such curtain wall), has historically been referred to as "partially reinforced" masonry. This term was introduced to make this type of reinforced masonry seem less intimidating to masons who were not used to using any reinforcement at all. As explained

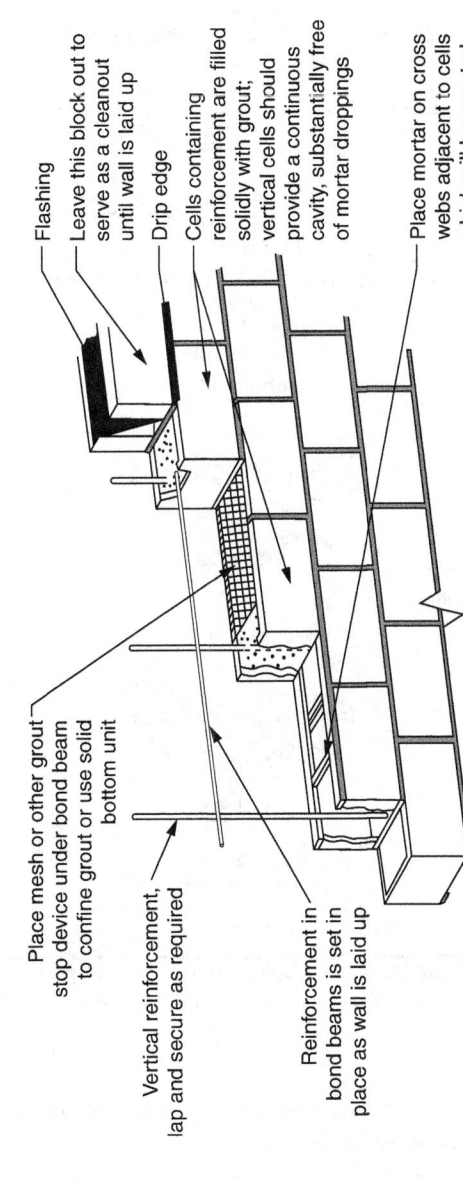

FIGURE 4.3 Example of placement of reinforcement in a masonry wall made of hollow units. (*Source:* Figure 1 of National Concrete Masonry Association TEK 3-2A typical reinforced concrete masonry wall section.)

126 Chapter Four

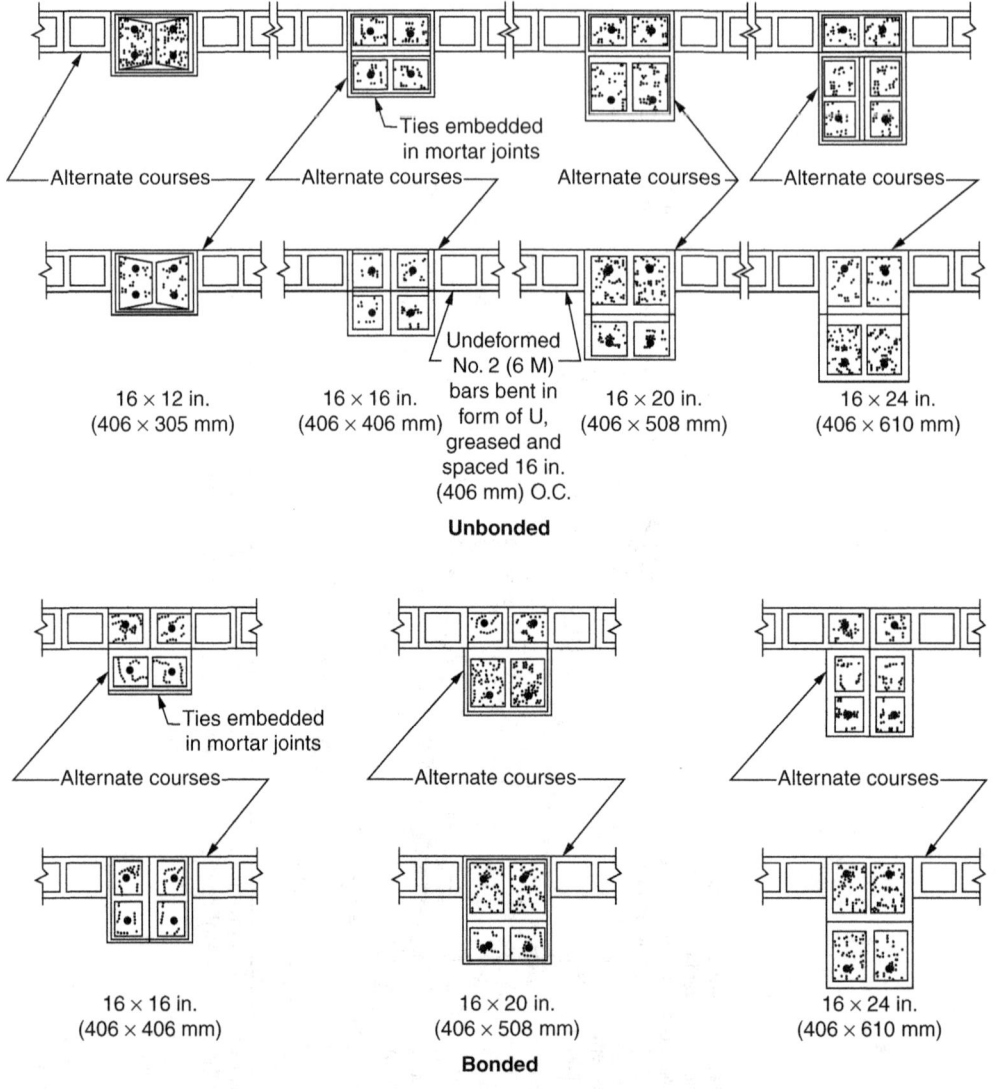

Figure 4.4 Examples of the placement of hollow units to form pilasters. (*Source*: Figure 1 of National Concrete Masonry Association TEK 17-4B 16 in. pilaster sections.)

below, this term has historically meant something completely different west of Denver.

- West of Denver, in regions where the *Uniform Building Code* formerly dominated, seismic resistance has traditionally been very important. After significant damage and deaths of school children in the Long Beach earthquake of 1933, masonry construction was revived only on the condition that it should be reinforced similarly to the reinforced concrete shear walls of the time. That new type of masonry construction, referred to as "reinforced masonry," required that the combined percentage of horizontal and vertical wall reinforcement be not less than 0.002, and that the percentage of reinforcement in each single direction (horizontal and vertical) be not less than one-third of this total, or 0.0007. That is,

$$\rho_h + \rho_v \geq 0.002$$

$$\rho_h \geq 0.0007$$

$$\rho_v \geq 0.0007$$

- West of Denver, masonry so reinforced has historically been referred to as "fully reinforced masonry."
- West of Denver, masonry with reinforcement not meeting those requirements has historically been referred to as "partially reinforced masonry."

When a relatively small amount of reinforcement is used, it can be placed in individually grouted cells, confining elements of reinforced concrete, bed joints, and bond beams. When relatively larger amounts of reinforcement are required, it is often more cost-effective to grout the wall completely.

4.7 How This Book Classifies Masonry Elements

In this book, masonry elements are distinguished first by strength design versus allowable-stress design, then by whether they are designed as unreinforced or reinforced, and finally by their structural function.

4.7.1 Elements Designed by the Strength Approach

Structural Design of Unreinforced Masonry Elements

Nonbearing Elements
- Unreinforced panel walls

Bearing Elements
- Unreinforced bearing walls with eccentric gravity load
- Unreinforced bearing walls with eccentric gravity load plus out-of-plane load

- Effect of openings on bearing walls
- Unreinforced shear walls

Structural Design of Reinforced Masonry Elements
- Reinforced beams and lintels
- Curtain walls with horizontal reinforcement only
- Reinforced bearing walls with eccentric gravity load
- Reinforced bearing walls with eccentric gravity load plus out-of-plane load
- Effect of openings on bearing walls
- Reinforced shear walls

4.7.2 Elements Designed by the Allowable-Stress Approach

Structural Design of Unreinforced Masonry Elements

Nonbearing Elements
- Unreinforced panel walls

Bearing Elements
- Unreinforced bearing walls with eccentric gravity load
- Unreinforced bearing walls with eccentric gravity load plus out-of-plane load
- Effect of openings on bearing walls
- Unreinforced shear walls

Structural Design of Reinforced Masonry Elements
- Reinforced beams and lintels
- Curtain walls with horizontal reinforcement only
- Reinforced bearing walls with eccentric gravity load
- Reinforced bearing walls with eccentric gravity load plus out-of-plane load
- Effect of openings on bearing walls
- Reinforced shear walls

In each case, we shall use both the strength provisions and the allowable-stress provisions of the MSJC Code (2008a). We shall compare the results of those provisions.

4.7.3 Design of Overall Buildings by the Strength Approach

Design of individual elements is followed by strength design of overall buildings, with one example of a high-rise building and one example of a low-rise building.

4.7.4 Design of Autoclaved Aerated Concrete Masonry

Finally, a chapter is devoted to autoclaved aerated concrete masonry, an innovative construction material.

CHAPTER 5
Strength Design of Unreinforced Masonry Elements

5.1 Strength Design of Unreinforced Panel Walls

5.1.1 Examples of Use of Unreinforced Panel Walls

Panel walls commonly comprise the masonry envelope surrounding reinforced concrete or steel frames. In the context of the 2008 MSJC *Code*, a panel wall would be termed a "multiwythe, noncomposite" wall. An example of an unreinforced panel wall is shown in Fig. 5.1.

The outer wythes of panel walls must span horizontally. They cannot span vertically, because of the open expansion joint under each shelf angle. Support conditions for the horizontally spanning outer wythe can be simple or continuous. An example of the connection of a panel wall to a column is shown in the horizontal section of Fig. 5.2. A simple support condition would be achieved by inserting a vertically oriented expansion joint in the clay masonry wythe on both sides of the column.

The inner wythes of panel walls can span horizontally and vertically. As a result, it is convenient to visualize panel walls as being composed of sets of vertical and horizontal crossing strips in each wythe. This is shown schematically in Fig. 5.3.

At the end of this section, it will be shown that

- Because of their aspect ratio, the inner wythe can almost always be considered to span in the vertical direction only.

134 Chapter Five

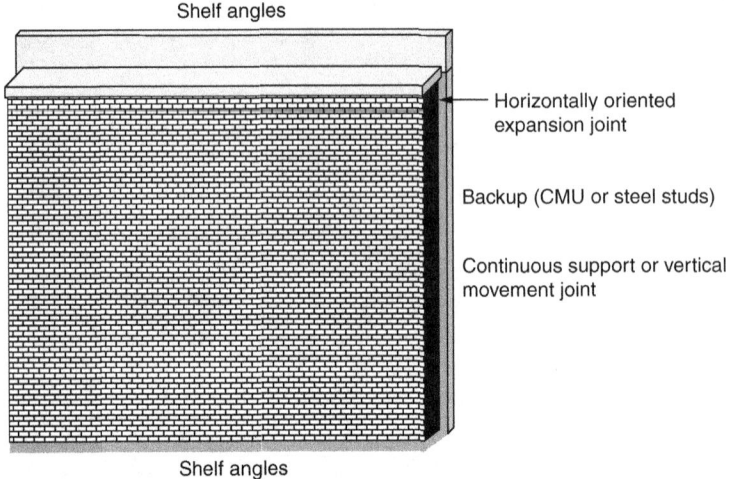

Figure 5.1 Example of an unreinforced panel wall.

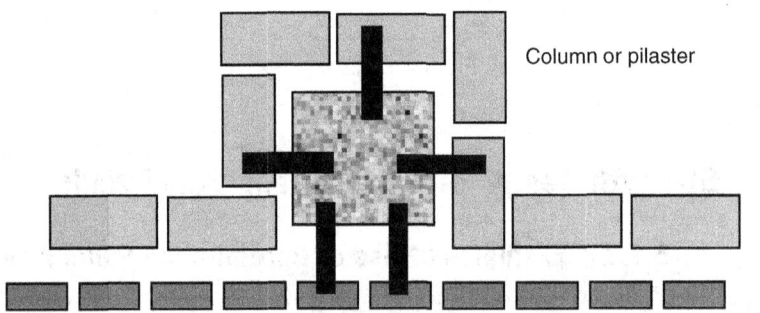

Figure 5.2 Horizontal section showing connection of a panel wall to a column.

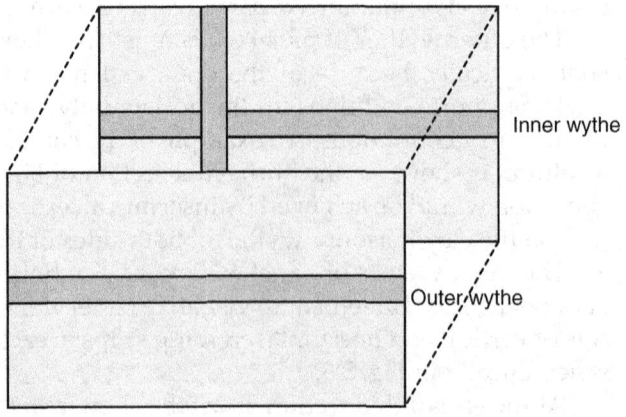

Figure 5.3 Schematic representation of an unreinforced, two-wythe panel wall as two sets of horizontal and vertical crossing strips.

- It is simple and only slightly conservative to design single-wythe panel walls as though the vertical strips resisted all out-of-plane load.
- It is simple and only slightly conservative to design two-wythe panel walls as though the vertical strips of the inner wythe resisted all out-of-plane load.

5.1.2 Flexural Design of Panel Walls Using Strength Provisions of 2008 MSJC *Code*

According to the strength provisions of the 2008 MSJC *Code*, nominal flexural capacity of unreinforced masonry is computed assuming linear stress-strain relationships. Nominal flexural capacity corresponds to a maximum flexural compressive stress of 0.80 f'_m, or a maximum flexural tensile stress equal to the modulus of rupture. Because the modulus of rupture is much lower than 0.80 f'_m it governs. Design actions are factored, and design capacities are computed using those nominal capacities and the appropriate strength-reduction factor.

Load Factors

Load factors are as discussed in Sec. 3.7.1. As prescribed in Sec. 1605.2 of the 2009 IBC, the two loading combinations involving wind are

1. $1.2D + 1.6W + f_1 L + 0.5\ (L_r\ \text{or}\ S\ \text{or}\ R)$
2. $0.9D + 1.6W + 1.6H$

Of these, the second will usually govern. Both combinations have a load factor for W of 1.6.

Modulus of Rupture

For strength design, the nominal flexural strength is to be computed using the modulus of rupture values given in Table 5.1 (Table 3.1.8.2.1 of the 2008 MSJC *Code*). Those values are intended to be 2.5 times the corresponding allowable stresses of the 2008 MSJC *Code*.

Strength-Reduction Factors

Strength-reduction factors are as discussed in Sec. 3.7.2. For combinations of flexure and axial load in unreinforced masonry, $\phi = 0.60$ (Sec. 3.1.4.2 of the 2008 MSJC *Code*).

5.1.3 Example of Strength Design of a Single-Wythe Panel Wall using Solid Units

Check the design of the panel wall shown in Fig. 5.4, for a wind load w of 20 lb/ft², using PCL *mortar*, Type N, and units with a nominal thickness of 8 in.

	Mortar types			
	PCL or mortar cement		Masonry cement or air-entrained PCL	
Masonry type	M or S	N	M or S	N
Normal to bed joints				
Solid units	100	75	60	38
Hollow units[1]				
Ungrouted	63	48	38	23
Fully grouted	163	158	153	145
Parallel to bed joints in running bond				
Solid units	200	150	120	75
Hollow units				
Ungrouted and partially grouted	125	95	75	48
Fully grouted	200	150	120	75
Parallel to bed joints in stack bond				
Continuous grout section parallel to bed joints	250	250	250	250
Other	0	0	0	0

Source: Table 3.1.8.2.1 of 2008 MSJC Code.

[1]For partially grouted masonry, allowable stresses shall be determined on the basis of linear interpolation between fully grouted hollow units and ungrouted hollow units based on amount (percentage) of grouting.

TABLE 5.1 Modulus of Rupture

The panel wall will be designed as unreinforced masonry. The design follows the following steps, using a nominal thickness of 8 in. The panel could be designed as a two-way panel. Nevertheless, because of its aspect ratio, the vertical strips will carry practically all the load. Therefore, design it as a one-way panel, consisting of a series of vertically spanning, simply supported strips.

Calculate the maximum factored design bending moment and corresponding factored design flexural tensile stress in a strip 1-ft wide, with a nominal thickness of 8 in.:

$$M_{u\max} = \frac{w_u l^2}{8} = \frac{1.6 \cdot 20 \text{ lb/ft } (8 \text{ ft})^2}{8} \times 12 \text{ in./ft} = 3072 \text{ lb-in.}$$

$$f_t = \frac{Mc}{I} = \frac{3072 \text{ lb-in.} \cdot (7.625/2) \text{ in.}}{[12 \text{ in.} \cdot (7.625 \text{ in.})^3/12]} = 26.4 \text{ lb/in.}^2$$

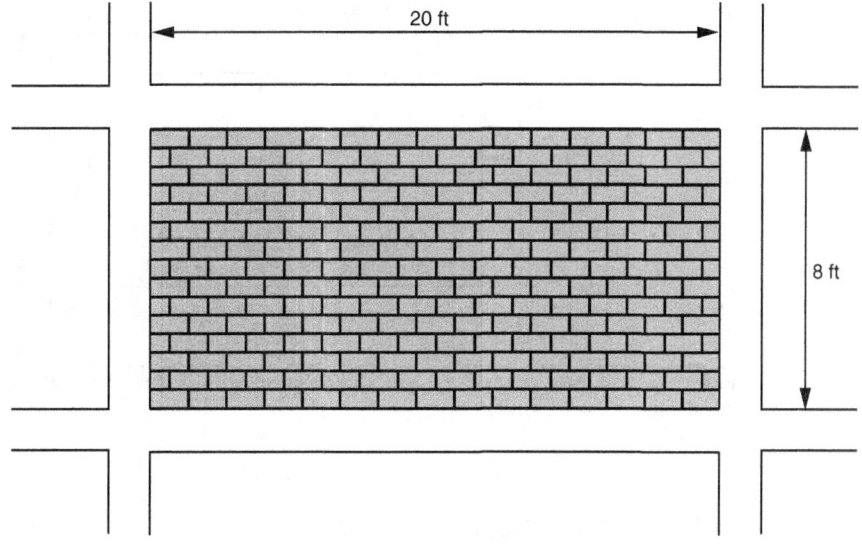

Figure 5.4 Example panel wall to be designed.

The factored flexural tensile stress, 26.4 lb/in.², is less than the modulus of rupture normal to bed joints for solid units and Type N PCL mortar (75 lb/in.²), reduced by a strength-reduction factor of 0.6, or 45.0 lb/in.². The design is therefore satisfactory. We should also check one-way (beam) shear. An example of this is given in Sec. 5.1.8.

5.1.4 Example of Strength Design of a Single-Wythe Panel Wall Using Hollow Units

Check the design of the panel wall of the example of Sec. 5.1.3 for a wind load w of 20 lb/ft², using PCL mortar, Type N, and assuming hollow units with face shells and cross-webs mortared.

The panel wall will be designed as unreinforced masonry. The design follows the following steps, using a nominal thickness of 8 in. The panel could be designed as a two-way panel. Nevertheless, because of its aspect ratio, the vertical strips will carry practically all the load. Therefore, design it as a one-way panel, consisting of a series of vertically spanning, simply supported strips.

Calculate the maximum bending moment and corresponding flexural tensile stress in a strip 1-ft wide, with a nominal thickness of 8 in. Assume that the head joints are only 1.25-in. thick. All length dimensions are in inches. These dimensions are shown in Fig. 5.5.

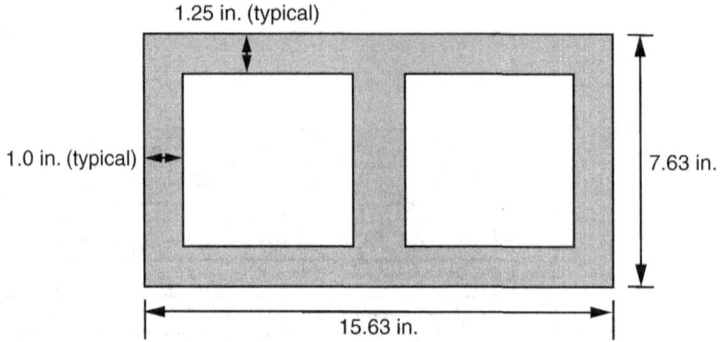

FIGURE 5.5 Idealized cross-sectional dimensions of a nominal 8 × 8 × 16 in. concrete masonry unit.

$$I = I_{solid} - I_{cells} - I_{head\,joint}$$

$$I = \frac{(16)(7.63)^3}{12} - \frac{(15.63 - 3 \cdot 1.00)(7.63 - 2 \cdot 1.25)^3}{12} - \frac{0.37(7.63 - 2 \cdot 1.25)^3}{12}$$

$$I = 592.3 - 142.1 - 4.2 = 446 \text{ in.}^4$$

$$c = \frac{7.63}{2} = 3.81 \text{ in.}$$

$$S = \frac{I}{c} = \frac{446.0}{3.81} = 117.1 \text{ in.}^3 \text{ per 16 in. of width}$$

$$S = \frac{117.1}{16.0} \times 12 = 87.8 \text{ in.}^3 \text{ for a 12-in. wide strip}$$

For the 12-in. wide strip,

$$M_{u\,max} = \frac{w_u l^2}{8} = \frac{1.6 \cdot 20 \text{ lb/ft } (8 \text{ ft})^2}{8} \times 12 \text{ in./ft} = 3072 \text{ lb-in.}$$

$$f_t = \frac{M}{S} = \frac{3072 \text{ lb-in.}}{87.8 \text{ in.}^3} = 35.0 \text{ lb/in.}^2$$

The factored flexural tensile stress, 35.0 lb/in.^{2}, exceeds the modulus of rupture normal to bed joints for hollow units and Type N PCL mortar (48 lb/in.^{2}), reduced by the strength-reduction factor of 0.6 (28.8 lb/in.^{2}). If the mortar is changed to Type S (modulus of rupture 63 lb/in.^{2}), the design will be satisfactory.

5.1.5 Example of Strength Design of a Single-Wythe Panel Wall Using Hollow Units, Face-Shell Bedding Only

Check the design of the panel wall of Sec. 5.1.3 assuming face-shell bedding only (mortar on the face shells of the units only).

The panel wall will be designed as unreinforced masonry. The design follows the following steps, using a nominal thickness of 8 in. The panel could be designed as a two-way panel. Nevertheless, because of its aspect ratio, the vertical strips will carry practically all the load. Therefore, design it as a one-way panel, consisting of a series of vertically spanning, simply supported strips.

The critical stresses will occur on the bed joint, which is the horizontal plane through the masonry where the section modulus is minimum. All length dimensions are in inches. The dimensions of this critical cross-section are shown in Fig. 5.6.

$$I = I_{faceshells} = 309.2 \text{ in.}^4 \text{ per ft of width}$$

$$S = \frac{I}{c} = \frac{309.2}{3.81} = 81.2 \text{ in.}^3 \text{ per ft of width}$$

For the 12-in. wide strip,

$$M_{u\,max} = \frac{w_u l^2}{8} = \frac{1.6 \cdot 20 \text{ lb/ft } (8 \text{ ft})^2}{8} \times 12 \text{ in./ft} = 3072 \text{ lb-in.}$$

$$f_t = \frac{M}{S} = \frac{3072 \text{ lb-in.}}{81.2 \text{ in.}^3} = 37.9 \text{ lb/in.}^2$$

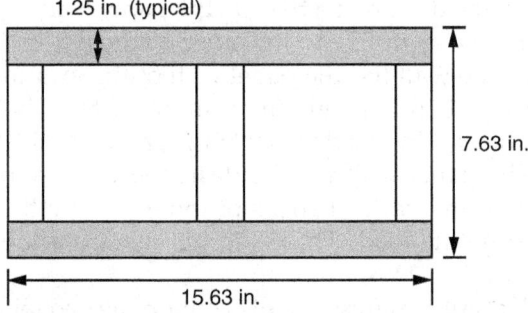

FIGURE 5.6 Idealized cross-sectional dimensions of a nominal 8 × 8 × 16 in. concrete masonry unit with face-shell bedding.

The calculated flexural tensile stress, 37.9 lb/in.², exceeds the modulus of rupture normal to bed joints for hollow units and Type N PCL mortar (48 lb/in.²), reduced by the strength-reduction factor of 0.6 (28.8 lb/in.²). If the mortar is changed to Type S modulus of rupture 63 lb/in.², the design will be satisfactory.

5.1.6 Example of Strength Design of a Single-Wythe Panel Wall Using Hollow Units, Fully Grouted

Check the design of the panel wall of Sec. 5.1.3 for a wind load w of 20 lb/ft², using PCL mortar, Type N, and assuming hollow units, fully grouted.

As in the example of Sec. 5.1.3, assume vertically spanning, simply supported strips. Calculate the maximum bending moment and corresponding flexural tensile stress in a strip 12-in. wide, with a nominal thickness of 8 in.:

$$M_{u\,max} = \frac{w_u l^2}{8} = \frac{1.6 \cdot 20 \text{ lb/ft } (8 \text{ ft})^2}{8} = 3072 \text{ lb-in.}$$

$$f_t = \frac{Mc}{I} = \frac{3072 \text{ lb-in.} \cdot (7.625/2) \text{ in.}}{[12 \text{ in.} \cdot (7.625 \text{ in.})^3 / 12]} = 26.4 \text{ lb/in.}^2$$

The calculated flexural tensile stress, 26.4 lb/in.², is less than the modulus of rupture normal to bed joints for fully grouted hollow units and Type N PCL mortar (150 lb/in.²), reduced by a strength-reduction factor of 0.6 (90 lb/in.²). The design is therefore satisfactory.

5.1.7 Example of Strength Design of a Two-Wythe Panel Wall Using Hollow Units, Face-Shell Bedding Only

Check the design of a two-wythe panel wall in which the outer wythe is modular clay units and the inner wythe is 8-in. CMU with face-shell bedding. The wall has the panel wall of the example of Sec. 5.1.5, assuming face-shell bedding only (mortar on the face shells of the units only). The wall has a wind load w of 20 lb/ft², and uses PCL mortar, Type N.

The panel wall will be designed as unreinforced masonry, assuming that the vertical strips of the inner wythe resist 100 percent of the out-of-plane load. The design is therefore identical to the example of Sec. 5.1.5.

The critical stresses will occur on the bed joint, which is the horizontal plane through the masonry where the section modulus is minimum. This idealized horizontal cross section is shown in Fig. 5.7.

Strength Design of Unreinforced Masonry Elements

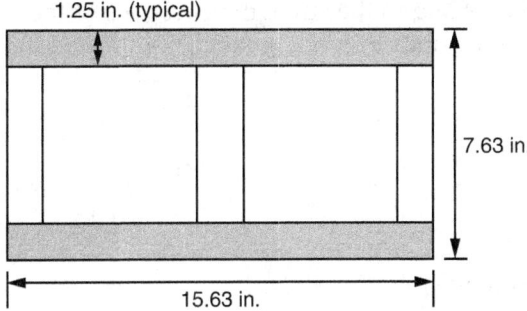

FIGURE 5.7 Idealized cross-sectional dimensions of a nominal 8 × 8 × 16 in. concrete masonry unit with face-shell bedding.

$$I = I_{faceshells} = 309.2 \text{ in.}^4 \text{ per ft of width}$$

$$S = \frac{I}{c} = \frac{309.2}{3.81} = 81.17 \text{ in.}^3 \text{ per ft of width}$$

For the 12-in. wide strip,

$$M_{u\max} = \frac{w_u l^2}{8} = \frac{1.6 \cdot 20 \text{ lb/ft } (8 \text{ ft})^2}{8} \times 12 \text{ in.ft} = 3072 \text{ lb-in.}$$

$$f_t = \frac{M}{S} = \frac{3072 \text{ lb-in.}}{81.17 \text{ in.}^3} = 37.9 \text{ lb/in.}^2$$

The calculated flexural tensile stress, 37.9 lb/in.², exceeds the modulus of rupture normal to bed joints for hollow units and Type N PCL mortar (48 lb/in.²), reduced by the strength-reduction factor of 0.6 (28.8 lb/in.²). If the mortar is changed to Type S (modulus of rupture 63 lb/in.²), the design will be satisfactory. Addition of the outer wythe does not change the design of the panel wall.

5.1.8 Strength Checks of One-Way Shear for Unreinforced Panel Walls

The examples of Sec. 5.1.3 through 5.1.7 in this chapter dealt with design of unreinforced panel walls for flexure. In theory, we should also check one-way shear. In practice, an example shows that shear does not come close to governing the design.

According to the strength provisions of the 2008 MSJC *Code* (Sec. 3.2.4), for hollow units in running bond, nominal shear strength V_n is the least of

$$\begin{cases} 3.8\sqrt{f'_m}\, A_n \\ 300 A_n \\ 56 A_n + 0.45 N \end{cases}$$

The third criterion is

$56A_n + 0.45N_u$ for running bond masonry not grouted solid
$56A_n + 0.45N_u$ for stack bond masonry with open-end units, grouted solid
$90A_n + 0.45N_u$ for running bond masonry grouted solid
$23A_n$ for other stack bond masonry

The strength-reduction factor for shear is 0.80 (2008 MSJC *Code*, Sec. 3.1.4.3).

5.1.9 Example of Strength Check of Shear Capacity for Unreinforced Panel Walls

Check the effect of shear in the example of Sec. 5.1.7.

With f'_m of 1500 lb/in.² and masonry in running bond, and conservatively neglecting the beneficial effects of axial load, the third of the above equations governs, and $V_n = 56 A_n$.

On a strip 1-ft wide, the factored wind load of 1.6 times 20 lb/ft² produces a factored design shear of

$$V_u = \frac{q_u L}{2} = \frac{1.6 \cdot 20 \text{ lb/ft} \cdot (8 \text{ ft})}{2} = 128 \text{ lb}$$

The design shear is the nominal capacity on two-face shells, each 1.25 in. in width, reduced by the strength-reduction factor for shear (0.8):

$$\phi V_n\, 0.8 \cdot 56 \text{ lb/in.}^2 \cdot (2 \cdot 1.25) \cdot 12 \text{ in.} = 1334 \text{ lb}$$

This is far greater than the factored design shear, and one-way shear does not govern the design.

5.1.10 Overall Comments on Strength Design of Unreinforced Panel Walls

- Nonload bearing masonry, without calculated reinforcement, can easily resist wind loads. It approaches its capacity only in the case of ungrouted hollow masonry. In this case, the lower allowable flexural tensile stress for masonry cement-mortar can be critical in design.

Unit	Area in.² per ft	Moment of inertia in.⁴ per ft
4-in. modular	43.6	47.8
6-in hollow CMU, fully bedded	32.2	139
6-in hollow CMU, face-shell bedded	24.0	130
8-in hollow CMU, fully bedded	41.5	334
8-in hollow CMU, face-shell bedded	30.0	309
12-in hollow CMU, fully bedded	57.8	1065
12-in hollow CMU, face-shell bedded	36.0	929

TABLE 5.2 Section Properties for Masonry Walls

- If noncalculated reinforcement is included, it will not act until the masonry has cracked.
- Elements such as the ones we have calculated in this section can be designed in many cases by prescription.

5.1.11 Section Properties for Masonry Walls

Section properties for masonry walls are summarized in Table 5.2.

5.1.12 Theoretical Derivation of the Strip Method (Hillerborg, 1996)

In the preceding examples, we have used the simplifications that single-wythe panel walls can be designed assuming that the entire load is resisted by vertically spanning strips, and that two-wythe panel walls can be designed assuming the entire load is resisted by vertically spanning strips of the inner wythe.

It is now appropriate to consider the theoretical basis for this simplification. We first consider the simplification of wythes as crossing strips using the strip method, and then derive additional simplifications based on the relative stiffnesses of those strips.

Consider the differential equation for the out-of-plane deflection of an elastic plate with a uniformly distributed, out-of-plane load:

$$\frac{\partial^4 w}{\partial x^4} + \frac{\partial^4 w}{\partial x^2 \partial y^2} + \frac{\partial^4 w}{\partial y^4} = \frac{-q}{D}$$

where w = out-of-plane deflection
q = uniformly distributed load

and

$$D = \frac{EI}{(1-v^2)} = \frac{Et^3}{12(1-v^2)}$$

(EI is calculated per unit width, and Poisson effects are included)
Conservatively, ignore twisting moments:

$$\frac{\partial^4 w}{\partial x^4} + \frac{\partial^4 w}{\partial y^4} = \frac{-q}{D} = -\left(\frac{q_x}{D} + \frac{q_y}{D}\right)$$

This is the differential equation for independent x and y strips (beams). The two sets of strips can be designed independently, provided that equilibrium is satisfied at every point

$$q_x + q_y = q$$

5.1.13 Distribution of Out-of-Plane Load to Vertical and Horizontal Strips of a Single-Wythe Panel Wall

In Sec. 5.1.1, it was stated that single-wythe panel walls can be designed as though out-of-plane load were carried by the vertical strips alone. Now let's show why that's true. Consider the single-wythe panel shown in Fig. 5.8.

Assume that the panel resists out-of-plane loading as an assemblage of crossing strips, in the x direction (horizontally on the page) and the y direction (vertically on the page). At the very end of this section, it will be shown that such an assumption is legitimate ("strip method").

Now impose compatibility of out-of-plane deflections on the strips—that is, the crossing x and y strips must have equal out-of-plane

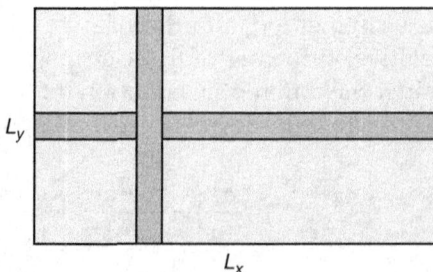

FIGURE 5.8 Idealization of a panel wall as an assemblage of crossing strips.

displacements. For a simply supported strip with uniformly distributed loading q, the center-line displacement is

$$\Delta_{center} = \frac{5qL^4}{384EI}$$

For equal deflections of the x and y strips,

$$\frac{5q_x L_x^4}{384EI} = \Delta_{center} = \frac{5q_y L_y^4}{384EI}$$

And because those strips have equal moduli and moments of inertia,

$$\frac{q_x}{q_y} = \frac{L_y^4}{L_x^4}$$

The span of the vertical strips, L_y, is the distance from the top of a floor slab to the underside of the slab or beam above, typically about 10 ft. The span of the horizontal strips, L_x, is the distance from the face of a column to the face of the adjacent column, typically about 20 ft. So

$$\frac{q_x}{q_y} = \frac{L_y^4}{L_x^4} = \left(\frac{10}{20}\right)^4 = 0.063$$

The horizontal strips will carry only about 6 percent of the out-of-plane load. If it is conservatively assumed that the vertical strips will carry 100 percent of the out-of-plane load, and the horizontal strip will carry zero load, the design work is halved, and the results will be conservative.

5.1.14 Distribution of Out-of-Plane Load to Vertical and Horizontal Strips of a Two-Wythe Panel Wall

The analysis of Sec. 5.1.13 can easily be extended to the case of a two-wythe panel wall, with an outer wythe of clay masonry and an inner wythe of concrete masonry shown in Fig. 5.9.

As before, assume that each wythe of the panel resists out-of-plane loading as an assemblage of crossing strips, in the x direction (horizontally on the page) and the y direction (vertically on the page). At the very end of this section, it will be shown that such an assumption is legitimate ("strip method").

Let's also assume that the inner wythe is of hollow, ungrouted, 8-in. CMU laid in face-shell bedding, and that the outer wythe is of solid modular units (nominal thickness of 4 in.).

From Sec. 5.1.6, we know that the moment of inertia of the hollow CMU is 444.9 in.^{4} per 16 in. of width, or 333.7 in.^{4} per foot of width. It is

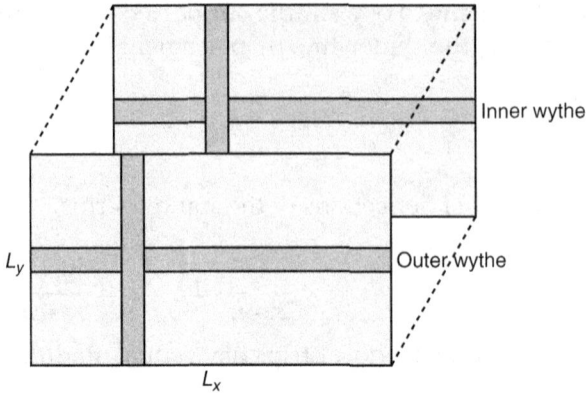

FIGURE 5.9 Idealization of a two-wythe panel wall as an assemblage of two sets of crossing strips.

proper to compute the average flexural stiffness of the wall using the moment of inertia of the hollow unit rather than the moment of the face-shell bedding only, because the bed joints occupy only a small portion of the volume of the wall.

The moment of inertia of the modular outer wythe, per foot of width, is

$$I = \frac{bt^3}{12} = \frac{(12 \text{ in.})(3.63 \text{ in.})^3}{12} = 47.9 \text{ in.}^4 \text{ per ft of width}$$

Assume that the ties between wythes are axially rigid, so that the two wythes have equal out-of-plane deflection.

The total load q must be equilibrated by the summation of the load resisted by each strip of each wythe:

$$(q_x + q_y)_{\text{exterior}} + (q_x + q_y)_{\text{interior}} = q$$

Because the vertical strips in the exterior wythe are simply supported at the bottom and free at the top (expansion joint), they carry no load:

$$(q_x)_{\text{exterior}} + (q_x + q_y)_{\text{interior}} = q$$

Now impose compatibility of out-of-plane deflections on the strips,

$$\Delta_{x\text{exterior}} = \frac{5q_{x\text{exterior}} L^4_{x\text{exterior}}}{384 E_{x\text{exterior}} I_{x\text{exterior}}} = \frac{5q_{x\text{exterior}} 20^4}{384 E_{x\text{exterior}} \cdot 47.9}$$

$$\Delta_{x\text{interior}} = \frac{5q_{x\text{interior}} L^4_{x\text{interior}}}{384 E_{x\text{interior}} I_{x\text{interior}}} = \frac{5q_{x\text{interior}} 20^4}{384 E_{x\text{interior}} \cdot 333.7}$$

$$\Delta_{y\text{interior}} = \frac{5q_{y\text{interior}} L^4_{y\text{interior}}}{384 E_{y\text{interior}} I_{y\text{interior}}} = \frac{5q_{y\text{interior}} 10^4}{384 E_{y\text{interior}} \cdot 333.7}$$

Equate those deflections, cancel out the common term of (5/384), and assume that all Es are the same:

$$\frac{q_{x\,exterior}\,20^4}{47.9} = \frac{q_{x\,interior}\,20^4}{333.7} = \frac{q_{y\,interior}\,10^4}{333.7}$$

$$3342\,q_{x\,exterior} = 479.4\,q_{x\,interior} = 29.97\,q_{y\,interior}$$

Express $q_{x\,exterior}$ and $q_{x\,interior}$ in terms of $q_{y\,interior}$:

$$q_{x\,exterior} = 0.00897\,q_{y\,interior}$$

$$q_{x\,interior} = 0.0625\,q_{y\,interior}$$

Now recall that the loads resisted by each strip must equilibrate the total load:

$$(q_x)_{exterior} + (q_x + q_y)_{interior} = q$$

$$(0.00897 + 0.0625 + 1.0)q_{y\,interior} = q$$

Solve for $q_{y\,interior}$ in terms of q:

$$q_{y\,interior} = \frac{q}{1.0715} = 0.933q$$

Finally, express the load carried by each set of strips in terms of q:

$$q_{y\,interior} = 0.93q$$

$$q_{x\,interior} = 0.06q$$

$$q_{x\,exterior} = 0.008q$$

Clearly, it is conservative and very reasonable to assume that the vertical strips of the interior wythe resist 100 percent of the out-of-plane load, and the other strips resist no load.

5.2 Strength Design of Unreinforced Bearing Walls

5.2.1 Basic Behavior of Unreinforced Bearing Walls

Load-bearing masonry (without calculated reinforcement) must be designed for the effects of

1. Gravity loads from self-weight, plus gravity loads from overlying roof or floor levels

2. Moments from eccentric gravity load, or out-of-plane wind or earthquake
3. In-plane shear

For now, we shall study Loadings (1) and (2). In Sec. 5.3, we shall study Loading (3), in the general context of design of masonry shear walls.

For Loadings (1) and (2), we shall design unreinforced, load-bearing masonry as a series of vertically spanning strips, (Fig. 5.10), subjected to gravity loads (possibly eccentric) and out-of-plane wind or earthquake.

The only aspect of behavior that we haven't studied so far is the effect of slenderness on the load-carrying capacity of a column or wall. This effect is shown in Fig. 5.11.

At low values of slenderness, a masonry column in compression exhibits material failure. At high values of slenderness, it exhibits stability failure.

For masonry design, the effective length coefficient, k, is usually equal to 1.

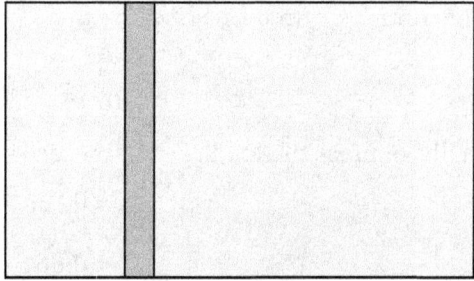

FIGURE 5.10 Idealization of bearing walls as vertically spanning strips.

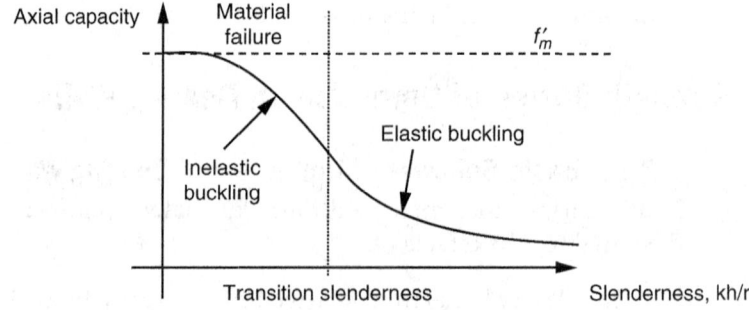

FIGURE 5.11 Effect of slenderness on the axial capacity of a column or wall.

5.2.2 Steps in Strength Design of Unreinforced Bearing Walls

In the 2008 MSJC *Code*, design of unreinforced bearing walls is similar to the design of panel walls, except that axial load must be considered. In Sec. 3.2.2 of the 2008 MSJC *Code*, no explicit equations are given for computing flexural strength. The usual assumption of plane sections is invoked, and tensile and compressive stresses in masonry are to be assumed proportional to strain.

Nominal capacities in masonry are reached at an extreme fiber tension equal to the modulus of rupture (Sec. 3.1.8.2 of the 2008 MSJC *Code*), and at a compressive stress of $0.80\, f'_m$.

Compressive capacity is given by Eqs. (3.11) and (3.12) of the 2008 MSJC *Code*:

For $kh/r = h/r \le 99$,

$$P_n = 0.80 \left\{ 0.80\, A_n f'_m \left[1 - \left(\frac{h}{140\, r} \right)^2 \right] \right\}$$

and for $kh/r = h/r > 99$,

$$P_n = 0.80 \left[0.80\, A_n f'_m \left(\frac{70\, r}{h} \right)^2 \right]$$

The strength reduction factor, ϕ, is equal to 0.60 (Sec. 3.1.4.2 of the 2008 MSJC *Code*). Unlike the allowable-stress provisions, the strength provisions of the 2008 MSJC *Code* require a direct stability check using a moment magnifier.

5.2.3 Example of Strength Design of Unreinforced Bearing Wall with Concentric Axial Load

The bearing wall shown in Fig. 5.12 has an unfactored, concentric axial load of 1050 lb/ft. Using hollow concrete masonry units with face-shell bedding, design the wall.

According to the 2009 IBC, and in the context of these example problems (dead load, wind load and roof live load), the following loading combinations must be checked for strength design:

1. $1.2D + 1.6W + f_1 L + 0.5\,(L_r \text{ or } S \text{ or } R)$
2. $0.9D + 1.6W + 1.6H$

The second of these is usually critical, because roof live load must be considered off as well as on.

To apply those loading combinations, let us assume that the total unfactored wall load of 1050 lb/ft represents 700 lb/ft of dead load and 350 lb/ft of live load.

150 Chapter Five

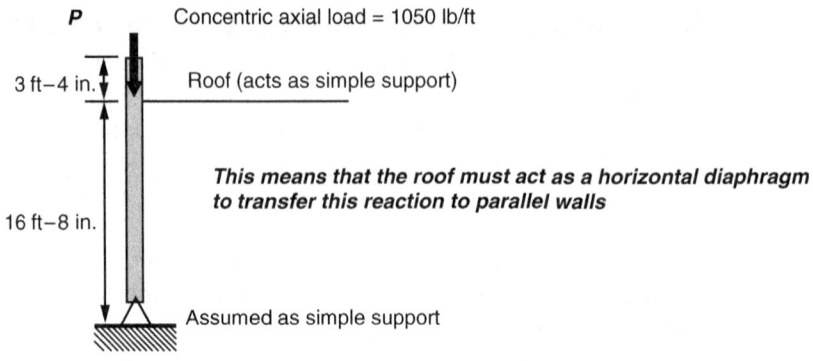

FIGURE 5.12 Unreinforced masonry bearing wall with concentric axial load.

Table 5.3, repeated from Sec. 5.1.11, gives section properties for masonry units:

Unit	Area in.² per ft	Moment of inertia in.⁴ per ft
4-in. modular	43.6	47.8
6-in hollow CMU, fully bedded	32.2	139
6-in hollow CMU, face-shell bedded	24.0	130
8-in hollow CMU, fully bedded	41.5	334
8-in hollow CMU, face-shell bedded	30.0	309
12-in hollow CMU, fully bedded	57.8	1065
12-in hollow CMU, face-shell bedded	36.0	929

TABLE 5.3 Section Properties for Masonry Units

Table 5.4, taken from NCMA TEK 2-1A, gives the self-weight of hollow masonry walls, assuming units with a density of 120 lb/ft³.

Nominal thickness, in.	Weight per ft²
4	27
6	40
8	48
10	56

TABLE 5.4 Weights of Hollow CMU Walls

Strength Design of Unreinforced Masonry Elements

At each horizontal plane through the wall, the following conditions must be met:

- Maximum compressive stress from factored axial loads must not exceed the slenderness-dependent values in Eqs. (3-12) or (3-13) as appropriate, reduced by a φ-factor of 0.60.
- Maximum compressive stress from factored loads (including a moment magnifiers) must not exceed 0.80 f'_m in the extreme compression fiber, reduced by a φ-factor of 0.60.
- Maximum tension stress from factored loads (including a moment magnifier) must not exceed the modulus of rupture in the extreme tension fiber, reduced by the φ-factor of 0.60.

For each condition, the more critical of the two possible loading combinations must be checked. Because there is no wind load, this example will be worked using that the loading combination $1.2D + 1.6L$.

In theory, we must check various points on the wall. In this problem, however, the wall has only axial load, which increases from top to bottom due to the wall's self-weight. Therefore we need to check only at the base of the wall.

Try 8-in. nominal units, and a specified compressive strength, f'_m, of 1500 lb/in.² This can be satisfied using units with a net-area compressive strength of 1900 lb/in.², and Type S PCL mortar. Work with a strip with a width of 1 ft (measured along the length of the wall in plan). Stresses are calculated using the critical section, consisting of the bedded area only (2008 MSJC, Sec. 1.9.1.1).

At the base of the wall, the factored axial force is

$$P_u = 1.2\,(700\text{ lb}) + 1.6\,(350\text{ lb}) + 1.2\,(20\text{ ft} \times 48\text{ lb/ft}) = 2552\text{ lb}$$

To calculate stiffness-related parameters for the wall, we use the average cross section, corresponding to the fully bedded section in the table (2008 MSJC, Sec. 1.9.3).

$$r = \sqrt{\frac{I}{A}} = \sqrt{\frac{334\text{ in.}^4}{41.5\text{ in.}^2}} = 2.84\text{ in.}$$

$$\frac{kh}{r} = \frac{16.67\text{ ft} \times 12\text{ in./ft}}{2.84\text{ in.}} = 70.5$$

This is less than the transition slenderness of 99, so the nominal axial capacity is based on the curve that is an approximation to inelastic buckling:

$$\phi P_n = \phi \, 0.80 \left\{ 0.80 \, A_n \, f'_m \left[1 - \left(\frac{h}{140\,r} \right)^2 \right] \right\}$$

$$\phi P_n = 0.60 \cdot 0.80 \cdot \left\{ 0.80 \times 30.0 \text{ in.}^2 \times 1500 \text{ lb/in.}^2 \left[1 - \left(\frac{16.67 \text{ ft} \times 12 \text{ in./ft}}{140 \times 2.84 \text{ in.}^2} \right)^2 \right] \right\}$$

$$\phi P_n = 17{,}280 \text{ lb} \times 0.746 = 12{,}898 \text{ lb}$$

The factored axial load, P_u, 2552 lb, is far less than this, and this part of the design is satisfactory.

Now check the net compressive stress. Because the load is concentric, there is no bending stress. At the base of the wall,

$$f_a = \frac{P_u}{A} = \frac{1.2\,(700 \text{ lb}) + 1.6\,(350 \text{ lb}) + 1.2\,(20 \text{ ft} \times 48 \text{ lb/ft})}{30 \text{ in.}^2} = \frac{2552 \text{ lb}}{30 \text{ in.}^2} = 85.7 \text{ lb/in.}^2$$

$$f_a = 85.7 \text{ lb/in.}^2$$

The maximum permitted compressive stress is

$$0.60 \cdot 0.80 f'_m = 0.60 \times 0.80 \times 1500 \text{ lb/in.}^2 = 720 \text{ lb/in.}^2$$

The maximum compressive stress is much less than this, and the design is satisfactory for this also. Clearly, because this example involves concentric axial loads only, the first criterion (axial capacity reduced by slenderness effects) is more severe than the second (maximum compressive stress from axial loads and bending moments).

Because the wall has concentric axial load, there is no net tensile stress, and that criterion does not have to be checked.

Because the axial load is concentric, there is no moment, and the magnified moment does not have to be checked. The design is satisfactory.

It would probably be possible to achieve a satisfactory design with a smaller nominal wall thickness. To maintain continuity in the example problems that follow, however, the design will stop at this point.

Although the 2008 MSJC *Code* has no explicit minimum eccentricity requirements for walls, the leading coefficient of 0.80 for nominal axial compressive capacity effectively imposes a minimum eccentricity of about $0.1t$.

5.2.4 Example of Strength Design of Unreinforced Bearing Wall with Eccentric Axial Load

Now consider the same bearing wall of the previous example, but make the gravity load eccentric. As before, suppose that the load is applied over a 4-in. bearing plate, and assume that bearing stresses vary linearly under the bearing plate as shown in Fig. 5.13.

Then the eccentricity of the applied load with respect to the centerline of the wall is

$$e = \frac{t}{2} - \frac{\text{plate}}{3} = \frac{7.63 \text{ in.}}{2} - \frac{4 \text{ in.}}{3} = 2.48 \text{ in.}$$

The wall is as shown in Fig. 5.14.

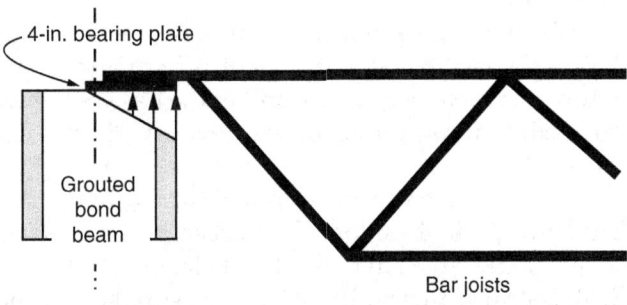

FIGURE 5.13 Assumed linear variation of bearing stresses under the bearing plate.

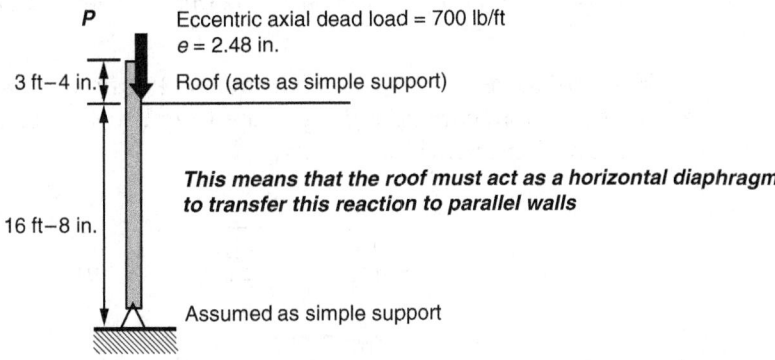

FIGURE 5.14 Unreinforced masonry bearing wall with eccentric axial load.

At each horizontal plane through the wall, the following conditions must be met:

- Maximum compressive stress from factored axial loads must not exceed the slenderness-dependent values in Eqs. (3-12) or (3-13) as appropriate, reduced by a ϕ-factor of 0.60.
- Maximum compressive stress from factored loads (including a moment magnifiers) must not exceed 0.80 f'_m in the extreme compression fiber, reduced by a ϕ-factor of 0.60.
- Maximum tension stress from factored loads (including a moment magnifier) must not exceed the modulus of rupture in the extreme tension fiber, reduced by the ϕ-factor of 0.60.

For each condition, the more critical of the two possible loading combinations must be checked. Because there is no wind load, this example will be worked using that the loading combination $1.2D + 1.6L$.

We must check various points on the wall. Critical points are just below the roof reaction (moment is high and axial load is low, so maximum tension may govern); and at the base of the wall (axial load is high, so maximum compression may govern). Check each of these locations on the wall.

As before, try 8-in. nominal units, and a specified compressive strength, f'_m, of 1500 lb/in.². This can be satisfied using units with a net-area compressive strength of 1900 lb/in.², and Type S PCL mortar. Work with a strip with a width of 1 ft (measured along the length of the wall in plan). Stresses are calculated using the critical section, consisting of the bedded area only (2008 MSJC, Sec. 1.9.1.1).

Just below the roof reaction, the axial force is

$$P_u = 1.2\,(700\text{ lb}) + 1.6\,(350\text{ lb}) + 1.2\,(3.33\text{ ft} \times 48\text{ lb/ft}) = 1592\text{ lb}$$

To calculate stiffness-related parameters for the wall, we use the average cross section, corresponding to the fully bedded section in the table (2008 MSJC Code, Sec. 1.9.3).

$$r = \sqrt{\frac{I}{A}} = \sqrt{\frac{334 \text{ in.}^4}{41.5 \text{ in.}^2}} = 2.84 \text{ in.}$$

$$\frac{kh}{r} = \frac{16.67 \cdot 12 \text{ in.}}{2.84 \text{ in.}} = 70.5$$

Strength Design of Unreinforced Masonry Elements

This is less than the transition slenderness of 99, so the nominal axial capacity is based on the curve that is an approximation to inelastic buckling:

$$\phi P_n = \phi\, 0.80\left\{0.80\, A_n f'_m\left[1-\left(\frac{h}{140\,r}\right)^2\right]\right\}$$

$$\phi P_n = 0.60 \times 0.80 \cdot \left\{0.80 \times 30.0 \text{ in.}^2 \times 1500 \text{ lb/in.}^2 \left[1-\left(\frac{16.67 \text{ ft} \times 12 \text{ in./ft}}{140 \times 2.84 \text{ in.}^2}\right)^2\right]\right\}$$

$$\phi P_n = 17{,}280 \text{ lb} \times 0.746 = 12{,}898 \text{ lb}$$

Because the factored axial force is much less than slenderness-dependent nominal capacity, reduced by the appropriate ϕ factor, the axial force check is satisfied.

Now check the net compressive stress. Because the loading is eccentric, there is bending stress:

$$f_{compression} = \frac{P_u}{A} + \frac{M_u c}{I}$$

The factored design axial load, P_u, is computed above. The factored design moment, M_u, is given by:

$$M_u = P_u e = (1.2 \times 700 + 1.6 \times 350) \text{ lb} \times 2.48 \text{ in.} = 3472 \text{ lb-in.}$$

$$f_{compression} = \frac{P_u}{A} + \frac{M_u c}{I}$$

$$f_{compression} = \frac{1592 \text{ lb}}{30 \text{ in.}^2} + \frac{3472 \text{ lb-in.} \,(7.63/2)}{309 \text{ in.}^4} = 53.1 + 42.9 \text{ lb/in.}^2 = 95.9 \text{ lb/in.}^2$$

$$0.60 \times 0.80 f'_m = 0.60 \times 0.80 \times 1500 \text{ lb/in.}^2 = 720 \text{ lb/in.}^2$$

The net compressive stress does not exceed the prescribed value. Clearly, because this example involves eccentric axial loads, the first criterion (axial load reduced by slenderness effects) is less severe than the second (maximum compressive stress from axial loads and bending moments).

Because the bending stress from factored design moments (42.9 lb/in.²) is less than the axial stress from factored axial load (53.1 lb/in.²), there is no net tension, and the third criterion (net tension) is automatically satisfied.

Section 3.2.2.4 of the 2008 MSJC *Code* requires that magnified moments be checked. For this wall, the ratio of (h/r) is 70.5, which exceeds 45. Therefore a magnifier must be calculated.

$$\delta = \frac{1}{1 - \dfrac{P_u}{A_n f'_m \left(\dfrac{70\,r}{h}\right)^2}} = \frac{1}{1 - \dfrac{1592\ \text{lb}}{30\ \text{in.}^2 \times 1500\,\dfrac{\text{lb}}{\text{in.}^2}\left(\dfrac{70}{70.5}\right)^2}} = 1.04$$

This is very close to 1.0 and does not affect the previous checks.

The other critical section could be at the base of the wall, where the checks of all three criteria are identical to those of example in Sec. 5.2.3, plus the moment magnifier. All are satisfied, and the design is therefore satisfactory.

5.2.5 Example of Strength Design of Unreinforced Bearing Wall with Eccentric Axial Load Plus Wind

Now consider the same bearing wall of the previous example, but add a uniformly distributed wind load of 25 lb/ft².

The wall is as shown in Fig. 5.15.

At each horizontal plane through the wall, the following conditions must be met:

- Maximum compressive stress from factored axial loads must not exceed the slenderness-dependent values in Eqs. (3-12) or (3-13) as appropriate, reduced by a ϕ-factor of 0.60.
- Maximum compressive stress from factored loads (including a moment magnifiers) must not exceed 0.80 f'_m in the extreme compression fiber, reduced by a ϕ-factor of 0.60.

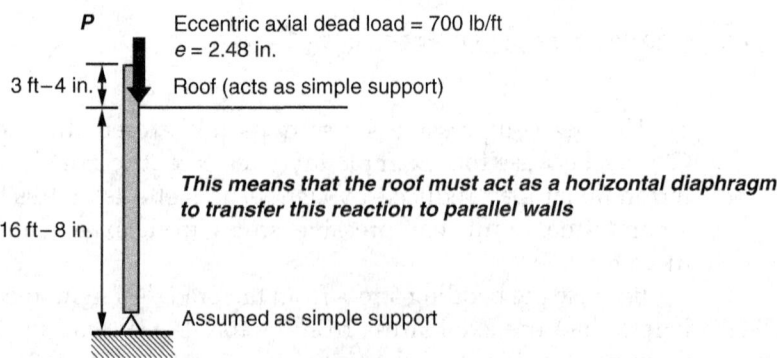

Figure 5.15 Unreinforced masonry bearing wall with eccentric axial load and wind load.

Strength Design of Unreinforced Masonry Elements

- Maximum tension stress from factored loads (including a moment magnifier) must not exceed the modulus of rupture in the extreme tension fiber, reduced by the φ-factor of 0.60.

For each condition, the more critical of the two possible loading combinations must be checked. Because there is wind load, and because the two previous examples showed little problem with the first two criteria, the third criterion (net tension) may well be critical. For this criterion, the critical loading condition could be either $1.2D + 1.6L$, or $0.9D + 1.6W$. Both loading conditions must be checked.

We must check various points on the wall. Critical points are just below the roof reaction (moment is high and axial load is low, so net tension may govern); at the mid-height of the wall, where moment from eccentric gravity load and wind load are highest; and at the base of the wall (axial load is high, so the maximum compressive stress may govern). Check each of these locations.

To avoid having to check a large number of loading combinations and potentially critical locations, it is worthwhile to assess them first, and check only the ones that will probably govern.

Due to wind only, the unfactored moment at the base of the parapet (roof level) is

$$M = \frac{qL_{parapet}^2}{2} = \frac{25 \text{ lb/ft} \times 3.33^2 \text{ ft}^2}{2} \times 12 \text{ in./ft} = 1663 \text{ lb-in.}$$

The maximum moment is close to that occurring at mid-height. The moment from wind load is the superposition of one-half moment at the upper support due to wind load on the parapet only, plus the midspan moment in a simply supported beam with that same wind load:

$$M_{midspan} = -\frac{1663}{2} + \frac{qL^2}{8} = -\frac{1663 \text{ lb-in.}}{2} + \frac{25 \text{ lb/ft} \times 16.67^2 \text{ ft}^2}{8}$$

$$\times 12 \text{ in./ft} = 9589 \text{ lb-in.}$$

The unfactored moment due to eccentric axial load is

$$M_{gravity} = Pe = 1050 \text{ lb} \times 2.48 \text{ in.} = 2604 \text{ lb-in.}$$

Unfactored moment diagrams due to eccentric axial load and wind are as shown in Fig. 5.16.

From the example of Sec. 5.2.4, we know that loading combination $1.2D + 1.6L$ was not close to critical directly underneath the roof. Because the wind-load moments directly underneath the roof are not very large, they will probably not be critical either. The critical location will probably be at mid-height; the critical loading condition will probably be $0.9D + 1.6W$; and the critical criterion will probably be net tension, because this masonry wall is unreinforced.

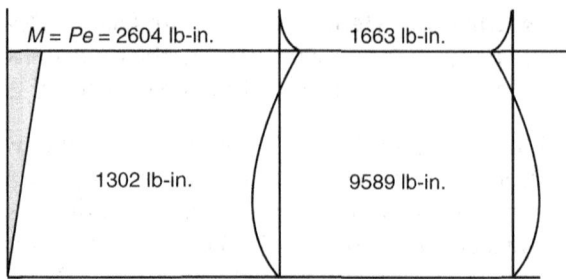

FIGURE 5.16 Unfactored moment diagrams due to eccentric axial load and wind.

As before, try 8-in. nominal units, and a specified compressive strength, f_m', of 1500 lb/in.². This can be satisfied using units with a net-area compressive strength of 1900 lb/in.², and Type S PCL mortar. Work with a strip with a width of 1 ft (measured along the length of the wall in plan). Stresses are calculated using the critical section, consisting of the bedded area only (2008 MSJC, Sec. 1.9.1.1).

Now check the net tensile stress. At the mid-height of the wall, the axial force due to 0.9D is

$$P_u = 0.9\,(700\text{ lb}) + 0.9\,(3.33\text{ ft} + 8.33\text{ ft}) \times 48\text{ lb/ft} = 1134\text{ lb}$$

At the mid-height of the wall, the factored design moment, M_u, is given by

$$M_u = P_{u\text{ eccentric}}\frac{e}{2} + M_{u\text{ wind}} = \left(\frac{1}{2}\right)0.9 \cdot 700\text{ lb} \times 2.48\text{ in.} + 1.6 \times 9589\text{ lb-in.} = 16,124\text{ lb-in.}$$

$$f_{\text{tension}} = -\frac{P_u}{A} + \frac{M_u c}{I}$$

$$f_{\text{tension}} = -\frac{1134\text{ lb}}{30\text{ in.}^2} + \frac{16,124\text{ lb-in. }(7.63/2)}{309\text{ in.}^4} = -37.8 + 199.1\text{ lb/in.}^2 = 161.3\text{ lb/in.}^2$$

$$0.60 f_r = 0.60 \times 63\text{ lb/in.}^2 = 37.8\text{ lb/in.}^2$$

The maximum tensile stress exceeds the prescribed value, and the design is not satisfactory. It will be necessary to grout the wall. The wall weight will increase somewhat, increasing the stress from factored dead load; the factored wind moments will remain unchanged; the section modulus of the wall will increase; and the modulus of rupture of the wall will increase from 63 to 163 lb/in.².

Recheck the wall as grouted:

$$P_u = 0.9\,(700\text{ lb}) + 0.9\,(3.33\text{ ft} + 8.33\text{ ft}) \times 76\text{ lb/ft} = 1428\text{ lb}$$

$$M_u = P_u \frac{e}{2} + M_{u\,\text{wind}} = \left(\frac{1}{2}\right) 0.9 \cdot 700\text{ lb} \times 2.48\text{ in.} + 1.6 \times 9589\text{ lb-in.} = 16{,}124\text{ lb-in.}$$

$$f_{\text{tension}} = -\frac{P_u}{A} + \frac{M_u c}{I}$$

$$f_{\text{tension}} = -\frac{1428\text{ lb}}{7.63 \times 12\text{ in.}^2} + \frac{16{,}124\text{ lb-in.}\,(7.63/2)\text{ in.}}{\left(\dfrac{12 \times 7.63^3\text{ in.}^4}{12}\right)} = -15.6 + 138.5\text{ lb/in.}^2 = 122.9\text{ lb/in.}^2$$

$$0.60 f_r = 0.60 \cdot 163\text{ lb/in.}^2 = 97.8\text{ lb/in.}^2$$

The wall would have to be thickened to 10 in. or reinforced.

Finally, the moment magnifier would have to be checked. Because we know that we have to thicken the wall, and the moment magnifier would make things worse, the calculation is not done here.

5.2.6 Comments on the Above Examples for Strength Design of Unreinforced Bearing Walls

1. In retrospect, it probably would not have been necessary to check all three criteria at all locations. With experience, a designer could realize that the location with highest wind moment would govern, and could therefore check only the mid-height of the wall.

2. The addition of wind load to the example of Sec. 5.2.4, to produce the example of Sec. 5.2.5, changes the critical location from just under the roof, to the mid-height of the simply supported section of the wall. The wind load of 25 lb/ft² in example of 5.2.5 produces maximum tensile stresses above the allowable values for ungrouted masonry, and makes it necessary to grout the wall, thicken it, or reinforce it.

5.2.7 Extension of the Above Concepts to Masonry Walls with Openings

In the previous examples, we have studied the behavior of bearing walls of unreinforced masonry, idealized as a series of vertical strips, simply supported at the level of the floor slab, and at the level of the roof. Let's see how this changes in the case of bearing walls with openings.

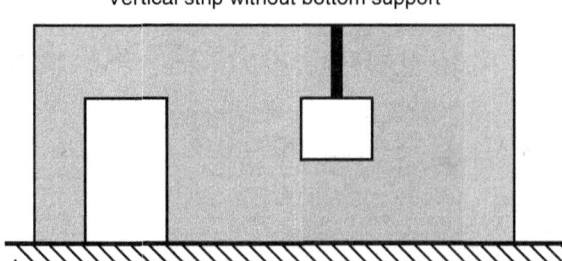

Figure 5.17 Hypothetical unstable resistance mechanism in a wall with openings, involving vertically spanning strips only.

In Fig. 5.17, load applied above the window and door openings clearly cannot be resisted by vertical strips, because those vertical strips have only one point of lateral support (at the roof level).

For that reason, the wall must be idealized as horizontal strips above and below the openings, supported by vertical strips on both sides of the openings, as shown in Fig. 5.18.

Each set of horizontal strips, idealized as simply supported, must be supported by the adjacent vertical strips. For example, the horizontal strips above the door are supported by Strip A and Strip B. The window and door are considered to transfer loads applied to them, via horizontal strips, to the vertical strips on either side of the openings.

Therefore, Strip A has to support, spanning vertically, the out-of-plane loads acting directly on it, plus the out-of-plane loads acting on the left

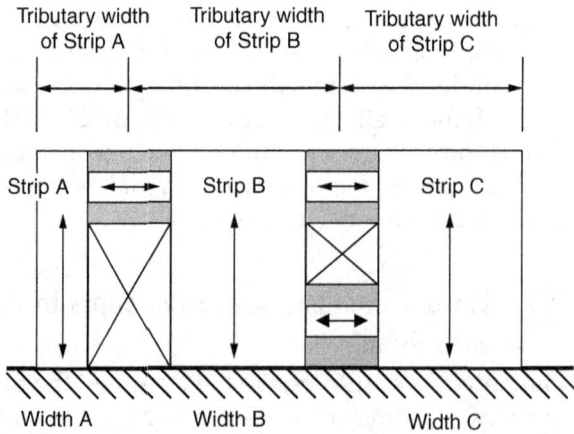

Figure 5.18 Stable resistance mechanism in a wall with openings, involving horizontally spanning strips in addition to vertically spanning strips.

half of the horizontal strips above the door. In other words, Strip A has to resist the out-of-plane loads acting on what might be termed a "tributary width," which extends from the left-hand edge of Strip A itself, to the midspan of the horizontal strips above the door. In the same way, Strips B and C have to resist the loads corresponding to Tributary Widths B and C, respectively. For example, if Strip B has to resist the loads acting over Tributary Width B, this represents an increase in the design loads on Strip B. That strip must resist the loads that normally would be applied to it (if no openings had existed), multiplied by the ratio of Tributary Width B, divided by Width B:

$$\text{Actions in Strip B} = \text{Initial actions} \left(\frac{\text{Tributary Width B}}{\text{Width B}} \right)$$

The same applies to vertical loads, because these also must be transferred from horizontal to vertical strips. In any event, the presence of openings can be considered to increase the initial actions in the vertical strips adjacent to the openings. Aside from this increase, the design of those elements proceeds exactly as before.

5.2.8 Final Comment on the Effect of Openings in Unreinforced Masonry Bearing Walls

As the summation of the plan lengths of openings in a bearing wall exceeds about one-half the plan length of the wall, even the higher allowable stresses (or moduli of rupture) corresponding to fully grouted walls will be exceeded, and it will generally become necessary to use reinforcement. Design of reinforced masonry bearing walls is addressed later in this book.

5.3 Strength Design of Unreinforced Shear Walls

5.3.1 Basic Behavior of Unreinforced Shear Walls

Box-type structures resist lateral loads as shown in Fig. 5.19.

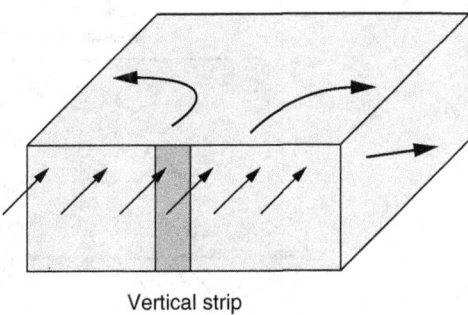

Vertical strip

FIGURE 5.19 Basic behavior of box-type buildings in resisting lateral loads.

This resistance mechanism involves three steps:

- Walls oriented perpendicular to the direction of lateral load transfer those loads to the level of the foundation and the levels of the horizontal diaphragms. The walls are idealized and designed as vertically oriented strips.
- The roof and floors act as horizontal diaphragms, transferring their forces to walls oriented parallel to the direction of lateral load.
- Walls oriented parallel to the direction of applied load must transfer loads from the horizontal diaphragms to the foundation. In other words, they act as shear walls.

As noted previously in the sections dealing with unreinforced bearing walls, this overall mechanism demands that the horizontal roof diaphragm have sufficient strength and stiffness to transfer the required loads. This is discussed again in a later section dealing with horizontal diaphragms.

The rest of this section addresses the design of shear walls. We shall see that in almost all cases, the design itself is very simple, because the cross-sectional areas of the masonry walls are so large that nominal stresses are quite low.

5.3.2 Design Steps for Unreinforced Shear Walls

Unreinforced masonry shear walls must be designed for the effects of

1. Gravity loads from self-weight, plus gravity loads from overlying roof or floor levels
2. Moments and shears from in-plane shear loads

Actions are shown in Fig. 5.20. Either allowable-stress design or strength design can be used.

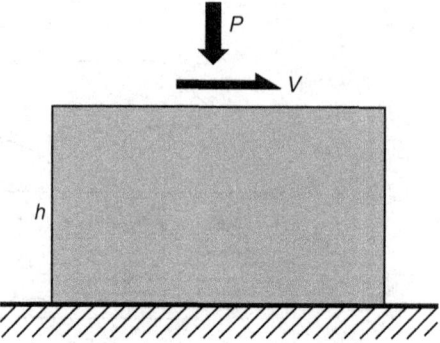

FIGURE 5.20 Design actions for unreinforced shear walls.

When strength design is used, the 2008 MSJC *Code* requires that maximum tensile stresses from in-plane flexure, alone or in combination with axial loads, not exceed the in-plane modulus of rupture from Table 5.1 (Table 3.1.8.2.1 of the 2008 MSJC *Code*).

Shear must also be checked. From Sec. 3.2.4 of the 2008 MSJC *Code*, for hollow units in running bond, nominal shear strength is the least of

$$\begin{cases} 3.8\sqrt{f'_m}\, A_n \\ 300 A_n \\ 56 A_n + 0.45 N \end{cases}$$

The third criterion is

$56 A_n + 0.45 N$	for running bond masonry not grouted solid
$56 A_n + 0.45 N$	for stack bond masonry with open-end units, grouted solid
$90 A_n + 0.45 N$	for running bond masonry grouted solid
$23 A_n$	for other stack bond masonry

The strength-reduction factor for shear is 0.80 (2008 MSJC, Sec. 3.1.4.3).

5.3.3 Example of Strength Design of Unreinforced Masonry Shear Wall

Consider the simple structure of Fig. 5.21, the same one whose bearing walls have been designed previously in this book. Use nominal 8-in. concrete masonry units, $f'_m = 1500$ lb/in.^{2}, and Type S PCL mortar. The roof applies a gravity load of 1050 lb/ft to the walls; the walls measure 16 ft, 8 in. height to the roof, and have an additional 3 ft, 4 in. parapet. The walls are loaded with a wind load of 20 lb/ft^2. The roof acts as a one-way system, transmitting gravity loads to the front and back walls. At this stage, all loads are unfactored; load factors will be applied later.

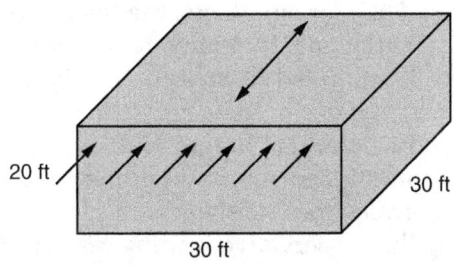

FIGURE 5.21 Example problem for strength design of unreinforced shear wall.

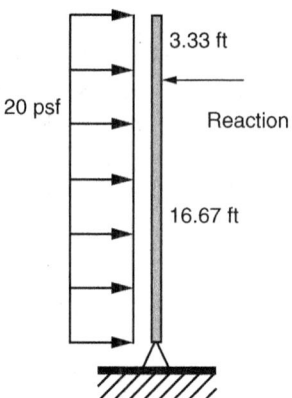

FIGURE 5.22 Calculation of reaction on roof diaphragm, strength design of unreinforced shear wall.

Now design the shear wall. Try an 8-in. wall with face-shell bedding only. The critical section for shear is just under the roof, where axial load in the shear walls is least, coming from the parapet only.

As a result of the wind loading, the reaction transmitted to the roof diaphragm is as calculated using Fig. 5.22.

$$\text{Reaction} = \frac{20 \text{ lb/ft}^2 \cdot \left(\frac{20^2 \text{ ft}^2}{2}\right)}{16.67 \text{ ft}} = 240 \text{ lb/ft}$$

Total roof reaction acting on one side of the roof is

$$\text{Reaction} = 240 \text{ lb/ft} \cdot 30 \text{ ft} = 7200 \text{ lb}$$

This is divided evenly between the two shear walls, so the shear per wall is 3600 lb.

In Fig. 5.23, for simplicity, the lateral load is shown as if it acted on the front wall alone. In reality, it also acts on the back wall, so that the structure is subjected to pressure on the front wall, and suction on the back wall.

The horizontal diaphragm reaction transferred to each shear wall is 240 lb/ft, multiplied by the building width of 30 ft, and then divided equally between the two shear walls, for a total of 3600 lb per shear wall.

Using the conservative loading case of $0.9D + 1.6W$,

$$V_u = 1.6 V_{\text{unfactored}} = 1.6 \cdot 3600 \text{ lb} = 5760 \text{ lb}$$

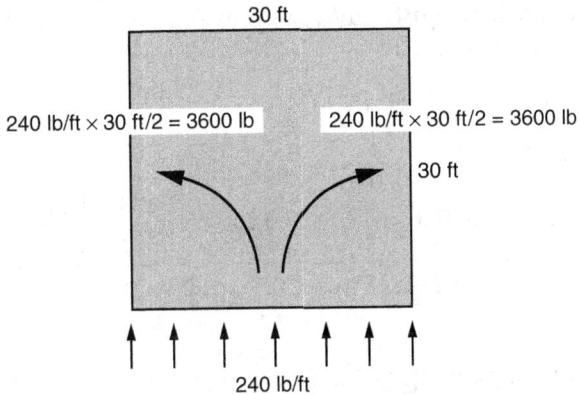

FIGURE 5.23 Transmission of forces from roof diaphragm to shear walls.

Compute the axial force in the wall at that level. To be conservative, use the loading combination $0.9D + 1.6W$. The force acting normal to the shear-transfer plane is

$$N_u = 0.9 \times 3.33 \text{ ft} \cdot 48 \text{ lb/ft}^2 \cdot 30 \text{ ft} = 4316 \text{ lb}$$

The nominal shear capacity at that level is

$$V_{nm} = \min \begin{cases} 3.8\sqrt{f'_m} A_n = 3.8 \cdot 38.73 \text{ lb/in.}^2 \cdot (30 \text{ ft} \cdot 30 \text{ in.}^2/\text{ft}) = 132,457 \text{ lb} \\ 300 A_n = 300 \cdot (30 \text{ ft} \cdot 30 \text{ in.}^2/\text{ft}) = 270,000 \text{ lb} \\ 56 A_n + 0.45 N_u = 56 \times (30 \text{ ft} \cdot 30 \text{ in.}^2/\text{ft}) + 0.45 \cdot 4316 \text{ lb} = 51,893 \text{ lb} \end{cases}$$

$$V_{nm} = 51,893 \text{ lb}$$

The design shear capacity is

$$\phi V_n = 0.80 \cdot 51,893 \text{ lb} = 41,514 \text{ lb}$$

The design shear capacity far exceeds the factored design shear of 5760 lb, and the wall is satisfactory for shear.

Now check for the net flexural tensile stress. The critical section is at the base of the wall, where in-plane moment is maximum. Because the roof spans between the front and back walls, the distributed gravity load on the roof does not act on the side walls, and their axial load comes from self-weight only. Again, use the conservative loading combination of $0.9D + 1.6W$:

$$f_{tension} = \frac{M_u c}{I} - \frac{P_u}{A} = \frac{V_u h c}{I} - \frac{P_u}{A} \leq \phi f_r$$

$$f_{tension} = \frac{1.6 \times 3600 \text{ lb} \cdot 16.67 \text{ ft} \times 12 \text{ in./ft} \left(\frac{30 \text{ ft} \cdot 12 \text{ in./ft}}{2}\right)}{\left[\frac{2 \cdot 1.25 \text{ in.} \times (30 \text{ ft} \times 12 \text{ in./ft})^3}{12}\right]} - \frac{0.9 \times 20 \text{ ft} \cdot 48 \text{ lb/ft}}{30 \text{ in.}^2}$$

$$f_{tension} = 1.6 \times 13.33 \text{ lb/in.}^2 - 0.9 \cdot 32.00 \text{ lb/in.}^2$$

$$f_{tension} = 21.33 \text{ lb/in.}^2 - 28.80 \text{ lb/in.}^2$$

$$f_{tension} = -7.47 \text{ lb/in.}^2$$

The net tension in the wall (actually a compressive stress) is less than the modulus of rupture for Type S PCL mortar and hollow units (63 lb/in.²) times the ϕ-factor of 0.6 (in other words, 37.8 lb/in.²), and the design is satisfactory.

When the wind blows against the side walls, these walls transfer their loads to the roof diaphragm, and the front and back walls act as shear walls. The side walls must be checked for this loading direction also, following the procedures of previous examples in this book.

In-plane, the (h/r) value for this shear wall is much less than the triggering value of 45, and the moment magnifier can be taken as 1.0 (2008 MSJC Code Sec. 3.2.2.4).

5.3.4 Comments on Example Problem with Strength Design of Unreinforced Shear Walls

Clearly, unreinforced masonry shear walls, whether designed by allowable stress or strength design procedures, have tremendous shear capacity because of their large cross-sectional area. If this area is reduced by openings, then shear capacities will decrease, and in-plane flexural capacities as governed by net flexural tension may decrease even faster.

5.3.5 Comments on Behavior and Design of Wall Buildings in General

Wall buildings are very efficient structurally, because the same element can act as part of the building envelope, as a vertically spanning structural element perpendicular to the direction of applied lateral load, and as a shear wall parallel to the direction of applied lateral load. If the wall building is made of a material that is aesthetically pleasing, like masonry, even more efficiency is achieved. Wall buildings are also very efficient from the viewpoint of design. The ultimate objective is design, not analysis.

The basic steps that are discussed here, in the context of simple, one-story shear wall buildings, can be applied to multistory shear wall buildings as well. At the roof level and at each floor level, horizontal diaphragms receive reactions from vertically spanning strips, and transfer those reactions to shear walls. Each shear wall acts essentially as a free-standing, statically determinate cantilever, with axial loads and in-plane lateral loads applied at each floor level. At each floor level, the shear wall must simply be designed for shear, and for combined axial force and moment.

Design of shear wall buildings is typically much easier than the design of frames, which are statically indeterminate and must usually be analyzed using computer programs.

5.3.6 Extension to Design of Unreinforced Masonry Shear Walls with Openings

Consider the structure shown in Fig. 5.24. The wall is identical to that addressed in the previous examples, with the exception of two openings, each measuring 9 ft in plan. These openings divide the wall into three smaller wall segments.

Assume that the applied shear is divided equally among the three wall segments; that points of inflection exist at the mid-height of each wall segment; and that axial forces in the wall are negligible. Then the moments and shears, can be determined by statics, where L is the 10-ft height of the wall segments. A free body of one wall segment is shown in Fig. 5.25.

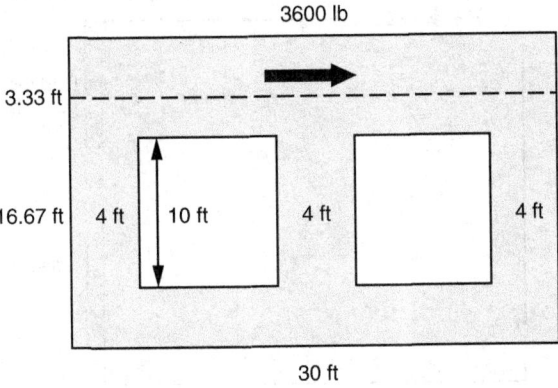

FIGURE 5.24 Shear wall with openings.

168 Chapter Five

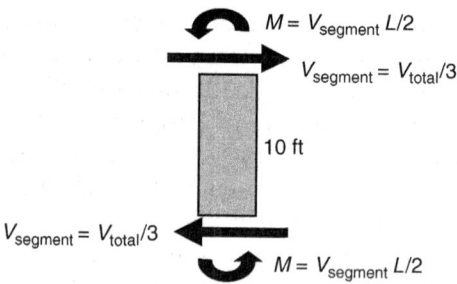

FIGURE 5.25 Free body of one wall segment.

The rest of the design proceeds as before. The shear area of the wall segments is reduced in proportion to the plan length of each segment, compared to the plan length of the original unperforated wall. The moment of inertia of the segments, however, is considerably less than the moment of inertia of the original unperforated wall.

5.4 Strength Design of Anchor Bolts

In masonry construction, anchor bolts are most commonly used to anchor roof or floor diaphragms to masonry walls. As shown in Fig. 5.26, vertically oriented anchor bolts can be placed along the top of a masonry wall to anchor a roof diaphragm resting on the top of the wall. Alternatively, horizontally oriented anchor bolts can be placed along the face of a masonry wall to anchor a diaphragm through a horizontal ledger. In these applications, anchor bolts are subjected to combinations of tension and shear. In this section, the behavior of anchors under those loadings is discussed, and 2008 MSJC strength design provisions are reviewed.

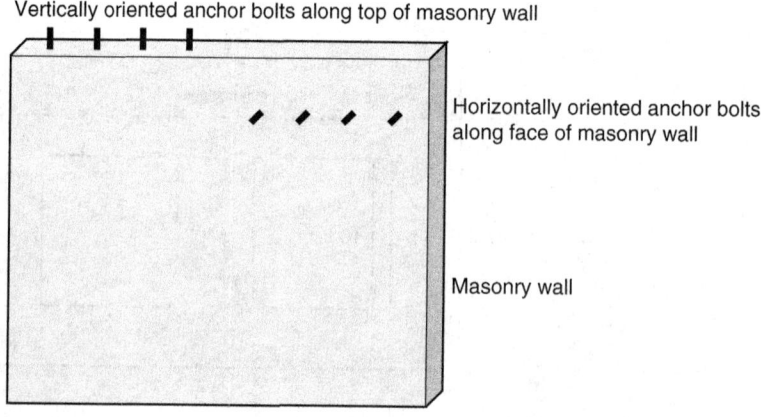

FIGURE 5.26 Common uses of anchor bolts in masonry construction.

5.4.1 Behavior and Design of Anchor Bolts Loaded in Tension

Anchor bolts loaded in tension can fail by breakout of a roughly conical body of masonry, or by yield and fracture of the anchor bolt steel. Bent-bar anchor bolts (such as J-bolts or L-bolts) can also fail by straightening of the bent portion of the anchor bolt, followed by pullout of the anchor bolt from the masonry. Nominal tensile capacity as governed by masonry breakout is evaluated using a design model based on a uniform tensile stress of $4\sqrt{f'_m}$ acting perpendicular to the inclined surface of an idealized breakout body consisting of a right circular cone (Fig. 5.27). The capacity associated with that stress state is identical with the capacity corresponding to a uniform tensile stress of $4\sqrt{f'_m}$ acting perpendicular to the projected area of the right circular cone. This design approach, while less sophisticated than that of ACI 318-08 App. D, has been shown to be user-friendly and safe for typical masonry applications.

Nominal tensile capacities for anchors as governed by masonry breakout are identical for headed and bent-bar anchors, and are given by Eqs. (3-1) and (3-3) of the 2008 MSJC *Code*.

$$B_{anb} = 4 A_{pt} \sqrt{f'_m} \quad \text{2008 MSJC } \textit{Code}, \text{ Eqs. (3-1) and (3-3)}$$

In Eqs. (3-1) and (3-3), the projected area A_{pt} is evaluated in accordance with Eq. (1-2) of the 2008 MSJC *Code*:

$$A_{pt} = \pi l_b^2 \quad \text{2008 MSJC } \textit{Code}, \text{ Eq. (1-2)}$$

As required by Sec. 1.16.4 of the 2008 MSJC *Code*, the effective embedment length, l_b, for headed anchors is the length of the embedment measured perpendicular from the masonry surface to the compression bearing surface of the anchor head. As required by Sec. 1.16.5 of the 2008 MSJC *Code*, the effective embedment for a bent-bar anchor bolt, l_b, is the length of embedment measured perpendicular from the masonry surface to the

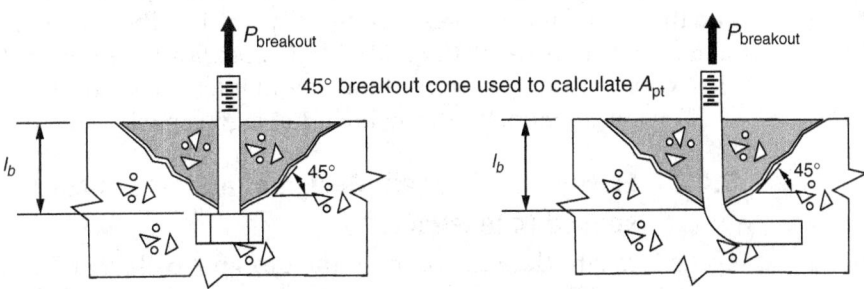

Figure 5.27 Idealized conical breakout cones for anchor bolts loaded in tension.

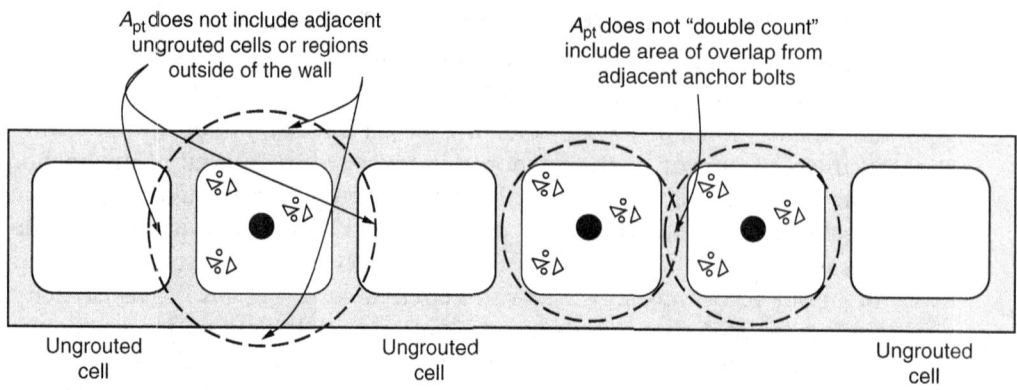

Figure 5.28 Modification of projected breakout area, A_{pt}, by void areas or adjacent anchors.

compression bearing surface of the bent end, minus one anchor bolt diameter. These are shown in Fig. 5.27. As shown in Fig. 5.28, the projected area must be reduced for the effect of overlapping projected circular areas, and for the effect of any portion of the project area falling in an open cell or core.

Nominal tensile capacities for anchors as governed by steel yield and fracture are also identical for headed and bent-bar anchors, and are given by Eqs. (3-2) and (3-5) of the 2008 MSJC *Code*. In those equations, A_b is the effective tensile stress area of the anchor bolt, including the effect of threads.

$$B_{ans} = A_b f_y \quad \text{2008 MSJC } Code, \text{ Eqs. (3-2) and (3-5)}$$

The nominal tensile capacity of bent-bar anchor bolts as governed by pullout is given by Eq. (3-4) of the 2008 MSJC *Code*.

$$B_{anp} = 1.5 f'_m e_b d_b + [300 \pi (l_b + e_b + d_b) d_b] \quad \text{2008 MSJC } Code, \text{ Eq. (3-4)}$$

In that equation, the first term represents capacity due to the hook, and the second term represents capacity due to adhesion along the anchor shank. Article 3.2A of the 2008 MSJC *Specification* requires that anchor shanks be cleaned of material that could interfere with that adhesion.

The failure mode with the lowest design capacity governs.

5.4.2 Example of Strength Design of a Single Anchor Loaded in Tension

Using strength design, compute the design tensile capacity of a 1/2-in. diameter, A307 bent-bar anchor with a 1-in. hook, embedded vertically in a grouted cell of a nominal 8-in. wall with a specified compressive

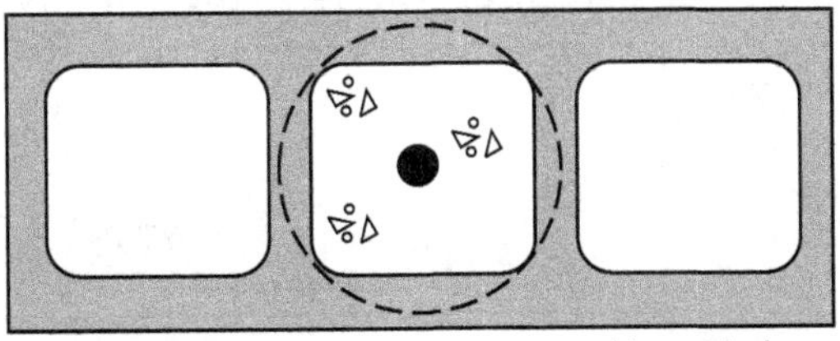

FIGURE 5.29 Example involving a single tensile anchor, placed vertically in a grouted cell.

*A*pt is not affected by adjacent ungrouted cells or regions outside of the wall

strength, f'_m, of 1500 lb/in.². Assume that the bottom of the anchor hook is embedded at a distance of 4.5 in. This example might represent a tensile anchor used to attach a roof diaphragm to a wall.

First, compute the effective embedment, l_b. In accordance with Sec. 1.16.5 of the 2008 MSJC *Code*, this is equal to the total embedment of 4.5 in., minus the diameter of the anchor (to get to the inside of the hook), and minus an additional anchor diameter, or 3.5 in. As shown in Fig. 5.29, the projected tensile breakout area has a radius of 3.5 in. (diameter of 7 in.). Because the masonry wall has a specified thickness of 7.63 in., the projected tensile breakout area is not affected by adjacent ungrouted cells or regions outside of the wall.

$$A_{pt} = \pi l_b^2 \quad \text{2008 MSJC } Code, \text{ Eq. (1-2)}$$

$$A_{pt} = \pi (3.5 \text{ in.})^2$$

$$A_{pt} = 38.5 \text{ in.}^2$$

Calculate the nominal capacity due to tensile breakout of masonry.

$$B_{anb} = 4 A_{pt} \sqrt{f'_m} \quad \text{2008 MSJC } Code, \text{ Eqs. (3-1) and (3-3)}$$

$$B_{anb} = 4 \times 38.5 \text{ in}^2 \sqrt{1500 \text{ lb/in.}^2}$$

$$B_{anb} = 5962 \text{ lb}$$

Now obtain the design capacity by multiplying the nominal capacity by the corresponding strength-reduction factor from Sec. 3.1.4.4 of the 2008 MSJC *Code*:

$$\phi B_{anb} = 0.5 \times 5962 \text{ lb} \quad \text{2008 MSJC } Code, \text{ Sec. 3.1.4.4}$$

$$\phi B_{anb} = 2981 \text{ lb}$$

Now compute the nominal tensile capacity as governed by steel yield. In this computation, A_b is the effective tensile stress area of the anchor bolt, including the effect of threads. According to ANSI/ASME B1.1,

$$A_b = \frac{\pi}{4}\left(d_o - \frac{0.9743}{n_t}\right)^2$$

where d_o = nominal anchor diameter, in.
n_t = number of threads per inch

For anchors with nominal diameters typically used in masonry, the effective tensile stress area can be approximated with sufficient accuracy as 0.75 times the nominal area. That approximation is used in this and other anchor bolt problems here. The minimum specified yield strength for A307 steel is 36 ksi.

$$B_{ans} = A_b f_y \qquad \text{2008 MSJC Code, Eqs. (3-2) and (3-5)}$$

$$B_{ans} = 0.75 \times 0.20 \text{ in.}^2 \times 60{,}000 \text{ lb/in.}^2$$

$$B_{ans} = 9000 \text{ lb}$$

Now obtain the design capacity by multiplying the nominal capacity by the corresponding strength-reduction factor from Sec. 3.1.4.4 of the 2008 MSJC Code:

$$\phi B_{ans} = 0.9 \times 9000 \text{ lb} \qquad \text{2008 MSJC Code, Sec. 3.1.4.4}$$

$$\phi B_{anb} = 8100 \text{ lb}$$

The nominal tensile capacity of bent-bar anchor bolts as governed by pullout is given by Eq. (3-4) of the 2008 MSJC Code.

$$B_{anp} = 1.5 f'_m e_b d_b + [300 \pi (l_b + e_b + d_b) d_b] \qquad \text{2008 MSJC Code, Eq. (3-4)}$$

$$B_{anp} = 1.5 \times 1500 \text{ lb/in.}^2 \times 1.0 \text{ in.} \times 0.5 \text{ in.}$$

$$+ \left[300 \pi (3.5 \text{ in.} + 1.0 \text{ in.} + 0.5 \text{ in.}) 0.5 \text{ in.}\right]$$

$$B_{anp} = 1125 \text{ lb} + 2356 \text{ lb}$$

$$B_{anp} = 3481 \text{ lb}$$

Now obtain the design capacity by multiplying the nominal capacity by the corresponding strength-reduction factor from Sec. 3.1.4.4 of the 2008 MSJC *Code*:

$\phi B_{anp} = 0.65 \times 3481$ lb   2008 MSJC *Code*, Sec. 3.1.4.4

$\phi B_{anp} = 2263$ lb

The governing design tensile capacity is the lowest of that governed by masonry breakout (2981 lb), yield of the anchor shank (8100 lb), and pullout (2263 lb). Pullout governs, and the design tensile capacity is 2263 lb.

If this problem had involved an anchor with deeper embedment (so that the projected tensile breakout area would have been affected by adjacent ungrouted cells or regions outside of the wall), only the anchor capacity as governed by tensile breakout would have been affected, due to a reduced projected tensile breakout area.

Similarly, if this problem had involved adjacent anchors with overlapping tensile breakout areas, only the anchor capacity as governed by tensile breakout would have been affected, again due to a reduced projected tensile breakout area.

5.4.3 Behavior and Design of Anchor Bolts Loaded in Shear

Anchor bolts loaded in shear, and located without a nearby free edge in the direction of load, can fail by local crushing of the masonry under bearing stresses from the anchor bolt; by pryout of the head of the anchor in a direction opposite to the direction of applied load, or by yield and fracture of the anchor bolt steel. Anchor bolts loaded in shear, and located near a free edge in the direction of load, can also fail by breakout of a roughly semi-conical volume of masonry in the direction of the applied shear. Pryout and shear breakout are shown in parts (a) and (b), respectively, of Fig. 5.30.

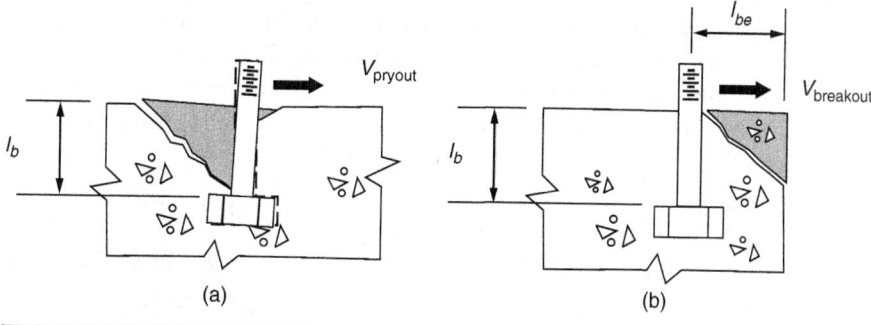

FIGURE 5.30 (a) Pryout failure and (b) shear breakout failure.

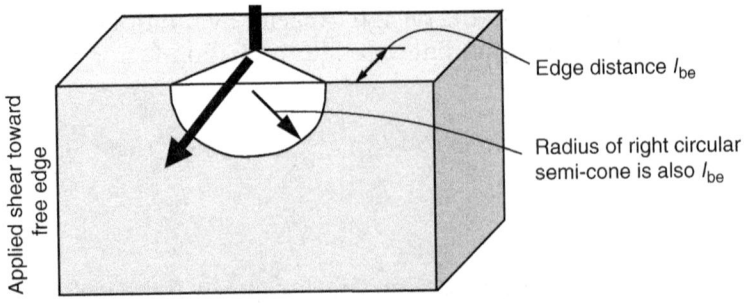

FIGURE 5.31 Design idealization associated with shear breakout failure.

Nominal shear capacity as governed by pryout is taken as twice the nominal tensile breakout capacity, based on the same empirical evidence used in ACI 318-08, App. D. Nominal shear capacity as governed by masonry breakout is evaluated using a design model based on a uniform tensile stress of $4\sqrt{f'_m}$ acting perpendicular to the inclined surface of an idealized breakout body consisting of a right circular semi-cone (Fig. 5.31).

The capacity associated with that stress state is identical with the capacity corresponding to a uniform tensile stress of $4\sqrt{f'_m}$ acting perpendicular to the projected area of the right circular semi-cone. This design approach, while less sophisticated than that of ACI 318-08 App. D, has been shown to be user-friendly and safe for typical masonry applications.

The nominal shear breakout capacity of an anchor is given by Eq. (3-6) of the 2008 MSJC *Code*. In evaluating that equation, the projected area of the breakout semi-cone is given by Eq. (1-3) of the 2008 MSJC *Code*.

$$B_{vnb} = 4 A_{pv} \sqrt{f'_m} \qquad \text{2008 MSJC } Code, \text{ Eq. (3-6)}$$

$$A_{pv} = \frac{\pi l_{be}^2}{2} \qquad \text{2008 MSJC } Code, \text{ Eq. (1-3)}$$

Nominal capacities of anchors loaded in shear are given by Eq. (3-7) of the 2008 MSJC *Code* for masonry crushing, by Eq. (3-8) of the 2008 MSJC *Code* for shear pryout, and by Eq. (3-9) of the 2008 MSJC *Code* for yield of the anchor in shear. In Eqs. (3-7) and (3-9) of the 2008 MSJC *Code* (masonry crushing and anchor yield, respectively), the effective tensile stress area of the bolt (including the effect of threads) is to be used, unless threads are excluded from the shear plane.

$$B_{vnc} = 1050 \sqrt[4]{f'_m A_b} \qquad \text{2008 MSJC } Code, \text{ Eq. (3-7)}$$

$$B_{vnpry} = 2.0 B_{anb} = 8 A_{pt} \sqrt{f'_m} \qquad \text{2008 MSJC } Code, \text{ Eq. (3-8)}$$

$$B_{vns} = 0.6 A_b f_y \qquad \text{2008 MSJC } Code, \text{ Eq. (3-9)}$$

The failure mode with the lowest design capacity governs.

5.4.4 Example of Strength Design of a Single Anchor Loaded in Shear

Using strength design, compute the design shear capacity of a 1/2-in. diameter, A307 bent-bar anchor with a 1-in. hook, embedded horizontally in a grouted cell of a nominal 8-in. wall with a specified compressive strength, f'_m, of 1500 lb/in.². Assume that the bottom of the anchor hook is embedded a distance of 4.5 in., and that the anchor is located far from free edges in the direction of applied shear. This might represent an anchor used to attach a ledger to a masonry wall. Because free edges are not a factor, shear breakout does not apply.

First, compute the effective embedment, l_b. In accordance with Sec. 1.16.5 of the 2008 MSJC *Code*, this is equal to the total embedment of 4.5 in., minus the diameter of the anchor (to get to the inside of the hook), and minus an additional anchor diameter, or 3.5 in. The projected tensile breakout area has a radius of 3.5 in. (diameter of 7 in.).

$$A_{pt} = \pi l_b^2 \qquad \text{2008 MSJC Code, Eq. (1-2)}$$

$$A_{pt} = \pi (3.5 \text{ in.})^2$$

$$A_{pt} = 38.5 \text{ in.}^2$$

First, compute the nominal capacity of the anchor as governed by masonry crushing.

$$B_{vnc} = 1050 \sqrt[4]{f'_m A_b} \qquad \text{2008 MSJC Code, Eq. (3-7)}$$

In this computation, A_b is the effective tensile stress area of the anchor bolt, including the effect of threads. According to ANSI/ASME B1.1,

$$A_b = \frac{\pi}{4}\left(d_o - \frac{0.9743}{n_t}\right)^2$$

where d_o = nominal anchor diameter, in.
n_t = number of threads per inch

For anchors with nominal diameters typically used in masonry, the effective tensile stress area can be approximated with sufficient accuracy as 0.75 times the nominal area. That approximation is used in this and other anchor bolt problems here.

$$B_{vnc} = 1050 \sqrt[4]{f'_m A_b} \qquad \text{2008 MSJC Code, Eq. (3-7)}$$

$$B_{vnc} = 1050 \sqrt[4]{1500 \text{ lb/in.}^2 \times (0.75 \times 0.20 \text{ in.}^2)}$$

$$B_{vnc} = 4067 \text{ lb/in.}^2$$

Now obtain the design capacity by multiplying the nominal capacity by the corresponding strength-reduction factor from Sec. 3.1.4.4 of the 2008 MSJC Code:

$$\phi B_{vnc} = 0.5 \cdot 4067 \text{ lb} \qquad \text{2008 MSJC Code, Sec. 3.1.4.4}$$

$$\phi B_{vnc} = 2033 \text{ lb}$$

Next, compute the nominal capacity of the anchor as governed by pryout.

$$B_{vnpry} = 2.0 \, B_{anb} = 8 \, A_{pt} \sqrt{f'_m} \qquad \text{2008 MSJC Code, Eq. (3-8)}$$

Because nominal pryout capacity is a multiple of the nominal tensile breakout capacity, we must compute the nominal tensile breakout capacity.

$$B_{anb} = 4 \, A_{pt} \sqrt{f'_m} \qquad \text{2008 MSJC Code, Eqs. (3-1) and (3-3)}$$

$$B_{anb} = 4 \cdot 38.5 \text{ in.}^2 \, \sqrt{1500 \text{ lb/in.}^2}$$

$$B_{anb} = 5962 \text{ lb}$$

Continue with the pryout calculation:

$$B_{vnpry} = 2.0 \, B_{anb} \qquad \text{2008 MSJC Code, Eq. (3-8)}$$

$$B_{vnpry} = 2.0 \cdot 5962 \text{ lb}$$

$$B_{vnpry} = 11,924 \text{ lb}$$

Now obtain the design capacity by multiplying the nominal capacity by the corresponding strength-reduction factor from Sec. 3.1.4.4 of the 2008 MSJC Code:

$$\phi B_{vnpry} = 0.5 \times 11,924 \text{ lb} \qquad \text{2008 MSJC Code, Sec. 3.1.4.4}$$

$$\phi B_{vnpry} = 5962 \text{ lb}$$

Next, compute the nominal capacity of the anchor as governed by yield and fracture of the anchor shank.

$$B_{vns} = 0.6 \, A_b f_y \qquad \text{2008 MSJC Code, Eq. (3-9)}$$

$$B_{vns} = 0.6 \times (0.75 \cdot 0.20 \text{ in.}^2) \cdot 60,000 \text{ lb/in.}^2$$

$$B_{vns} = 5400 \text{ lb}$$

Now obtain the design capacity by multiplying the nominal capacity by the corresponding strength-reduction factor from Sec. 3.1.4.4 of the 2008 MSJC *Code*:

$$\phi B_{vns} = 0.9 \cdot 5400 \text{ lb} \quad \text{2008 MSJC } Code, \text{ Sec. 3.1.4.4}$$

$$\phi B_{vns} = 4860 \text{ lb}$$

The governing design shear capacity is the lowest of that governed by masonry crushing (2033 lb), pryout (5962 lb), and yield of the anchor shank (4860 lb). Because the anchor is not close to a free edge, shear breakout does not apply. Masonry crushing governs, and the design shear capacity is 2033 lb.

If this problem had involved an anchor loaded toward a free edge, then shear breakout would have had to be checked.

5.4.5 Behavior and Design of Anchor Bolts Loaded in Combined Tension and Shear

Design capacities of anchor bolts in combined tension and shear are given by the linear interaction equation of Eq. (3-10) of the 2008 MSJC *Code*. While an elliptical or trilinear interaction equation would be slightly more accurate, a linear interaction is conservative and simple for design.

$$\frac{b_{af}}{\phi B_{an}} + \frac{b_{vf}}{\phi B_{vn}} \leq 1 \quad \text{2008 MSJC } Code, \text{ Eq. (3-10)}$$

5.5 Required Details for Unreinforced Bearing Walls and Shear Walls

Bearing walls that resist out-of-plane lateral loads, and shear walls, must be designed to transfer lateral loads to the floors above and below. Examples of such connections are shown below. These connections would have to be strengthened for regions subject to strong earthquakes or strong winds. Section 1604.8.2 of the 2009 IBC has additional requirements for anchorage of diaphragms to masonry walls. Section 12.11 of ASCE 7-05 has additional requirements for anchorage of structural walls for structures assigned to Seismic Design Categories C and higher.

5.5.1 Wall-to-Foundation Connections

As shown in Fig. 5.32, CMU walls (or the inner CMU wythe of a drainage wall) must be connected to the concrete foundation. Bond breaker should be used only between the outer veneer wythe and the foundation.

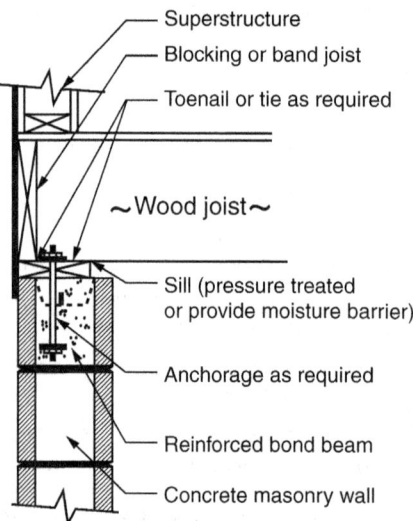

FIGURE 5.32 Example of wall-to-foundation connection (*Source*: Figure 1 of National Concrete Masonry Association TEK 05-07A.)

5.5.2 Wall-to-Floor Details

Example of a wall-to-floor detail are shown in Figs. 5.33 and 5.34. In the latter detail (floor or roof planks oriented parallel to walls), the planks are actually cambered. They are shown on the outside of the walls so that this camber does not interfere with the coursing of the units. Some designers object to this detail because it could lead to spalling of the cover. If it is modified so that the planks rest on the face shells of the walls, then the thickness of the topping must vary to adjust for the camber, and form boards must be used against both sides of the wall underneath the planks, so that the concrete or grout that is cast into the bond beam does not run out underneath the cambered beam.

5.5.3 Wall-to-Roof Details

An example of a wall-to-roof detail is shown in Fig. 5.35.

5.5.4 Typical Details of Wall-to-Wall Connections

Typical details of wall-to-wall connections are shown in Fig. 5.36.

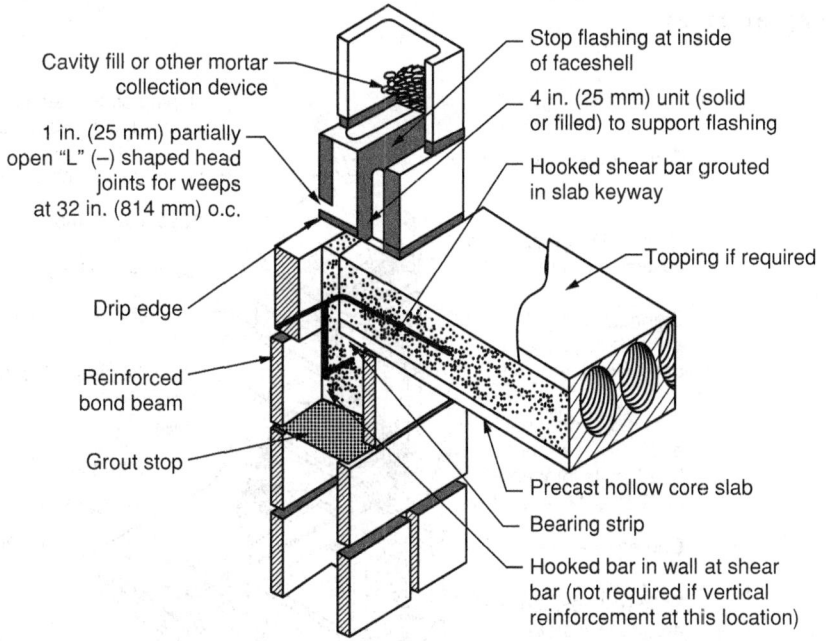

Figure 5.33 Example of wall-to-floor connection, planks perpendicular to wall (*Source*: Figure 14 of National Concrete Masonry Association TEK 05-07A.)

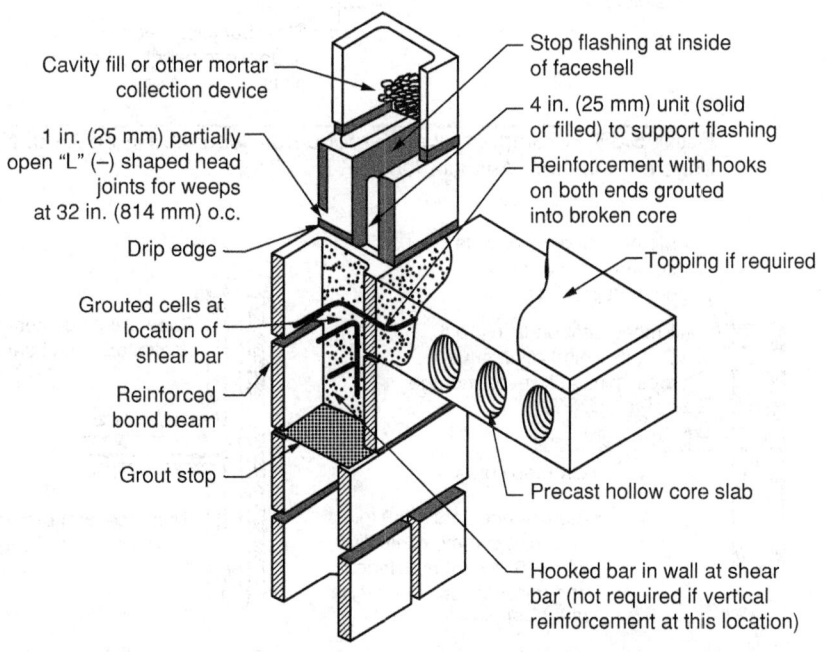

Figure 5.34 Example of wall-to-floor connection, planks parallel to wall (*Source*: Figure 15 of National Concrete Masonry Association TEK 05-07A.)

179

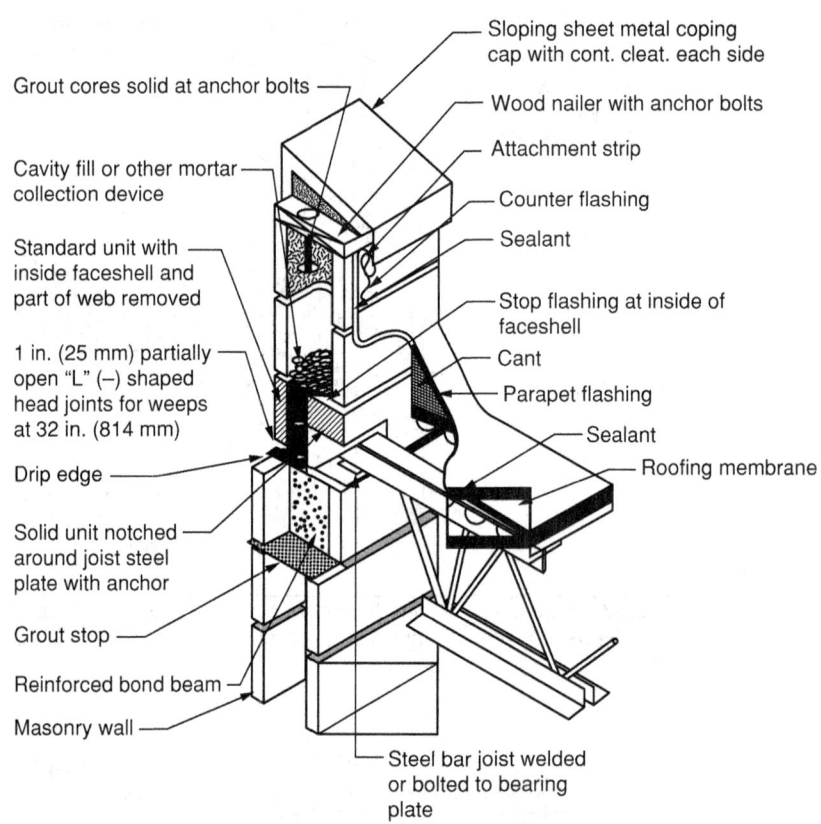

Figure 5.35 Example of wall-to-roof detail (*Source*: Figure 11 of National Concrete Masonry Association TEK 05-07A.)

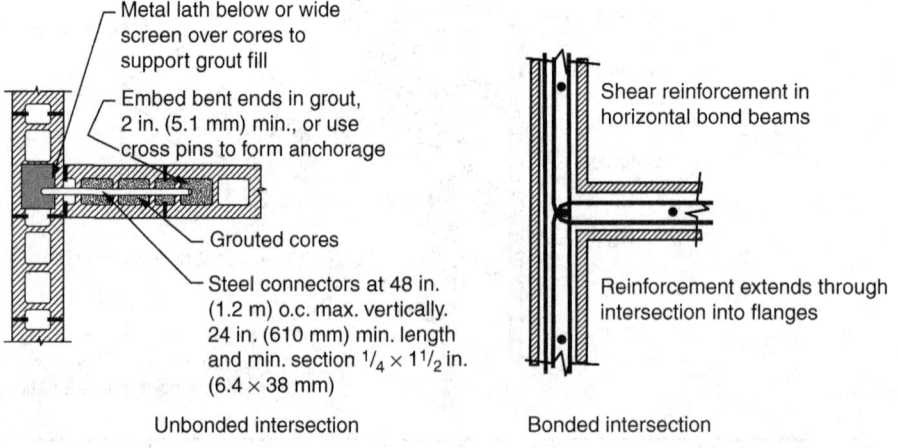

Figure 5.36 Examples of wall-to-wall connection details. (*Source*: Figure 2 of National Concrete Masonry Association TEK 14-08B.)

CHAPTER 6
Strength Design of Reinforced Masonry Elements

6.1 Strength Design of Reinforced Beams and Lintels

6.1.1 Background on Strength Design of Reinforced Masonry Beams for Flexure

Strength design of reinforced masonry beams follows the same steps used for reinforced concrete beams. The basic assumptions are shown in Fig. 6.1.

Strain in the masonry is assumed to have a maximum useful value of 0.0025 for concrete masonry and 0.0035 for clay masonry. Tension reinforcement is assumed to be somewhere on the yield plateau. Because axial load is zero, flexural capacity is equal to either the tension force or the compression force on the cross-section, multiplied by the internal lever arm as shown in Fig. 6.2 (the distance between the tensile and compressive forces).

$$M_n = A_s f_y \left(d - \frac{\beta_1 c}{2} \right)$$

But the depth of the compressive stress block is known from equilibrium of axial forces:

$$T = C$$
$$A_s f_y = 0.80 f'_m \beta_1 cb$$
$$\beta_1 c = \frac{A_s f_y}{0.80 f'_m b}$$

Chapter Six

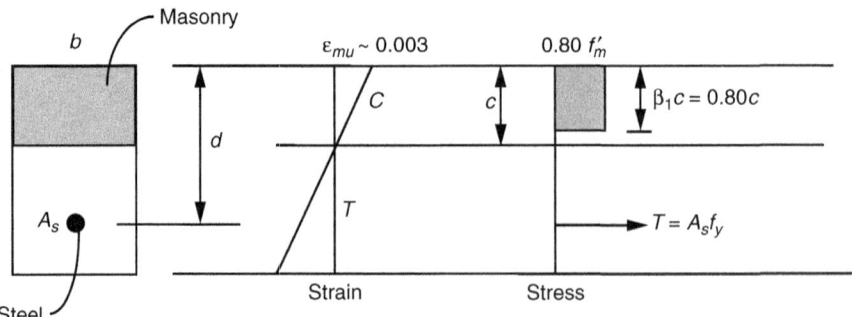

FIGURE 6.1 Assumptions used in strength design of reinforced masonry for flexure.

Now define $\rho \equiv \dfrac{A_s}{bd}$ and $\omega \equiv \rho\left(\dfrac{f_y}{f'_m}\right)$. Then

$$M_n = A_s f_y \left(d - \dfrac{\beta_1 c}{2}\right)$$

$$M_n = A_s f_y \left(d - \dfrac{A_s f_y}{2 \cdot 0.80 f'_m b}\right)$$

$$M_n = \rho b d f_y \left(d - \dfrac{\rho b d f_y}{2 \cdot 0.80 f'_m b}\right)$$

$$M_n = \rho b d f_y \left(\dfrac{f'_m}{f'_m}\right)\left(d - \dfrac{\rho b d f_y}{2 \cdot 0.80 f'_m b}\right)$$

$$M_n = \omega b d f'_m \left(d - \dfrac{\omega d}{1.6}\right)$$

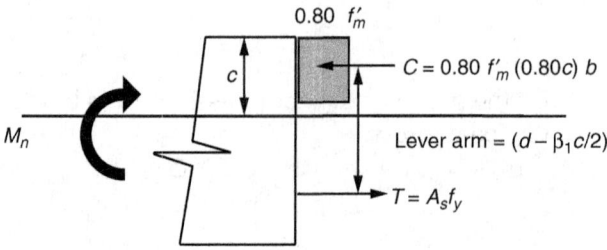

FIGURE 6.2 Equilibrium of internal stresses and external nominal moment for strength design of reinforced masonry for flexure.

Strength Design of Reinforced Masonry Elements

and finally,

$$M_n = \omega b d^2 f'_m (1 - 0.63\omega)$$

This closed-form expression permits solving for the required dimensions if the steel percentage is known. The variable ω is sometimes referred to as the "tensile reinforcement index." For design of masonry beams, where dimensions are known, it is usually easier simply to use $M_n = A_s f_y \left(d - \dfrac{\beta_1 c}{2} \right)$. For additional simplicity, the internal lever arm $\left(d - \dfrac{\beta_1 c}{2} \right)$ can be approximated as $0.9\,d$.

The steel percentage is constrained by the requirement that the steel be on the yield plateau when the masonry reaches its maximum useful strain. For this condition to be satisfied, the steel must yield before the masonry reaches its maximum useful strain. In other words, the steel percentage must be less than the balanced steel percentage, at which the steel yields just as the masonry reaches its maximum useful strain.

The balanced steel percentage for strength design can be derived based on the strains in steel and masonry (Fig. 6.3).

First, locate the neutral axis under balanced conditions:

$$\frac{\varepsilon_{mu}}{\varepsilon_y} = \frac{c}{d-c}$$

$$c\varepsilon_y = \varepsilon_{mu}(d-c)$$

$$c(\varepsilon_y + \varepsilon_{mu}) = \varepsilon_{mu} d$$

$$c = d\left(\frac{\varepsilon_{mu}}{\varepsilon_y + \varepsilon_{mu}} \right)$$

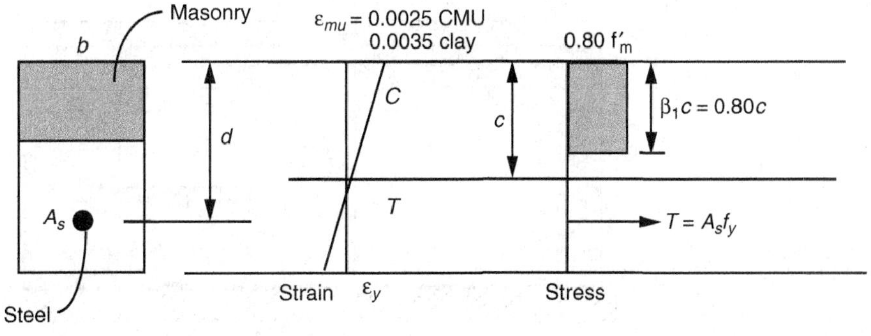

Figure 6.3 Conditions corresponding to balanced reinforcement percentage for strength design.

Next, compute the compressive force under those conditions, and compute balanced steel area as the steel area, acting at yield, that is necessary to equilibrate that compressive force:

$$T = C$$

$$A_{sb} f_y = 0.80 f'_m \beta_1 cb$$

$$A_{sb} f_y = 0.80 f'_m \beta_1 bd \left(\frac{\varepsilon_{mu}}{\varepsilon_y + \varepsilon_{mu}} \right)$$

$$\rho_b = 0.80 \left(\frac{f'_m}{f_y} \right) \beta_1 \left(\frac{\varepsilon_{mu}}{\varepsilon_y + \varepsilon_{mu}} \right)$$

6.1.2 Steps in Strength Design of Reinforced Beams and Lintels

The most common reinforced masonry beam is a lintel. Lintels are beams that support masonry over openings. Strength design of reinforced beams and lintels follows the steps given below:

1. Shear design
    a. Calculate the design shear, and compare it with the corresponding resistance. Revise the lintel depth if necessary.
2. Flexural design
    a. Calculate the design moment.
    b. Calculate the required flexural reinforcement. Check that it fits within minimum and maximum reinforcement limitations.

In many cases, the depth of the lintel is determined by architectural considerations. In other cases, it is necessary to determine the number of courses of masonry that will work as a beam. For example, consider the lintel in Fig. 6.4.

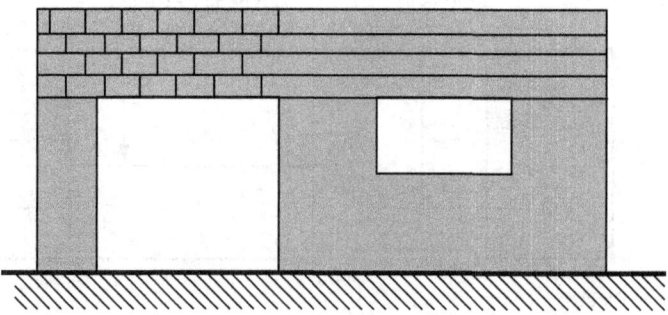

FIGURE 6.4 Example of masonry lintel.

The depth of the beam, and hence the area that is effective in resisting shear, is determined by the number of courses that we consider to comprise it. Because it is not very practical to put shear reinforcement in masonry beams, the depth of the beam may be determined by this. In other words, the beam design may start with the number of courses that are needed to that shear can be resisted by masonry alone.

6.1.3 Physical Properties of Steel Reinforcing Wire and Bars

Physical properties of steel reinforcing wire and bars are given in Table 6.1.

Cover requirements are given in Sec. 1.15.4 of the 2008 MSJC *Code*. Minimum cover for joint reinforcement (exterior exposure) is 5/8 in.

6.1.4 Example of Lintel Design According to Strength Provisions

Suppose that we have a uniformly distributed load of 1050 lb/ft, applied at the level of the roof of the structure shown in Fig. 6.5. Design the lintel.

According to Table 2 of the 2008 MSJC *Specification*, for Type M or S mortar and concrete units with a specified strength of 1900 psi (the minimum specified strength for ASTM C90 units), the compressive strength of the masonry can conservatively be taken as 1500 psi (the so-called "unit strength method"). If the compressive strength is evaluated by

Designation	Diameter, in.	Area, in.^{2}
Wire		
W1.1 (11 gage)	0.121	0.011
W1.7 (9 gage)	0.148	0.017
W2.1 (8 gage)	0.162	0.020
W2.8 (3/16 wire)	0.187	0.027
W4.9 (1/4 wire)	0.250	0.049
Bars		
#3	0.375	0.11
#4	0.500	0.20
#5	0.625	0.31
#6	0.750	0.44
#7	0.875	0.60
#8	1.000	0.79
#9	1.128	1.00
#10	1.270	1.27
#11	1.410	1.56

TABLE 6.1 Physical Properties of Steel Reinforcing Wire and Bars

prism testing, a higher value can probably be used. Take the specified compressive strength of the masonry as $f'_m = 1500$ psi.

Assume fully grouted concrete masonry with a nominal thickness of 8 in., a weight of 80 lb/ft², and a specified compressive strength of 1500 lb/in.². Use Type S PCL mortar. The lintel has a span of 10 ft, and a total depth (height of parapet plus distance between the roof and the lintel) of 4 ft. These are shown in the schematic figure in Fig. 6.5. Assume that 700 lb/ft of the roof load is D, and the remaining 350 lb/ft is L. The governing loading combination is $1.2D + 1.6L$. Our design presumes that entire height of the lintel is grouted.

First check whether the depth of the lintel is sufficient to avoid the use of shear reinforcement. Because the opening may have a movement joint on either side, again use a span equal to the clear distance, plus one-half of a half-unit on each side. So the span is 10 ft plus 8 in., or 10.67 ft.

$$M_u = \frac{w_u l^2}{8} = \frac{[(700 + 4 \text{ ft} \cdot 80 \text{ lb/ft}) \cdot 1.2 + 350 \text{ lb/ft} \cdot 1.6] \cdot 10.67^2 \text{ ft}^2 \cdot 12 \text{ in./ft}}{8}$$

$$= 304,660 \text{ in.-lb}$$

$$V_u = \frac{w_u l}{2} = \frac{[(700 + 4 \text{ ft} \cdot 80 \text{ lb/ft}) \cdot 1.2 + 350 \text{ lb/ft} \cdot 1.6] \cdot 10.67 \text{ ft}}{2} = 9518 \text{ lb}$$

The bars in the lintel will probably be placed in the lower part of an inverted bottom course.

The effective depth d is calculated using the minimum cover of 1.5 in. (Sec. 1.15.4.1 of the 2008 MSJC *Code*), plus one-half the diameter of an assumed #8 bar.

Because this is a reinforced element, shearing capacity is calculated using Sec. 3.3.4.1.2.1 of the 2008 MSJC *Code*.

$$V_{nm} = \left[4.0 - 1.75\left(\frac{M_u}{V_u d_v}\right)\right] A_n \sqrt{f'_m} + 0.25 P_u$$

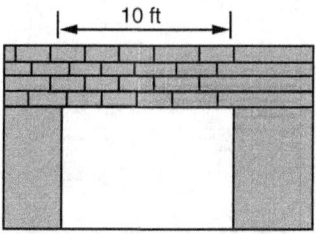

FIGURE 6.5 Example for strength design of a lintel.

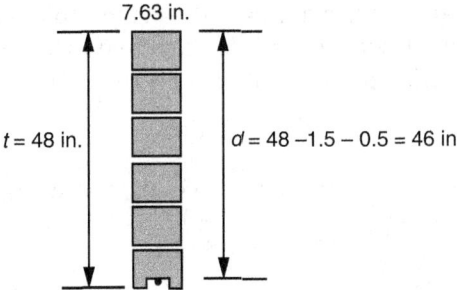

FIGURE 6.6 Example showing placement of bottom reinforcement in lowest course of lintel.

As $(M_u/V_u d_v)$ increases, V_{nm} decreases. Because $(M_u/V_u d_v)$ need not be taken greater than 1.0 (2008 MSJC *Code* Sec. 3.3.4.1.2.1), the most conservative (lowest) value of V_{nm} is obtained with $(M_u/V_u d_v)$ equal to 1.0. Also, factored design axial load, P_u, is zero:

$$V_{nm} = [4.0 - 1.75\,(1.0)] A_n \sqrt{f'_m}$$

$$V_{nm} = 2.25\, A_n \sqrt{f'_m}$$

$V_u = 9518 \text{ lb} \le \phi V_n = 0.8 \times 2.25 \times 7.63 \text{ in.} \times 46 \text{ in.} \times \sqrt{1500} \text{ lb/in.}^2 = 24{,}468 \text{ lb}$

Also, according to Eq. 3-20,

$$V_n \le 4\sqrt{f'_m}\, A_n$$

This does not govern, and the shear design is acceptable.
Now check the required flexural reinforcement:

$$M_n = A_s f_y \text{ (lever arm)}$$

$$M_n \approx A_s f_y \cdot 0.9\, d$$

In our case,

$$M_n^{\text{required}} = \frac{M_u}{\phi} = \frac{M_u}{0.9} = \frac{304{,}660 \text{ lb-in.}}{0.9} = 338{,}511 \text{ lb-in.}$$

$$A_s^{\text{required}} \approx \frac{M_n^{\text{required}}}{0.9 d\, f_y} = \frac{338{,}511 \text{ lb-in.}}{0.9 \times 46 \text{ in.} \times 60{,}000 \text{ lb/in.}^2} = 0.14 \text{ in.}^2$$

Because of the depth of the beam, this can easily be satisfied with a #4 bar in the lowest course (of concrete masonry units). The corresponding nominal flexural capacity is approximately

$$M_n \approx A_s f_y (0.9\, d)$$

$$M_n \approx 0.20 \text{ in.}^2 \cdot 60{,}000 \text{ lb/in.}^2 \cdot 0.9 \cdot 46 \text{ in.}$$

$$M_n \approx 496{,}800 \text{ lb-in.}$$

Also include two #4 bars at the level of the roof (bond beam reinforcement). The flexural design is quite simple.

Section 3.3.4.2.2.2 of the 2008 MSJC *Code* does require that the nominal flexural strength of a beam not be less than 1.3 times the nominal cracking capacity, calculated using the modulus of rupture from *Code* Sec. 3.1.8.2. In our case, the nominal cracking moment for the 4-ft deep section is

$$M_{cr} = S f_r = \frac{bt^2}{6} f_r = \frac{7.63 \text{ in.} \cdot 48^2 \text{ in.}^2}{6} \times 200 \text{ lb/in.}^2 = 585{,}984 \text{ lb-in.}$$

This value, multiplied by 1.3, is 761,779 lb-in., which exceeds the nominal capacity of this lintel with the provided #4 bar. Flexural reinforcement must be increased to

$$A_s \approx 0.20 \text{ in.}^2 \left(\frac{761{,}779}{496{,}800 \text{ lb-in.}} \right) = 0.31 \text{ in.}^2$$

Use two #4 bars.

Section 3.3.4.2.2.2 of the 2008 MSJC *Code* need not be met if the amount of tensile reinforcement is at least one-third greater than required by analysis (*Code* Sec. 3.3.4.2.2.3).

Finally, Sec. 3.3.3.5 of the 2008 MSJC *Code* imposes maximum flexural reinforcement limitations that are based on a series of critical strain gradients. These generally do not govern for members with little or no axial load, like this lintel. They may govern for members with significant axial load, such as tall shear walls.

6.1.5 Comments on Arching Action

1. Using the traditional assumption that distributed loads act only within a beam length defined by 45° lines from the ends of the distributed load, it would have been possible to take advantage of so-called "arching action" to reduce the gravity load for which the lintel must be designed. Nevertheless, this measure is hardly necessary, because the required area of reinforcement is quite small in any case.

2. Even though it would have been possible to refine the flexural design (e.g., by reducing the required depth of the lintel, or including the mid-depth reinforcement in the bond beam in the calculation of flexural resistance), this additional design effort would not have been cost-effective. The goal is to simplify the design process, and the final layout of reinforcement.

6.2 Strength Design of Reinforced Curtain Walls

6.2.1 Background on Curtain Walls

In the first part of the structural design section of this book, we began with the design of panel walls, which can be designed as unreinforced masonry, and which span primarily in the vertical direction to transmit out-of-plane loads to the structural system. Panel walls are nonload bearing masonry, because they support gravity loads from self-weight only.

At this point, it is appropriate for us to study another type of nonload bearing masonry, the curtain wall. Like panel walls, curtain walls carry gravity load from self-weight only, and transmit out-of-plane loads to a structural frame. Unlike panel walls, however, curtain walls can be more than one story high, and span horizontally rather than vertically. Typical curtain wall construction is shown in Fig. 6.7.

A single wythe of masonry spans horizontally between columns, which support the roof. This type of construction can be used for industrial buildings, gymnasiums, theaters, and other buildings of similar configuration.

In previous sections dealing with panel walls, we have seen that because those walls are unreinforced, their design is governed by the flexural tensile strength of masonry. In previous example, using reasonable unfactored wind loads of about 20 lb/ft², the flexural tensile stresses in vertically spanning panel walls were comfortably within allowable values.

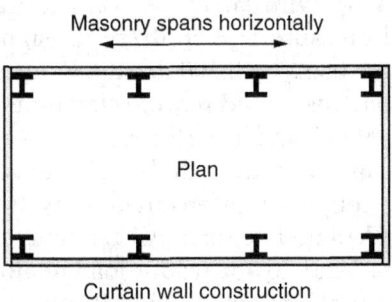

FIGURE 6.7 Plan view of typical curtain wall construction.

If we tried to use the same principles to design horizontally spanning curtain walls, however, they wouldn't work. In the Fig. 6.7, the horizontal span between columns is at least 20 ft, about twice the typical vertical span of panel walls. Since moments increase as the square of the span, doubling the span would increase the flexural tensile stresses by a factor of 4. Even considering that allowable flexural tensile stresses parallel to bed joints in running bond are about twice as high as those normal to the bed joints (reflecting the interlocking nature of running bond), the calculated flexural tensile stresses in the direction of span would exceed the allowable values.

The most reasonable solution to this problem is to reinforce the masonry horizontally. Single-wythe curtain walls are commonly used for industrial buildings, where water-penetration resistance is not a primary design consideration.

6.2.2 Examples of Use of Curtain Walls—Clay Masonry

Examples of use of curtain walls with clay masonry are shown in Fig. 6.8.

6.2.3 Example of Use of Curtain Walls—Concrete Masonry

Examples of use of curtain walls with concrete masonry are shown in Fig. 6.9.

6.2.4 Structural Action of Curtain Walls

Curtain walls act as horizontal strips to transfer out-of-plane loads to vertical supporting members such as steel or reinforced concrete columns, or masonry pilasters (masonry columns partially embedded in the wall).

6.2.5 Example of Strength Design of a Reinforced Curtain Wall

A curtain wall of standard modular clay units spans 20 ft between columns, and is simply supported at each column. It has reinforcement consisting of W4.9 wire each face, every course. The curtain wall is subjected to a wind pressure $w = 20$ lb/ft^2. Design the curtain wall. As an initial assumption, use $f_m' = 2{,}500$ lb/in.^{2}. Referring to Table 1 of the 2008 MSJC *Specification*, this would require clay units with a compressive strength of at least 6600 psi, and Type S mortar.

In actuality, the strength design provisions of the 2008 MSJC *Code* permit the use of joint reinforcement only as prescriptive reinforcement, not to resist calculated actions. All references to "reinforcement" in Chap. 3 are to "bars," and do not include joint reinforcement. The following example ignores that prohibition, because its purpose is simply to show differences between strength design and allowable stress design for flexure.

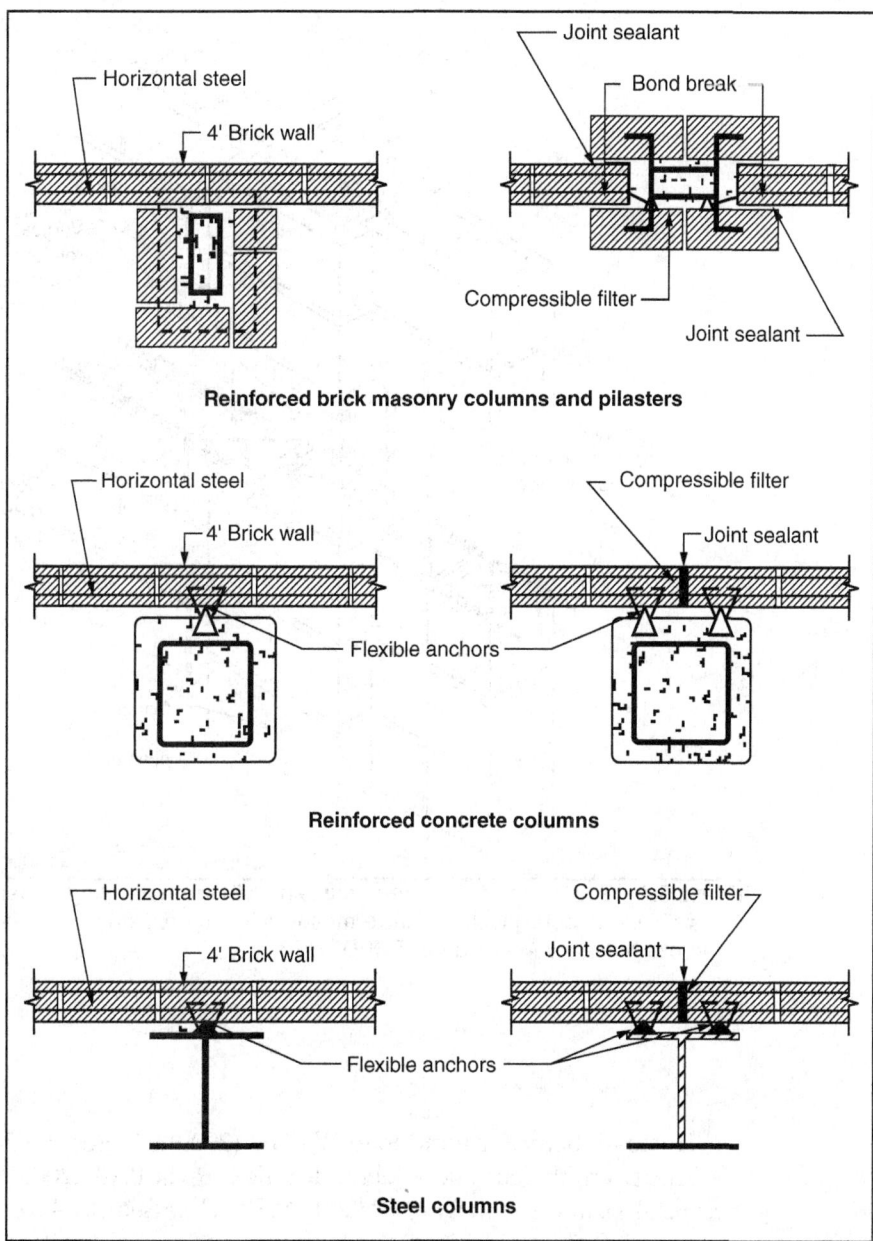

FIGURE 6.8 Examples of use of curtain walls of clay masonry. (*Source*: Figure 6 of Technical Notes on Brick Association 17L. © Brick Industry Association.)

194 Chapter Six

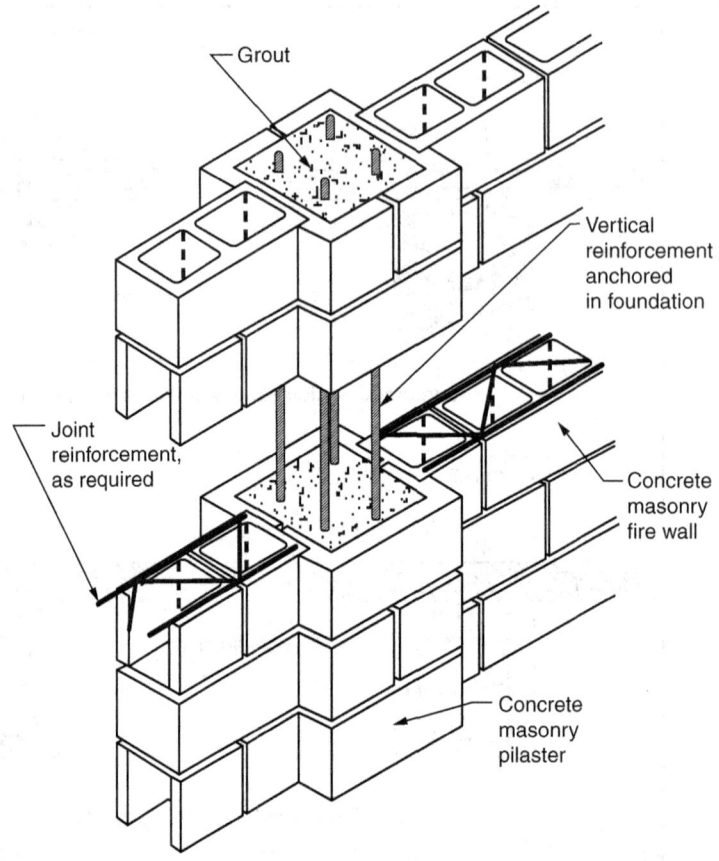

FIGURE 6.9 Examples of the use of curtain walls (freestanding or cantilevered fire wall with pilaster) with concrete masonry. (*Source*: Figure 1 of National Concrete Masonry Association TEK 05-08B.)

The load factor for wind load W is 1.6 (2009 IBC, Sec. 1605.2.1).

The strength-reduction factor for flexure is 0.90 (2008 MSJC *Code* Sec. 3.1.4.1), and for shear, 0.80 (2008 MSJC *Code* Sec. 3.1.4.3).

For a 1-ft strip,

$$M_u = \frac{q_u L^2}{8} = \frac{1.6 \cdot 20 \text{ lb/ft} \cdot (20 \text{ ft})^2}{8} \times 12 \text{ in./ft} = 19,200 \text{ lb-in.}$$

$$V_u = \frac{q_u L}{2} = \frac{1.6 \cdot 20 \text{ lb/ft} \cdot (20 \text{ ft})}{2} = 320 \text{ lb}$$

Flexural Design
Approximate the internal lever arm as 0.9 d.

$$d - d' = t - 2 \cdot \text{cover} - 2 \cdot \frac{d_b}{2} = 3.63 - 2 \cdot 0.63 - 0.25 = 2.13 \text{ in.}$$

$$M_n \approx A_s f_y (0.9\,d) = 0.049 \text{ in.}^2 \cdot \left(\frac{12 \text{ in.}}{2.67 \text{ in.}}\right) \cdot 60{,}000 \text{ lb/in.}^2 \cdot 0.9 \cdot 2.87 \text{ in.}$$

$$= 34{,}130 \text{ lb-in.}$$

$$M_u = 19{,}200 \text{ lb-in.} \leq \phi M = 0.9 \cdot 34{,}130 \text{ lb-in.} = 30{,}717 \text{ lb-in.}$$

Shear Design
From the 2008 MSJC *Code*, Sec. 3.3.4.1.2.1,

$$V_{nm} = \left[4.0 - 1.75\left(\frac{M_u}{V_u d_v}\right)\right] A_n \sqrt{f'_m} + 0.25 P_u$$

As $(M_u/V_u d_v)$ increases, V_{nm} decreases. Because $(M_u/V_u d_v)$ need not be taken greater than 1.0 (MSJC *Code* Sec. 3.3.4.1.2.1), the most conservative (lowest) value of V_{nm} is obtained with $(M_u/V_u d_v)$ equal to 1.0. Also, the factored design axial load, P_u, is zero:

$$V_{nm} = [4.0 - 1.75(1.0)] A_n \sqrt{f'_m}$$

$$V_{nm} = 2.25 A_n \sqrt{f'_m}$$

$$V_u = 320 \text{ lb} \leq \phi V_{nm} = 0.8 \cdot 2.25 \cdot 12 \text{ in.} \times 3.63 \text{ in.} \times \sqrt{2500} \text{ lb/in.}^2 = 3924 \text{ lb}$$

and the design is acceptable.

For this example, strength design is probably a little simpler than allowable-stress design. Note the above prohibition, in the strength-design provisions of the 2008 MSJC *Code*, from using joint reinforcement to resist calculated actions. The purpose of this example has been to show how strength calculations differ from allowable-stress calculations for design of curtain walls. Strength design of flexural members using deformed reinforcement, which is permitted by the 2008 MSJC *Code*, is presented in later sections.

6.2.6 Design of Anchors for Curtain Wall
The design of the curtain wall would have to finish with the design of anchors holding the ends of the curtain wall strips, to the columns (Fig. 6.10).

FIGURE 6.10 Ten anchors holding the ends of curtain wall strips to columns.

For example, if anchors are spaced at 12 in. vertically, the load per anchor is

$$\text{Anchor load} = \frac{qL}{2} = \frac{20 \text{ lb} \cdot 20 \text{ ft}}{2} = 200 \text{ lb}$$

6.3 Strength Design of Reinforced Bearing Walls

6.3.1 Introduction to Strength Design of Reinforced Bearing Walls

In this section, we shall study the behavior and design of reinforced masonry wall elements subjected to combinations of axial force and out-of-plane flexure. In the context of engineering mechanics, they are beam-columns. In the context of the MSJC *Code*, however, a "column" is an isolated masonry element rarely found in real masonry construction.

Masonry beam-columns, like those of reinforced concrete, are designed using moment-axial force interaction diagrams. Combinations of axial force and moment lying inside the diagram represent permitted designs; combinations lying outside, prohibited ones.

Unlike reinforced concrete, however, reinforced masonry beam-columns rarely take the form of isolated rectangular elements with four longitudinal bars and transverse ties. The most common form for a reinforced masonry beam-column is a wall, loaded out-of-plane by eccentric gravity load, alone or in combination with wind.

For example, Fig. 6.11 shows a portion of a wall, with a total effective width of 6*t* prescribed by Sec. 1.9.6.1 of the 2008 MSJC *Code*.

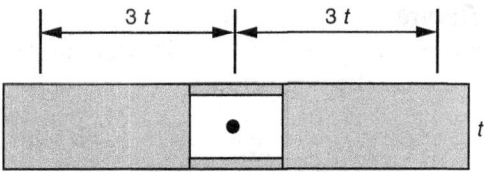

FIGURE 6.11 Effective width of a reinforced masonry bearing wall.

6.3.2 Background on Moment-Axial Force Interaction Diagrams by the Strength Approach

Using the strength approach, we seek to construct interaction diagrams that represent combinations of axial and flexural capacity. This can be done completely by hand, or with the help of a spreadsheet.

6.3.3 Background on Strength Interaction Diagrams by Hand

By hand, we can compute three points (pure compression, pure flexure, and the balance point). We then draw a straight line between pure compression and the balance point, and either a straight line or an appropriate curve between the balance point and pure flexure. This approach is commonly applied to reinforced concrete columns.

Pure Compression

$$P_0 = 0.80 \cdot 0.80 f'_m (A_c - A_{st}) + A_{st} f_y$$

As in ACI 318-08, the leading factor of 0.80 in effect imposes a minimum design eccentricity.

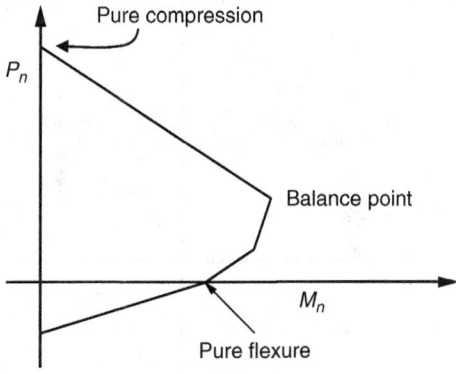

FIGURE 6.12 Idealized moment-axial force interaction diagram using strength design.

Pure Flexure

As before, the possible contribution of compressive reinforcement is small, and can be neglected.

$$\rho = \frac{A_s}{bd}$$

$$\omega = \rho\left(\frac{f_y}{f'_m}\right)$$

$$M_n = A_s f_y \left(d - \frac{\beta_1 c}{2}\right) = \omega b d^2 f'_m (1 - 0.63\omega)$$

Balance Point

First, locate the neutral axis

$$\frac{c}{d-c} = \frac{\varepsilon_{mu}}{\varepsilon_y}$$

$$c = d\left(\frac{\varepsilon_{mu}}{\varepsilon_{mu} + \varepsilon_y}\right)$$

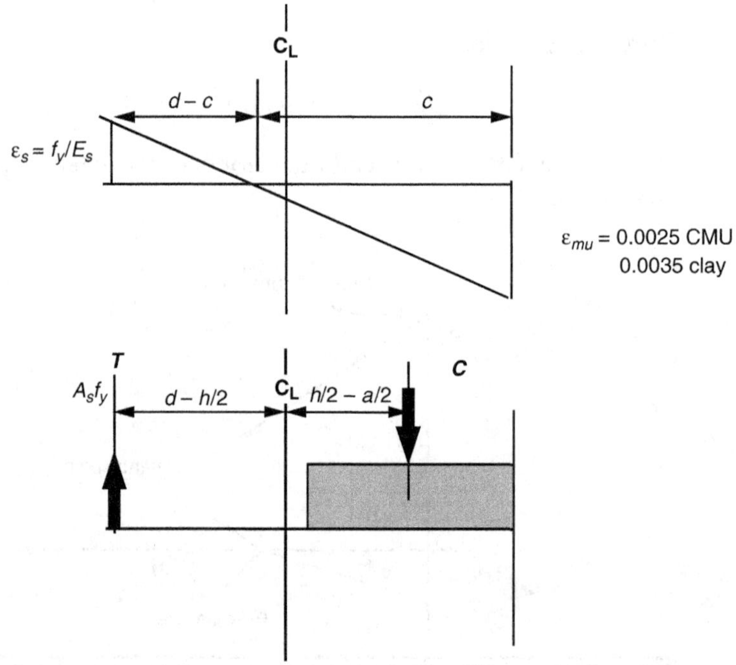

$\varepsilon_{mu} = 0.0025$ CMU
   0.0035 clay

FIGURE 6.13 Location of neutral axis under balanced conditions, strength design.

Next, calculate the corresponding tensile and compressive forces:

$$T = A_s f_y$$
$$C = 0.80 f'_m (\beta_1 c) b$$
$$P_n = C - T$$
$$M_n = T\left(d - \frac{h}{2}\right) + C\left(\frac{h}{2} - \frac{\beta_1 c}{2}\right)$$

6.3.4 Example of Moment-Axial Force Interaction Diagram by the Strength Approach (Hand Calculation)

Construct the moment-axial force interaction diagram by the strength approach for a nominal 8-in. CMU wall, fully grouted, with $f'_m = 1500$ lb/in.² and reinforcement consisting of #5 bars at 48 in., placed in the center of the wall. Compute the interaction diagram per foot of wall length.

For the case of a wall with reinforcement at mid-depth, the reference axis for moment is located at the plastic centroid (geometric centroid) of the cross-section, which is also at mid-depth. This leads to results which appear considerably different from what we are used to for a symmetrically reinforced column. For example, using the geometric centroid (the level of the reinforcement) as the reference axis, the contribution of the reinforcement to the moment is always zero. In what is apparently even stranger, the balanced-point axial force does not coincide with the maximum moment capacity. As a result, hand calculations are useful for some reinforced masonry beam-columns, but not all.

Pure Compression

Because the compressive reinforcement in the wall is not supported laterally, it is not counted in computed the capacity.

$$P_0 = 0.80 \times 0.80 f'_m (A_c - A_{st}) + A_{st} f_y$$

$P_n = 0.80 \times 0.80 \cdot 1500 \text{ lb/in.}^2 \cdot (7.63 \text{ in.} \cdot 48 \text{ in.} - 0.31 \text{ in.}^2) + 0.31 \text{ in.}^2 \cdot 0 \text{ lb/in.}^2$

$P_n = 351{,}293$ lb

Per foot of wall length, the design capacity will be the above value, divided by 4 (the length of the wall in feet), and multiplied by the strength reduction factor of 0.90:

$$\phi P_n = 79{,}041 \text{ lb}$$

Pure Flexure

As before, neglect the influence of compressive reinforcement:

$$d = \left(\frac{1}{2}\right) 7.63 \text{ in.} = 3.81 \text{ in.}$$

$$\rho = \frac{A_s}{bd} = \frac{0.31 \text{ in.}^2}{48 \text{ in.} \times 3.81 \text{ in.}} = 1.70 \cdot 10^{-3}$$

$$\omega = \rho\left(\frac{f_y}{f'_m}\right) = (1.70 \cdot 10^{-3}) \cdot \left(\frac{60{,}000}{1500}\right) = 0.0678$$

$$M_n = \omega b d^2 f'_m (1 - 0.63\,\omega)$$
$$M_n = 0.0678 \times 48 \text{ in.} \times 3.81^2 \text{ in.}^2 \times 1500 \text{ lb/in.}^2 \times (1 - 0.63 \times 0.0678)$$
$$M_n = 67{,}838 \text{ lb-in.}$$

Per foot of wall length, the design capacity is the above value, divided by 4 and multiplied by the strength reduction factor of 0.90:

$$\phi M_n = 15{,}264 \text{ lb-in.}$$

Balance Point

First, locate the neutral axis

$$c = d\left(\frac{\varepsilon_{mu}}{\varepsilon_{mu} + \varepsilon_y}\right)$$

$$c = d\left(\frac{0.0025}{0.0025 + 0.00207}\right)$$

$$c = 0.55\,d$$

$$c = 0.55 \times 3.81 \text{ in.} = 2.08 \text{ in.}$$

The value of β_1 is prescribed in Section 3.3.2 (g) of the 2008 MSJC *Code* as 0.80:

$$T = A_s f_y = 0.31 \text{ in.}^2 \cdot 60{,}000 \text{ lb/in.}^2 = 18{,}600 \text{ lb}$$
$$C = 0.80\,f'_m(\beta_1 c)b = 0.80\,f'_m(0.80 \times 0.55\,d)\,b$$
$$C = 0.80 \times 1500 \text{ lb/in.}^2 (0.80 \times 0.55 \times 3.81 \text{ in.}) \times 48 \text{ in.} = 96{,}561 \text{ lb}$$
$$P_n = C - T = 77{,}961 \text{ lb}$$

$$M_n = T\left(d - \frac{h}{2}\right) + C\left(\frac{h}{2} - \frac{\beta_1 c}{2}\right)$$

$$M_n = 18{,}600 \text{ lb} (3.81 - 3.81) \text{ in.} + 96{,}561 \text{ lb} \left(\frac{7.63}{2} - \frac{0.80 \times 0.55 \times 3.81}{2}\right) \text{in.}$$

$$M_n = 0 \text{ lb-in.} + 287{,}442 \text{ lb-in.} = 287{,}442 \text{ lb-in.}$$

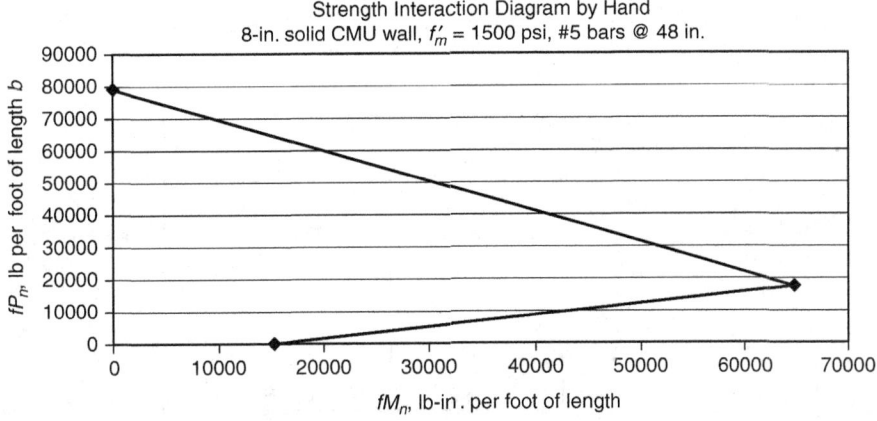

FIGURE 6.14 Moment-axial force interaction (strength basis), calculated by hand.

Per foot of wall length, the design capacities are the above values, divided by 4 and multiplied by the strength reduction factor of 0.90:

$$\phi P_n = 17{,}541 \text{ lb}$$
$$\phi M_n = 64{,}674 \text{ lb-in.}$$

Plot of Strength Interaction Diagram by Hand

The strength-basis moment-axial force interaction diagram calculated above is plotted in Fig. 6.14.

As we shall shortly see, the points that we have calculated are correct. The form of the diagram is misleading, however, because the balance point is actually not the point of maximum moment. It is incorrect to draw the diagram with a straight line from the balance point to the pure-compression point. The balance point becomes the point of maximum moment as the reinforcement is placed farther apart than about 70 percent of the thickness of the wall.

6.3.5 Strength Interaction Diagrams by Spreadsheet

To calculate strength interaction diagrams using a spreadsheet, we first calculate the position of the neutral axis corresponding to the balance point (Fig. 6.15).

$$\frac{c}{d-c} = \frac{\varepsilon_{mu}}{\varepsilon_y}$$

$$c = d\left(\frac{\varepsilon_{mu}}{\varepsilon_{mu} + \varepsilon_y}\right)$$

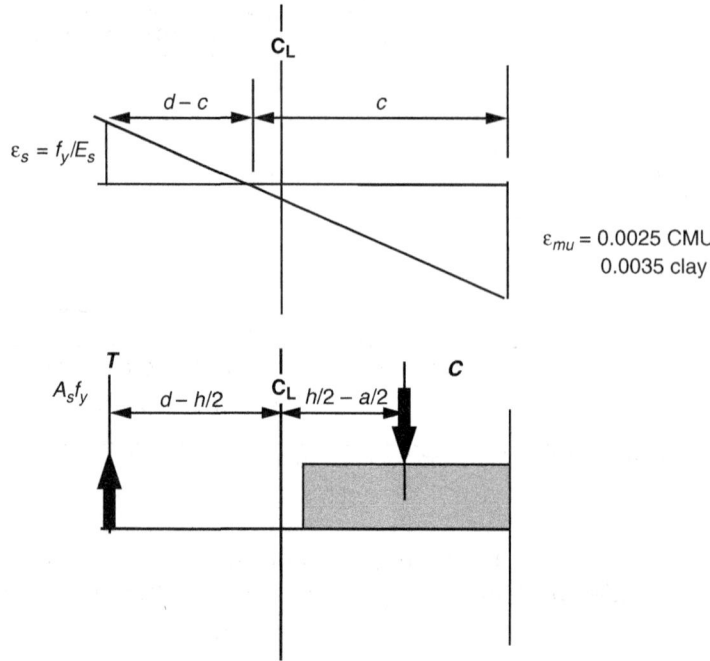

FIGURE 6.15 Position of neutral axis at balanced conditions, strength calculation of moment-axial force interaction diagram by spreadsheet.

For values of c less than that balanced value, the steel will yield before the masonry reaches its maximum useful strain. Combinations of axial force and moment corresponding to nominal capacity can then be calculated, as can the corresponding moment.

Given the position of the neutral axis, c, less than or equal to the balance-point value:

$$C = 0.80c\,(0.80\,f'_m)b$$

$$T = A_s f_y$$

$$P_n = C - T$$

$$M_n = T\left(d - \frac{h}{2}\right) + C\left(\frac{h}{2} - \frac{\beta_1 c}{2}\right)$$

Similarly, for values of c greater than the balance-point value, the steel will still be elastic when the masonry reaches its maximum useful strain. Compute the strain (and corresponding stress) in the steel by proportion, and find combinations of axial force and moment corresponding to each position of the neutral axis.

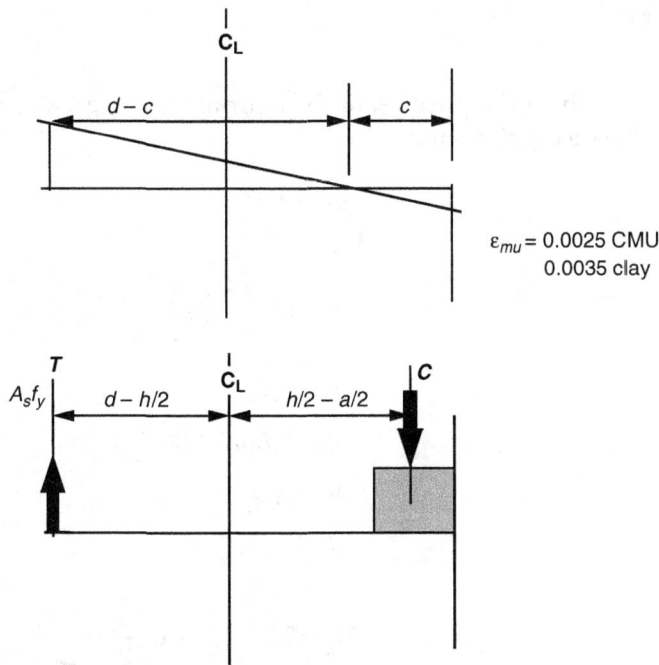

FIGURE 6.16 Position of the neutral axis for axial loads less than the balance-point axial load, strength design.

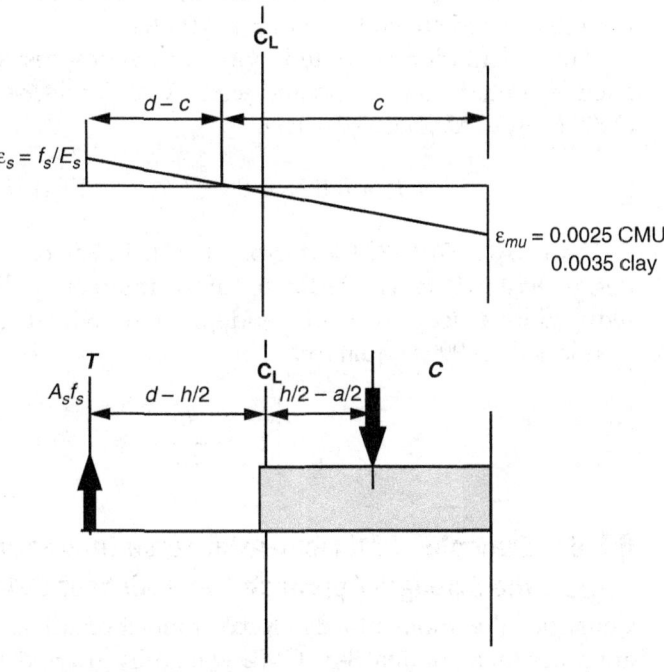

FIGURE 6.17 Position of the neutral axis for axial loads greater than the balance-point axial load, strength design.

203

Given the position of the neutral axis, c, greater than or equal to the balance-point value:

$$\frac{\varepsilon_s}{\varepsilon_{mu}} = \frac{d-c}{c}$$

$$\varepsilon_s = \varepsilon_{mu}\left(\frac{d-c}{c}\right)$$

$$f_s = E_s \varepsilon_s$$

$$C = 0.80c\,(0.80\,f'_m)b$$

$$T = A_s f_s$$

$$P_n = C - T$$

$$M_n = T\left(d - \frac{h}{2}\right) + C\left(\frac{h}{2} - \frac{\beta_1 c}{2}\right)$$

In the above expression, β_1 is the ratio between the depth of the compressive stress block and the distance from the neutral axis to the extreme compression fiber. It is also denoted as "a" in technical references dealing with reinforced concrete, so that $a = 0.80\,c$.

This calculation is limited by the pure-compression resistance, calculated as noted above, and reduced by a slenderness-dependent factor (2008 MSJC *Code*, Sec. 3.2.4.1.1):

$$P_0 = 0.80 \times 0.80\, f'_m (A_c - A_{st}) + A_{st} f_y$$

As in ACI 318-08, the factor of 0.80 in effect imposes a minimum design eccentricity. The ordinates of the interaction diagram must also be reduced by a slenderness-dependent factor, which for values of slenderness less than 99 is equal to

$$\left[1 - \left(\frac{h}{140\,r}\right)^2\right]$$

6.3.6 Example of Moment-Axial Force Interaction Diagram by the Strength Approach (Spreadsheet Calculation)

Construct the moment-axial force interaction diagram by the strength approach for a nominal 8-in. CMU wall, fully grouted, with $f'_m = 1500\ \text{lb/in.}^2$ and reinforcement consisting of #5 bars at 48 in., placed in the center of the wall.

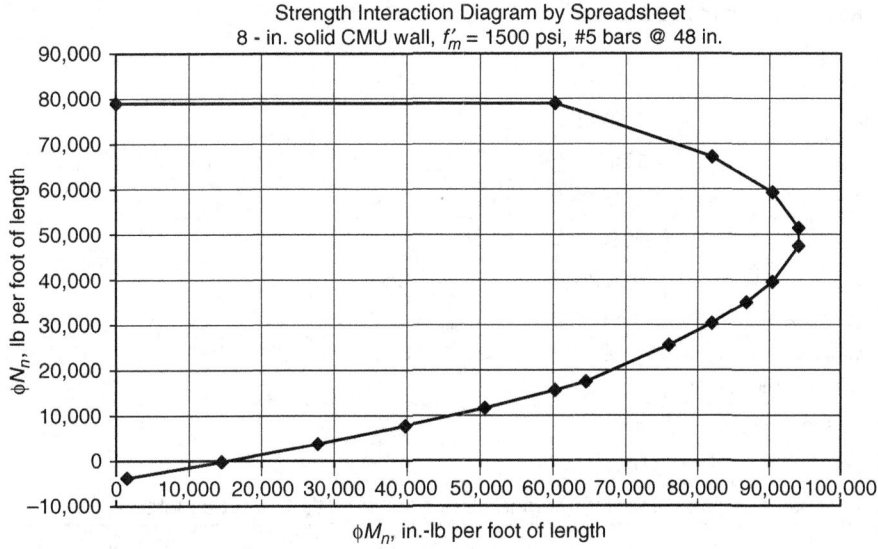

FIGURE 6.18 Moment-axial force interaction diagram (strength approach), spreadsheet calculation.

The effective width of the wall is $6t$, or 48 in. The spreadsheet and corresponding interaction diagram are shown in Fig. 6.18. As noted above in the example of Sec. 6.3.3, the results are interesting. Because the reinforcement is located at the geometric centroid of the section, the balance-point axial load (about 100,000 lb) does not correspond to the maximum moment capacity.

Plot of Strength Interaction Diagram by Spreadsheet

The moment-axial force interaction diagram (strength basis) as plotted by spreadsheet is shown in Fig. 6.18.

Relevant cells from the spreadsheet are reproduced in Table 6.2.

6.3.7 Example of Strength Design of Masonry Walls Loaded Out-of-Plane

Once we have developed the moment-axial force interaction diagram by the strength approach, the actual design simply consists of verifying that the combination of factored design axial force and moment lies within the diagram of nominal axial and flexural capacity, reduced by strength-reduction factors.

Consider the bearing wall designed previously as unreinforced, shown in Fig. 6.19. It has an eccentric axial load plus out-of-plane wind load of 25 lb/ft².

Example of spreadsheet for calculating strength moment-axial force interaction diagram for solid masonry wall

Reinforcement at mid-depth						
Specified thickness	7.625					
e_{mu}	0.0025					
f'_m	1500					
f_y	60000					
E_s	29000000					
d	3.8125					
(c/d) balanced	0.54717					
Tensile reinforcement area	0.31					
Effective width	48					
phi	0.9					
Because compression reinforcement is not supported, it is not counted						
	C/d	c	C_{mas}	f_s	Moment	Axial force
Points controlled by steel	0.01	0.038125	1757	−60000	1501	−3790
	0.1	0.38125	17568	−60000	14467	−232
	0.2	0.7625	35136	−60000	27729	3721
	0.3	1.14375	52704	−60000	39785	7673
	0.4	1.525	70272	−60000	50635	11626
	0.5	1.90625	87840	−60000	60280	15579
	0.54717	2.086085	96127	−60000	64411	17444
Points controlled by masonry	0.54717	2.086085	96127	−60000	64411	17444
	0.7	2.66875	122976	−31071	75953	25502
	0.8	3.05	140544	−18125	81981	30358
	0.9	3.43125	158112	−8056	86803	35013
	1	3.8125	175680	0	90420	39528
	1.2	4.575	210816	0	94037	47434
	1.3	4.95625	228384	0	94037	51386
	1.5	5.71875	263520	0	90420	59292
	1.7	6.48125	298656	0	81981	67198
	2	7.625	351360	0	60280	79056
Pure axial load					0	78989

TABLE 6.2 Spreadsheet for Computing Moment-Axial Force Interaction Diagram (Strength Approach)

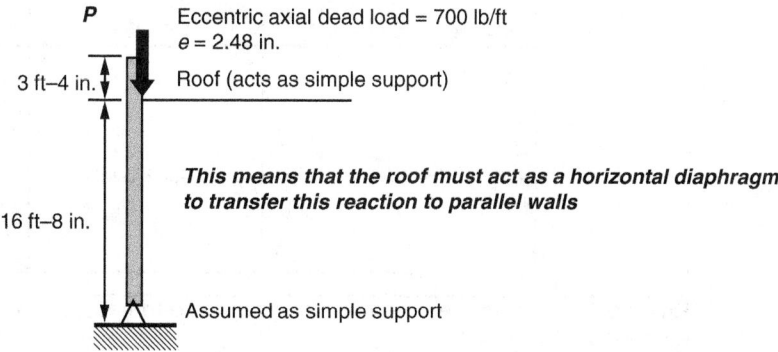

FIGURE 6.19 Reinforced masonry wall loaded by eccentric gravity axial load plus out-of-plane wind load.

At each horizontal plane through the wall, the following condition must be met:

- Combinations of factored axial load and moment must lie within the moment-axial force interaction diagram, reduced by strength-reduction factors.

Because flexural capacity increases with increasing axial load, the critical loading combination is probably $0.9D + 1.6W$.

From our previous experience, we know that the critical point on the wall is at the midspan of the lower portion.

Due to wind only, the unfactored moment at the base of the parapet (roof level) is

$$M = \frac{qL_{parapet}^2}{2} = \frac{25 \text{ lb/ft} \times 3.33^2 \text{ ft}^2}{2} \times 12 \text{ in./ft} = 1663 \text{ lb-in.}$$

The maximum moment is close to that occurring at mid-height. The moment from wind load is the superposition of one-half moment at the upper support due to wind load on the parapet only, plus the midspan moment in a simply supported beam with that same wind load:

$$M_{midspan} = -\frac{1663}{2} + \frac{qL^2}{8} = -\frac{1663}{2} + \frac{25 \text{ lb/ft} \cdot 16.67^2 \text{ ft}^2}{8} \times 12 \text{ in./ft} = 9589 \text{ lb-in.}$$

The unfactored moment due to eccentric axial load is

$$M_{gravity} = Pe = 1050 \text{ lb} \times 2.48 \text{ in.} = 2604 \text{ lb-in.}$$

Unfactored moment diagrams due to eccentric axial load and wind are as shown in Fig. 6.20.

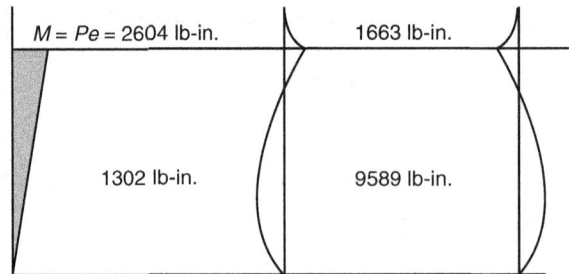

Figure 6.20 Unfactored moment diagrams due to eccentric axial load plus wind load.

Check the adequacy of the wall with 8-in. nominal units, a specified compressive strength, f_m', of 1500 lb/in.², and #5 bars spaced at 48 in. All design actions are calculated per foot of width of the wall.

At the mid-height of the wall, the axial force due to 0.9D is

$$P_u = 0.9(700 \text{ lb}) + 0.9(3.33 \text{ ft} + 8.33 \text{ ft}) \times 48 \text{ lb/ft} = 1134 \text{ lb}$$

At the mid-height of the wall, the factored design moment, M_u, is given by

$$M_u = P_u \frac{e}{2} + M_{u\,\text{wind}} = \left(\frac{1}{2}\right) 0.9 \times 700 \text{ lb} \times 2.48 \text{ in.} + 1.6 \times 9589 \text{ lb-in.} = 16,124 \text{ lb-in.}$$

In each foot of wall, the design actions are $P_u = 1134$ lb and $M_u = 16,124$ lb-in. That combination lies within the interaction diagram of design capacities (Fig. 6.18), and the design is satisfactory.

Because this out-of-plane wall is checked for magnified moments in accordance Sec. 3.3.5.2 of the 2008 MSJC *Code*, the slenderness-dependant reduction factor is not applied to the moment-axial force interaction diagram.

The strength provisions of the 2008 MSJC *Code* also require a check of the possible effects of secondary moments for reinforced walls loaded out of plane (MSJC *Code* Section 3.3.5.3).

In accordance with those sections, MSJC *Code* Eq. 3-25 is used to calculate the maximum moment, including possible secondary moments. That maximum moment is then compared with the interaction diagram. MSJC *Code* Eq. 3-25 is based on a member simply supported at top and bottom, which is the case here:

$$M_u = \frac{w_u h^2}{8} + P_{uf}\left(\frac{e_u}{2}\right) + P_u \delta_u$$

As calculated above, for each feet of wall length, the first two terms in this equation total 16,124 lb in., and P_u equals 1134 lb. In accordance with MSJC *Code* Sec. 3.3.5.3, δ_u is to be calculated using *Code* Eqs. 3-31 and 3-32, replacing M_{ser} with M_u.

Because the cracking moment used in those equations is calculated without strength-reduction factors, it might exceed the factored design moment. Nevertheless, it is believed prudent to assume that reinforced masonry is cracked at bed joints.

For this problem, the cracked moment of inertia I_{cr} for use in MSJC *Code* Sec. 3.3.5.3 is approximately and conservatively be taken as 40 percent of the gross moment of inertia. This relationship between cracked and gross inertia is commonly used for lightly reinforced concrete or masonry sections. Its use here is consistent with the assumption that the entire wall is initially cracked on the bed joints before any load is applied. Section properties are per foot of plan length.

$$\delta_u = \frac{5M_u h^2}{48 E_m I_{cr}}$$

$$I_{cr} = 0.40\, I_g = 0.40\left(\frac{bt^3}{12}\right) = 0.40\left[\frac{12 \text{ in.} \times (7.625 \text{ in.})^3}{12}\right]$$

$$= 0.40 \times 443.3 \text{ in.}^4 = 177.3 \text{ in.}^4$$

$$\delta_{u1} = \frac{5 \times 16{,}124 \text{ lb-in.} \times (16.67 \text{ ft} \times 12 \text{ in./ft})^2}{48\,(900 \times 1500 \text{ lb/in.}^2)(177.3 \text{ in.}^4)} = 0.281 \text{ in.}$$

$M_{u2} = 16{,}124$ lb-in. $+ 1134(0.281)$ lb-in. $= 16{,}442$ lb-in.

Check convergence:

$$\delta_{u2} = \frac{5\,(16{,}442 \text{ lb-in.})(16.67 \text{ ft} \times 12 \text{ in./ft})^2}{48\,(900 \times 1500 \text{ lb/in.}^2)(177.3 \text{ in.}^4)} = 0.286 \text{ in.}$$

$M_{u3} = 16{,}124$ lb-in. $+ 1134(0.286)$ lb-in. $= 16{,}449$ lb-in.

Because the moment is changing by less than 0.04 percent, it can be assumed to have converged.

The combination of factored axial force and factored moment (including secondary moments) remains within the moment-axial force interaction diagram, and the design is still satisfactory. Finally, the strength provisions of the 2008 MSJC *Code* also require a check of out-of-plane deflections for reinforced walls loaded out of plane (MSJC *Code* Sec. 3.3.5.4).

In accordance with those sections, MSJC *Code* Eqs. 3-31 or 3-32 is used to calculate the mid-height deflection. Those equations are based on a member simply supported at top and bottom, which is the case here. Conservatively assuming the section to be cracked, the out-of-plane deflection is given by the converged δ_{u2} from above (0.286 in.). That deflection is less than $0.007\,h$ (equal to 0.007 times 16.67 ft, or 1.40 in.). The out-of-plane deflection requirement is satisfied.

6.3.8 Minimum and Maximum Reinforcement Ratios for Out-of-Plane Flexural Design of Masonry Walls by the Strength Approach

The strength design provisions of the 2008 MSJC *Code* include requirements for minimum and maximum flexural reinforcement. In this section, the implications of those requirements for the out-of-plane flexural design of masonry walls are addressed.

Minimum Flexural Reinforcement by 2008 MSJC *Code*

The 2008 MSJC *Code* has no requirements for minimum flexural reinforcement for out-of-plane design of masonry walls.

Maximum Flexural Reinforcement by 2008 MSJC *Code*

The 2008 MSJC *Code* has a maximum reinforcement requirement (Sec. 3.3.3.5) that is intended to ensure ductile behavior over a range of axial loads. As compressive axial load increases, the maximum permissible reinforcement percentage decreases. For compressive axial loads above a critical value, the maximum permissible reinforcement percentage drops to zero, and design is impossible unless the cross-sectional area of the element is increased.

For walls subjected to out-of-plane forces, for columns, and for beams, the provisions of the 2008 MSJC *Code* set the maximum permissible reinforcement based on a critical strain condition in which the masonry is at its maximum useful strain, and the extreme tension reinforcement is set at 1.5 times the yield strain.

The critical strain condition for walls with a single layer of concentric reinforcement and loaded out-of-plane is shown in Fig. 6.21 along with the corresponding stress state. The parameters for the equivalent rectangular stress block are the same as those used for conventional flexural design. The height of the equivalent rectangular stress block is $0.80\,f_m'$, and the depth is $0.80\,c$. The tensile reinforcement is assumed to be at f_y.

Locate the neutral axis using the critical strain condition

$$\frac{\varepsilon_{mu}}{1.5\,\varepsilon_y} = \frac{c}{d-c} \qquad c = d\left(\frac{\varepsilon_{mu}}{1.5\,\varepsilon_y + \varepsilon_{mu}}\right)$$

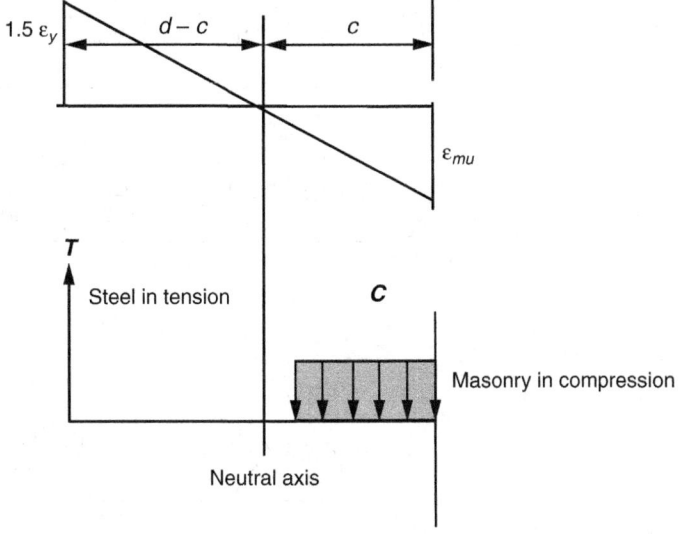

FIGURE 6.21 Critical strain condition for a masonry wall loaded out of plane.

Compute the tensile and compressive forces acting on the section, assuming concentric reinforcement with a percentage of reinforcement $\rho = \dfrac{A_s}{bd}$, where $d = \dfrac{t}{2}$.

The compressive force in the masonry is given by

$$C_{masonry} = 0.80\, f'_m\, 0.80\, cb$$

The tensile force in the reinforcement is given by

$$T_{steel} = \rho b d f_y$$

Equilibrium of axial forces requires

$$N_n = C - T$$

$$\frac{N_u}{\phi} = C - T$$

$$\frac{N_u}{\phi} = 0.80\, f'_m\, 0.80\, cb - \rho db f_y$$

$$\frac{N_u}{\phi} = 0.80\, f'_m\, 0.80\, d\left(\frac{\varepsilon_{mu}}{1.5\,\varepsilon_y + \varepsilon_{mu}}\right)b - \rho db f_y$$

$$\rho = \frac{0.80 f'_m \, 0.80 d \left(\dfrac{\varepsilon_{mu}}{1.5\varepsilon_y + \varepsilon_{mu}} \right) b - \dfrac{N_u}{\phi}}{bd f_y}$$

$$\rho = \frac{0.80 f'_m \, 0.80 \left(\dfrac{\varepsilon_{mu}}{1.5\varepsilon_y + \varepsilon_{mu}} \right) - \dfrac{N_u}{bd\phi}}{f_y}$$

So

$$\rho_{max} = \frac{0.64 f'_m \left(\dfrac{\varepsilon_{mu}}{1.5\varepsilon_y + \varepsilon_{mu}} \right) - \dfrac{N_u}{bd\phi}}{f_y}$$

6.4 Strength Design of Reinforced Shear Walls

6.4.1 Introduction to Strength Design of Reinforced Shear Walls

In this section, we shall study the behavior and design of reinforced masonry shear walls. The discussion follows the same approach used previously for unreinforced masonry shear walls.

6.4.2 Design Steps for Strength Design of Reinforced Shear Walls

Reinforced masonry shear walls must be designed for the effects of:

1. Gravity loads from self-weight, plus gravity loads from overlying roof or floor levels.
2. Moments and shears from in-plane shear loads.

Actions are shown in Fig. 6.22.

Flexural capacity of reinforced shear walls using strength procedures is calculated using moment-axial force interaction diagrams as discussed in the section on masonry walls loaded out-of-plane. In contrast to the elements addressed in that section, a shear wall is subjected to flexure in its own plane rather than out-of-plane. It therefore usually has multiple layers of flexural reinforcement. Computation of moment-axial force interaction diagrams for shear walls is much easier using a spreadsheet.

Strength Design of Reinforced Masonry Elements

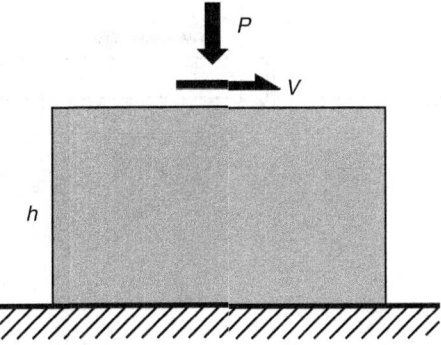

FIGURE 6.22 Design actions for reinforced masonry shear walls.

From the 2008 MSJC *Code*, Sec. 3.3.4.1.2, nominal shear strength is the summation of shear strength from masonry and shear strength from shear reinforcement:

$$V_n = V_{nm} + V_{ns}$$

From the 2008 MSJC *Code*, Sec. 3.3.4.1.2.1,

$$V_{nm} = \left[4.0 - 1.75\left(\frac{M_u}{V_u d_v}\right)\right] A_n \sqrt{f'_m} + 0.25P$$

As $(M_u/V_u d_v)$ increases, V_{nm} decreases. Because $(M_u/V_u d_v)$ need not be taken greater than 1.0 (2008 MSJC *Code*, Sec. 3.3.4.1.2.1), the most conservative (lowest) value of V_{nm} is obtained with $(M_u/V_u d_v)$ equal to 1.0.

Just as in reinforced concrete design, this model assumes that shear is resisted by reinforcement crossing a hypothetical failure surface oriented at 45°, as shown in Fig. 6.24.

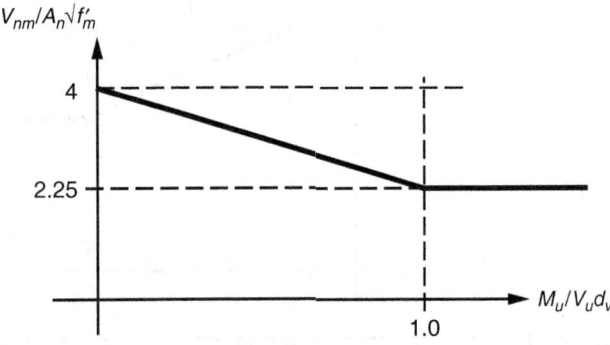

FIGURE 6.23 V_{nm} as a function of $(M_u/V_u d_v)$.

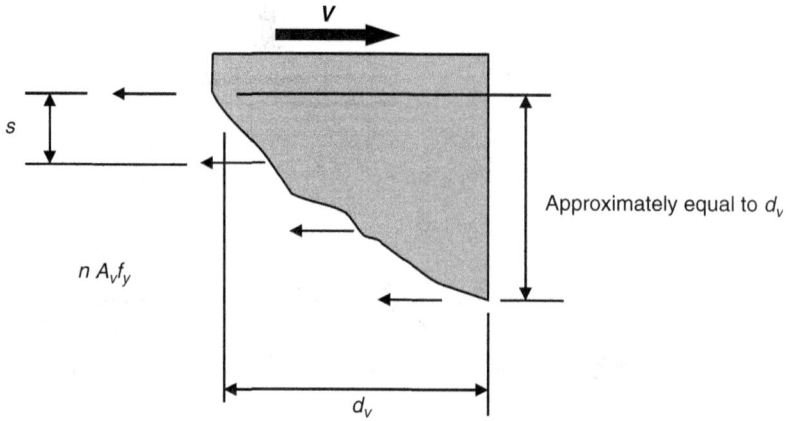

FIGURE 6.24 Idealized model used in evaluating the resistance due to shear reinforcement.

The nominal resistance from reinforcement is taken as the area associated with each set of shear reinforcement, multiplied by the number of sets of shear reinforcement crossing the hypothetical failure surface. Because the hypothetical failure surface is assumed to be inclined at 45°, its projection along the length of the member is approximately equal to d_v, and number of sets of shear reinforcement crossing the hypothetical failure surface can be approximated by (d_v/s):

$$V_{ns} = A_v f_y n$$

$$V_{ns} = A_v f_y \left(\frac{d_v}{s}\right)$$

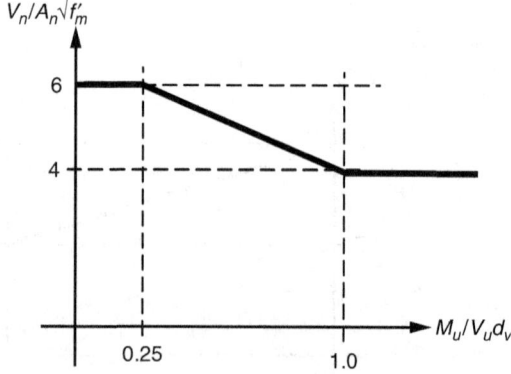

FIGURE 6.25 Maximum permitted nominal shear capacity as a function of $(M_u/V_u d_v)$.

The actual failure surface may be inclined at a larger angle with respect to the axis of the wall, however. Also, all reinforcement crossing the failure surface may not yield. For both these reasons, the assumed resistance is decreased by an efficiency factor of 0.5. From the 2008 MSJC Code, Sec. 3.3.4.1.2.2,

$$V_{ns} = 0.5 \left(\frac{A_v}{s}\right) f_y d_v$$

Finally, because shear resistance really comes from a truss mechanism in which horizontal reinforcement is in tension, and diagonal struts in the masonry are in compression, crushing of the diagonal compressive struts is controlled by limiting the total shear resistance V_n, regardless of the amount of shear reinforcement.

For $(M_u/V_u d_v) < 0.25$,

$$V_n = 6 A_n \sqrt{f'_m}$$

and for $(M_u/V_u d_v) > 1.00$,

$$V_n = 4 A_n \sqrt{f'_m}$$

Interpolation is permitted between these limits.

If these upper limits on V_n are not satisfied, the cross-sectional area of the section must be increased.

6.4.3 Example of Strength Design of Reinforced Clay Masonry Shear Wall

Consider the masonry shear wall shown in Fig. 6.26.

Design the wall. Unfactored in-plane lateral loads at each floor level are due to earthquake, and are shown in Fig. 6.27, along with the corresponding shear and moment diagrams.

Assume an 8-in. nominal clay masonry wall, grouted solid, with Type S PCL mortar. The total plan length of the wall is 24 ft (288 in.), and its thickness is 7.5 in. Assume an effective depth d of 285 in.

	Clay masonry
Unit Strength	6600
Mortar	Type S
f'_m or f_g (psi)	2500
Reinforcement = Grade 60; $E_s = 29 \times 10^6$ psi	

Chapter Six

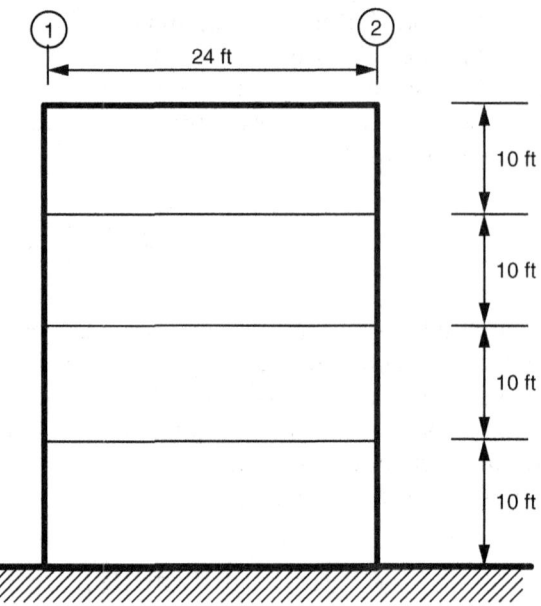

FIGURE 6.26 Reinforced masonry shear wall to be designed.

Unfactored axial loads on the wall are given in the table below.

Level (Top of wall)	DL (kips)	LL (kips)
4	90	15
3	180	35
2	270	55
1	360	75

Use 2009 IBC SD load Combination 7: $0.9D + 1.0E$

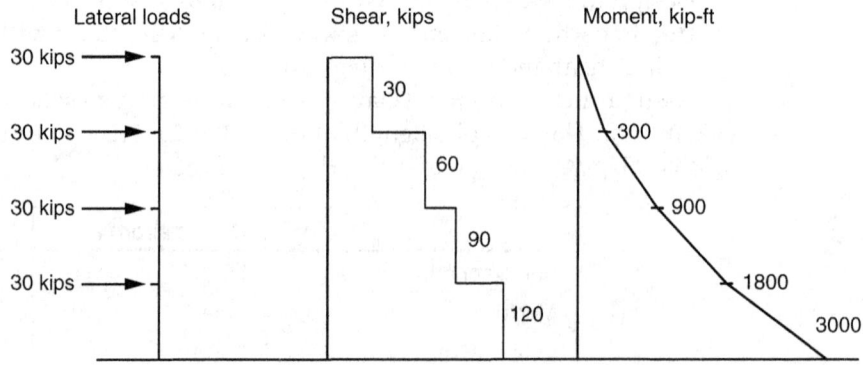

FIGURE 6.27 Unfactored in-plane lateral loads, shear and moment diagrams for reinforced masonry shear wall.

Check shear for assumed wall thickness. By Sec. 3.3.4.1.2 of the 2008 MSJC *Code*,

$$V_n = V_{nm} + V_{ns}$$

$$M_u = 3000 \times 12 \times 1000 \text{ in.-lb} = 36.0 \times 10^6 \text{ in.-lb}$$

$$V_u d_v = 120,000 \text{ lb} \qquad d_v = 285 \text{ in.}$$

$$M_u/V_u d_v = \frac{36 \times 10^6 \text{ in.-lb}}{120,000 \text{ lb} (285 \text{ in.})} = 1.05$$

$$V_{nm} = \left[4.0 - 1.75 \left(\frac{M_u}{V_u d_v}\right)\right] A_n \sqrt{f'_m} + 0.25 P_u$$

$$V_{nm} = [4.0 - 1.75(1.0)] 7.5 \text{ in.} \times 285 \text{ in.} (\sqrt{2500} \text{ psi}) + 0.25(0.9 \times 360,000 \text{ lb})$$

$$V_{nm} = 240.4 \text{ kips} + 81.0 \text{ kips} = 321.4 \text{ kips}$$

$$\phi V_n > V_u \qquad \phi = 0.80 \qquad V_n = V_{nm} = 321.4 \text{ kips}$$

$$0.80(321.4 \text{ kips}) = 257.2 \text{ kips} \geq V_u = 120 \text{ kips}$$

Shear design is satisfactory so far, even without shear reinforcement. *Code* Sec. 1.17.3.2.6.1 will be checked later.

Now check flexural capacity using a spreadsheet-generated moment-axial force interaction diagram. Try #5 bars @ 4 ft. Neglecting slenderness effects, the diagram is shown in Fig. 6.28.

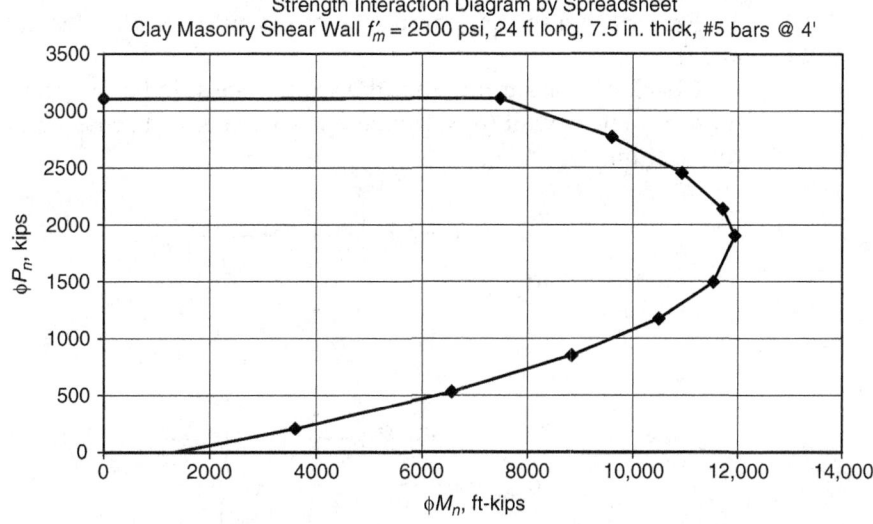

FIGURE 6.28 Moment-axial force interaction (strength basis) for reinforced shear wall, neglecting slenderness effects.

At a factored axial load of 0.9 D, or 0.9×360 kips = 324 kips, the design flexural capacity of this wall is about 4000 ft-kips, and the design is satisfactory for flexure.

We have designed the wall for the calculated design shear, which is normally sufficient. Now suppose that the wall is a special reinforced masonry shear wall (required in areas of high seismic risk), so that the capacity design requirements of *Code* Sec. 1.17.3.2.6.1.1 apply. First try to meet the capacity design provisions of that section. At an axial load of 324 kips, the nominal flexural capacity of this wall is the design flexural capacity of 4000 ft-kips, divided by the strength reduction factor of 0.9, or 4444 ft-kips. The ratio of this nominal flexural capacity to the factored design moment is 4444 divided by 3000, or 1.48. Including the additional factor of 1.25, that gives a ratio of 1.85.

$$\phi V_n \geq 1.85 V_u$$

$$V_n \geq \frac{1.85}{\phi} V_u = \frac{1.85}{0.8} V_u = 2.31 V_u = 2.31 \times 120 = 277.5 \text{ kips}$$

The wall is satisfactory without shear reinforcement. Prescriptive seismic reinforcement for seismic design category D will probably require #5 bars horizontally @ 24 in.

$$V_n = V_{nm} + V_{ns} = 321.4 \text{ kips} + \left(\frac{1}{2}\right) A_v f_y \frac{d}{s} = 321.4 + \left(\frac{1}{2}\right) 0.31 \text{ in.}^2 \times 60,000 \text{ psi} \times \left(\frac{285 \text{ in.}}{24 \text{ in.}}\right)$$

$$V_n = 321.4 \text{ kips} + 110.4 \text{ kips} = 431.8 \text{ kips}$$

Prescriptive seismic reinforcement is sufficient for shear. Use #5 bars at 2 ft.

Check ρ_{max}, assuming that the wall is classified as a special reinforced masonry shear wall ($\alpha = 4$). See derivation and discussion at the end of this section.

$$\rho_{max} = \frac{0.64 f'_m \left(\frac{\varepsilon_{mu}}{\alpha \varepsilon_y + \varepsilon_{mu}}\right) - \frac{N_u}{bd\phi}}{f_y \left(\frac{\alpha \varepsilon_y - \varepsilon_{mu}}{\alpha \varepsilon_y + \varepsilon_{mu}}\right)}$$

$$\rho_{max} = \frac{0.64 f'_m \left(\frac{\varepsilon_{mu}}{4\varepsilon_y + \varepsilon_{mu}}\right) - \frac{N_u}{bd\phi}}{f_y \left(\frac{4\varepsilon_y - \varepsilon_{mu}}{4\varepsilon_y + \varepsilon_{mu}}\right)}$$

In accordance with MSJC *Code* Section 3.3.3.5.1(d), the governing axial load combination is $D + 0.75\,L + 0.525\,Q_E$, and the axial load is $(360{,}000 + 0.75 \times 75{,}000\text{ lb})$, or 416,250 lb.

$$\rho_{max} = \frac{0.64\,f'_m \left(\dfrac{\varepsilon_{mu}}{4\varepsilon_y + \varepsilon_{mu}} \right) - \dfrac{N_u}{bd\phi}}{f_y \left(\dfrac{4\varepsilon_y - \varepsilon_{mu}}{4\varepsilon_y + \varepsilon_{mu}} \right)}$$

$$\rho_{max} = \frac{0.64\,(2500\text{ psi})\left[\dfrac{0.0035}{4(0.00207) + 0.0035} \right] - \dfrac{416{,}250\text{ lb}}{(7.50\text{ in.})(285\text{ in.})(0.9)}}{(60{,}000\text{ psi}) \left[\dfrac{4(0.00207) - 0.0035}{4(0.00207) + 0.0035} \right]}$$

$\rho_{max} = 0.01064$

Check maximum area of flexural reinforcement per 48 in. of wall length

$$A_{s\,max} = \rho_{max}\,b \times 48\text{ in.} = 0.01064\,(7.50\text{ in.})\,48\text{ in.} = 3.83\text{ in.}^2$$

We have 0.31 in.² every 48 in., and the design is satisfactory.
Summary: Use #5 @ 4 ft vertically, #5 @ 2 ft horizontally.

6.4.4 Minimum and Maximum Reinforcement Ratios for Flexural Design of Masonry Shear Walls by the Strength Approach

Minimum Flexural Reinforcement by 2008 MSJC *Code*
The 2008 MSJC *Code* has no global requirements for minimum flexural reinforcement for shear walls.

Maximum Flexural Reinforcement by 2008 MSJC *Code*
The 2008 MSJC *Code* has a maximum reinforcement requirement (Sec. 3.3.3.5) that is intended to ensure ductile behavior over a range of axial loads. As compressive axial load increases, the maximum permissible reinforcement percentage decreases. For compressive axial loads above a critical value, the maximum permissible reinforcement percentage drops to zero, and design is impossible unless the cross-sectional area of the element is increased.

For walls subjected to in-plane forces, for columns, and for beams, the provisions of the 2008 MSJC *Code* set the maximum permissible reinforcement based on a critical strain condition in which the masonry is at its maximum useful strain, and the extreme tension reinforcement is set at a multiple of the yield strain, where the multiple depends on the expected curvature ductility demand on the wall. For "special" reinforced masonry shear walls, the multiple is 4; for "intermediate" walls, it is 3. For walls not required to undergo inelastic deformations, no upper limit is imposed.

The critical strain condition for walls loaded in-plane, and for columns and beams, is shown in Fig. 6.29, along with the corresponding stress state. The multiple is termed "α." The parameters for the equivalent rectangular stress block are the same as those used for conventional flexural design. The height of the stress block is $0.80 f'_m$, and the depth is $0.80c$. The stress in yielded tensile reinforcement is assumed to be f_y. Compression reinforcement is included in the calculation, based on the assumption that protecting the compression toe will permit the masonry there to provide lateral support to the compression reinforcement. This assumption, while perhaps reasonable, is not consistent with that used for calculation of moment-axial force interaction diagrams.

Locate the neutral axis using the critical strain condition:

$$\frac{\varepsilon_{mu}}{\alpha \varepsilon_y} = \frac{c}{d-c}$$

$$c = d \left(\frac{\varepsilon_{mu}}{\alpha \varepsilon_y + \varepsilon_{mu}} \right)$$

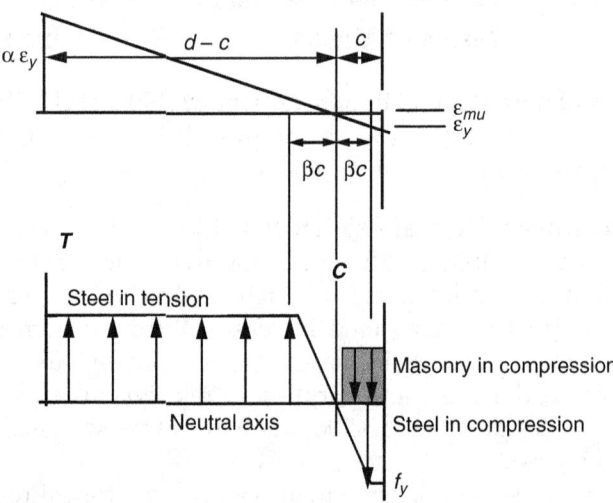

Figure 6.29 Critical strain condition for strength design of masonry walls loaded in-plane, and for columns and beams.

Strength Design of Reinforced Masonry Elements

Compute the tensile and compressive forces acting on the section, assuming uniformly distributed flexural reinforcement, with a percentage of reinforcement $\rho = A_s/bd$. On each side of the neutral axis, the distance over which the reinforcement, in the elastic range, is βc, where β is given by proportion as $\beta = \varepsilon_y/\varepsilon_{mu}$.

The compressive force in the masonry is given by

$$C_{masonry} = 0.80\, f'_m\; 0.80\, cb$$

The compressive force in the reinforcement is given by

$$C_{steel} = \rho\beta cb f_y \left(\frac{1}{2}\right) + \rho(1-\beta)\, cb f_y$$

$$C_{steel} = \rho\left(\frac{\varepsilon_y}{\varepsilon_{mu}}\right) cb f_y \left(\frac{1}{2}\right) + \rho\left[1 - \left(\frac{\varepsilon_y}{\varepsilon_{mu}}\right)\right] cb f_y$$

The tensile force in the reinforcement is given by

$$T_{steel} = \rho\beta cb f_y \left(\frac{1}{2}\right) + \rho(d - c - \beta c)\, b f_y$$

$$T_{steel} = \rho\left(\frac{\varepsilon_y}{\varepsilon_{mu}}\right) cb f_y \left(\frac{1}{2}\right) + \rho\left[d - c - \left(\frac{\varepsilon_y}{\varepsilon_{mu}}\right) c\right] b f_y$$

Equilibrium of axial forces requires

$$N_n = C - T$$

$$\frac{N_u}{\phi} = C - T$$

$$\frac{N_u}{\phi} = 0.80\, f'_m\; 0.80\, cb$$

$$+ \rho\left(\frac{\varepsilon_y}{\varepsilon_{mu}}\right) cb f_y \left(\frac{1}{2}\right) + \rho\left[1 - \left(\frac{\varepsilon_y}{\varepsilon_{mu}}\right)\right] cb f_y$$

$$- \rho\left(\frac{\varepsilon_y}{\varepsilon_{mu}}\right) cb f_y \left(\frac{1}{2}\right) - \rho\left[d - c - \left(\frac{\varepsilon_y}{\varepsilon_{mu}}\right) c\right] b f_y$$

Chapter Six

$$\frac{N_u}{\phi} = 0.80 f'_m \, 0.80 \, cb + \rho \left[1 - \left(\frac{\varepsilon_y}{\varepsilon_{mu}}\right)\right] cb f_y - \rho \left[d - c - \left(\frac{\varepsilon_y}{\varepsilon_{mu}}\right)c\right] b f_y$$

$$\frac{N_u}{\phi} = 0.80 f'_m \, 0.80 \, cb + \rho(2 - d) cb f_y$$

$$\frac{N_u}{\phi} = 0.80 f'_m \, 0.80 \, d \left(\frac{\varepsilon_{mu}}{\alpha \varepsilon_y + \varepsilon_{mu}}\right) b + 2\rho cb f_y - \rho db f_y$$

$$\frac{N_u}{\phi} = 0.80 f'_m \, 0.80 \, d \left(\frac{\varepsilon_{mu}}{\alpha \varepsilon_y + \varepsilon_{mu}}\right) b + 2\rho d \left(\frac{\varepsilon_{mu}}{\alpha \varepsilon_y + \varepsilon_{mu}}\right) b f_y - \rho db f_y$$

$$\frac{N_u}{\phi} = 0.80 f'_m \, 0.80 \, d \left(\frac{\varepsilon_{mu}}{\alpha \varepsilon_y + \varepsilon_{mu}}\right) b + \rho bd f_y \left[\left(\frac{2\varepsilon_{mu}}{\alpha \varepsilon_y + \varepsilon_{mu}}\right) - 1\right]$$

$$\rho bd f_y \left[1 - \left(\frac{2\varepsilon_{mu}}{\alpha \varepsilon_y + \varepsilon_{mu}}\right)\right] = 0.80 f'_m \, 0.80 \, d \left(\frac{\varepsilon_{mu}}{\alpha \varepsilon_y + \varepsilon_{mu}}\right) b - \frac{N_u}{\phi}$$

$$\rho = \frac{0.80 f'_m \, 0.80 \, d \left(\dfrac{\varepsilon_{mu}}{\alpha \varepsilon_y + \varepsilon_{mu}}\right) b - \dfrac{N_u}{\phi}}{bd f_y \left[1 - \left(\dfrac{2\varepsilon_{mu}}{\alpha \varepsilon_y + \varepsilon_{mu}}\right)\right]} = \frac{0.80 f'_m \, 0.80 \left(\dfrac{\varepsilon_{mu}}{\alpha \varepsilon_y + \varepsilon_{mu}}\right) - \dfrac{N_u}{bd\phi}}{f_y \left[1 - \left(\dfrac{2\varepsilon_{mu}}{\alpha \varepsilon_y + \varepsilon_{mu}}\right)\right]}$$

$$\rho = \frac{0.80 f'_m \, 0.80 \left(\dfrac{\varepsilon_{mu}}{\alpha \varepsilon_y + \varepsilon_{mu}}\right) - \dfrac{N_u}{bd\phi}}{f_y \left[\left(\dfrac{\alpha \varepsilon_y + \varepsilon_{mu}}{\alpha \varepsilon_y + \varepsilon_{mu}}\right) - \left(\dfrac{2\varepsilon_{mu}}{\alpha \varepsilon_y + \varepsilon_{mu}}\right)\right]} = \frac{0.80 f'_m \, 0.80 \left(\dfrac{\varepsilon_{mu}}{\alpha \varepsilon_y + \varepsilon_{mu}}\right) - \dfrac{N_u}{bd\phi}}{f_y \left(\dfrac{\alpha \varepsilon_y - \varepsilon_{mu}}{\alpha \varepsilon_y + \varepsilon_{mu}}\right)}$$

So

$$\rho_{max} = \frac{0.64 f'_m \left(\dfrac{\varepsilon_{mu}}{\alpha \varepsilon_y + \varepsilon_{mu}}\right) - \dfrac{N_u}{bd\phi}}{f_y \left(\dfrac{\alpha \varepsilon_y - \varepsilon_{mu}}{\alpha \varepsilon_y + \varepsilon_{mu}}\right)}$$

6.4.5 Additional Comments of the Design of Reinforced Shear Walls

Reinforced masonry shear walls, like unreinforced ones, are relatively easy to design by either strength or allowable-stress approaches. Although shear capacities per unit area is small, the available area is large.

With either strength or allowable-stress approaches, it is rarely necessary to use shear reinforcement. In this sense, the best shear design strategy for shear walls is like that for shear design of beams—use enough cross-sectional area to eliminate the need for shear reinforcement. Seismic requirements may still dictate some shear reinforcement, however.

6.5 Required Details for Reinforced Bearing Walls and Shear Walls

Bearing walls that resist out-of-plane lateral loads, and shear walls, must be designed to transfer lateral loads to the floors above and below. Examples of such connections are shown in Figs. 6.30 through 6.34. These connections would have to be strengthened for regions subject to strong earthquakes or strong winds. Section 1604.8.2 of the 2009 IBC has additional requirements for anchorage of diaphragms to masonry walls. Section 12.11 of ASCE 7-05 has additional requirements for anchorage of structural walls for structures assigned to seismic design categories C and higher.

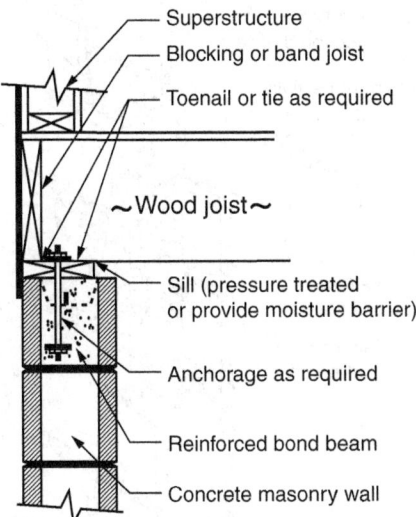

Figure 6.30 Example of wall-to-foundation connection. (*Source:* Figure 1 of National Concrete Masonry Association TEK 05-07A.)

224 Chapter Six

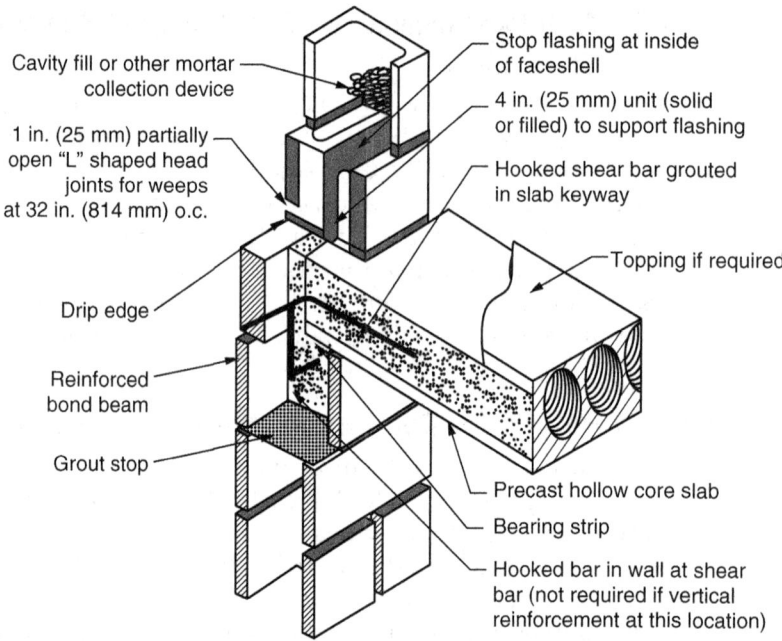

Figure 6.31 Example of wall-to-floor connection, planks perpendicular to wall. (*Source*: Figure 14 of National Concrete Masonry Association TEK 05-07A.)

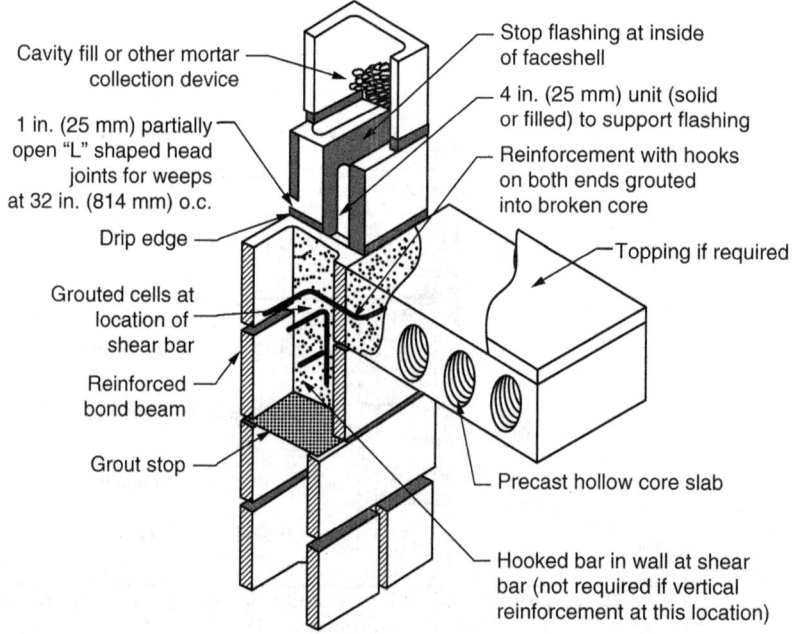

Figure 6.32 Example of wall-to-floor connection, planks parallel to wall. (*Source*: Figure 15 of National Concrete Masonry Association TEK 05-07A.)

Strength Design of Reinforced Masonry Elements 225

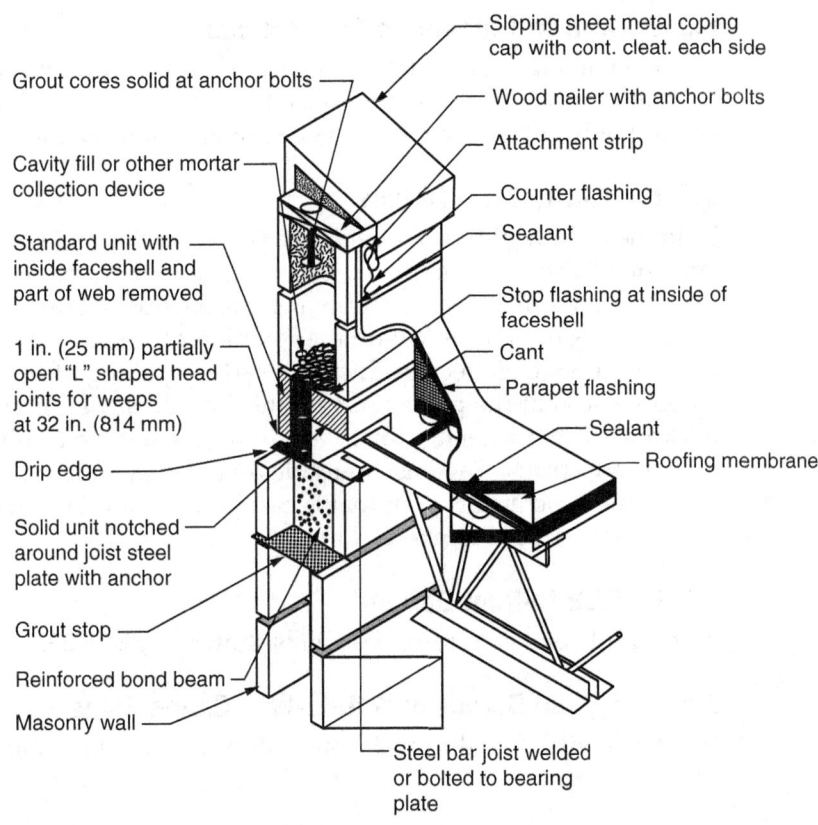

FIGURE 6.33 Example of wall-to-roof detail. (*Source*: Figure 11 of National Concrete Masonry Association TEK 05-07A.)

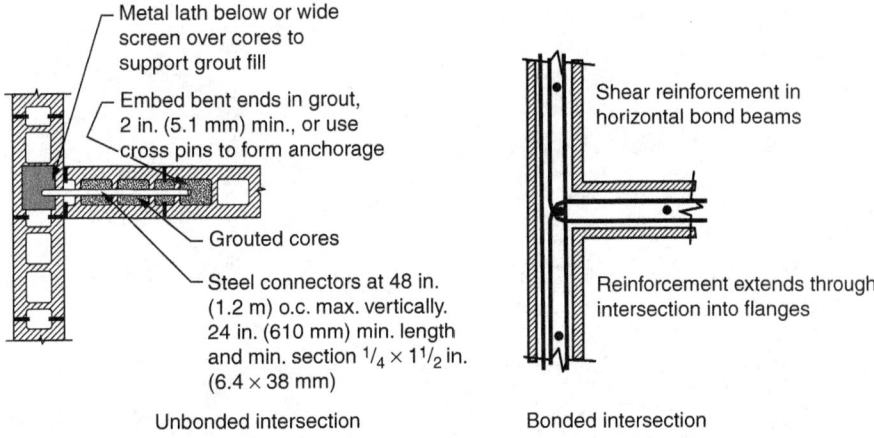

FIGURE 6.34 Examples of wall-to-wall unbonded and bonded intersections connection details. (*Source*: Figure 2 of National Concrete Masonry Association TEK 14-08B.)

6.5.1 Wall-to-Foundation Connections

As shown in Fig. 6.31, CMU walls (or the inner CMU wythe of a drainage wall) must be connected to the concrete foundation. Bond breaker should be used only between the outer veneer wythe and the foundation.

6.5.2 Wall-to-Floor Details

Examples of a wall-to-floor detail are shown in Figs. 6.31 and 6.32. In the latter detail (floor or roof planks oriented parallel to walls), the planks are actually cambered. They are shown on the outside of the walls so that this camber does not interfere with the coursing of the units. Some designers object to this detail because it could lead to spalling of the cover. If it is modified so that the planks rest on the face shells of the walls, then the thickness of the topping must vary to adjust for the camber, and form boards must be used against both sides of the wall underneath the planks, so that the concrete or grout that is cast into the bond beam does not run out underneath the cambered beam.

6.5.3 Wall-to-Roof Details

An example of a wall-to-roof detail is shown in Fig. 6.33.

6.5.4 Typical Details of Wall-to-Wall Connections

Typical details of wall-to-wall connections are shown in Fig. 6.34.

CHAPTER 7
Allowable-Stress Design of Unreinforced Masonry Elements

7.1 Allowable-Stress Design of Unreinforced Panel Walls

7.1.1 Examples of Use of Unreinforced Panel Walls

Panel walls commonly comprise the masonry envelope surrounding reinforced concrete or steel frames. In the context of the 2008 MSJC *Code*, a panel wall would be termed a "multiwythe, noncomposite" wall. An example of an unreinforced panel wall is shown in Fig. 7.1.

The outer wythes of panel walls must span horizontally. They cannot span vertically, because of the open expansion joint under each shelf angle. Support conditions for the horizontally spanning outer wythe can be simple or continuous. An example of the connection of a panel wall to a column is shown in the horizontal section of Fig. 7.2. A simple support condition would be achieved by inserting a vertically oriented expansion joint in the clay masonry wythe on both sides of the column. The inner wythes of panel walls can span horizontally and vertically. As a result, it is convenient to visualize panel walls as being composed of sets of vertical and horizontal crossing strips in each wythe. This is shown schematically in Fig. 7.3.

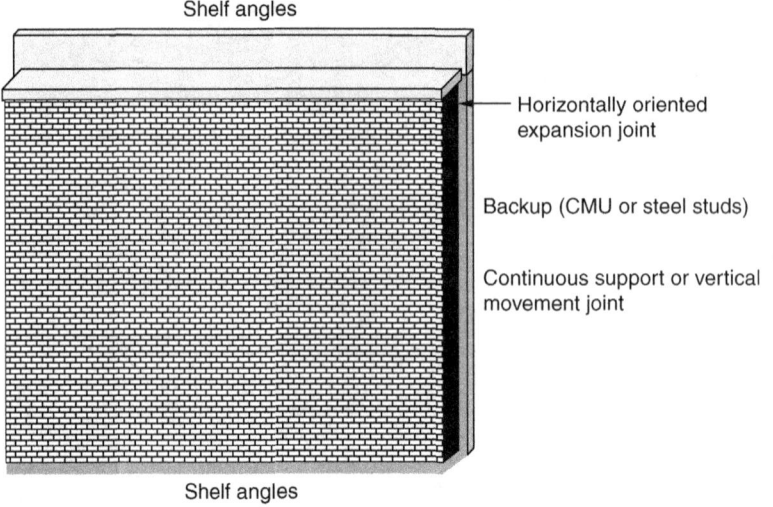

Figure 7.1 Example of an unreinforced panel wall.

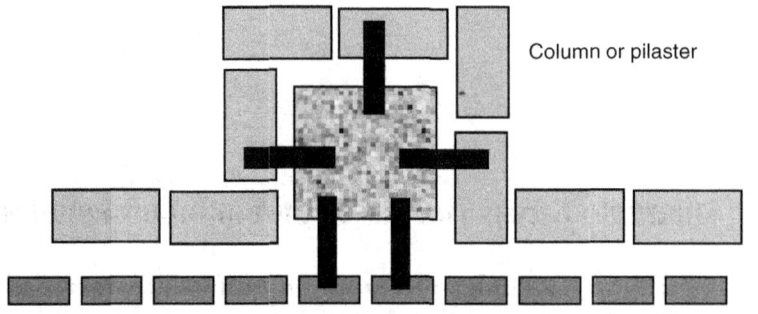

Figure 7.2 Horizontal section showing connection of a panel wall to a column.

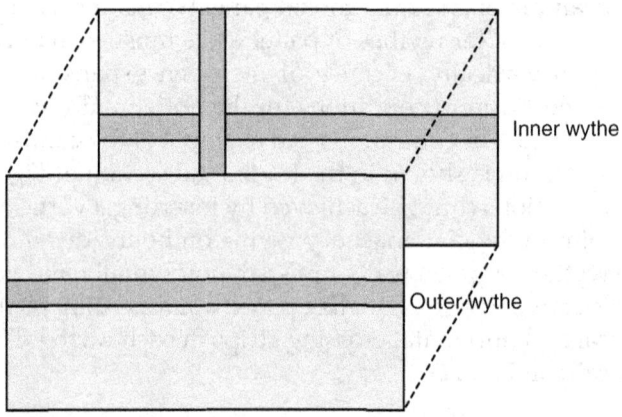

Figure 7.3 Schematic representation of an unreinforced, two-wythe panel wall as two sets of horizontal and vertical crossing strips.

At the end of this section, it will be shown that

- Because of their aspect ratio, the inner wythe can almost always be considered to span in the vertical direction only
- It is simple and only slightly conservative to design single-wythe panel walls as though the vertical strips resisted all out-of-plane load
- It is simple and only slightly conservative to design two-wythe panel walls as though the vertical strips of the inner wythe resisted all out-of-plane load

7.1.2 Flexural Design of Panel Walls Using Allowable-Stress Provisions of 2008 MSJC *Code*

According to the allowable-stress provisions of the 2008 MSJC *Code*, the allowable stress in flexural compression is $(1/3)f'_m$ [Code Sec. 2.2.3.1(c)]. The allowable stress in flexural tension is much lower, and governs in design as shown in Table 7.1.

Direction of flexural tensile stress and masonry type	Mortar types			
	PCL or mortar cement		Masonry cement or air-entrained PCL	
	M or S	N	M or S	N
Normal to bed joints				
Solid units	40	30	24	15
Hollow units[1]				
Ungrouted	25	19	15	9
Fully grouted	65	63	61	58
Parallel to bed joints in running bond				
Solid units	80	60	48	30
Hollow units				
Ungrouted and partially grouted	50	38	30	19
Fully grouted	80	60	48	30
Parallel to bed joints in stack bond				
Continuous grout section parallel to bed joints	100	100	100	100
Other	0	0	0	0

[1] For partially grouted masonry, allowable stresses shall be determined on the basis of linear interpolation between fully grouted hollow units and ungrouted hollow units based on amount (percentage) of grouting.

Source: Table 2.2.3.2 of 2008 MSJC *Code*.

TABLE 7.1 Allowable Flexural Tension for Clay and Concrete Masonry, psi

The 2008 MSJC *Code* provides different allowable stresses for flexural tension normal to bed joints, and parallel to bed joints in running bond. Allowable stresses are higher parallel to bed joints in running bond, because of the interlocking of units laid in that bond pattern. Allowable flexural tension is zero parallel to bed joints in stack bond, unless they are resisted by a continuous grout section parallel to the bed joints.

The allowable stresses in Table 2.2.3.2 of 2008 MSJC *Code* apply to out-of-plane and in-plane bending.

7.1.3 Example of Allowable-Stress Design of a Single-Wythe Panel Wall Using Solid Units

Check the design of the panel wall shown in Fig. 7.4, for a wind load w of 20 lb/ft^2, using PCL mortar, Type N, and units with a nominal thickness of 8 in.

The panel wall will be designed as unreinforced masonry. The design follows the steps below, using a nominal thickness of 8 in. The panel could be designed as a two-way panel. Nevertheless, because of its aspect ratio, the vertical strips will carry practically all the load. Therefore, design it as a one-way panel, consisting of a series of vertically spanning, simply supported strips.

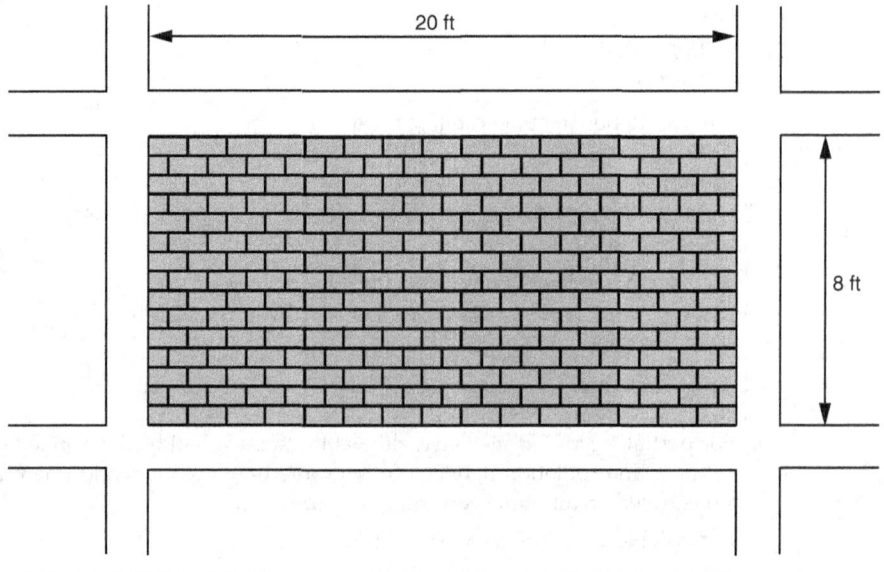

FIGURE 7.4 Example panel wall to be designed.

Calculate the maximum bending moment and corresponding flexural tensile stress in a strip 1-ft wide, with a nominal thickness of 8 in.:

$$M_{max} = \frac{w\ell^2}{8} = \frac{20 \text{ lb/ft }(8 \text{ ft})^2}{8} \times 12 \text{ in./ft} = 1920 \text{ lb-in.}$$

$$f_t = \frac{Mc}{I} = \frac{1920 \text{ lb-in.} \cdot \left(7.625/2\right) \text{ in.}}{\left[12 \text{ in.} \cdot (7.625 \text{ in.})^3 / 12\right]} = 16.5 \text{ lb/in.}^2$$

The calculated flexural tensile stress, 16.5 lb/in.², is less than the allowable flexural tensile stress normal to bed joints for solid units and Type N PCL mortar (30 lb/in.²). The design is therefore satisfactory. We should also check one-way (beam) shear. An example of this is given in Sec. 7.1.8.

7.1.4 Example of Allowable-Stress Design of a Single-Wythe Panel Wall Using Hollow Units

Check the design of the panel wall of the example of Sec. 7.1.3 for a wind load w of 20 lb/ft², using PCL mortar, Type N, and assuming hollow units.

The panel wall will be designed as unreinforced masonry. The design follows the following steps, using a nominal thickness of 8 in. The panel could be designed as a two-way panel. Nevertheless, because of its aspect ratio, the vertical strips will carry practically all the load. Therefore, design it as a one-way panel, consisting of a series of vertically spanning, simply supported strips.

Calculate the maximum bending moment and corresponding flexural tensile stress in a strip 1-ft wide, with a nominal thickness of 8 in. Assume that the head joints are only 1.25-in. thick. All length dimensions are in inches. These dimensions are shown in Fig. 7.5.

$$I = I_{solid} - I_{cells} - I_{head\,joint}$$

$$I = \frac{(16)(7.63)^3}{12} - \frac{(15.63 - 3 \cdot 1.00)(7.63 - 2 \cdot 1.25)^3}{12} - \frac{0.37(7.63 - 2 \cdot 1.25)^3}{12}$$

$$I = 592.3 - 142.1 - 4.2 = 446 \text{ in.}^4$$

$$c = \frac{7.63}{2} = 3.81 \text{ in.}$$

$$S = \frac{I}{c} = \frac{446.0}{3.81} = 117.1 \text{ in.}^3 \text{ per 16 in. of width}$$

$$S = \frac{117.1}{16.0} \times 12 = 87.8 \text{ in.}^3 \text{ for a 12-in. wide strip}$$

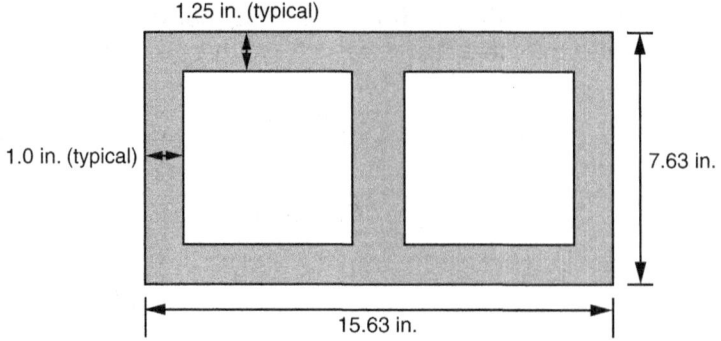

FIGURE 7.5 Idealized cross-sectional dimensions of a nominal 8 × 8 × 16 in. concrete masonry unit.

For the 12-in. wide strip,

$$M_{max} = \frac{w\ell^2}{8} = \frac{20 \text{ lb/ft } (8 \text{ ft})^2}{8} \times 12 \text{ in./ft} = 1920 \text{ lb-in.}$$

$$f_t = \frac{M}{S} = \frac{1920 \text{ lb-in.}}{87.8 \text{ in.}^3} = 21.9 \text{ lb/in.}^2$$

The calculated flexural tensile stress, 21.9 lb/in.², exceeds the allowable flexural tensile stress normal to bed joints for hollow units and Type N PCL mortar (19 lb/in.²). If the mortar is changed to Type S (allowable stress 25 lb/in.²), the design will be satisfactory.

7.1.5 Example of Design of a Single-Wythe Panel Wall Using Hollow Units, Face-Shell Bedding Only

Check the design of the panel wall of the previous example assuming face-shell bedding only (mortar on the face shells of the units only).

The panel wall will be designed as unreinforced masonry. The design follows the following steps, using a nominal thickness of 8 in. The panel could be designed as a two-way panel. Nevertheless, because of its aspect ratio, the vertical strips will carry practically all the load. Therefore, design it as a one-way panel, consisting of a series of vertically spanning, simply supported strips.

The critical stresses will occur on the bed joint, which is the horizontal plane through the masonry where the section modulus is minimum. All length dimensions are in inches. The dimensions of this critical cross section are shown in Fig. 7.6.

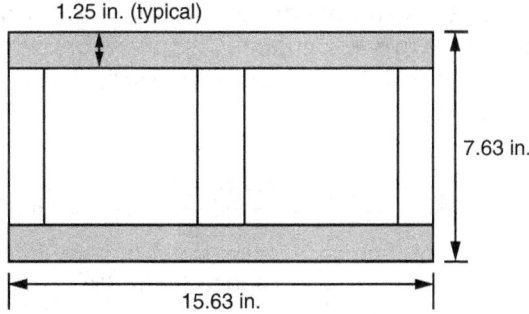

FIGURE 7.6 Idealized cross-sectional dimensions of a nominal 8 × 8 × 16 in. concrete masonry unit with face-shell bedding.

$I = I_{faceshells}$

$I = \dfrac{2(12)(1.25)^3}{12} + 2(12 \cdot 1.25)\left(\dfrac{7.63 - 1.25}{2}\right)^2 = 309.2 \text{ in.}^4 \text{ per ft of width}$

$c = \dfrac{7.63}{2} = 3.81 \text{ in.}$

$S = \dfrac{I}{c} = \dfrac{309.2}{3.81} = 81.2 \text{ in.}^3 \text{ per ft of width}$

For the 1-ft wide strip,

$M_{max} = \dfrac{w\ell^2}{8} = \dfrac{20 \text{ lb/ft } (8 \text{ ft})^2}{8} \times 12 \text{ in./ft} = 1920 \text{ lb-in.}$

$f_t = \dfrac{M}{S} = \dfrac{1920 \text{ lb-in.}}{81.2 \text{ in.}^3} = 23.7 \text{ lb/in.}^2$

The calculated flexural tensile stress, 23.7 lb/in.², exceeds the allowable flexural tensile stress normal to bed joints for hollow units and Type N PCL mortar (19 lb/in.²). If the mortar is changed to Type S (allowable stress 25 lb/in.²), the design will be satisfactory.

7.1.6 Example of Allowable-Stress Design of a Single-Wythe Panel Wall Using Hollow Units, Fully Grouted

Check the design of the panel wall of the example of Sec. 7.1.3 for a wind load w of 20 lb/ft², using PCL mortar, Type N, and assuming hollow units, fully grouted.

As in the example of Sec. 7.1.3, assume vertically spanning, simply supported strips. Calculate the maximum bending moment and corresponding flexural tensile stress in a strip 1-ft wide, with a nominal thickness of 8 in.:

$$M_{max} = \frac{w\ell^2}{8} = \frac{20 \text{ lb/ft } (8 \text{ ft})^2}{8} \times 12 \text{ in./ft} = 1920 \text{ lb-in.}$$

$$f_t = \frac{Mc}{I} = \frac{1920 \text{ lb-in.} \cdot (7.625/2) \text{ in.}}{\left[12 \text{ in.} \cdot (7.625 \text{ in.})^3 / 12\right]} = 16.5 \text{ lb/in.}^2$$

The calculated flexural tensile stress, 16.5 lb/in.², is less than the allowable flexural tensile stress normal to bed joints for fully grouted hollow units and Type N PCL mortar (63 lb/in.²). The design is therefore satisfactory.

7.1.7 Example of Design of a Two-Wythe Panel Wall Using Hollow Units, Face-Shell Bedding Only

Check the design of a two-wythe panel wall in which the outer wythe is modular clay units and the inner wythe is 8-in. CMU with face-shell bedding. The wall has the panel wall of Sec. 7.1.6 assuming face-shell bedding only (mortar on the face shells of the units only). The wall has a wind load w of 20 lb/ft², and uses PCL mortar, Type N.

The panel wall will be designed as unreinforced masonry, assuming that the vertical strips of the inner wythe resist 100 percent of the out-of-plane load. The design is therefore identical to the design of Sec. 7.1.5. This idealized horizontal cross section is shown in Fig. 7.7.

$$I = I_{faceshells} = 309.2 \text{ in.}^4 \text{ per ft of width}$$

$$S = \frac{I}{c} = \frac{309.2}{3.81} = 81.2 \text{ in.}^3 \text{ per ft of width}$$

For the 12-in. wide strip,

$$M_{max} = \frac{w\ell^2}{8} = \frac{20 \text{ lb/ft } (8 \text{ ft})^2}{8} \times 12 \text{ in./ft} = 1920 \text{ lb-in.}$$

$$f_t = \frac{M}{S} = \frac{1920 \text{ lb-in.}}{81.2 \text{ in.}^3} = 23.7 \text{ lb/in.}^2$$

The calculated flexural tensile stress, 23.7 lb/in.², exceeds the allowable flexural tensile stress normal to bed joints for hollow units and Type N

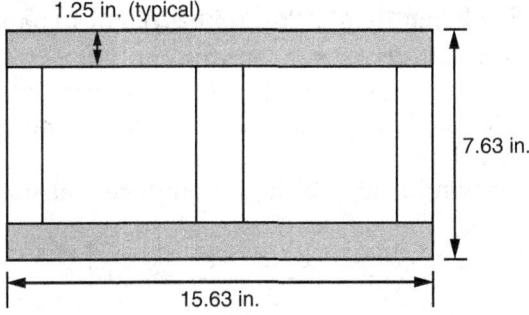

FIGURE 7.7 Idealized cross-sectional dimensions of a nominal 8 × 8 × 16 in. concrete masonry unit with face-shell bedding.

PCL mortar (19 lb/in.²). If the mortar is changed to Type S (allowable stress 25 lb/in.²), the design will be satisfactory. Addition of the outer wythe does not change the design of the panel wall.

7.1.8 Allowable-Stress Checks of One-Way Shear

The examples of Sec. 7.1.3 through Sec. 7.1.7 dealt with design of panel walls for flexure. In theory, we should also check one-way shear. In practice, an example shows that shear does not come close to governing the design.

According to the 2008 MSJC *Code*, allowable one-way shear for unreinforced masonry flexural elements is prescribed by Sec. 2.2.5:

$$f_v = \frac{VQ}{I_n b}$$

and allowable in-plane shear stresses, F_v, shall not exceed any of

$$F_v = \begin{cases} 1.5\sqrt{f'_m} \\ 120\,\text{psi} \\ v + 0.45(N_v/A_v) \end{cases}$$

in the third equation,

> v = 37 psi for masonry in running bond that is not grouted solid
> = 37 psi for masonry in other than running bond with open-end units grouted solid
> = 60 psi for masonry in running bond that is grouted solid

7.1.9 Example of Allowable-Stress Check of Shear Stress

Check the effect of shear in the example of Sec. 7.1.7. With f'_m of 1500 lb/in.² and masonry in running bond, and conservatively neglecting the beneficial effects of axial load, the third of the above equations governs, and $F_v = 37$ psi.

The wind load of 20 lb/ft.² produces shears on a strip 12-in. wide, of

$$V = \frac{qL}{2} = \frac{20 \text{ lb/ft} \cdot (8 \text{ ft})}{2} = 80.0 \text{ lb}$$

$$f_v = \frac{VQ}{Ib} = \left(\frac{3}{2}\right)\left(\frac{V}{A}\right) = \left(\frac{3}{2}\right)\left(\frac{80.0 \text{ lb}}{2 \cdot 12 \text{ in.} \cdot 1.25 \text{ in.}}\right) = 4.0 \text{ lb/in.}^2$$

This is far less than the allowable shear stress, and one-way shear does not govern the design.

7.1.10 Overall Comments on Allowable-Stress Design of Unreinforced Panel Walls

- Nonload bearing masonry, without calculated reinforcement, can easily resist wind loads. It approaches its capacity only in the case of ungrouted hollow masonry. In this case, the lower allowable flexural tensile stress for masonry cement mortar can be critical in design.
- If noncalculated reinforcement is included, it will not act until the masonry has cracked.
- Elements such as the ones we have calculated in this section can be designed in many cases by prescription.

7.1.11 Section Properties for Masonry Walls

Section properties for masonry walls are summarized in Table 7.2.

7.1.12 Theoretical Derivation of the Strip Method (Hillerborg, 1996)

In the preceding examples, we have used the simplifications that single-wythe panel walls can be designed assuming that entire load is resisted by vertically spanning strips, and that two-wythe panel walls can be designed assuming that entire load is resisted by vertically spanning strips of the inner wythe.

It is now appropriate to consider the theoretical basis for this simplification. We first consider the simplification of wythes as crossing strips using the strip method, and then derive additional simplifications based on the relative stiffnesses of those strips.

Unit	Area in.² per ft	Moment of inertia in.⁴ per ft
4-in. modular	43.6	47.8
6-in hollow CMU, fully bedded	32.2	139
6-in hollow CMU, face-shell bedded	24.0	130
8-in hollow CMU, fully bedded	41.5	334
8-in hollow CMU, face-shell bedded	30.0	309
12-in hollow CMU, fully bedded	57.8	1065
12-in hollow CMU, face-shell bedded	36.0	929

TABLE 7.2 Section Properties for Masonry Walls

Consider the differential equation for the out-of-plane deflection of an elastic plate with a uniformly distributed, out-of-plane load:

$$\frac{\partial^4 w}{\partial x^4} + \frac{\partial^4 w}{\partial x^2 \partial y^2} + \frac{\partial^4 w}{\partial y^4} = \frac{-q}{D}$$

where w = out-of-plane deflection
q = uniformly distributed load

and

$$D = \frac{EI}{(1-v^2)} = \frac{Et^3}{12(1-v^2)}$$

(EI is calculated per unit width, and Poisson effects are included) Conservatively, ignore twisting moments:

$$\frac{\partial^4 w}{\partial x^4} + \frac{\partial^4 w}{\partial y^4} = \frac{-q}{D} = -\left(\frac{q_x}{D} + \frac{q_y}{D}\right)$$

This is the differential equation for independent x and y strips (beams). The two sets of strips can be designed independently, provided that equilibrium is satisfied at every point:

$$q_x + q_y = q$$

7.1.13 Distribution of Out-of-Plane Load to Vertical and Horizontal Strips of a Single-Wythe Panel Wall

In Sec. 7.1.1, it was stated that single-wythe panel walls can be designed as though out-of-plane load were carried by the vertical strips alone.

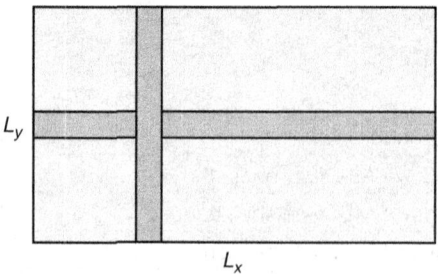

Figure 7.8 Idealization of a panel wall as an assemblage of crossing strips.

Now let's show why that's true. Consider the single-wythe panel shown in Fig. 7.8.

Assume that the panel resists out-of-plane loading as an assemblage of crossing strips, in the x direction (horizontally on the page) and the y direction (vertically on the page). At the very end of this section, it will be shown that such an assumption is legitimate ("strip method").

Now impose compatibility of out-of-plane deflections on the strips—that is, the crossing x and y strips must have equal out-of-plane displacements. For a simply supported strip with uniformly distributed loading q, the center-line displacement is

$$\Delta_{center} = \frac{5qL^4}{384EI}$$

For equal deflections of the x and y strips,

$$\frac{5q_x L_x^4}{384EI} = \Delta_{center} = \frac{5q_y L_y^4}{384EI}$$

And because those strips have equal moduli and moments of inertia,

$$\frac{q_x}{q_y} = \frac{L_y^4}{L_x^4}$$

The span of the vertical strips, L_y, is the distance from the top of a floor slab to the underside of the slab or beam above, typically about 10 ft. The span of the horizontal strips, L_x, is the distance from the face of a column to the face of the adjacent column, typically about 20 ft. So

$$\frac{q_x}{q_y} = \frac{L_y^4}{L_x^4} = \left(\frac{10}{20}\right)^4 = 0.063$$

The horizontal strips will carry only about 6 percent of the out-of-plane load. If it is conservatively assumed that the vertical strips will carry 100 percent of the out-of-plane load, and the horizontal strip will carry zero load, the design work is halved, and the results will be conservative.

7.1.14 Distribution of Out-of-Plane Load to Vertical and Horizontal Strips of a Two-Wythe Panel Wall

The analysis of Sec. 7.1.13 can easily be extended to the case of a two-wythe panel wall, with an outer wythe of clay masonry and an inner wythe of concrete masonry shown in Fig. 7.9.

As before, assume that each wythe of the panel resists out-of-plane loading as an assemblage of crossing strips, in the x direction (horizontally on the page) and the y direction (vertically on the page). At the very end of this section, it will be shown that such an assumption is legitimate ("strip method").

Also assume that the inner wythe is of hollow, ungrouted, 8-in. CMU laid in face-shell bedding, and that the outer wythe is of solid modular units (nominal thickness of 4 in.).

From Sec. 7.1.6, we know that the moment of inertia of the hollow CMU is 444.9 in.^{4} per 16 in. of width, or 333.7 in.^{4} per foot of width. It is proper to compute the average flexural stiffness of the wall using the moment of inertia of the hollow unit rather than the moment of the face-shell bedding only, because the bed joints occupy only a small portion of the volume of the wall.

The moment of inertia of the modular outer wythe, per foot of width, is

$$I = \frac{bt^3}{12} = \frac{(12 \text{ in.})(3.63 \text{ in.})^3}{12} = 47.9 \text{ in.}^4 \text{ per ft of width}$$

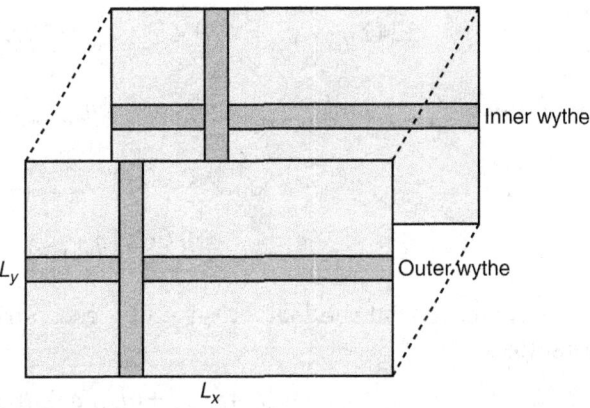

FIGURE 7.9 Idealization of a two-wythe panel wall as an assemblage of two sets of crossing strips.

Assume that the ties between wythes are axially rigid, so that the two wythes have equal out-of-plane deflection.

The total load q must be equilibrated by the summation of the load resisted by each strip of each wythe:

$$(q_x + q_y)_{exterior} + (q_x + q_y)_{interior} = q$$

Because the vertical strips in the exterior wythe are simply supported at the bottom and free at the top (expansion joint), they carry no load:

$$(q_x)_{exterior} + (q_x + q_y)_{interior} = q$$

Now impose compatibility of out-of-plane deflections on the strips,

$$\Delta_{x\,exterior} = \frac{5 q_{x\,exterior} L^4_{x\,exterior}}{384 E_{x\,exterior} I_{x\,exterior}} = \frac{5 q_{x\,exterior} 20^4}{384 E_{x\,exterior}} \cdot 47.9$$

$$\Delta_{x\,interior} = \frac{5 q_{x\,interior} L^4_{x\,interior}}{384 E_{x\,interior} I_{x\,interior}} = \frac{5 q_{x\,interior} 20^4}{384 E_{x\,interior}} \cdot 333.7$$

$$\Delta_{y\,interior} = \frac{5 q_{y\,interior} L^4_{y\,interior}}{384 E_{y\,interior} I_{y\,interior}} = \frac{5 q_{y\,interior} 10^4}{384 E_{y\,interior}} \cdot 333.7$$

Equate those deflections, cancel out the common term of (5/384), and assume that all Es are the same:

$$\frac{q_{x\,exterior} 20^4}{47.9} = \frac{q_{x\,interior} 20^4}{333.7} = \frac{q_{y\,interior} 10^4}{333.7}$$

$$3342\, q_{x\,exterior} = 479.4\, q_{x\,interior} = 29.97\, q_{y\,interior}$$

Express $q_{x\,exterior}$ and $q_{x\,interior}$ in terms of $q_{y\,interior}$:

$$q_{x\,exterior} = 0.00897\, q_{y\,interior}$$

$$q_{x\,interior} = 0.0625\, q_{y\,interior}$$

Now recall that the loads resisted by each strip must equilibrate the total load:

$$(q_x)_{exterior} + (q_x + q_y)_{interior} = q$$

$$(0.00897 + 0.0625 + 1.0) q_{y\,interior} = q$$

Solve for $q_{y \text{ interior}}$ in terms of q:

$$q_{y \text{ interior}} = \frac{q}{1.0715} = 0.933q$$

Finally, express the load carried by each set of strips in terms of q:

$$q_{y \text{ interior}} = 0.93q$$
$$q_{x \text{ interior}} = 0.06q$$
$$q_{x \text{ exterior}} = 0.008q$$

Clearly, it is conservative and very reasonable to assume that the vertical strips of the interior wythe resist 100 percent of the out-of-plane load, and the other strips resist no load.

7.2 Allowable-Stress Design of Unreinforced Bearing Walls

7.2.1 Basic Behavior of Unreinforced Bearing Walls

Load-bearing masonry (without calculated reinforcement) must be designed for the effects of

1. Gravity loads from self-weight, plus gravity loads from overlying roof or floor levels
2. Moments from eccentric gravity load, or out-of-plane wind or earthquake
3. In-plane shear

For now, we shall study Loadings (1) and (2). In Sec. 7.3, we shall study Loading (3), in the general context of design of masonry shear walls.

For Loadings (1) and (2), we shall design unreinforced, load-bearing masonry as a series of vertically spanning strips (Fig. 7.10), subjected to gravity loads (possibly eccentric) and out-of-plane wind or earthquake.

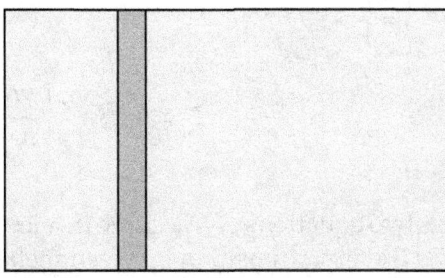

FIGURE 7.10 Idealization of bearing walls as vertically spanning strips.

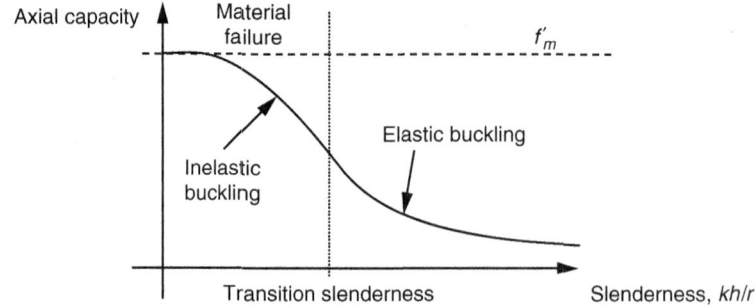

FIGURE 7.11 Effect of slenderness on the axial capacity of a column or wall.

The only aspect of behavior that we haven't studied so far is the effect of slenderness on the load-carrying capacity of a column or wall. This effect is shown in Fig. 7.11.

At low values of slenderness, a masonry column in compression exhibits material failure. At high values of slenderness, it exhibits stability failure.

For masonry design, the effective length coefficient, k, is usually equal to 1.

7.2.2 Steps in Allowable-Stress Design of Unreinforced Bearing Walls

Allowable-stress codes usually address slenderness effects by decreasing the allowable axial stress with increasing slenderness. Equations (2-15) and (2-16) (Sec. 2.2.3.1) of the 2008 MSJC *Code*, for example, require that for $\dfrac{kh}{r} = \dfrac{h}{r} \leq 99$,

$$F_a = 0.25 f'_m \left[1 - \left(\dfrac{h}{140r} \right)^2 \right]$$

and for $\dfrac{kh}{r} = \dfrac{h}{r} > 99$,

$$F_a = 0.25 f'_m \left(\dfrac{70}{h/r} \right)^2$$

These two equations give a curve that looks very much like that shown in Fig. 7.11, with a transition between inelastic and elastic buckling at a slenderness value of 99.

Combinations of axial force and bending are addressed by a so-called "unity equation" [2008 MSJC *Code*, Eq. (2-13)]:

$$\frac{f_a}{F_a} + \frac{f_b}{F_b} \leq 1$$

where f_a = calculated axial stress (P/A)
F_a = allowable axial stress as specified above
f_b = calculated bending stress (M/S)
F_b = allowable bending stress, ($f_m'/3$)

Next, the 2008 MSJC *Code* requires that the extreme-fiber tensile stress not exceed the out-of-plane allowable stress from Table 7.1 (Table 2.2.3.2 of 2008 MSJC *Code*):

$$f_t = \frac{Mc}{I} - \frac{P}{A} \leq F_t$$

Finally, in addition to the slenderness-dependent allowable axial stress in the unity equation, the 2008 MSJC *Code* imposes another stability requirement. The axial load in the bearing wall must not exceed one-quarter of the Euler buckling load, decreased by a penalty factor that becomes very significant at high eccentricities [2008 MSJC *Code*, Eq. (2-18)]:

$$P \leq \frac{P_e}{4}$$

$$P_e = \frac{\pi^2 E_m I}{h^2}\left(1 - 0.577\frac{e}{r}\right)^3$$

where e is effective eccentricity, equal to the moment at the point under consideration, divided by the axial force at that same point. For this calculation, moments and axial forces are from gravity loads only.

This equation is effectively equivalent to assuming that the masonry units are stacked one on top of the other without any mortar. For units with a rectangular cross section, for which $r = t/\sqrt{12}$, at an eccentricity of ($t/2$) (load applied at the edge of the units), the penalty factor becomes equal to zero, and so does the allowable axial load. This equation was developed as a conservative alternative to moment magnifier methods, which were regarded as too complex. The MSJC is working to develop moment magnifier methods that will be user-friendly and not so conservative.

246 Chapter Seven

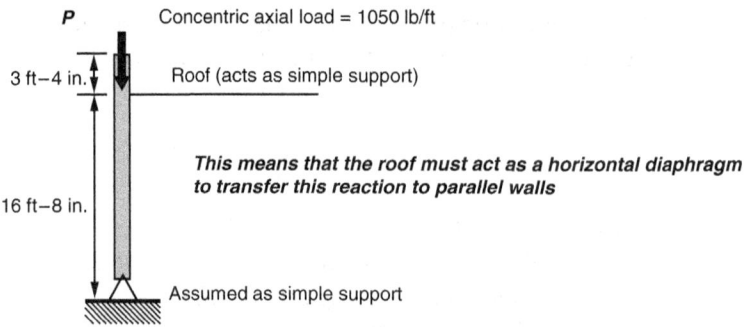

FIGURE 7.12 Unreinforced masonry bearing wall with concentric axial load.

7.2.3 Example of Allowable Stress Design of Unreinforced Bearing Wall with Concentric Axial Load

The bearing wall shown in Fig. 7.12 has a concentric axial load of 1050 lb/ft, due to dead plus live load. Using hollow concrete masonry units with face-shell bedding, design the wall. The governing load combination for allowable-stress design is $D + L$.

Table 7.3 repeated from Sec. 7.1.11, gives section properties for masonry units.

Table 7.4 taken from NCMA TEK 2-1A, gives the self-weight of hollow masonry walls, assuming units with a density of 120 lb/ft^3.

At each horizontal plane through the wall, the following conditions must be met:

- Combination of axial and flexural compressive stresses must not violate the unity equation
- Net tension stress must not exceed the allowable flexural tension
- Separate stability check must be satisfied

Unit	Area in.^{2} per ft	Moment of inertia in.^{4} per ft
4-in. modular	43.6	47.8
6-in hollow CMU, fully bedded	32.2	139
6-in hollow CMU, face-shell bedded	24.0	130
8-in hollow CMU, fully bedded	41.5	334
8-in hollow CMU, face-shell bedded	30.0	309
12-in hollow CMU, fully bedded	57.8	1065
12-in hollow CMU, face-shell bedded	36.0	929

TABLE 7.3 Section Properties for Masonry Units

Allowable-Stress Design of Unreinforced Masonry Elements

Nominal thickness, in.	Weight per ft²
4	27
6	40
8	48
10	56

TABLE 7.4 Weights of Hollow CMU Walls

In theory, we must check various points on the wall. In this problem, however, the wall has only axial load, which increases from top to bottom due to the wall's self-weight. Therefore we need to check only at the base of the wall.

Try 8-in. nominal units, and a specified compressive strength, f'_m of 1500 lb/in.². This can be satisfied using units with a net-area compressive strength of 1900 lb/in.², and Type S PCL mortar. Work with a strip with a width of 1 ft (measured along the length of the wall in plan). Stresses are calculated using the critical section, consisting of the bedded area only (2008 MSJC Code, Sec. 1.9.1.1):

$$f_a = \frac{P}{A} = \frac{1050 \text{ lb} + 20 \text{ ft} \cdot 48 \text{ lb/ft}}{30 \text{ in.}^2} = \frac{2010 \text{ lb}}{30 \text{ in.}^2} = 68.6 \text{ lb/in.}^2$$

To calculate stiffness-related parameters for the wall, we use the average cross section, corresponding to the fully bedded section in the table (2008 MSJC Code, Sec. 1.9.3).

$$r = \sqrt{\frac{I}{A}} = \sqrt{\frac{334 \text{ in.}^4}{41.5 \text{ in.}^2}} = 2.84 \text{ in.}$$

$$\frac{kh}{r} = \frac{16.67 \cdot 12 \text{ in.}}{2.84 \text{ in.}} = 70.5$$

This is less than the transition slenderness of 99, so the allowable stress is based on the curve that is an approximation to inelastic buckling:

$$F_a = 0.25 f'_m \left[1 - \left(\frac{h}{140\,r}\right)^2\right]$$

$$F_a = 0.25 \times 1500 \text{ lb/in.}^2 \left[1 - \left(\frac{16.67 \text{ ft} \times 12 \text{ in./ft}}{140 \cdot 2.84 \text{ in.}}\right)^2\right]$$

$$= 375 \text{ lb/in.}^2 \times 0.746 = 280 \text{ lb/in.}^2$$

Now check the unity equation. Because the load is concentric, there is no bending stress:

$$\frac{f_a}{F_a} + \frac{f_b}{F_b} \leq 1$$

$$\frac{f_a}{F_a} = \frac{68.6 \text{ lb/in.}^2}{280 \text{ lb/in.}^2} = 0.25 \leq 1$$

and the unity equation is satisfied. Because the wall has concentric axial load, there is no net tensile stress, and that equation does not have to be checked.

Now check the stability equation. Because the load is concentric, the eccentricity is zero, and the penalty term has a value of 1.

$$P_e = \frac{\pi^2 E_m I}{h^2}\left(1 - 0.577\frac{e}{r}\right)^3$$

$$P_e = \frac{\pi^2 \times 900 \times 1500 \text{ lb/in.}^2 \times 334 \text{ in.}^4}{(16.67 \text{ ft} \times 12 \text{ in./ft})^2} = 111{,}211 \text{ lb}$$

$$\frac{P_e}{4} = 27{,}803 \text{ lb}$$

Because the calculated axial force per foot of length, 2010 lb, is much less than the allowable value of 27,803 lb, the design is satisfactory.

It would probably be possible to achieve a satisfactory design with a smaller nominal wall thickness. To maintain continuity in the example problems that follow, however, the design will stop at this point.

The 2008 MSJC *Code* has no minimum eccentricity requirements for walls. It does require that for columns, the minimum design eccentricity in each direction be taken as 0.1 times the specified cross-sectional dimension in that direction (2008 MSJC *Code*, Sec. 2.1.6.3).

7.2.4 Example of Allowable-Stress Design of Unreinforced Bearing Wall with Eccentric Axial Load

Now consider the same bearing wall of the previous example, but make the gravity load eccentric. As before, the governing load combination for allowable-stress design is $D + L$.

First, review the calculation of the gravity load itself. Suppose that it comes from a uniformly distributed roof dead load of 50 lb/ft² and live

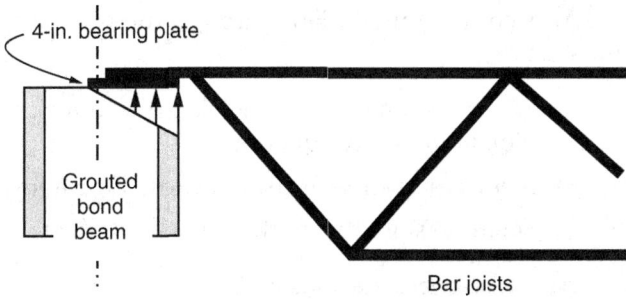

FIGURE 7.13 Assumed linear variation of bearing stresses under the bearing plate.

load of 20 lb/ft², acting on a 30-ft span. The reaction on the wall per foot of plan length is then:

$$\text{Wall load} = \frac{ql}{2} = \frac{(50+20)\text{ lb/ft} \times 30 \text{ ft}}{2} = 1050 \text{ lb/ft}$$

Now consider how it is applied to the wall. Suppose that the load is applied over a 4-in. bearing plate, and assume that bearing stresses vary linearly under the bearing plate as shown in Fig. 7.13.

Then the eccentricity of the applied load with respect to the centerline of the wall is

$$e = \frac{t}{2} - \frac{\text{plate}}{3} = \frac{7.63 \text{ in.}}{2} - \frac{4 \text{ in.}}{3} = 2.48 \text{ in.}$$

The wall is as shown in Fig. 7.14.

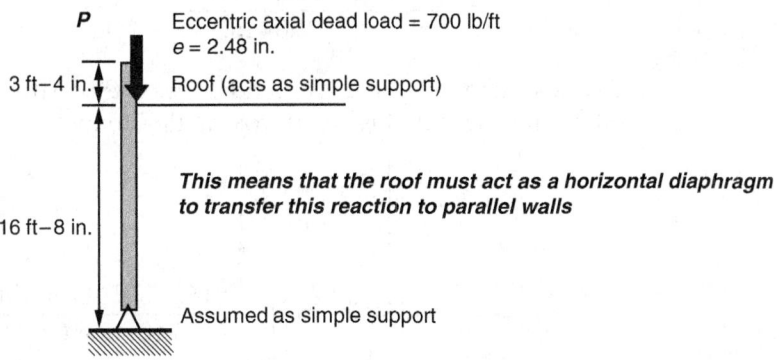

FIGURE 7.14 Unreinforced masonry bearing wall with eccentric axial load.

250 Chapter Seven

At each horizontal plane through the wall, the following conditions must be met:

- Combination of axial and flexural compressive stresses must not violate the unity equation
- Net tension stress must not exceed the allowable flexural tension
- Separate stability check must be satisfied

We must check various points on the wall. Critical points are just below the roof reaction (moment is high and axial load is low, so net tension may govern); and at the base of the wall (axial load is high, so the unity equation or the stability equation may govern). Check each of these locations on the wall.

As before, try 8-in. nominal units, and a specified compressive strength, f'_m, of 1500 lb/in.². This can be satisfied using units with a net-area compressive strength of 1900 lb/in.², and Type S PCL mortar. Work with a strip with a width of 1 ft (measured along the length of the wall in plan). Stresses are calculated using the critical section, consisting of the bedded area only (2008 MSJC *Code*, Sec. 1.9.1.1).

Just below the roof reaction,

$$f_a = \frac{P}{A} = \frac{1050 \text{ lb} + 3.33 \text{ ft} \times 48 \text{ lb/ft}}{30 \text{ in.}^2} = \frac{1210 \text{ lb}}{30 \text{ in.}^2} = 40.3 \text{ lb/in.}^2$$

To calculate stiffness-related parameters for the wall, we use the average cross section, corresponding to the fully bedded section in the table (2008 MSJC *Code*, Sec. 1.9.3).

$$r = \sqrt{\frac{I}{A}} = \sqrt{\frac{334 \text{ in.}^4}{41.5 \text{ in.}^2}} = 2.84 \text{ in.}$$

$$\frac{kh}{r} = \frac{16.67 \text{ ft} \times 12 \text{ in./ft}}{2.84 \text{ in.}} = 70.5$$

This is less than the transition slenderness of 99, so the allowable stress is based on the curve that is an approximation to inelastic buckling:

$$F_a = 0.25 f'_m \left[1 - \left(\frac{h}{140r} \right)^2 \right]$$

$$F_a = 0.25 \cdot 1500 \text{ lb/in.}^2 \left[1 - \left(\frac{16.67 \text{ ft} \times 12 \text{ in./ft}}{140 \times 2.84 \text{ in.}} \right)^2 \right]$$

$$= 375 \text{ lb/in.}^2 \cdot 0.746 = 280 \text{ lb/in.}^2$$

Now there is bending stress:

$$M = P_e = 1050 \text{ lb} \times 2.48 \text{ in.} = 2604 \text{ lb-in.}$$

$$f_b = \frac{Mc}{I} = \frac{2604 \text{ lb-in.} \times \left(7.63/2\right) \text{ in.}}{309 \text{ in.}^4} = 32.15 \text{ lb/in.}^2$$

$$F_b = \frac{f'_m}{3} = \frac{1500 \text{ lb/in.}^2}{3} = 500 \text{ lb/in.}^2$$

Now check the unity equation.

$$\frac{f_a}{F_a} + \frac{f_b}{F_b} \leq 1$$

$$\frac{40.3 \text{ lb/in.}^2}{280 \text{ lb/in.}^2} + \frac{32.15 \text{ lb/in.}^2}{500 \text{ lb/in.}^2} = 0.144 + 0.064 = 0.208 \leq 1$$

and the unity equation is satisfied.

Because the bending stress (32.15 lb/in.²) is less than the axial stress (40.3 lb/in.²), there is no net tensile stress, and that equation does not have to be checked.

Now check the stability equation. Because the load is eccentric, the penalty term has a value less than 1. The effective eccentricity is the moment at the section under consideration, divided by the axial force there:

$$e = \frac{M}{P} = \frac{2604 \text{ lb-in.}}{1210 \text{ lb}} = 2.15 \text{ in.}$$

$$P_e = \frac{\pi^2 E_m I}{h^2}\left(1 - 0.577\frac{e}{r}\right)^3$$

$$P_e = \frac{\pi^2 \cdot 900 \cdot 1500 \text{ lb/in.}^2 \cdot 334 \text{ in.}^4}{(16.67 \cdot 12 \text{ in.})^2}\left(1 - 0.577 \cdot \frac{2.15}{2.84}\right)^3$$

$$= 111{,}211 \cdot 0.178 = 19821 \text{ lb}$$

$$\frac{P_e}{4} = 4955 \text{ lb}$$

Because the calculated axial force per foot of length, 1210 lb, is less than the allowable value of 4955 lb, the stability check is also satisfied.

The other critical section is at the base of the wall. The checks of the unity equation, net flexural tensile stress, and stability are identical to those of the previous example, and are satisfied. The design is therefore satisfactory. The increase in effective eccentricity from the previous example to this example makes a significant difference in this problem.

7.2.5 Example of Allowable-Stress Design of Unreinforced Bearing Wall with Eccentric Axial Load plus Wind

Now consider the same bearing wall of the previous example, but add a uniformly distributed wind load of 25 lb/ft^2. The governing allowable-stress loading combination from the 2009 IBC is $0.6 D + W$. Assume that 700 lb/ft of the 1050 lb/ft is due to D, and 350 to L.

The wall is as shown in Fig. 7.15.

At each horizontal plane through the wall, the following conditions must be met:

- Combination of axial and flexural compressive stresses must not violate the unity equation
- Net tension stress must not exceed the allowable flexural tension
- Separate stability check must be satisfied

We must check various points on the wall. Critical points are just below the roof reaction (moment is high and axial load is low, so net tension may govern); and at the base of the wall (axial load is high, so the unity equation or the stability equation may govern). Check each of these locations on the wall.

To avoid having to check a large number of loading combinations and potentially critical locations, it is worthwhile to assess them first, and check only the ones that will probably govern.

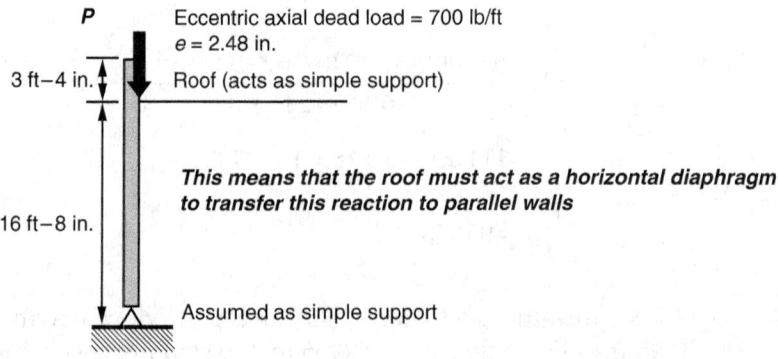

Figure 7.15 Unreinforced masonry bearing wall with eccentric axial load and wind load.

Allowable-Stress Design of Unreinforced Masonry Elements

Due to wind only, the unfactored moment at the base of the parapet (roof level) is

$$M = \frac{qL^2}{2} = \frac{25 \text{ lb/ft} \times 3.33^2 \text{ ft}^2}{2} \times 12 \text{ in./ft} = 1663 \text{ lb-in.}$$

The maximum moment is close to that occurring at mid-height. The moment from wind load is the superposition of one-half moment at the upper support due to wind load on the parapet only, plus the midspan moment in a simply supported beam with that same wind load:

$$M_{midspan} = -\frac{1663}{2} + \frac{qL^2}{8} = -\frac{1663}{2} + \frac{25 \text{ lb/ft} \cdot 16.67^2 \text{ ft}^2}{8} \cdot 12 \text{ in./ft}$$

$$= 9589 \text{ lb-in.}$$

The unfactored moment due to eccentric axial load is

$$M_{gravity} = Pe = 1050 \text{ lb} \times 2.48 \text{ in.} = 2604 \text{ lb-in.}$$

Unfactored moment diagrams due to eccentric axial load and wind are as shown in Fig. 7.16.

As before, try 8-in. nominal units, and a specified compressive strength, f'_m, of 1500 lb/in.² This can be satisfied using units with a net-area compressive strength of 1900 lb/in.², and Type S PCL mortar. Work with a strip of 1 ft width (measured along the length of the wall in plan). Stresses are calculated using the critical section, consisting of the bedded area only (2008 MSJC *Code*, Sec. 1.9.1.1):

Just below the roof reaction,

$$f_a = \frac{0.6 \, P}{A} = \frac{0.6 \times 700 \text{ lb} + 0.6 \times 3.33 \text{ ft} \cdot 48 \text{ lb/ft}}{30 \text{ in.}^2} = \frac{516 \text{ lb}}{30 \text{ in.}^2} = 17.20 \text{ lb/in.}^2$$

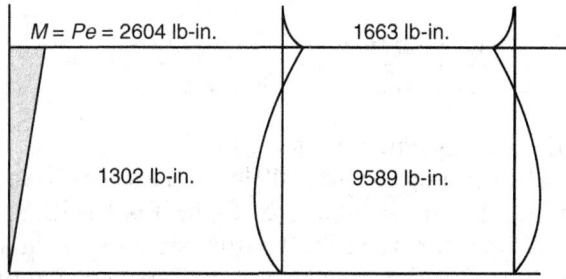

FIGURE 7.16 Unfactored moment diagrams due to eccentric axial load and wind.

To calculate stiffness-related parameters for the wall, we use the average cross section, corresponding to the fully bedded section in the table (2008 MSJC *Code*, Sec. 1.9.3).

$$r = \sqrt{\frac{I}{A}} = \sqrt{\frac{334 \text{ in.}^4}{41.5 \text{ in.}^2}} = 2.84 \text{ in.}$$

$$\frac{kh}{r} = \frac{16.67 \text{ ft} \times 12 \text{ in.}/\text{ft}}{2.84 \text{ in.}} = 70.5$$

This is less than the transition slenderness of 99, so the allowable stress is based on the curve that is an approximation to inelastic buckling:

$$F_a = 0.25 f'_m \left[1 - \left(\frac{h}{140r}\right)^2\right]$$

$$F_a = 0.25 \cdot 1500 \text{ lb/in.}^2 \left[1 - \left(\frac{16.67 \text{ ft} \times 12 \text{ in.}/\text{ft}}{140 \times 2.84 \text{ in.}}\right)^2\right]$$

$$= 375 \text{ lb/in.}^2 \times 0.746 = 280 \text{ lb/in.}^2$$

Bending stress comes from wind load plus eccentric gravity load:

$$M = Pe + \text{wind} = 0.6 \times 700 \text{ lb} \times 2.48 \text{ in.} + 1663 \text{ lb-in.} = 2705 \text{ lb-in.}$$

$$f_b = \frac{Mc}{I} = \frac{2705 \text{ lb-in.} \cdot (7.63/2) \text{ in.}}{309 \text{ in.}^4} = 33.40 \text{ lb/in.}^2$$

$$F_b = \frac{f'_m}{3} = \frac{1500 \text{ lb/in.}^2}{3} = 500 \text{ lb/in.}^2$$

Now check the unity equation.

$$\frac{f_a}{F_a} + \frac{f_b}{F_b} \leq 1$$

$$\frac{17.20 \text{ lb/in.}^2}{280 \text{ lb/in.}^2} + \frac{33.40 \text{ lb/in.}^2}{500 \text{ lb/in.}^2} = 0.061 + 0.067 = 0.128 \leq 1$$

and the unity equation is satisfied.

The bending stress (33.40 lb/in.²) exceeds the axial stress (17.20 lb/in.²). The net tensile stress (16.20 lb/in.²) is less than the allowable stress of 25 lb/in.² for flexural tensile stresses normal to the bed joint in ungrouted hollow masonry, with Type S PCL mortar. Net tensile stresses are satisfactory.

Now check the stability equation. Moments and axial forces from lateral loads are not included in this check. Therefore, it proceeds exactly as in the previous example:

$$e = \frac{M}{P} = \frac{2604 \text{ lb-in.}}{1210 \text{ lb}} = 2.15 \text{ in.}$$

$$P_e = \frac{\pi^2 E_m I}{h^2}\left(1 - 0.577\frac{e}{r}\right)^3$$

$$P_e = \frac{\pi^2 \cdot 900 \cdot 1500 \text{ lb/in.}^2 \cdot 334 \text{ in.}^4}{(16.67 \cdot 12 \text{ in.})^2}\left(1 - 0.577 \cdot \frac{2.15}{2.84}\right)^3$$

$$= 111{,}211 \cdot 0.178 = 19{,}821 \text{ lb}$$

$$\frac{P_e}{4} = 4955 \text{ lb}$$

Because the calculated axial force per foot of length, 1210 lb, is less than the allowable value of 4955 lb, the stability check is also satisfied.

Now check the wall at mid-height:

$$f_a = \frac{P}{A} = \frac{0.6 \times 700 \text{ lb} + 0.6 \times (3.33 + 8.33) \text{ ft} \cdot 48 \text{ lb/ft}}{30 \text{ in.}^2}$$

$$= \frac{755.8 \text{ lb}}{30 \text{ in.}^2} = 25.19 \text{ lb/in.}^2$$

To calculate stiffness-related parameters for the wall, we use the average cross section, corresponding to the fully bedded section in the table (2008 MSJC *Code*, Sec. 1.9.3).

$$r = \sqrt{\frac{I}{A}} = \sqrt{\frac{334 \text{ in.}^4}{41.5 \text{ in.}^2}} = 2.84 \text{ in.}$$

$$\frac{kh}{r} = \frac{16.67 \cdot 12 \text{ in.}}{2.84 \text{ in.}} = 70.5$$

This is less than the transition slenderness of 99, so the allowable stress is based on the curve that is an approximation to inelastic buckling:

$$F_a = 0.25 f'_m \left[1 - \left(\frac{h}{140r}\right)^2\right]$$

$$F_a = 0.25 \cdot 1500 \text{ lb/in.}^2 \left[1 - \left(\frac{16.67 \cdot 12 \text{ in.}}{140 \cdot 2.84 \text{ in.}}\right)^2\right]$$

$$= 375 \text{ lb/in.}^2 \cdot 0.746 = 280 \text{ lb/in.}^2$$

Bending stress comes from the net moment shown above, assuming that the wind is directed so that moments from eccentric gravity load are added to moments from wind:

$$M = P\left(\frac{e}{2}\right) + \text{wind} = 0.6 \times 868 + 9589 \text{ lb-in.} = 10,110 \text{ lb-in.}$$

$$f_b = \frac{Mc}{I} = \frac{10,110 \text{ lb-in.} \times (7.63/2) \text{ in.}}{309 \text{ in.}^4} = 124.8 \text{ lb/in.}^2$$

$$F_b = \frac{f'_m}{3} = \frac{1500 \text{ lb/in.}^2}{3} = 500 \text{ lb/in.}^2$$

Now check the unity equation.

$$\frac{f_a}{F_a} + \frac{f_b}{F_b} \leq 1$$

$$\frac{25.19 \text{ lb/in.}^2}{280 \text{ lb/in.}^2} + \frac{124.8 \text{ lb/in.}^2}{500 \text{ lb/in.}^2} = 0.090 + 0.250 = 0.340 \leq 1$$

and the unity equation is satisfied.

The bending stress (124.8 lb/in.²) exceeds the axial stress (25.19 lb/in.²). The net tensile stress (99.61 lb/in.²) exceeds the allowable stress of 25 lb/in.² for flexural tensile stresses normal to the bed joint in ungrouted hollow masonry, with Type S PCL mortar. It will be necessary to grout the wall. The section modulus will increase, and the allowable stress will also increase from 25 lb/in.² to 65 lb/in.². Net tensile stresses will be satisfactory. Recheck the wall as grouted.

$$f_a = \frac{P}{A} = \frac{0.6 \times 700 \text{ lb} + 0.6 \times (3.33 + 8.33) \text{ ft} \times 76 \text{ lb/ft}}{12 \times 7.63 \text{ in.}^2}$$

$$= \frac{951.7 \text{ lb}}{91.6 \text{ in.}^2} = 10.39 \text{ lb/in.}^2$$

$$f_b = \frac{Mc}{I} = \frac{10,110 \text{ lb-in.} \times (7.63/2) \text{ in.}}{\left(\frac{12 \times 7.63^3}{12}\right) \text{ in.}^4} = 86.83 \text{ lb/in.}^2$$

The bending stress (86.83 lb/in.²) exceeds the axial stress (10.39 lb/in.²). The net tensile stress (76.44 lb/in.²) exceeds the allowable stress of 65 lb/in.² for flexural tensile stresses normal to the bed joint in solid-grouted masonry, with Type S PCL mortar. It would be necessary either to increase the thickness of the wall to 10 in., or to reinforce the wall.

Now check the stability equation. Moments and axial forces from lateral loads are not included in this check. Therefore, it proceeds exactly as in the previous example. Because the calculated axial force per foot of length at mid-height, 1610 lb, is less than the allowable value of 4955 lb, the stability check is also satisfied.

The other critical section is at the base of the wall. The checks of the unity equation, net flexural tensile stress, and stability are identical to those of the example in Sec. 7.2.4, and are satisfied.

7.2.6 Comments on the Above Examples for Allowable-Stress Design of Unreinforced Bearing Walls

1. In retrospect, it probably would not have been necessary to check all three criteria at all locations. With experience, a designer could realize that the location with highest wind moment would govern, and could therefore check only the mid-height of the wall.

2. The addition of wind load to the example of Sec. 7.2.4, to produce the example of Sec. 7.2.5, changes the critical location from just under the roof, to the mid-height of the simply supported section of the wall. The wind load of 25 lb/ft^2 in the third example produces maximum tensile stresses above the allowable values for ungrouted masonry, and makes it necessary to grout the wall, thicken it, or reinforce it.

7.2.7 Extension of the Above Concepts to Masonry Walls with Openings

In the previous examples, we have studied the behavior of bearing walls of unreinforced masonry, idealized as a series of vertical strips, simply supported at the level of the floor slab, and at the level of the roof. Let's see how this changes in the case of bearing walls with openings.

In Fig. 7.17, load applied above the window and door openings clearly cannot be resisted by vertical strips, because those vertical strips have only one point of lateral support (at the roof level).

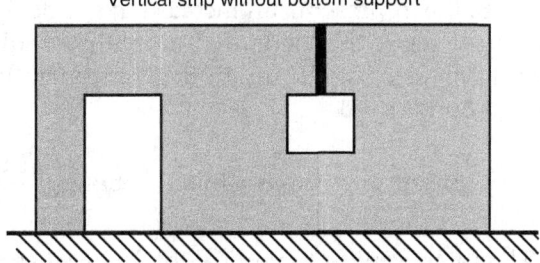

FIGURE 7.17 Hypothetical unstable resistance mechanism in a wall with openings, involving vertically spanning strips only.

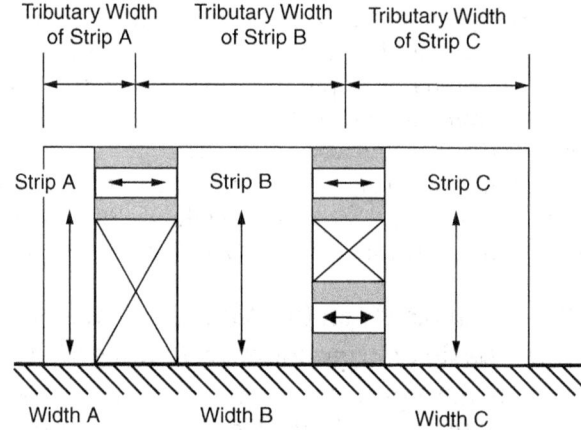

Figure 7.18 Stable resistance mechanism in a wall with openings, involving horizontally spanning strips in addition to vertically spanning strips.

For that reason, the wall must be idealized as horizontal strips above and below the openings, supported by vertical strips on both sides of the openings, as shown in Fig. 7.18.

Each set of horizontal strips, idealized as simply supported, must be supported by the adjacent vertical strips. For example, the horizontal strips above the door are supported by Strip A and Strip B. The window and door are considered to transfer loads applied to them, via horizontal strips, to the vertical strips on either side of the openings.

Therefore, Strip A has to support, spanning vertically, the out-of-plane loads acting directly on it, plus the out-of-plane loads acting on the left half of the horizontal strips above the door. In other words, Strip A has to resist the out-of-plane loads acting on what might be termed a "tributary width," which extends from the left-hand edge of Strip A itself, to the midspan of the horizontal strips above the door. In the same way, Strips B and C have to resist the loads corresponding to Tributary Widths B and C, respectively.

For example, if Strip B has to resist the loads acting over Tributary Width B, this represents an increase in the design loads on Strip B. That strip must resist the loads that normally would be applied to it (if no openings had existed), multiplied by the ratio of Tributary Width B, divided by Width B:

$$\text{Actions in Strip B} = \text{Initial actions} \left(\frac{\text{Tributary Width B}}{\text{Width B}} \right)$$

The same applies to vertical loads, because these also must be transferred from horizontal to vertical strips.

In any event, the presence of openings can be considered to increase the initial actions in the vertical strips adjacent to the openings. Aside from this increase, the design of those elements proceeds exactly as before.

7.2.8 Final Comment on the Effect of Openings in Unreinforced Masonry Bearing Walls

As the summation of the plan lengths of openings in a bearing wall exceeds about one-half the plan length of the wall, even the higher allowable stresses (or moduli of rupture) corresponding to fully grouted walls will be exceeded, and it will generally become necessary to use reinforcement. Design of reinforced masonry bearing walls is addressed later in this book.

7.3 Allowable-Stress Design of Unreinforced Shear Walls

7.3.1 Basic Behavior of Unreinforced Shear Walls

Box-type structures resist lateral loads as shown in Fig. 7.19.

This resistance mechanism involves three steps:

- Walls oriented perpendicular to the direction of lateral load transfer those loads to the level of the foundation and the levels of the horizontal diaphragms. The walls are idealized and designed as vertically oriented strips.

- The roof and floors act as horizontal diaphragms, transferring their forces to walls oriented parallel to the direction of lateral load.

- Walls oriented parallel to the direction of applied load must transfer loads from the horizontal diaphragms to the foundation. In other words, they act as shear walls.

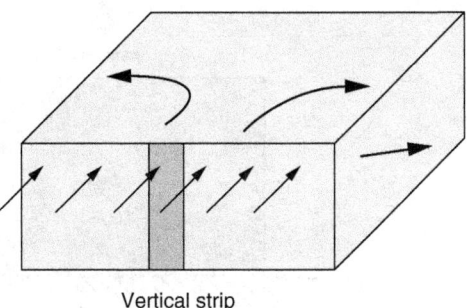

Vertical strip

Figure 7.19 Basic behavior of box-type buildings in resisting lateral loads.

As noted previously in the sections dealing with unreinforced bearing walls, this overall mechanism demands that the horizontal roof diaphragm have sufficient strength and stiffness to transfer the required loads. This is discussed again in a later section of this book dealing with horizontal diaphragms.

The rest of this section addresses the design of shear walls. We shall see that in almost all cases, the design itself is very simple, because the cross-sectional areas of the masonry walls so large that nominal stresses are quite low.

7.3.2 Design Steps for Unreinforced Shear Walls

Unreinforced masonry shear walls must be designed for the effects of:

1. Gravity loads from self-weight, plus gravity loads from overlying roof or floor levels
2. Moments and shears from in-plane shear loads

Actions are shown in Fig. 7.20. Either allowable-stress design or strength design can be used.

The 2008 MSJC *Code* specifies allowable flexural tensile stresses for unreinforced masonry in Table 7.1 (Table 2.2.3.2 of 2008 MSJC *Code*), which applies equally to in-plane and out-of-plane bending.

$$f_{\text{tension}} = \frac{M}{S} - \frac{P}{A} = \frac{Vh}{S} - \frac{P}{A} \leq F_t$$

Shearing capacity is calculated using Sec. 2.2.5 of the 2008 MSJC *Code*. Shear stresses are calculated by:

$$f_v = \frac{VQ}{I_n b}$$

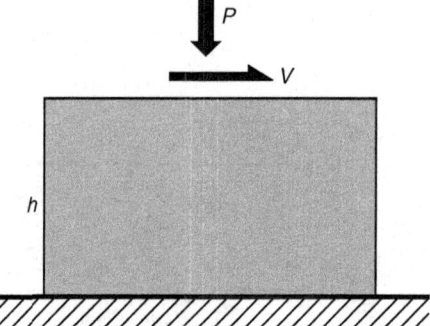

Figure 7.20 Design actions for unreinforced shear walls.

Allowable-Stress Design of Unreinforced Masonry Elements

and allowable in-plane shear stresses, F_v, shall not exceed any of:

$$F_v = \begin{cases} 1.5\sqrt{f'_m} \\ 120\,\text{psi} \\ v + 0.45(N_v/A_v) \end{cases}$$

in the third equation,

v = 37 psi for masonry in running bond that is not grouted solid
  = 37 psi for masonry in other than running bond with open-end units grouted solid
  = 60 psi for masonry in running bond that is grouted solid

7.3.3 Design Example of Allowable-Stress Design of Unreinforced Masonry Shear Wall

Consider the simple structure of Fig. 7.21, the same one whose bearing walls have been designed previously in this book. Use nominal 8-in. concrete masonry units, f'_m = 1500 lb/in.², and Type S PCL mortar. The roof applies a gravity load of 1050 lb/ft to the walls; the walls measure 16 ft, 8 in. height to the roof, and have an additional 3 ft, 4 in. parapet. The walls are loaded with a wind load of 20 lb/ft.². The roof acts as a one-way system, transmitting gravity loads to the front and back walls.

Now design the shear wall. Try an 8-in. wall with face-shell bedding only. The critical section for shear is just under the roof, where axial load in the shear walls is least, coming from the parapet only.

As a result of the wind loading, the reaction transmitted to the roof diaphragm is as calculated using Fig. 7.22.

$$\text{Reaction} = \frac{20\,\text{lb/ft}^2 \cdot \left(\dfrac{20^2\,\text{ft}^2}{2}\right)}{16.67\,\text{ft}} = 240\,\text{lb/ft}$$

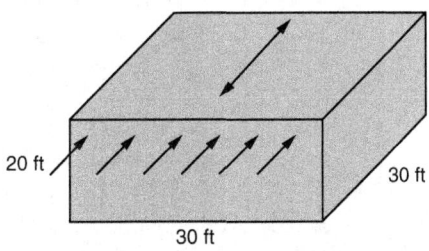

FIGURE 7.21 Example problem for strength design of unreinforced shear wall.

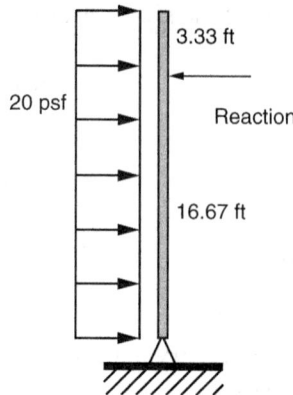

FIGURE 7.22 Calculation of reaction on roof diaphragm, strength design of unreinforced shear wall.

Total roof reaction acting on one side of the roof is

$$\text{Reaction} = 240 \text{ lb/ft} \cdot 30 \text{ ft} = 7200 \text{ lb}$$

This is divided evenly between the two shear walls, so the shear per wall is 3600 lb.

In Fig. 7.23, for simplicity, the lateral load is shown as if it acted on the front wall alone. In reality, it also acts on the back wall, so that the structure is subjected to pressure on the front wall, and suction on the back wall.

Now design the shear wall. Try an 8-in. wall with face-shell bedding only. The critical section for shear is just under the roof, where axial load in the shear walls is least, coming from the parapet only.

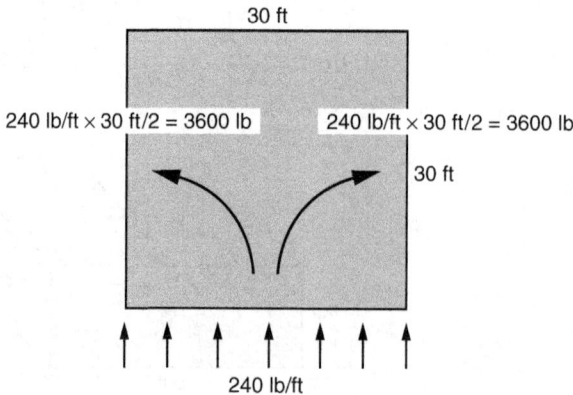

FIGURE 7.23 Transmission of forces from roof diaphragm to shear walls.

Compute the axial force in the wall at that level. To be conservative, assume that only dead load act. The force acting normal to the shear-transfer plane is

$$N_v = 3.33 \text{ ft} \times 48 \text{ lb/ft}^2 = 160 \text{ lb/ft}$$

The maximum shear stress at that level is

$$f_v = \frac{VQ}{I_n b} = \left(\frac{3}{2}\right)\frac{V}{A} = \left(\frac{3}{2}\right)\frac{3600 \text{ lb}}{30 \text{ in.}^2/\text{ft} \times 30 \text{ ft}} = 6.00 \text{ lb/in.}^2$$

and allowable in-plane shear stresses, F_v, is the smallest of:

$$F_v = \begin{cases} 1.5\sqrt{f'_m} = 1.5\sqrt{1500} = 58.1 \text{ lb/in.}^2 \\ 120 \text{ psi} \\ v + 0.45(N_v/A_v) = 37 \text{ lb/in.}^2 + 0.45\left(\frac{160 \text{ lb}}{30 \text{ in.}^2}\right) = 39.4 \text{ lb/in.}^2 \end{cases}$$

The lowest allowable shear stress, 39.4 lb/in.², far exceeds the maximum shear stress of 6.0 lb/in.², and the design is satisfactory for shear.

Now check the net flexural tensile stress. The critical section is at the base of the wall, where in-plane moment is maximum. Because the roof spans between the front and back walls, the distributed gravity load on the roof does not act on the side walls, and their axial load comes from self-weight only:

$$f_{tension} = \frac{Mc}{I} - \frac{P}{A} = \frac{Vhc}{I} - \frac{P}{A} \leq F_t$$

$$f_{tension} = \frac{3600 \text{ lb} \times 16.67 \text{ ft} \cdot 12 \text{ in./ft}\left(\frac{30 \text{ ft} \times 12 \text{ in./ft}}{2}\right)}{\left[\frac{2 \times 1.25 \text{ in.} \cdot (30 \text{ ft} \cdot 12 \text{ in./ft})^3}{12}\right]} - \frac{20 \text{ ft} \times 48 \text{ lb/ft}}{30 \text{ in.}^2}$$

$$f_{tension} = 13.33 \text{ lb/in.}^2 - 32.00 \text{ lb/in.}^2 = -18.67 \text{ lb/in.}^2$$

The net flexural tension is actually compression, and is certainly less than the allowable flexural tension from Table 7.1 (Table 2.2.3.2 of the 2008 MSJC *Code*). The design is satisfactory.

When the wind blows against the side walls, these walls transfer their loads to the roof diaphragm, and the front and back walls act as shear walls. The side walls must be checked for this loading direction also, following the procedures of previous examples.

7.3.4 Comments on Example Problem with Allowable-Stress Design of Unreinforced Shear Walls

Clearly, unreinforced masonry shear walls, whether designed by allowable stress or strength design procedures, have tremendous shear capacity because of their large cross-sectional area. If this area is reduced by openings, then shear capacities will decrease, and in-plane flexural capacities as governed by net flexural tension may decrease even faster.

7.3.5 Comments on Behavior and Design of Wall Buildings in General

Wall buildings are very efficient structurally, because the same element can act as part of the building envelope, as a vertically spanning structural element perpendicular to the direction of applied lateral load, and as a shear wall parallel to the direction of applied lateral load. If the wall building is made of a material that is aesthetically pleasing, like masonry, even more efficiency is achieved. Wall buildings are also very efficient from the viewpoint of design. The ultimate objective of design is design, not analysis.

The basic steps that are discussed here, in the context of simple, one-story shear wall buildings, can be applied to multistory shear wall buildings as well. At the roof level and at each floor level, horizontal diaphragms receive reactions from vertically spanning strips, and transfer those reactions to shear walls. Each shear wall acts essentially as a free-standing, statically determinate cantilever, with axial loads and in-plane lateral loads applied at each floor level. At each floor level, the shear wall must simply be designed for shear, and for combined axial force and moment.

Design of shear wall buildings is typically much easier than the design of frames, which are statically indeterminate and must usually be analyzed using computer programs.

7.3.6 Extension to Design of Unreinforced Masonry Shear Walls with Openings

Consider the structure shown in Fig. 7.24. The wall is identical to that addressed in the previous examples, with the exception of two openings, each measuring 9 ft in plan. These openings divide the wall into three smaller wall segments.

Assume that the applied shear is divided equally among the three wall segments; that points of inflection exist at the mid-height of each wall segment; and that axial forces in the wall are negligible. Then the moments and shears, can be determined by statics, where L is the 10-ft height of the wall segments. A free body of one wall segment is shown in Fig. 7.25.

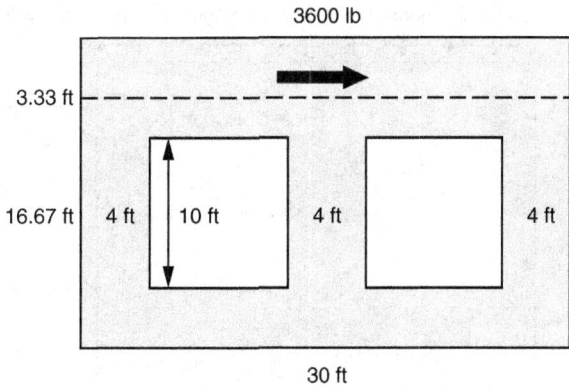

FIGURE 7.24 Shear wall with openings.

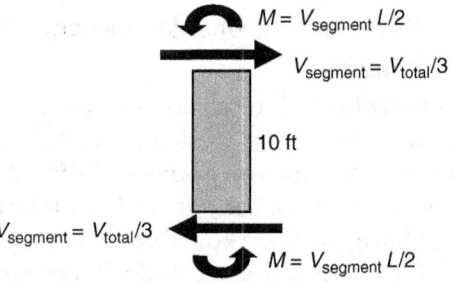

FIGURE 7.25 Free body of one wall segment.

The rest of the design proceeds as before. The shear area of the wall segments is reduced in proportion to the plan length of each segment, compared to the plan length of the original unperforated wall. The moment of inertia of the segments, however, is considerably less than the moment of inertia of the original unperforated wall.

7.4 Allowable-Stress Design of Anchor Bolts

In masonry construction, anchor bolts are most commonly used to anchor roof or floor diaphragms to masonry walls. As shown in Fig. 7.26, vertically oriented anchor bolts can be placed along the top of a masonry wall to anchor a roof diaphragm resting on the top of the wall. Alternatively, horizontally oriented anchor bolts can be placed along the face of a masonry wall to anchor a diaphragm through a horizontal ledger. In these applications, anchor bolts are subjected to combinations of tension and shear. In this section, the behavior of anchors under those loadings is discussed, and 2008 MSJC allowable-stress design provisions are reviewed.

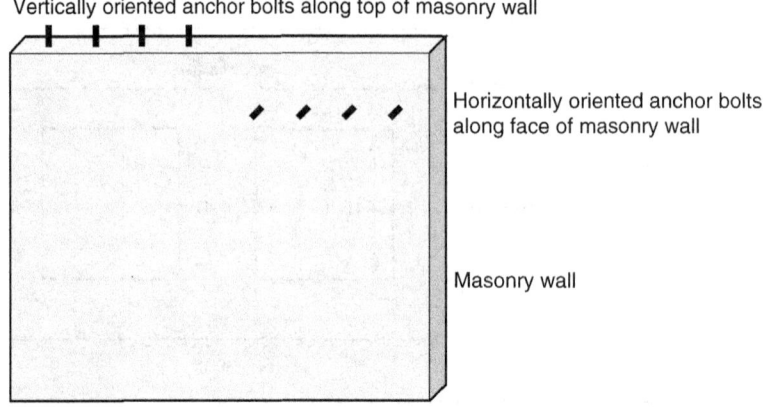

FIGURE 7.26 Common uses of anchor bolts in masonry construction.

7.4.1 Behavior and Design of Anchor Bolts Loaded in Tension

Anchor bolts loaded in tension can fail by breakout of a roughly conical body of masonry, or by yield and fracture of the anchor bolt steel. Bent-bar anchor bolts (such as J-bolts or L-bolts) can also fail by straightening of the bent portion of the anchor bolt, followed by pullout of the anchor bolt from the masonry. Allowable tensile capacity as governed by masonry breakout is evaluated using a design model based on a uniform tensile stress of $1.25\sqrt{f'_m}$ acting perpendicular to the inclined surface of an idealized breakout body consisting of a right circular cone (Fig. 7.27). The capacity associated with that stress state is identical with the capacity corresponding to a uniform tensile stress of $1.25\sqrt{f'_m}$ acting perpendicular to the projected area of the right circular cone. This design approach, while less sophisticated than that of ACI318-08 App. D, has been shown to be user-friendly and safe for typical masonry applications.

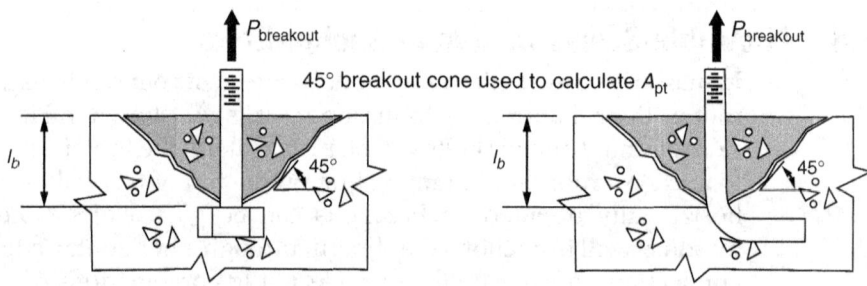

FIGURE 7.27 Idealized conical breakout cones for anchor bolts loaded in tension.

Allowable tensile capacities for anchors as governed by masonry breakout are identical for headed and bent-bar anchors, and are given by Eqs. (3-1) and (3-3) of the 2008 MSJC *Code*.

$$B_{ab} = 1.25 A_{pt} \sqrt{f'_m} \qquad \text{2008 MSJC } Code\text{, Eqs. (2-1) and (2-3)}$$

In Eqs. (2-1) and (2-3), the projected area A_{pt} is evaluated in accordance with Eq. (1-2) of the 2008 MSJC *Code*:

$$A_{pt} = \pi l_b^2 \qquad \text{2008 MSJC } Code\text{, Eq. (1-2)}$$

As required by Sec. 1.16.4 of the 2008 MSJC *Code*, the effective embedment length, l_b, for headed anchors is the length of the embedment measured perpendicular from the masonry surface to the compression bearing surface of the anchor head. As required by Sec. 1.16.5 of the 2008 MSJC *Code*, the effective embedment for a bent-bar anchor bolt, l_b, is the length of embedment measured perpendicular from the masonry surface to the compression bearing surface of the bent end, minus one anchor bolt diameter. These are shown in Fig. 7.27. As shown in Fig. 7.28, the projected area must be reduced for the effect of overlapping projected circular areas, and for the effect of any portion of the project area falling in an open cell or core.

Allowable tensile capacities for anchors as governed by steel yield and fracture are also identical for headed and bent-bar anchors, and are given by Eqs. (2-2) and (2-5) of the 2008 MSJC *Code*. In those equations, A_b is the effective tensile stress area of the anchor bolt, including the effect of threads.

$$B_{as} = 0.6 A_b f_y \qquad \text{2008 MSJC } Code\text{, Eqs. (2-2) and (2-5)}$$

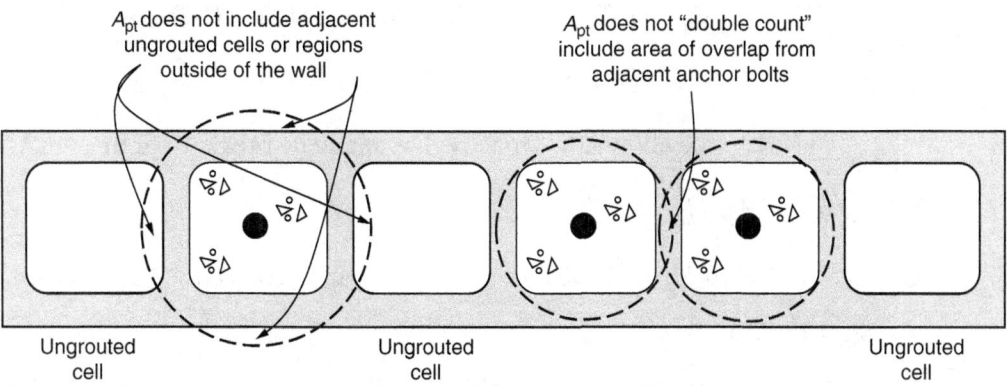

FIGURE 7.28 Modification of projected breakout area, A_{pt}, by void areas or adjacent anchors.

The allowable tensile capacity of bent-bar anchor bolts as governed by pullout is given by Eq. (2-4) of the 2008 MSJC *Code*.

$$B_{anp} = 0.6 f'_m e_b d_b + [120 \pi (l_b + e_b + d_b) d_b] \quad \text{2008 MSJC } Code, \text{ Eq. (2-4)}$$

In that equation, the first term represents capacity due to the hook, and the second term represents capacity due to adhesion along the anchor shank. Article 3.2A of the 2008 MSJC *Specification* requires that anchor shanks be cleaned of material that could interfere with that adhesion.

The failure mode with the lowest allowable capacity governs.

7.4.2 Example of Allowable-Stress Design of a Single Anchor Loaded in Tension

Using allowable-stress design, compute the allowable tensile capacity of a 1/2-in. diameter, A307 bent-bar anchor with a 1-in. hook, embedded vertically in a grouted cell of a nominal 8-in. wall with a specified compressive strength, f'_m, of 1500 lb/in.² Assume that the bottom of the anchor hook is embedded a distance of 4.5 in. This example might represent a tensile anchor used to attach a roof diaphragm to a wall.

First, compute the effective embedment, l_b, in accordance with Sec. 1.16.5 of the 2008 MSJC *Code*, this is equal to the total embedment of 4.5 in., minus the diameter of the anchor (to get to the inside of the hook), and minus an additional anchor diameter, or 3.5 in. As shown in Fig. 7.29, the projected tensile breakout area has a radius of 3.5 in. (diameter of 7 in.). Because the masonry wall has a specified thickness of 7.63 in., the projected tensile breakout area is not affected by adjacent ungrouted cells or regions outside of the wall.

$$A_{pt} = \pi l_b^2$$

$$A_{pt} = \pi (3.5 \text{ in.})^2 \quad \text{2008 MSJC } Code, \text{ Eq. (1-2)}$$

$$A_{pt} = 38.5 \text{ in.}^2$$

Calculate the allowable capacity due to tensile breakout of masonry.

$$B_{ab} = 1.25 A_{pt} \sqrt{f'_m}$$

$$B_{ab} = 1.25 \times 38.5 \text{ in.}^2 \sqrt{1500 \text{ lb/in.}^2} \quad \text{2008 MSJC } Code, \text{ Eqs. (2-1) and (2-3)}$$

$$B_{ab} = 1863 \text{ lb}$$

Now compute the allowable tensile capacity as governed by steel yield. In this computation, A_b is the effective tensile stress area of

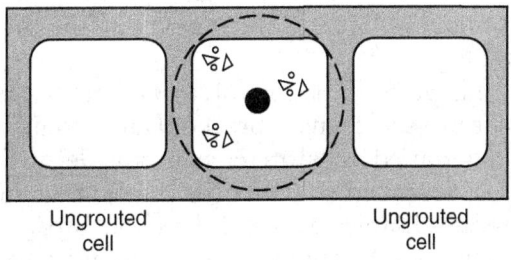

FIGURE 7.29 Example involving a single tensile anchor, placed vertically in a grouted cell.

the anchor bolt, including the effect of threads. According to ANSI/ASME B1.1,

$$A_b = \frac{\pi}{4}\left(d_o - \frac{0.9743}{n_t}\right)^2$$

where d_o = nominal anchor diameter, in.
n_t = number of threads per inch

For anchors with nominal diameters typically used in masonry, the effective tensile stress area can be approximated with sufficient accuracy as 0.75 times the nominal area. That approximation is used in this and other anchor bolt problems here. The minimum specified yield strength for A307 steel is 36 ksi.

$$B_{as} = 0.6\, A_b f_y$$

$B_{ans} = 0.6 \times 0.75 \times 0.20 \text{ in.}^2 \times 60{,}000 \text{ lb/in.}^2$ 2008 MSJC *Code*, Eqs. (2-2) and (2-5)

$B_{ans} = 5400 \text{ lb}$

The allowable tensile capacity of bent-bar anchor bolts as governed by pullout is given by Eq. (2-4) of the 2008 MSJC *Code*.

$$B_{ap} = 0.6\, f'_m e_b d_b + [120\, \pi(l_b + e_b + d_b)d_b]$$

$B_{ap} = 0.6 \times 1500 \text{ lb/in.}^2 \times 1.0 \text{ in.} \times 0.5 \text{ in.}$

$\quad + [120\, \pi(3.5 \text{ in.} + 1.0 \text{ in.} + 0.5 \text{ in.})\, 0.5 \text{ in.}]$ 2008 MSJC *Code*, Eq. (2-4)

$B_{ap} = 450 \text{ lb} + 942 \text{ lb}$

$B_{ap} = 1392 \text{ lb}$

The governing allowable tensile capacity is the lowest of that governed by masonry breakout (1863 lb), yield of the anchor shank (5400 lb), and pullout (1392 lb). Pullout governs, and the allowable tensile capacity is 1392 lb.

If this problem had involved an anchor with deeper embedment (so that the projected tensile breakout area would have been affected by adjacent ungrouted cells or regions outside of the wall), only the anchor capacity as governed by tensile breakout would have been affected, due to a reduced projected tensile breakout area.

Similarly, if this problem had involved adjacent anchors with overlapping tensile breakout areas, only the anchor capacity as governed by tensile breakout would have been affected, again due to a reduced projected tensile breakout area.

7.4.3 Behavior and Design of Anchor Bolts Loaded in Shear

Anchor bolts loaded in shear, and located without a nearby free edge in the direction of load, can fail by local crushing of the masonry under bearing stresses from the anchor bolt; by pryout of the head of the anchor in a direction opposite to the direction of applied load, or by yield and fracture of the anchor bolt steel. Anchor bolts loaded in shear, and located near a free edge in the direction of load, can also fail by breakout of a roughly semi-conical volume of masonry in the direction of the applied shear. Pryout and shear breakout are shown in parts (a) and (b), respectively, of Fig. 7.30.

Allowable shear capacity as governed by pryout is taken as twice the allowable tensile breakout capacity, based on the same empirical evidence used in ACI318-08, App. D. Allowable shear capacity as governed by masonry breakout is evaluated using a design model based on a uniform tensile stress of $1.25\sqrt{f'_m}$ acting perpendicular to the inclined surface of an idealized breakout body consisting of a right circular semi-cone (Fig. 7.31).

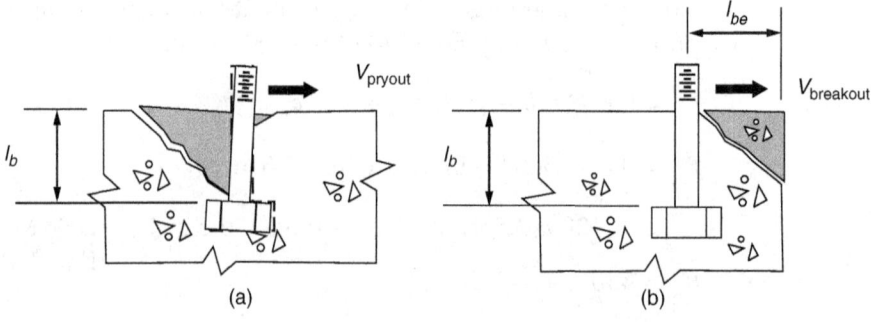

Figure 7.30 (a) Pryout failure and (b) shear breakout failure.

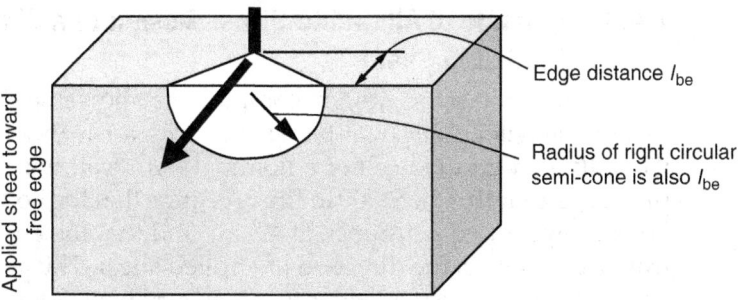

FIGURE 7.31 Design idealization associated with shear breakout failure.

The capacity associated with that stress state is identical with the capacity corresponding to a uniform tensile stress of $1.25\sqrt{f'_m}$ acting perpendicular to the projected area of the right circular semi-cone. This design approach, while less sophisticated than that of ACI318-08 App. D, has been shown to be user-friendly and safe for typical masonry applications.

The nominal shear breakout capacity of an anchor is given by Eq. (2-6) of the 2008 MSJC *Code*. In evaluating that equation, the projected area of the breakout semi-cone is given by Eq. (1-3) of the 2008 MSJC *Code*.

$$B_{vb} = 1.25\, A_{pv}\sqrt{f'_m} \qquad \text{2008 MSJC } Code,\text{ Eq. (2-6)}$$

$$A_{pv} = \frac{\pi l_{be}^2}{2} \qquad \text{2008 MSJC } Code,\text{ Eq. (1-3)}$$

Allowable capacities of anchors loaded in shear are given by Eq. (2-7) of the 2008 MSJC *Code* for masonry crushing, by Eq. (2-8) of the 2008 MSJC *Code* for shear pryout, and by Eq. (2-9) of the 2008 MSJC *Code* for yield of the anchor in shear. In Eqs. (2-7) and (2-9) of the 2008 MSJC *Code* (masonry crushing and anchor yield, respectively), the effective tensile stress area of the bolt (including the effect of threads) is to be used, unless threads are excluded from the shear plane.

$$B_{vc} = 350\sqrt[4]{f'_m A_b} \qquad \text{2008 MSJC } Code,\text{ Eq. (2-7)}$$

$$B_{vpry} = 2.0\, B_{nb} = 2.5\, A_{pt}\sqrt{f'_m} \qquad \text{2008 MSJC } Code,\text{ Eq. (2-8)}$$

$$B_{vs} = 0.36\, A_b f_y \qquad \text{2008 MSJC } Code,\text{ Eq. (2-9)}$$

The failure mode with the lowest allowable capacity governs.

7.4.4 Example of Allowable-Stress Design of a Single Anchor Loaded in Shear

Using allowable-stress design, compute the allowable shear capacity of a 1/2-in. diameter, A307 bent-bar anchor with a 1-in. hook, embedded horizontally in a grouted cell of a nominal 8-in. wall with a specified compressive strength, f'_m, of 1500 lb/in.^{2}. Assume that the bottom of the anchor hook is embedded a distance of 4.5 in., and that the anchor is located far from free edges in the direction of applied shear. This might represent an anchor used to attach a ledger to a masonry wall. Because free edges are not a factor, shear breakout does not apply.

First, compute the effective embedment, l_b. In accordance with Sec. 1.16.5 of the 2008 MSJC *Code*, this is equal to the total embedment of 4.5 in., minus the diameter of the anchor (to get to the inside of the hook), and minus an additional anchor diameter, or 3.5 in. The projected tensile breakout area has a radius of 3.5 in. (diameter of 7 in.).

$$A_{pt} = \pi l_b^2$$

$$A_{pt} = \pi (3.5 \text{ in.})^2 \quad \text{2008 MSJC } \textit{Code}, \text{ Eq. (1-2)}$$

$$A_{pt} = 38.5 \text{ in.}^2$$

First, compute the allowable capacity of the anchor as governed by masonry crushing.

$$B_{vc} = 350 \sqrt[4]{f'_m A_b} \quad \text{2008 MSJC } \textit{Code}, \text{ Eq. (2-7)}$$

In this computation, A_b is the effective tensile stress area of the anchor bolt, including the effect of threads. According to ANSI/ASME B1.1,

$$A_b = \frac{\pi}{4}\left(d_o - \frac{0.9743}{n_t}\right)^2$$

where d_o = nominal anchor diameter, in.
n_t = number of threads per inch

For anchors with nominal diameters typically used in masonry, the effective tensile stress area can be approximated with sufficient accuracy as 0.75 times the nominal area. That approximation is used in this and other anchor bolt problems here.

$$B_{vc} = 350 \sqrt[4]{f'_m A_b}$$

$$B_{vc} = 350 \sqrt[4]{1500 \text{ lb/in.}^2 \times (0.75 \times 0.20 \text{ in.}^2)} \quad \text{2008 MSJC } \textit{Code}, \text{ Eq. (2-7)}$$

$$B_{vc} = 1356 \text{ lb/in.}^2$$

Allowable-Stress Design of Unreinforced Masonry Elements

Next, compute the allowable capacity of the anchor as governed by pryout.

$$B_{vpry} = 2.0\, B_{ab} = 2.5\, A_{pt}\sqrt{f'_m} \qquad \text{2008 MSJC Code, Eq. (2-8)}$$

Because allowable pryout capacity is a multiple of the allowable tensile breakout capacity, we must compute the allowable tensile breakout capacity.

$$B_{nb} = 1.25\, A_{pt}\sqrt{f'_m}$$

$$B_{ab} = 1.25 \times 38.5\,\text{in.}^2 \sqrt{1500\,\text{lb/in.}^2} \qquad \text{2008 MSJC Code, Eqs. (2-1) and (2-3)}$$

$$B_{anb} = 1863\,\text{lb}$$

Continue with the pryout calculation:

$$B_{vpry} = 2.0\, B_{ab}$$

$$B_{vpry} = 2.0 \times 1863\,\text{lb} \qquad \text{2008 MSJC Code, Eq. (2-8)}$$

$$B_{vpry} = 3726\,\text{lb}$$

Next, compute the allowable capacity of the anchor as governed by yield and fracture of the anchor shank.

$$B_{vs} = 0.36\, A_b f_y$$

$$B_{vs} = 0.36 \times (0.75 \times 0.20\,\text{in.}^2) \times 60{,}000\,\text{lb/in.}^2 \qquad \text{2008 MSJC Code, Eq. (2-9)}$$

$$B_{vs} = 3240\,\text{lb}$$

The governing allowable shear capacity is the lowest of that governed by masonry crushing (1356 lb), pryout (3726 lb), and yield of the anchor shank (3240 lb). Because the anchor is not close to a free edge, shear breakout does not apply. Masonry crushing governs, and the allowable shear capacity is 1356 lb.

If this problem had involved an anchor loaded toward a free edge, then shear breakout would have had to be checked.

7.4.5 Behavior and Design of Anchor Bolts Loaded in Combined Tension and Shear

Design capacities of anchor bolts in combined tension and shear are given by the linear interaction equation of Eq. (2-10) of the 2008 MSJC

Code. While an elliptical or trilinear interaction equation would be slightly more accurate, a linear interaction is conservative and simple for design.

$$\frac{b_a}{B_a} + \frac{b_v}{B_v} \leq 1 \qquad \text{2008 MSJC } Code, \text{ Eq. (2-10)}$$

7.5 Required Details for Unreinforced Bearing Walls and Shear Walls

Bearing walls that resist out-of-plane lateral loads, and shear walls, must be designed to transfer lateral loads to the floors above and below. Examples of such connections are shown below. These connections would have to be strengthened for regions subjected to strong earthquakes or strong winds. Section 1604.8.2 of the 2009 IBC has additional requirements for anchorage of diaphragms to masonry walls. Section 12.11 of ASCE 7-05 has additional requirements for anchorage of structural walls for structures assigned to Seismic Design Categories C and higher.

7.5.1 Wall-to-Foundation Connections

As shown in Fig. 7.32, CMU walls (or the inner CMU wythe of a drainage wall) must be connected to the concrete foundation. Bond breaker should be used only between the outer veneer wythe and the foundation.

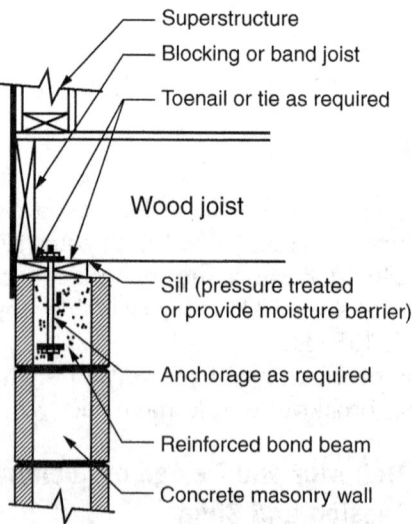

Figure 7.32 Example of wall-to-foundation connection. (*Source*: Figure 1 of National Concrete Masonry Association TEK 05-07A.)

7.5.2 Wall-to-Floor Details

Example of a wall-to-floor detail are shown in Figs. 7.33 and 7.34. In the latter detail (floor or roof planks oriented parallel to walls), the planks are actually cambered. They are shown on the outside of the walls so that this camber does not interfere with the coursing of the units. Some designers object to this detail because it could lead to spalling of the cover. If it is modified so that the planks rest on the face shells of the walls, then the thickness of the topping must vary to adjust for the camber, and form boards must be used against both sides of the wall underneath the planks, so that the concrete or grout that is cast into the bond beam does not run out underneath the cambered beam.

7.5.3 Wall-to-Roof Details

An example of a wall-to-roof detail is shown in Fig. 7.35.

7.5.4 Typical Details of Wall-to-Wall Connections

Typical details of wall-to-wall connections are shown in Fig. 7.36.

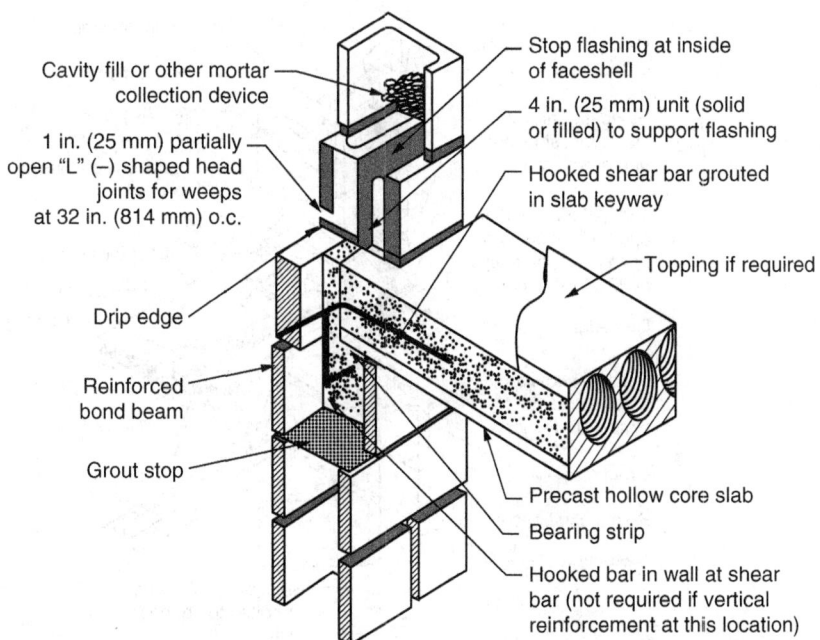

Figure 7.33 Example of wall-to-floor connection, planks perpendicular to wall. (*Source*: Figure 14 of National Concrete Masonry Association TEK 05-07A.)

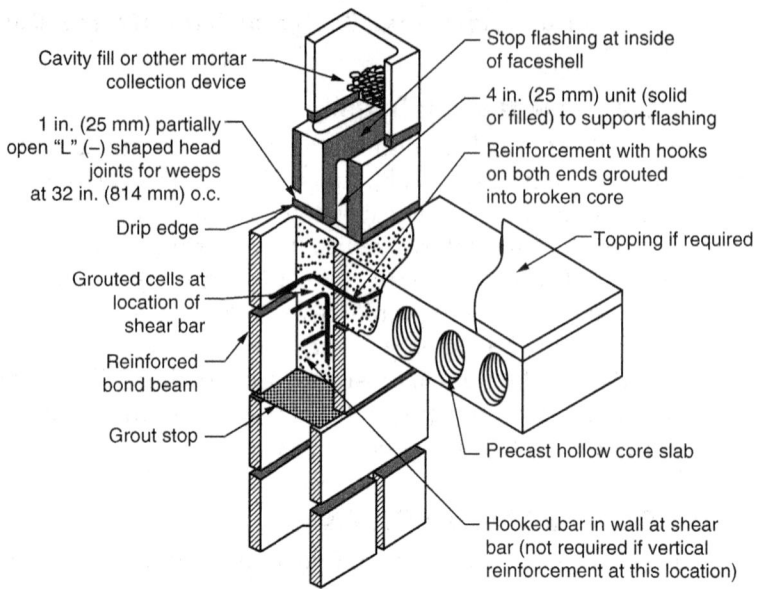

Figure 7.34 Example of wall-to-floor connection, planks parallel to wall. (*Source*: Figure 15 of National Concrete Masonry Association TEK 05-07A.)

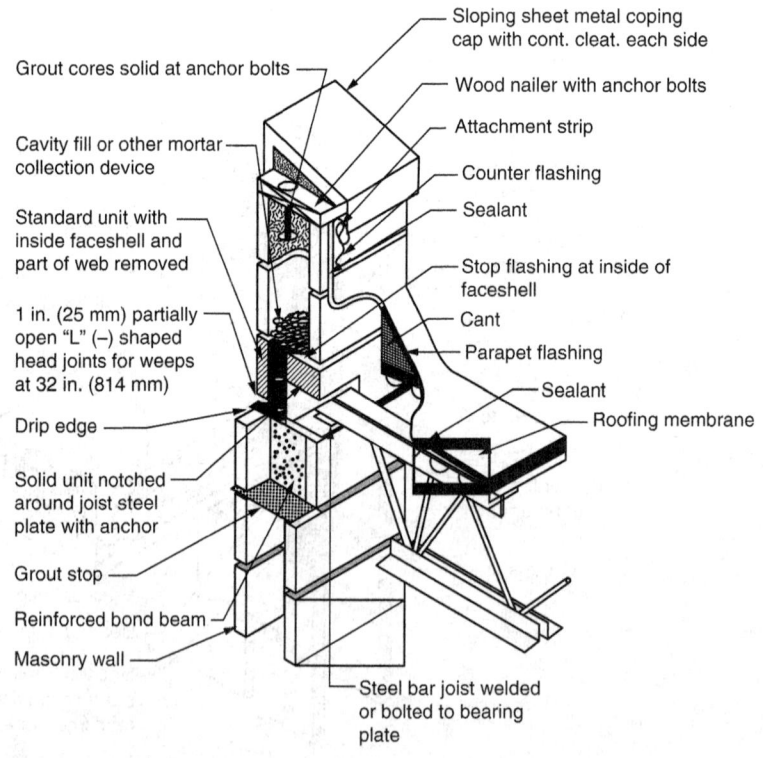

Figure 7.35 Example of wall-to-roof detail. (*Source*: Figure 11 of National Concrete Masonry Association TEK 05-07A.)

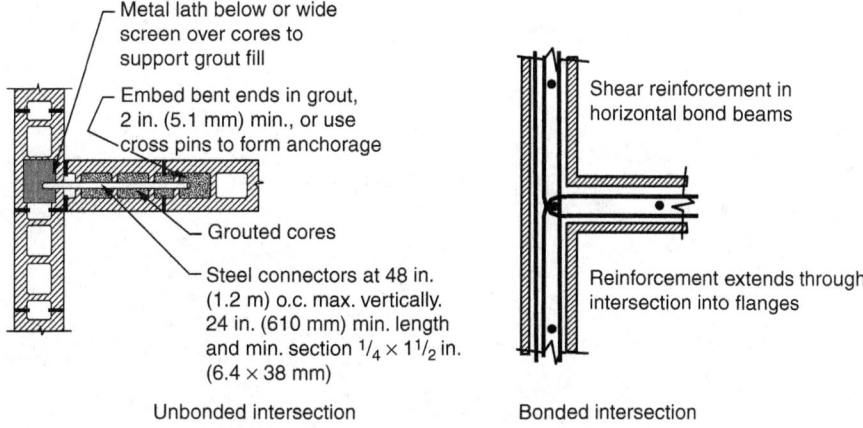

FIGURE 7.36 Examples of wall-to-wall connection details.

CHAPTER 8
Allowable-Stress Design of Reinforced Masonry Elements

Before studying the allowable-stress design of reinforced masonry elements, it is useful to review the behavior of reinforced masonry elements in the linear elastic range. This is accomplished in Sec. 8.1.1. Later sections in this chapter then address the allowable-stress design of beams, lintels, reinforced bearing walls, and reinforced shear walls.

8.1 Review: Behavior of Cracked, Transformed Sections

8.1.1 Review: Flexural Behavior of Cracked, Transformed Sections

The basic principles of the flexural behavior of cracked, transformed sections are developed using kinematics, stress-strain relations and equilibrium, as shown in Fig. 8.1.

The cracked masonry section must satisfy kinematics (plane sections remain plane), stress-strain relationships, and statics.

Kinematics	Stress-strain relation	Statics
$\varepsilon_m = \varepsilon_s$	$f_m = E_m \varepsilon_m$	$\sum f dA = 0$
$\varepsilon = \phi y$	$f_s = E_s \varepsilon_s$	$\sum f y dA = M$

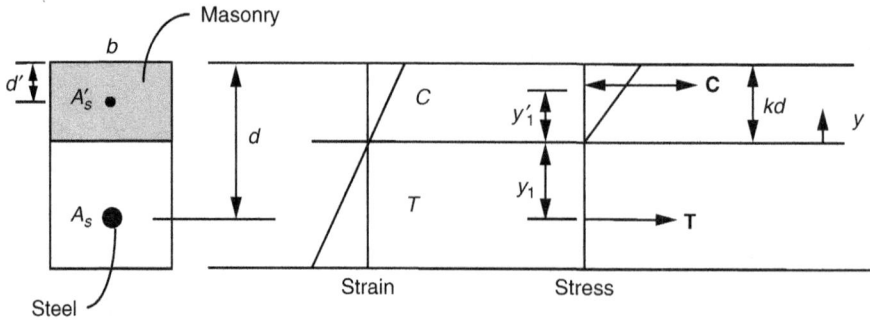

FIGURE 8.1 States of strain and stress in a cracked masonry section.

Axial equilibrium of the section locates the neutral axis of the cracked, transformed section as shown below:

$$b\int_0^{kd} f_m dy + A'_s f'_s + A_s f_s = 0$$

$$b\int_0^{kd} \phi E_m y dy + A'_s E_s \phi y'_1 + A_s E_s \phi y_1 = 0$$

$$E_m \phi \left[b\int_0^{kd} y dy + (n-1)A'_s y'_1 + nA_s y_1 \right] = 0$$

$$n \equiv \frac{E_s}{E_m} \quad \text{(Modular ratio)}$$

$$E_m \phi \neq 0$$

$$b\int_0^{kd} y dy + (n-1)A'_s y'_1 + nA_s y_1 = 0$$

In other words, the neutral axis is located at the geometric centroid of the cracked, transformed section. In Sec. 8.1.2 we shall develop closed-form expressions for that location.

A derivation quite similar to the foregoing can be carried out for moment equilibrium:

$$b\int_0^{kd} f_m y dy + A'_s y'_1 f'_s + A_s y_1 f_s = M_o$$

$$b\int_0^{kd} \phi E_m y^2 dy + A'_s E_s \phi {y'_1}^2 + A_s E_s \phi y_1^2 = M_o$$

$$E_m \phi \left[b\int_0^{kd} y^2 dy + (n-1)A'_s {y'_1}^2 + nA_s y_1^2 \right] = M_o$$

The quantity in brackets represents the centroidal moment of inertia of the cracked, transformed section. Continuing,

$$E_m \phi I_{c,t} = M_o$$

$$\phi = \frac{M}{EI_{c,t}}$$

but

$$\varepsilon = y\phi$$

$$f_s = E_s \varepsilon = E_s y\phi = nE_m y\phi = \frac{nM_o y}{I_{c,t}} = f_s$$

$$f_m = E_m \varepsilon = E_m y\phi = \frac{M_o y}{I_{c,t}} = f_m$$

and in summary,

$$f_s = \frac{nM_o y}{I_{c,t}}$$

$$f_m = \frac{M_o y}{I_{c,t}}$$

8.1.2 Location of the Neutral Axis for Particular Cases

Now examine the location of the neutral axis. The neutral axis is located using the information shown in Fig. 8.2.

$$\rho \equiv \frac{A_s}{bd} \qquad \rho' \equiv \frac{A'_s}{bd}$$

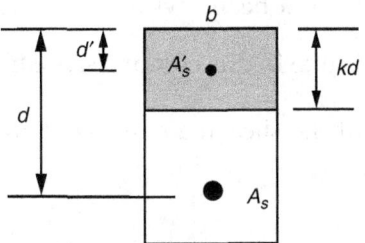

FIGURE 8.2 Location of the neutral axis for particular cases.

The neutral axis is located at the centroid of the cracked, transformed section:

$$(bkd)\left(\frac{kd}{2}\right) + (n-1)A'_s(kd - d') = nA_s(d - kd)$$

$$\left(\frac{d}{2}\right)k^2 + (n-1)\rho'(kd - d') = n\rho(d - kd)$$

$$\left(\frac{d}{2}\right)k^2 + [n\rho + (n-1)\rho']dk - \left[n\rho + (n-1)\rho'\left(\frac{d'}{d}\right)\right]d = 0$$

$$k^2 + 2[n\rho + (n-1)\rho']k - 2\left[n\rho + (n-1)\rho'\left(\frac{d'}{d}\right)\right] = 0$$

Using the quadratic formula, and noting that only positive areas of reinforcement have physical meaning:

$$k = \frac{-2[n\rho + (n-1)\rho'] \pm \sqrt{4[n\rho + (n-1)\rho']^2 + 8\left[n\rho + (n-1)\rho'\left(\frac{d'}{d}\right)\right]}}{2}$$

$$k = -[n\rho + (n-1)\rho'] + \sqrt{[n\rho + (n-1)\rho']^2 + 2\left[n\rho + (n-1)\rho'\left(\frac{d'}{d}\right)\right]}$$

Neglecting the compressive reinforcement,

$$k = -n\rho + \sqrt{n^2\rho^2 + 2n\rho}$$

8.1.3 Review: Shear Behavior of Cracked, Transformed Sections

Now examine the shear behavior of a cracked, transformed section. Consider a slice of a beam, with a maximum compressive stress f_b on one side, and a maximum compressive stress $f_b + \frac{df_b}{dx}dx$ on the other side (Fig. 8.3).

Now cut the slice at a distance y_1 from the neutral axis, as shown in Fig. 8.4.

$$\tau dx b = \int_{y_1}^{y_{max}} \frac{df_b}{dx} dx b dy$$

Allowable-Stress Design of Reinforced Masonry Elements

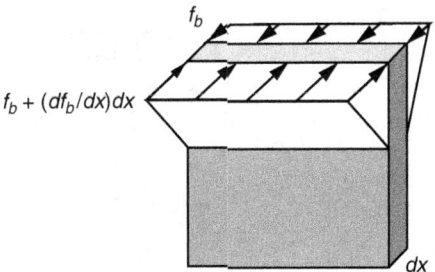

FIGURE 8.3 Slice of a cracked, transformed section showing triangular compressive stress blocks.

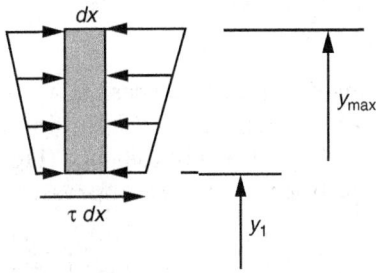

FIGURE 8.4 Slice of a cracked, transformed section showing equilibrium between shear forces and difference in shear forces.

but

$$f_b = \frac{My}{I}$$

so

$$\frac{df_b}{dx} = \frac{dM}{dx}\frac{y}{I} = \frac{Vy}{I}$$

and

$$\tau \, dx \, b = \int_{y_1}^{y_{max}} b \, y \, dy \, \frac{V}{I}$$

finally,

$$\tau = \frac{V \int_{y_1}^{y_{max}} b \, y \, dy}{Ib}$$

$$\tau = \frac{VQ}{Ib}$$

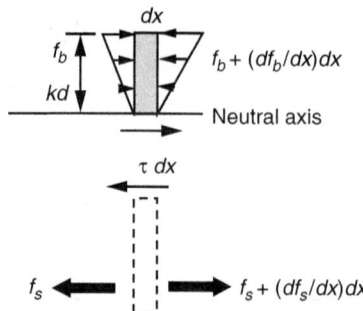

FIGURE 8.5 Slice of a cracked, transformed section showing equilibrium of the compressive and tensile portions of the slice.

Now consider the special case of a cracked, transformed section. Shear is greatest at the neutral axis. Examine the equilibrium of the compressive and tensile portions of the slice (Fig. 8.5).

Axial equilibrium of the compressive block requires

$$\tau b\, dx = b\int_0^{kd} \frac{df_b}{dx} dx\, dy$$

but

$$b\int_0^{kd} \frac{df_b}{dx} dx\, dy = \frac{df_s}{dx} dx\, A_s$$

So

$$\tau b\, dx = \frac{df_s}{dx} dx\, A_s = \frac{d\left(\dfrac{M}{A_s jd}\right)}{dx} dx\, A_s = \left(\frac{dM}{dx}\right)\frac{dx}{jd} = \frac{V}{jd} dx$$

and finally

$$\tau = \frac{V}{bjd}$$

This derivation implies maximum shear at the neutral axis, where cracked and uncracked masonry meet. This result is theoretically correct but practically suspect.

8.1.4 Review Bond Behavior of Cracked, Transformed Sections

Now examine the bond behavior of a cracked, transformed section. Consider a slice of a beam, with a maximum steel stress f_s on the far

Allowable-Stress Design of Reinforced Masonry Elements

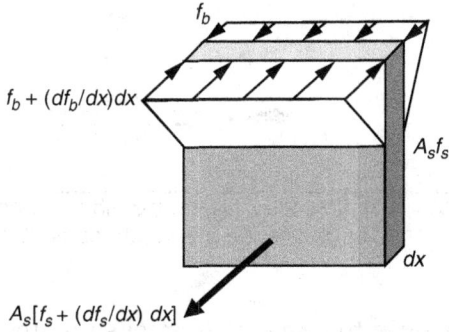

FIGURE 8.6 Slice of a cracked, transformed section, showing equilibrium of difference in compressive force and difference in tensile force.

side, and a maximum steel stress $f_s + \dfrac{df_s}{dx}dx$ on the other side, as shown in Fig. 8.6.

The far side of the slice has a moment M; the near side has a moment $M + \dfrac{dM}{dx}dx$.

Far side of slice	Near side of slice
$A_s f_s = \dfrac{M}{jd}$	$A_s\left(f_s + \dfrac{df_s}{dx}dx\right) = \dfrac{M + \dfrac{dM}{dx}dx}{jd}$

For the near side of the slice,

$$A_s\left(f_s + \frac{df_s}{dx}dx\right) = \frac{\left(M + \dfrac{dM}{dx}dx\right)}{jd}$$

$$A_s f_s + A_s \frac{df_s}{dx}dx = \frac{M}{jd} + \frac{dM}{dx}\frac{dx}{jd}$$

The first term on the left-hand side is equal to the first term on the right-hand side, so they cancel:

$$A_s \frac{df_s}{dx}dx = \frac{dM}{dx}\frac{dx}{jd}$$

$$\frac{df_s}{dx} = \frac{V}{A_s jd}$$

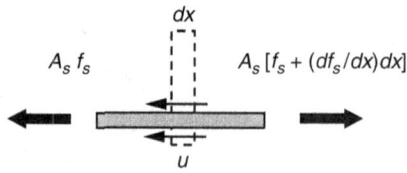

FIGURE 8.7 Tensile portion of a slice, showing equilibrium between bond force and difference in tensile force in reinforcement.

However, this change in steel stress requires a bond stress u, acting over the perimeter of the bar, Σ_o, times the length, dx. This is shown in Fig. 8.7.

Imposing horizontal equilibrium,

$$A_s f_s + A_s \frac{df_s}{dx} dx = A_s f_s + u \Sigma_o dx$$

$$A_s \frac{df_s}{dx} dx = u \Sigma_o dx$$

$$u = \frac{A_s}{\Sigma_o} \frac{df_s}{dx} = \frac{A_s}{\Sigma_o} \frac{dM}{dx} \frac{1}{A_s jd}$$

so

$$u = \frac{V}{\Sigma_o jd}$$

8.1.5 Physical Properties of Steel Reinforcing Wire and Bars

Physical properties of steel reinforcing wire and bars are given in Table 8.1.

Cover requirements are given in Sec. 1.15.4 of the 2008 MSJC *Code*. Minimum cover for joint reinforcement (exterior exposure) is 5/8 in.

8.1.6 Example of Location of the Neutral Axis (Cracked, Transformed Section)

For a 4-in. modular clay masonry wall with two W1.7 wires at every course, loaded out-of-plane, what is the effect on the location of the neutral axis if compressive reinforcement is neglected? Assume Type S mortar, and units with a compressive strength of 6600 psi (see Fig. 8.8). Because the section is loaded out-of-plane, the depth of the section is measured horizontally on the page, and the width is measured vertically. The effective width per bar is 2.67 in.

Designation	Diameter, in.	Area, in.^{2}
Wire		
W1.1 (11 gage)	0.121	0.011
W1.7 (9 gage)	0.148	0.017
W2.1 (8 gage)	0.162	0.020
W2.8 (3/16 wire)	0.187	0.027
W4.9 (1/4 wire)	0.250	0.049
Bars		
#3	0.375	0.11
#4	0.500	0.20
#5	0.625	0.31
#6	0.750	0.44
#7	0.875	0.60
#8	1.000	0.79
#9	1.128	1.00
#10	1.270	1.27
#11	1.410	1.56

TABLE 8.1 Physical Properties of Steel Reinforcing Wire and Bars

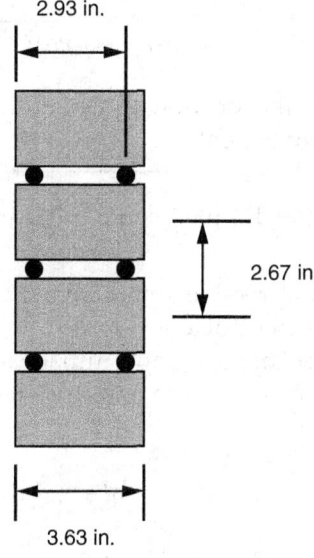

FIGURE 8.8 Example of calculation of the position of the neutral axis.

According to Table 1 of the 2008 MSJC *Specification*, for Type M or S mortar and clay units with a strength of 6600 psi, the compressive strength of the masonry can conservatively be taken as 2500 psi (the so-called "unit strength method"). If the compressive strength is evaluated by prism testing, a higher value can probably be used. Take the specified compressive strength of the masonry as $f'_m = 2500$ psi.

Then according to Sec. 1.8.2.2.1 of the 2008 MSJC *Code*, for clay masonry,

$$E_m = 700 f'_m = 700 \cdot 2500 \text{ lb/in.}^2 = 1.75 \times 10^6 \text{ lb/in.}^2$$

The modular ratio n is given by

$$n = \frac{E_s}{E_m} = \frac{29 \times 10^6}{1.75 \times 10^6} = 16.6$$

$$t = 3.63$$

$$d = 3.63 - \frac{5}{8} - \frac{0.148}{2} = 2.93 \text{ in.}$$

$$d' = \frac{5}{8} + \frac{0.148}{2} = 0.70 \text{ in.}$$

$$\rho = \rho' = \frac{A_s}{bd} = \frac{0.0172}{2.93 \times 8/3} = 0.00220$$

First, compute the location of the neutral axis neglecting the effect of compressive reinforcement.

$$k = -n\rho + \sqrt{(n\rho)^2 + 2n\rho} = 0.236$$

Now compute the location of the neutral axis including the effect of compressive reinforcement:

$$k = -[n\rho + (n-1)\rho'] + \sqrt{[n\rho + (n-1)\rho']^2 + 2\left[n\rho + (n-1)\rho'\left(\frac{d'}{d}\right)\right]} = 0.237$$

Therefore, compressive reinforcement does not significantly affect the location of the neutral axis.

For most sections, a good initial assumption is that k (the location of the neutral axis) is close to 3/8, and therefore j (the internal lever arm) is close to 7/8.

$$kd \approx \frac{3}{8}d$$

$$jd = d - \frac{kd}{3} \approx \frac{7}{8}d$$

8.1.7 Example of Allowable Flexural Capacity of the Cross-Section

Now compute the allowable flexural capacity of the cross-section considered in Sec. 8.1.6. The allowable flexural capacity could be governed by the maximum flexural tensile stress in the reinforcement, or by the maximum flexural compressive stress in the masonry.

Using the allowable-stress approach, computed stress are denoted by the lowercase letter f. For example, f_s is the computed stress in the reinforcement, and f_b is the computed bending stress in the masonry. Allowable stresses are denoted by the uppercase letter F. For example, F_s is the allowable stress in the reinforcement, and F_b is the allowable flexural compressive stress in the masonry.

In accordance with Sec. 2.3.2 of the 2008 MSJC *Code*, the allowable stress in drawn wire reinforcement is 30,000 lb/in.². In accordance with Sec. 2.3.3.2.2 of the 2008 MSJC *Code*, the allowable flexural compressive stress in masonry is $(1/3) f'_m$.

First consider the allowable moment capacity as governed by the allowable stress in the reinforcement, F_s (Fig. 8.9). The moment is the allowable tensile force in the reinforcement, multiplied by the internal lever arm.

$$M = A_s F_s j d$$

In our case,

$k = 0.24$

$$j = 1 - \frac{k}{3} = 1 - \frac{0.24}{3} = 0.92$$

$M = A_s F_s j d = 0.0172 \text{ in.}^2 \times 30{,}000 \text{ lb/in.}^2 \times 0.92 \times 2.93 \text{ in.}$

$M = 1391$ in.-lb/course

$M = 1391$ in.-lb/course $\cdot \left(\dfrac{3 \text{ courses}}{8 \text{ in.}} \right) \cdot 12$ in./ft

$M = 6259$ in.-lb per foot of width

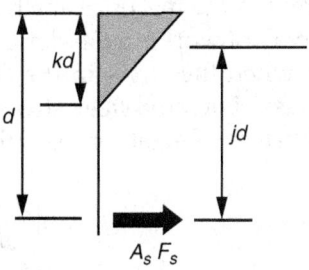

FIGURE 8.9 Equilibrium of forces in the cross section corresponding to allowable stress in the reinforcement.

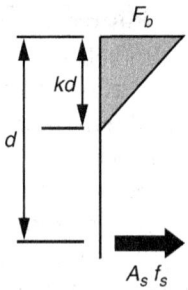

FIGURE 8.10 Equilibrium of forces in the cross section corresponding to allowable stress in the masonry.

Now consider the allowable moment capacity as governed by the allowable stress in the masonry, F_b (Fig. 8.10). The moment is the allowable compressive force in the masonry, multiplied by the internal lever arm:

$$M = \left(\frac{1}{2} \times F_b bkd\right)(jd)$$

$$M = \frac{1}{2} F_b jkbd^2$$

In our case,

$$M = \frac{1}{2} bjkd^2 F_b$$

$$M = \frac{1}{2} \times 12 \text{ in.} \times 0.92 \times 0.24 \times 2.93 \text{ in.} \times 2500 \text{ psi}$$

$$M = 9704 \text{ in.-lb per foot of width}$$

Because the allowable moment as governed by the allowable stress in the reinforcement is less than the allowable moment as governed by the allowable stress in the masonry, the former governs.

Another way of working the same problem is to compute the stress in the masonry when the stress in the allowable reinforcement equals the allowable stress. For any steel stress f_s (not necessarily the allowable stress), equilibrium of axial forces in the beam gives:

$$(f_b bkd)\left(\frac{1}{2}\right) = A_s f_s$$

$$f_b = \frac{2 A_s f_s}{bkd}$$

But from above,

$$A_s = \frac{M}{jdf_s}$$

so

$$f_b = \frac{2\left(M/f_s jd\right)f_s}{bkd} = \frac{2M}{jkbd^2}$$

In our case,

$$f_b = \frac{2M}{jkbd^2} = \frac{2 \cdot 6259 \text{ in.-lb}}{0.24 \times 0.92 \times 12 \text{ in.} \times 2.93^2 \text{ in.}^2} = 550 \text{ lb/in.}^2$$

This is less than the allowable stress of $(1/3)\,f'_m$, and the design is satisfactory.

8.1.8 Allowable-Stress Balanced Reinforcement

For allowable-stress design, the concept of balanced reinforcement exists, but has little physical significance. The allowable-stress balanced steel area is simply the steel area at which the masonry and steel reach their respective allowable stresses simultaneously. Because the factors of safety are different for steel and masonry, and because the assumed stress distribution in masonry is linear rather than an equivalent rectangular stress block, the allowable-stress balanced steel area does not mark a transition between ductile and brittle behavior. It is a reference point, however, between behavior governed by reinforcement and behavior governed by masonry.

The balanced steel percentage for allowable-stress design can be derived based on the strains in steel and masonry, as shown in Fig. 8.11.

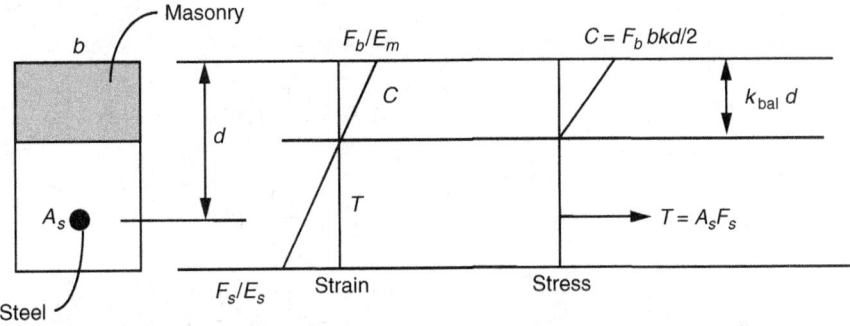

FIGURE 8.11 Conditions of stress and strain corresponding to allowable-stress balanced reinforcement.

First, locate the neutral axis under allowable-stress balanced conditions:

$$\frac{\varepsilon_m}{\varepsilon_s} = \frac{(F_b/E_m)}{(F_s/E_s)} = \frac{k_b d}{d - k_b d}$$

$$\left(\frac{F_b}{E_m}\right)(d - k_b d) = \left(\frac{F_s}{E_s}\right) k_b d$$

$$\frac{E_s}{E_m} \equiv n$$

$$\left(\frac{F_b}{E_m}\right) d - \left(\frac{F_b}{E_m}\right) k_b d = \left(\frac{F_s}{nE_m}\right) k_b d$$

$$k_b \left(\frac{F_b}{E_m} + \frac{F_s}{nE_m}\right) = \left(\frac{F_b}{E_m}\right)$$

$$k_b = \frac{F_b}{F_b + F_s/n}$$

and finally,

$$k_b = \frac{n}{(F_s/F_b) + n}$$

The corresponding steel percentage, ρ_b, is given by:

$$T = C$$

$$A_s F_s = \left(\frac{1}{2}\right) F_b b k d$$

$$\rho_b b d F_s = \left(\frac{1}{2}\right) F_b b k d$$

$$\rho_b = \left(\frac{1}{2}\right) k_b \left(\frac{F_b}{F_s}\right)$$

8.2 Allowable-Stress Design of Reinforced Beams and Lintels

8.2.1 Steps in Allowable-Stress Design of Reinforced Beams and Lintels

The most common reinforced masonry beam is a lintel, as shown in Fig. 8.12. Lintels are beams that support masonry over openings.

Allowable-stress design of reinforced beams and lintels follows the basic steps given below:

1. Shear design:
   a. Calculate the design shear, and compare it with the corresponding resistance. Revise the lintel depth if necessary.
2. Flexural design:
   a. Calculate the design moment.
   b. Calculate the required flexural reinforcement. Check that it fits within minimum and maximum reinforcement limitations.

In many cases, the depth of the lintel is determined by architectural considerations. In other cases, it is necessary to determine the number of courses of masonry that will work as a beam.

The depth of the beam, and hence the area that is effective in resisting shear, is determined by the number of courses that we consider to comprise it. Because it is not very practical to put shear reinforcement in masonry beams, the depth of the beam may be determined by this. In other words, the beam design may start with the number of courses that are needed so that shear can be resisted by masonry alone.

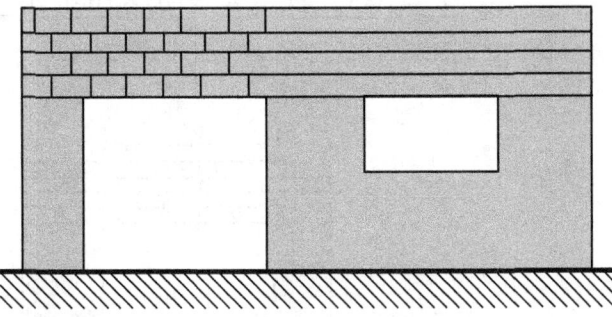

FIGURE 8.12 Example of masonry lintel.

8.2.2 Example of Lintel Design according to Allowable-Stress Provisions

Suppose that we have a uniformly distributed load of 1050 lb/ft, applied at the level of the roof of the structure shown in Fig. 8.13. Design the lintel.

According to Table 2 of the 2008 MSJC *Specification*, for Type M or S mortar and concrete masonry units with a specified strength of 1900 psi (the minimum specified strength for ASTM C90 units), the compressive strength of the masonry can conservatively be taken as 1500 psi (the so-called "unit strength method"). If the compressive strength is evaluated by prism testing, a higher value can probably be used. Take the specified compressive strength of the masonry as $f'_m = 1500$ psi.

Assume fully grouted concrete masonry with a nominal thickness of 8 in., a weight of 80 lb/ft², and a specified compressive strength of 1500 lb/in.². Use Type S PCL mortar. The lintel has a span of 10 ft, and a total depth (height of parapet plus distance between the roof and the lintel) of 4 ft. These are shown in the schematic Fig. 8.13.

Our design presumes that entire height of the lintel is grouted.

First check whether the depth of the lintel is sufficient to avoid the use of shear reinforcement. The opening may have a movement joint placed on either side, at a distance of one-half the unit length from the opening. The lintel therefore bears on two bearing areas, each of length 8 in. Conservatively, the span of the lintel is taken as the clear distance, plus one-half of 8 in. on each side. So the span is 10 ft plus 8 in., or 10.67 ft.

$$M = \frac{wl^2}{8} = \frac{(1050 + 4 \text{ ft} \times 80 \text{ lb/ft}) \times 10.67^2 \text{ lb-ft} \times 12 \text{ in./ft}}{8}$$

$$= 233,959 \text{ in.-lb}$$

$$V = \frac{wl}{2} = \frac{(1050 + 4 \text{ ft} \times 80 \text{ lb/ft}) \times 10.67 \text{ lb}}{2} = 7309 \text{ lb}$$

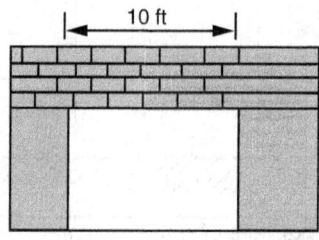

FIGURE 8.13 Example for allowable-stress design of a lintel.

Because this is a reinforced element subject to flexural tension, shearing capacity is calculated using Sec. 2.3.5.2.2 of the 2008 MSJC *Code*. Shear stresses are calculated by:

$$f_v = \frac{V}{bd}$$

and the allowable in-plane shear stress, F_v, shall not exceed either of:

$$F_v = \begin{cases} \sqrt{f'_m} \\ 50\,\text{psi} \end{cases}$$

In our case, the bars in the lintel will probably be placed in the lower part of an inverted bottom course as shown in Fig. 8.14.

The effective depth d is calculated using the minimum cover of 1.5 in. (Sec. 1.15.4.1 of the 2008 MSJC *Code*), plus one-half the diameter of an assumed #8 bar.

$$f_v = \frac{V}{bd} = \frac{7309\,\text{lb}}{7.63 \times 46\,\text{in.}^2} = 20.8\,\text{lb/in.}^2$$

$$F_v = \begin{cases} \sqrt{f'_m} = \sqrt{1500} = 38.7\,\text{lb/in.}^2 \\ 50\,\text{psi} \end{cases}$$

The 38.7 lb/in.² governs, and the depth is satisfactory to avoid the use of shear reinforcement.

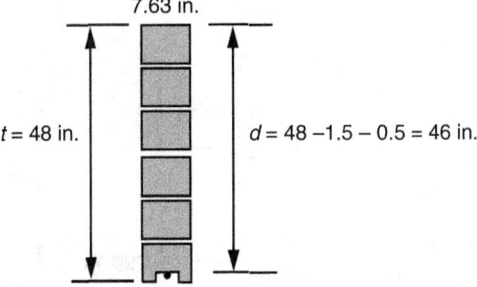

FIGURE 8.14 Example showing placement of bottom reinforcement in lowest course of lintel.

Now check the required flexural reinforcement:

$$A_s^{required} = \frac{M}{F_s jd} = \frac{233{,}959 \text{ lb-in.}}{24{,}000 \text{ lb/in.}^2 \times \left(\frac{7}{8} \times 46 \text{ in.}\right)} = 0.24 \text{ in.}^2$$

Because of the depth of the beam, this can easily be satisfied with a #5 bar. Also include two #4 bars at the level of the roof (bond beam reinforcement). The flexural design is quite simple. Because j is approximated as 0.9, the required steel area calculated above is not exact. The calculation could be refined by solving for the actual value of j.

The 2008 MSJC Code has no minimum reinforcement requirements for allowable-stress design of flexural members, and its maximum reinforcement requirements apply only to special reinforced masonry shear walls (2008 MSJC Code Sec. 2.3.3.4). This will probably be addressed in future editions of the MSJC Code.

Although the compressive stress in the masonry will probably not govern if the beam is deep enough not to need shear reinforcement, it is checked here for completeness. Equilibrium of forces on the cross-section is shown in Fig. 8.15.

First calculate the position of the neutral axis. Neglecting compressive reinforcement, the position of the neutral axis is given by:

$$n \equiv \frac{E_s}{E_m} \quad \text{and} \quad \rho \equiv \frac{A_s}{bd}$$

$$k = -n\rho + \sqrt{n^2\rho^2 + 2n\rho} \approx \frac{3}{8}$$

$$j = 1 - \frac{k}{3} \approx \frac{7}{8}$$

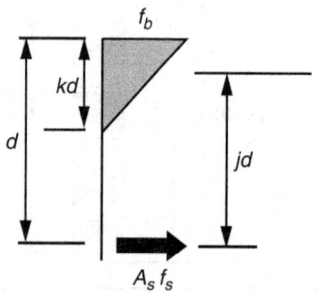

FIGURE 8.15 Equilibrium of forces on cross-section.

Allowable-Stress Design of Reinforced Masonry Elements

For a given moment, the tensile stress in the tensile reinforcement is

$$M = A_s f_s jd$$

$$f_s = \frac{M}{A_s jd}$$

and the maximum compressive stress in the masonry is given as follows:

$$(f_b bkd)\left(\frac{1}{2}\right) = A_s f_s$$

$$f_b = \frac{2 A_s f_s}{bkd}$$

$$A_s f_s = \frac{M}{jd}$$

$$f_b = \frac{2M}{bjdkd} = \frac{2M}{bjkd^2}$$

In this case, which involves concrete masonry,

$$n \equiv \frac{E_s}{E_m} = \frac{E_s}{900 f'_m} = \frac{29 \times 10^6 \text{ psi}}{900 \times 1500 \text{ psi}} = 21.48$$

$$\rho \equiv \frac{A_s}{bd} = \frac{0.31 \text{ in.}^2}{7.63 \times 46 \text{ in.}^2} = 8.83 \times 10^{-4}$$

$$n\rho = 0.190$$

$$k = -n\rho + \sqrt{n^2 \rho^2 + 2n\rho} = 0.177$$

$$j = 1 - \frac{k}{3} = 0.941$$

$$f_s = \frac{M}{A_s jd} = \frac{233{,}959 \text{ lb-in.}}{0.31 \text{ in.}^2 \times 0.941 \times 46 \text{ in.}} = 17{,}435 \text{ psi} \leq 24{,}000 \text{ psi} \quad \text{OK}$$

$$f_b = \frac{2M}{bjkd^2} = \frac{2 \times 233{,}959 \text{ lb-in.}}{7.63 \text{ in.} \times 0.177 \times 0.941 \times (46 \text{ in.})^2} = 174 \text{ psi} \leq \frac{f'_m}{3} = 500 \text{ psi} \quad \text{OK}$$

The design is therefore satisfactory. Note also that the approximate values of 3/8 and 7/8 could have been used for k and j, respectively.

8.2.3 Comments on Arching Action

Using the traditional assumption that distributed loads act only within a beam length defined by 45° lines from the ends of the distributed load, it would have been possible to take advantage of so-called "arching action" to reduce the gravity load for which the lintel must be designed. Nevertheless, this measure is hardly necessary, because the required area of reinforcement is quite small in any case.

Even though it would have been possible to refine the flexural design (e.g., by reducing the required depth of the lintel, or including the mid-depth reinforcement in the bond beam in the calculation of flexural resistance), this additional design effort would not have been cost-effective. The goal is to simplify the design process and the final layout of reinforcement.

8.3 Allowable-Stress Design of Curtain Walls

8.3.1 Background on Curtain Walls

In the first part of the structural design section of this book, we began with the design of panel walls, which can be designed as unreinforced masonry, and which span primarily in the vertical direction to transmit out-of-plane loads to the structural system. Panel walls are nonload bearing masonry, because they support gravity loads from self-weight only.

At this point, it is appropriate for us to study another type of nonload bearing masonry, the curtain wall. Like panel walls, curtain walls carry gravity load from self-weight only, and transmit out-of-plane loads to a structural frame. Unlike panel walls, however, curtain walls can be more than one story high and span horizontally rather than vertically. Typical curtain wall construction is shown in Fig. 8.16.

A single wythe of masonry spans horizontally between columns, which support the roof. This type of construction can be used for industrial buildings, gymnasiums, theaters, and other buildings of similar configuration.

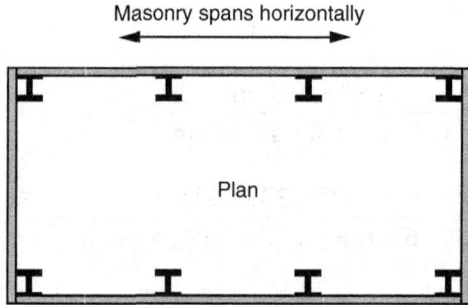

Figure 8.16 Plan view of typical curtain wall construction.

In previous sections dealing with panel walls, we have seen that because those walls are unreinforced, their design is governed by the flexural tensile strength of masonry. In the example of Sec. 7.1, using reasonable unfactored wind loads of about 20 lb/ft², the flexural tensile stresses in vertically spanning panel walls were comfortably within allowable values.

If we tried to use the same principles to design horizontally spanning curtain walls, however, they wouldn't work. In the Fig. 8.16, the horizontal span between columns is at least 20 ft, about twice the typical vertical span of panel walls. Since moments increase as the square of the span, doubling the span would increase the flexural tensile stresses by a factor of 4. Even considering that allowable flexural tensile stresses parallel to bed joints in running bond are about twice as high as those normal to the bed joints (reflecting the interlocking nature of running bond), the calculated flexural tensile stresses in the direction of span would exceed the allowable values.

The most reasonable solution to this problem is to reinforce the masonry horizontally. Single-wythe curtain walls are commonly used for industrial buildings, where water-penetration resistance is not a primary design consideration.

8.3.2 Examples of Use of Curtain Walls—Clay Masonry

Examples of use of curtain walls with clay masonry are shown in Fig. 8.17.

8.3.3 Example of Use of Curtain Walls—Concrete Masonry

Examples of use of curtain walls with concrete masonry are shown in Fig. 8.18.

8.3.4 Structural Action of Curtain Walls

Curtain walls act as horizontal strips to transfer out-of-plane loads to vertical supporting members such as steel or reinforced concrete columns or masonry pilasters (masonry columns partially embedded in the wall).

8.3.5 Example of Allowable-Stress Design of a Reinforced Curtain Wall

A curtain wall of standard modular clay units spans 20 ft between columns, and is simply supported at each column. It has reinforcement consisting of W4.9 wire each face, every course. The curtain wall is subjected to a wind pressure $w = 20$ lb/ft². Design the curtain wall. As an initial assumption, use $f'_m = 2500$ lb/in.². Referring to Table 1 of the 2008 MSJC *Specification*, this would require clay units with a compressive strength of at least 6600 psi, and Type S mortar.

302 Chapter Eight

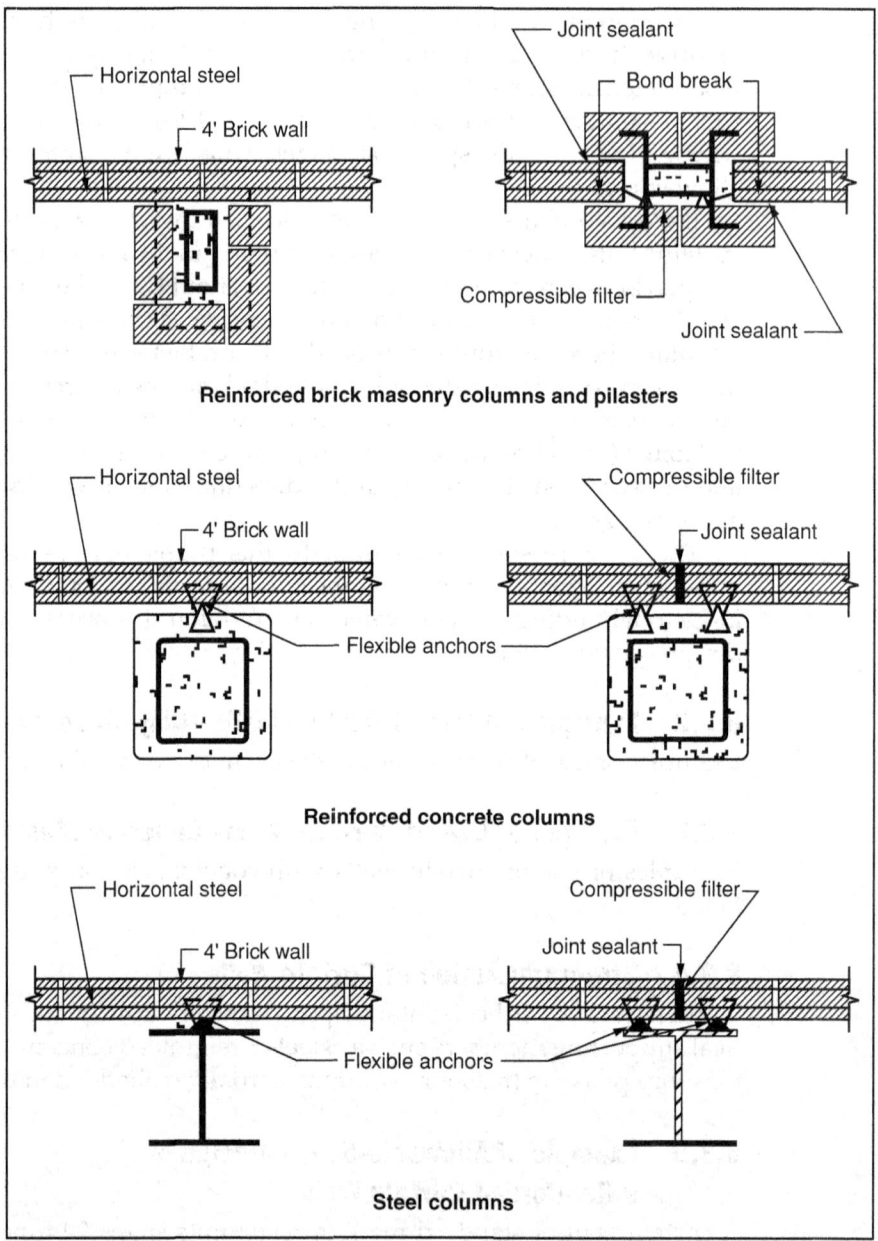

FIGURE 8.17 Examples of use of curtain walls of clay masonry. (*Source*: Technical-Note 17L, "Panel and Curtain Walls."© Brick Industry Association.)

Allowable-Stress Design of Reinforced Masonry Elements

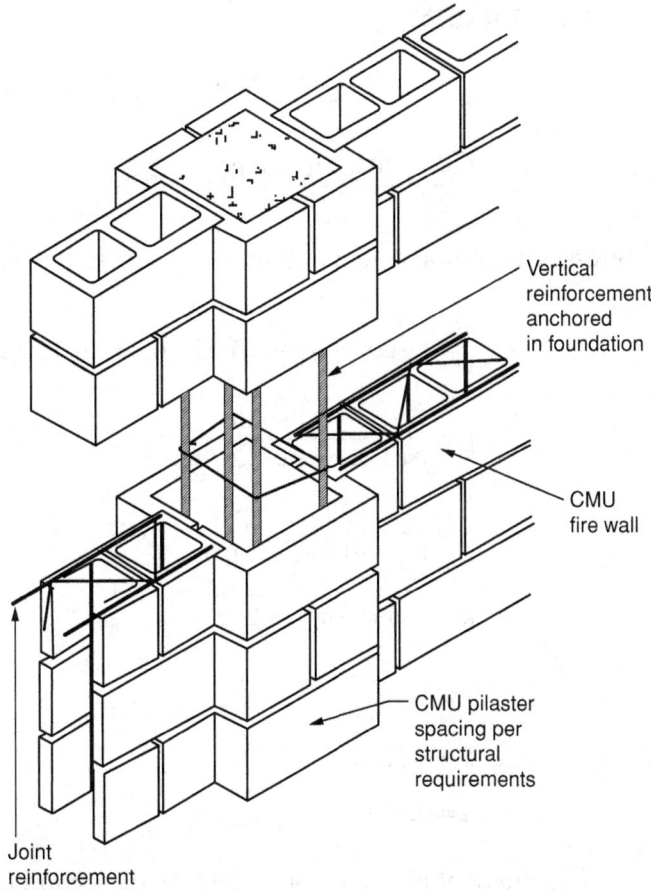

FIGURE 8.18 Examples of the use of curtain walls with concrete masonry. (*Source*: Figure 1 of National Concrete Masonry Association TEK-05-08A, "Details for Concrete Masonry Fire Walls.")

Maximum bending moment:

$$M = \frac{qL^2}{8}$$

Maximum shear:

$$V = \frac{qL}{2}$$

For a 1-ft strip,

$$M = \frac{qL^2}{8} = \frac{20 \text{ lb/ft} \cdot (20 \text{ ft})^2}{8} \times 12 \text{ in./ft} = 12{,}000 \text{ lb-in.}$$

$$V = \frac{qL}{2} = \frac{20 \text{ lb/ft} \cdot (20 \text{ ft})}{2} = 200 \text{ lb}$$

Stresses in reinforcement and masonry due to flexure:

$$d = t - \text{cover} - \frac{d_b}{2} = 3.63 - 0.63 - 0.13 = 2.87 \text{ in.}$$

$$\rho = \frac{A_s}{bd} = \frac{0.049 \text{ in.}^2}{2.67 \text{ in.} \cdot 2.87 \text{ in.}} = 0.00639$$

$$n = \frac{E_s}{E_m} = \frac{29 \times 10^6 \text{ psi}}{700 \cdot 2500 \text{ psi}} = 16.57$$

$$k = -n\rho + \sqrt{(n\rho)^2 + 2n\rho}$$

$$j = 1 - \frac{k}{3}$$

$$k = 0.366$$

$$j = 0.878$$

The moment above is for a 12-in. wide section of masonry. For flexural reinforcement spaced at 2.67 in., it is convenient to compute the moment on a 2.67-in. wide section.

$$f_s = \frac{M}{A_s jd} = \frac{12{,}000 \text{ lb-in.} \times \left(\frac{2.67 \text{ in.}}{12 \text{ in.}}\right)}{0.049 \text{ in.}^2 \times 0.878 \times 2.87 \text{ in.}} = 21{,}624 \text{ lb/in.}^2$$

$$f_b = \frac{2M}{jkbd^2} = \frac{2 \cdot 12{,}000 \text{ lb-in.} \left(\frac{2.67}{12}\right)}{0.878 \times 0.366 \times 2.67 \text{ in.} \times 2.87^2 \text{ in.}^2} = 755 \text{ lb/in.}^2$$

This analysis shows that the stress in reinforcement, 21.6 kips/in.², is less than or equal to the allowable stress of 30 kips/in.² [2008 MSJC *Code*, Sec. 2.3.2.1(c)]. The stress in the masonry, 755 lb/in.², is less than or equal to the allowable flexural compressive stress of $(f'_m/3)$, or 833 lb/in.² (2008 MSJC *Code*, Sec. 2.3.3.2.2).

Now check one-way shear, using the provisions of Sec. 2.3.5 of the 2008 MSJC *Code* for reinforced masonry:

$$f_v = \frac{V}{bd} = \frac{200 \text{ lb}}{12 \times 2.87 \text{ in.}^2} = 5.81 \text{ lb/in.}^2$$

This is considerably less than the allowable stress [Sec. 2.3.5.2.2.(a)],

$$F_v = \sqrt{f'_m} \leq 50 \text{ lb/in.}^2$$

and the design is acceptable.

8.3.6 Note on Simplification of Allowable-Stress Design for Flexure

From the example in Sec. 8.3.5, it is apparent that the flexural check requires most of the design time. For most practical combinations of f'_m and element dimensions, the values of k and j can be assumed rather than calculated:

$$k \approx \left(\frac{3}{8}\right)$$

$$j \approx \left(\frac{7}{8}\right)$$

This considerably decreases the time required for the flexural check.

8.3.7 Design of Anchors for Curtain Wall

The design of the curtain wall would have to finish with the design of anchors holding the ends of the curtain wall strips, to the columns (Fig. 8.19).

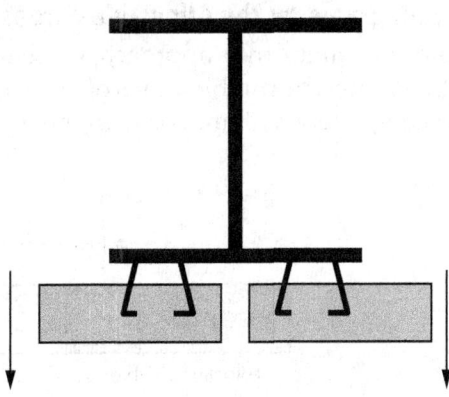

FIGURE 8.19 Anchors holding the ends of curtain wall strips to columns.

For example, if anchors are spaced at 12 in. vertically, the load per anchor is

$$\text{Anchor load} = \frac{qL}{2} = \frac{20 \text{ lb} \cdot 20 \text{ ft}}{2} = 200 \text{ lb}$$

8.4 Allowable-Stress Design of Reinforced Bearing Walls

8.4.1 Introduction to Allowable-Stress Design of Reinforced Bearing Walls

In this section, we shall study the behavior and design of reinforced masonry wall elements subjected to combinations of axial force and out-of-plane flexure. In the context of engineering mechanics, they are beam-columns. In the context of the MSJC *Code*, however, a "column" is an isolated masonry element rarely found in real masonry construction.

Masonry beam-columns, like those of reinforced concrete, are designed using moment-axial force interaction diagrams. Combinations of axial force and moment lying inside the diagram represent permitted designs; combinations lying outside, prohibited ones.

Unlike reinforced concrete, however, reinforced masonry beam-columns rarely take the form of isolated rectangular elements with four longitudinal bars and transverse ties. The most common form for a reinforced masonry beam-column is a wall, loaded out-of-plane by eccentric gravity load, alone or in combination with wind.

For example, Fig. 8.20 shows a portion of a wall, with a total effective width of 6*t* prescribed by Sec. 1.9.6.1 of the 2008 MSJC *Code*.

8.4.2 Background on Moment-Axial Force Interaction Diagrams by the Allowable-Stress Approach

Using the allowable-stress approach, we seek to construct interaction diagrams that represent combinations of axial and flexural capacity. This can be done completely by hand, or with the help of a spreadsheet.

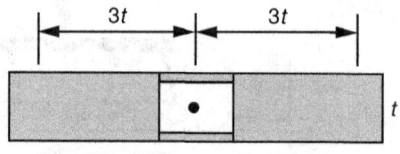

Effective width on each side of bar

FIGURE 8.20 Effective width of a reinforced masonry bearing wall.

8.4.3 Allowable-Stress Interaction Diagrams by Hand

By hand, we can compute three points (pure compression, pure flexure, and the allowable-stress balance point). We then connect those points by straight lines. As with the strength approach, if the balance point does not correspond to the maximum moment, the interaction diagram formed by connecting those points by straight lines may differ considerably from that obtained using more points and calculated by spreadsheet.

Pure Compression

For members with slenderness (h/r) less than or equal to 99 (the most common case),

$$P_a = (0.25 f'_m A_n + 0.65 A_{st} F_s)\left[1 - \left(\frac{h}{140r}\right)^2\right]$$

Pure Flexure

As before, the possible contribution of compressive reinforcement is small, and can be neglected. For most cases, flexural capacity is governed by the allowable stress in reinforcement.

$$jd = d - \frac{kd}{3}$$

$$jd \approx \frac{7}{8} d$$

$$M = A_s F_s jd$$

Balance Point

First, locate the neutral axis under allowable-stress balanced conditions, as shown in Fig. 8.21.

Because strains vary linearly over the depth of the cross section,

$$\frac{\varepsilon_m}{\varepsilon_s} = \frac{(F_b/E_m)}{(F_s/E_s)} = \frac{k_b d}{d - k_b d}$$

$$\left(\frac{F_b}{E_m}\right)(d - k_b d) = \left(\frac{F_s}{E_s}\right) k_b d$$

$$\frac{E_s}{E_m} \equiv n$$

$$\left(\frac{F_b}{E_m}\right)d - \left(\frac{F_b}{E_m}\right)k_b d = \left(\frac{F_s}{nE_m}\right)k_b d$$

$$k_b\left(\frac{F_b}{E_m} + \frac{F_s}{nE_m}\right) = \left(\frac{F_b}{E_m}\right)$$

$$k_b = \frac{F_b}{F_b + F_s/n}$$

so,

$$k_b = \frac{n}{\left(F_s/F_b\right) + n}$$

Next, we calculate the corresponding tensile and compressive forces:

$$T = A_s F_s$$
$$C = k_b db F_b/2$$
$$P = C - T$$
$$M = T\left(d - \frac{h}{2}\right) + C\left(\frac{h}{2} - \frac{k_b d}{3}\right)$$

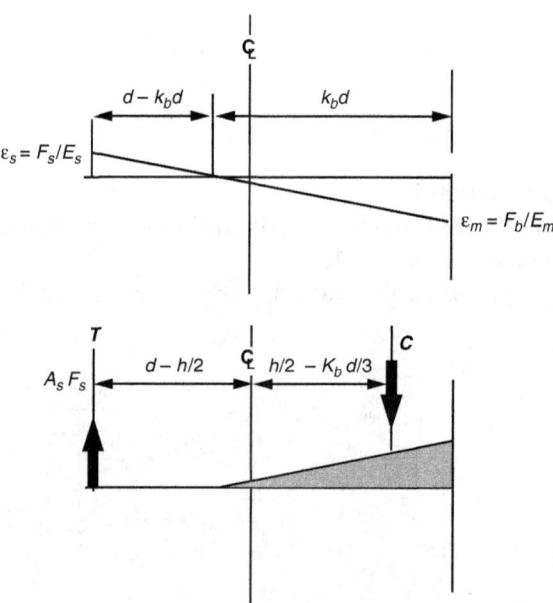

Figure 8.21 Location of neutral axis under allowable-stress balanced conditions.

8.4.4 Example of Moment-Axial Force Interaction Diagram by the Allowable-Stress Approach (Hand Calculation)

Construct the moment-axial force interaction diagram by the allowable-stress approach for a nominal 8-in. wall, fully grouted, with f'_m = 1500 lb/in.² and reinforcement consisting of #5 bars at 48 in., placed in the center of the wall.

Beam-columns with centrally located reinforcement can have moment-axial force interaction diagrams of that differ in appearance from those of conventional columns. This is even more so for allowable-stress interaction diagrams. For example, the balance point is located far below the point of maximum moment. Because of this, hand calculations are useful for some reinforced masonry beam-columns, but not all. In particular, they are not useful for most masonry walls loaded out-of-plane.

Pure Compression

Neglect slenderness effects (assume $h = 0$). Because reinforcement is not supported laterally, neglect it in compression.

$$P_a = (0.25 f'_m A_n + 0.65 A_{st} F_s)\left[1 - \left(\frac{h}{140\,r}\right)^2\right]$$

$P_a = 0.25 \times 1500 \text{ lb/in.}^2 (7.63 \text{ in.} \times 48 \text{ in.} - 0.31 \text{ in.}^2) + 0.65 \times 0.31 \text{ in.}^2 \times 0 \text{ lb/in.}^2$

$P_a = 137{,}224 \text{ lb}$

The capacity per foot of wall length will be the above value, divided by 4:

$$P_a = 34{,}306 \text{ lb}$$

Pure Flexure

$$jd = d - \frac{kd}{3}$$

$$jd \approx \frac{7}{8} d$$

$$M = A_s F_s jd$$

$$M = 0.31 \text{ in.}^2 \times 24{,}000 \text{ lb/in.}^2 \times \left(\frac{7}{8}\right) \times 3.81 \text{ in.}$$

$$M = 24{,}803 \text{ lb-in.}$$

The capacity per foot of wall length will be the above value, divided by 4:

$$M = 6201 \text{ lb}$$

Balance Point

First, locate the neutral axis. Assume concrete masonry:

$$E_s = 29,000,000 \text{ lb/in.}^2$$

$$E_m = 900 \, f'_m = 900 \times 1500 \text{ lb/in.}^2 = 1,350,000 \text{ lb/in.}^2$$

$$n = \frac{E_e}{E_m} = \frac{29}{1.35} = 21.48$$

$$k_b = \frac{n}{\left(\frac{F_s}{F_b}\right) + n} = \frac{21.48}{\left(\frac{24000}{500}\right) + 21.48} = 0.309$$

Now calculate the corresponding axial force and moment:

$$T = A_s F_s$$

$$T = 0.31 \text{ in.}^2 \times 24,000 \text{ lb/in.}^2 = 7440 \text{ lb}$$

$$C = 1/2 \, (k_b d b F_b)$$

$$C = 1/2(0.309 \times 3.81 \text{ in.} \times 48 \text{ in.} \times 500 \text{ psi}) = 14,127 \text{ lb}$$

$$P = C - T$$

$$P = 14,127 \text{ lb} - 7440 \text{ lb} = 6687 \text{ lb}$$

$$M = T\left(d - \frac{h}{2}\right) + C\left(\frac{h}{2} - \frac{k_b d}{3}\right)$$

$$M = 7440 \text{ lb} \left(3.81 \text{ in.} - \frac{7.63 \text{ in.}}{2}\right) + 14,127 \text{ lb} \left(\frac{7.63 \text{ in.}}{2} - \frac{0.309 \times 3.81 \text{ in.}}{3}\right)$$

$$M = 0 + 48,351 \text{ lb-in.} = 48,351 \text{ lb-in.}$$

The capacities per foot of wall length will be the above values, divided by 4:

$$P = 1672 \text{ lb}$$

$$M = 12,088 \text{ lb}$$

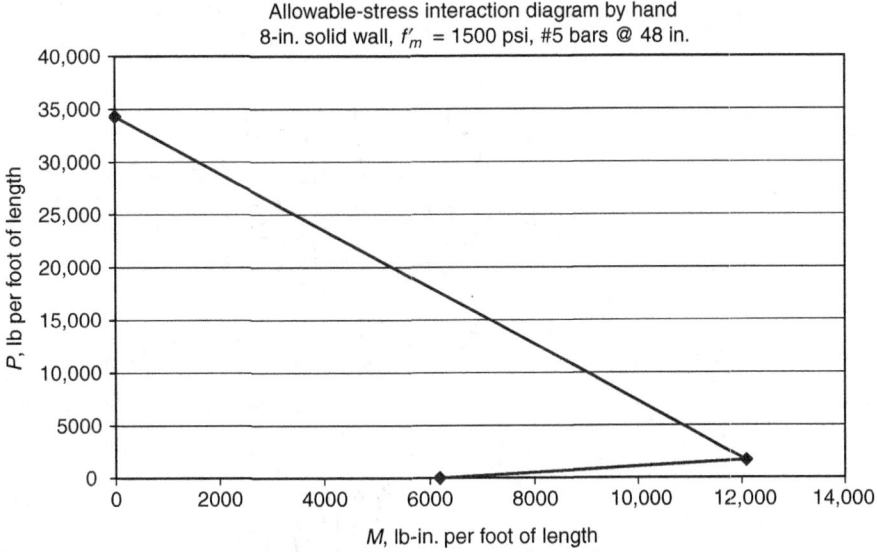

FIGURE 8.22 Plot of allowable-stress moment-axial force interaction diagram calculated by hand.

8.4.5 Plot of Allowable-Stress Interaction Diagram by Hand

The allowable-stress moment-axial force interaction diagram calculated in Sec. 8.4.3 is plotted in Fig. 8.22.

As we shall shortly see, the points that we have calculated are correct. The form of the diagram is misleading, however, because the balance point is actually not the point of maximum moment. It is incorrect but very conservative to draw the diagram with a straight line from the balance point to the pure-compression point.

8.4.6 Allowable-Stress Interaction Diagrams by Spreadsheet

To calculate allowable-stress interaction diagrams using a spreadsheet, first calculate the position of the neutral axis corresponding to the balance point (Fig. 8.23).

The location of the neutral axis can be determined:

$$\frac{\varepsilon_m}{\varepsilon_s} = \frac{\left(F_b/E_m\right)}{\left(F_s/E_s\right)} = \frac{k_b d}{d - k_b d}$$

$$\left(\frac{F_b}{E_m}\right)(d - k_b d) = \left(\frac{F_s}{E_s}\right) k_b d$$

$$\frac{E_s}{E_m} \equiv n$$

$$\left(\frac{F_b}{E_m}\right)d - \left(\frac{F_b}{E_m}\right)k_b d = \left(\frac{F_s}{nE_m}\right)k_b d$$

$$k_b\left(\frac{F_b}{E_m} + \frac{F_s}{nE_m}\right) = \left(\frac{F_b}{E_m}\right)$$

$$k_b = \frac{F_b}{F_b + F_s/n}$$

so,

$$k_b = \frac{n}{\left(F_s/F_b\right) + n}$$

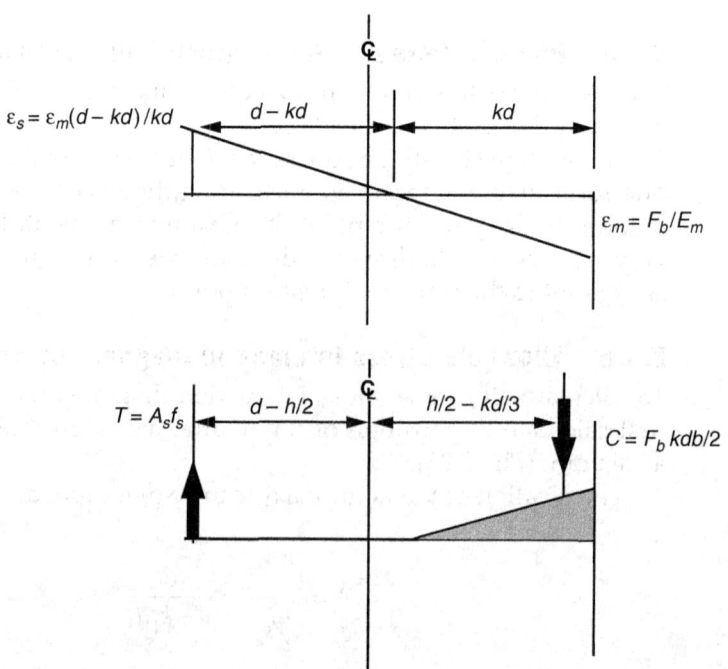

Figure 8.23 Conditions of strain and stress at allowable-stress balanced conditions.

For values of k less than that balanced value, the steel will reach its allowable stress before the masonry reaches its allowable stress. Combinations of axial force and moment can then be calculated, as can the corresponding moment (Fig. 8.24).

$$\varepsilon_m = \varepsilon_s \left(\frac{kd}{d - kd} \right)$$

$$f_b = \varepsilon_m E_m$$

$$C = kd \cdot b \cdot f_b / 2$$

$$T = A_s F_s$$

$$P = C - T$$

$$M = T\left(d - \frac{h}{2}\right) + C\left(\frac{h}{2} - \frac{kd}{3}\right)$$

Similarly, for values of kd greater than the allowable-stress balance-point value, the steel will not have reached its allowable stress when the masonry is at its allowable stress. Compute the strain (and corresponding

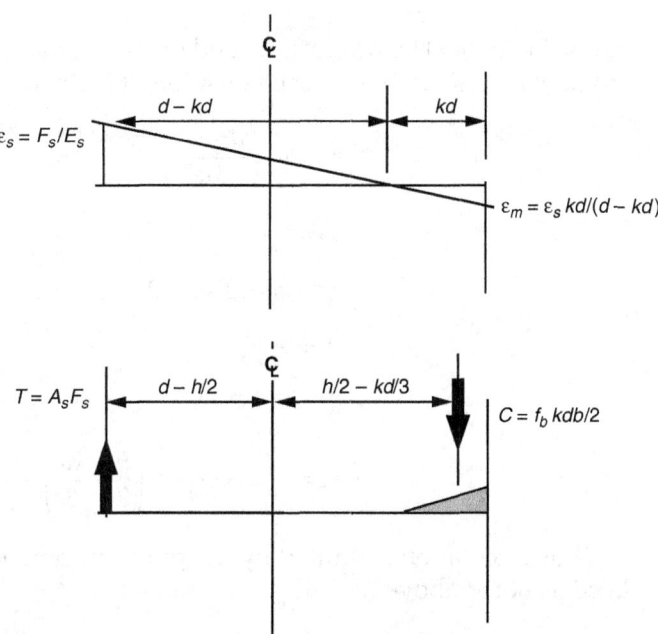

FIGURE 8.24 Conditions of strain and stress for values of k less than the allowable-stress balanced value.

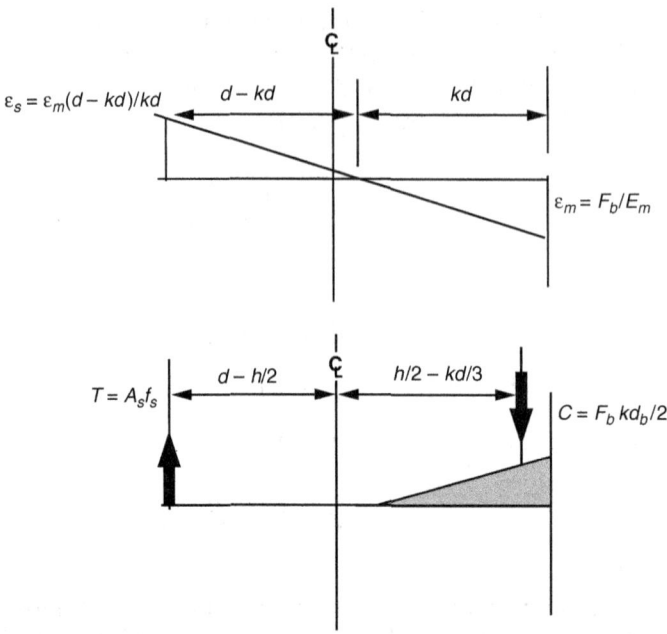

FIGURE 8.25 Conditions of strain and stress for values of k greater than the allowable-stress balanced value.

stress) in the steel by proportion, and find combinations of axial force and moment corresponding to each position of the neutral axis (Fig. 8.25).

$$\varepsilon_s = \varepsilon_m \left(\frac{d - kd}{kd} \right)$$

$$f_s = \varepsilon_s E_s$$

$$C = kd \times b \times F_b / 2$$

$$T = A_s f_s$$

$$P = C - T$$

$$M = T\left(d - \frac{h}{2}\right) + C\left(\frac{h}{2} - \frac{kd}{3}\right)$$

This calculation is limited by the pure compression resistance, calculated as noted above:

$$P_a = (0.25 f'_m A_n + 0.65 A_{st} F_s)\left[1 - \left(\frac{h}{140\, r}\right)^2\right]$$

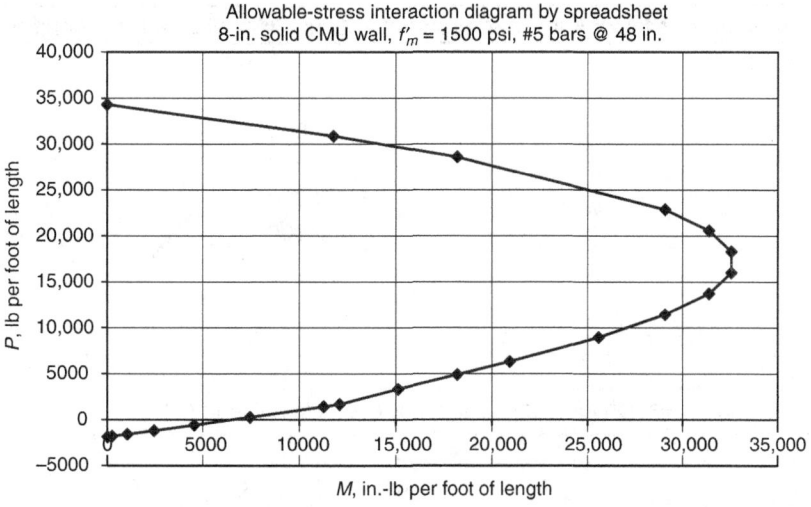

FIGURE 8.26 Plot of allowable-stress interaction calculated by spreadsheet.

8.4.7 Example of Moment-Axial Force Interaction Diagram by the Allowable-Stress Approach (Spreadsheet Calculation)

Construct the moment-axial force interaction diagram by the allowable-stress approach for a nominal 8-in. CMU wall, fully grouted, with f'_m = 1500 lb/in.² and reinforcement consisting of #5 bars at 48 in., placed in the center of the wall.

The effective width of the wall is $6t$, or 48 in. The spreadsheet and corresponding interaction diagram are shown in Fig. 8.26. As noted previously in this section, the results are interesting. Because the reinforcement is located at the geometric centroid of the section, and because the allowable-stress balance point does not reflect real behavior, the allowable-stress balance-point axial load (about 9000 lb) does not correspond to the maximum moment capacity.

8.4.8 Plot of Allowable-Stress Interaction Diagram by Spreadsheet

The allowable-stress interaction diagram, calculated by spreadsheet, is shown in Fig. 8.26.

Relevant cells from the spreadsheet are reproduced in Table 8.2.

8.4.9 Example of Allowable-Stress Design of Masonry Walls Loaded Out-of-Plane

Once we have developed the moment-axial force interaction diagram by the allowable-stress approach, the actual design simply consists of verifying

Chapter Eight

Spreadsheet for calculating allowable-stress M-N diagram for solid masonry wall	
Total depth	7.625
f'_m	1500
E_m	1350000
F_b	500
E_s	29000000
F_s	24000
d	3.8125
$k_{balanced}$	0.309168
Tensile reinforcement	0.31
Width	48
Because compression reinforcement is not tied, it is not counted	

	k	kd	f_b	C_{mas}	f_s	Moment	Axial force
Pure compression						0	34283
Points controlled by masonry	3	11.44	500	137250	0	0	34313
	2.7	10.29	500	123525	0	11773	30881
	2.5	9.53	500	114375	0	18169	28594
	2	7.63	500	91500	0	29070	22875
	1.8	6.86	500	82350	0	31396	20588
	1.6	6.10	500	73200	0	32559	18300
	1.4	5.34	500	64050	0	32559	16013
	1.2	4.58	500	54900	0	31396	13725
	1	3.81	500	45750	0	29070	11438
	0.8	3.05	500	36600	−2685	25582	8942
	0.6	2.29	500	27450	−7160	20931	6308
	0.5	1.91	500	22875	−10741	18169	4886
	0.4	1.53	500	18300	−16111	15117	3326
	0.309168	1.18	500	14144	−24000	12092	1676
Points controlled by steel	0.309168	1.18	500	14144	−24000	12092	1676
	0.3	1.14	479	13144	−24000	11275	1426
	0.25	0.95	372	8519	−24000	7443	270
	0.2	0.76	279	5111	−24000	4547	−582
	0.15	0.57	197	2706	−24000	2450	−1183
	0.1	0.38	124	1136	−24000	1047	−1576
	0.05	0.19	59	269	−24000	252	−1793
	0.01	0.04	11	10	−24000	10	−1857

TABLE 8.2 Spreadsheet for Calculating Allowable-Stress Interaction Diagram for Wall Loaded Out-of-Plane

Allowable-Stress Design of Reinforced Masonry Elements

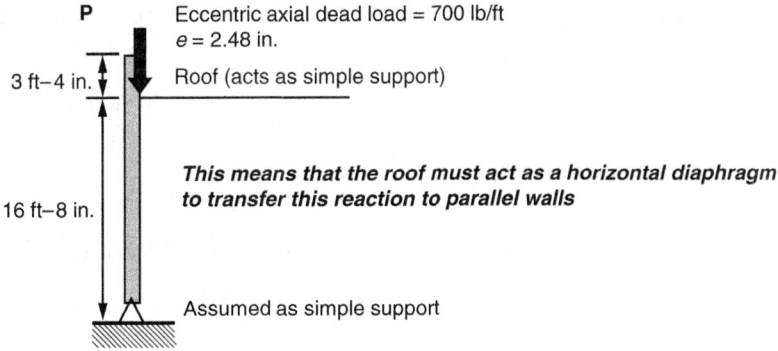

FIGURE 8.27 Example of reinforced bearing wall loaded out-of-plane.

that the combination of required axial and flexural capacity lies within the diagram. No load factors or ϕ factors are used.

Consider the bearing wall designed previously as unreinforced. It has an eccentric axial load plus out-of-plane wind load of 25 lb/ft².

The wall is as shown in Fig. 8.27.

At each horizontal plane through the wall, the following conditions must be met:

- Combination of axial and flexural compressive stresses must not violate the unity equation
- Net tension stress must not exceed the allowable flexural tension
- Separate stability check must be satisfied

We must check various points on the wall. Critical points are just below the roof reaction (moment is high and axial load is low, so net tension may govern); and at the base of the wall (axial load is high, so the unity equation or the stability equation may govern). Check each of these locations.

To avoid having to check a large number of loading combinations and potentially critical locations, it is worthwhile to assess them first, and check only the ones that will probably govern.

Due to wind only, the unfactored moment at the base of the parapet (roof level) is

$$M = \frac{qL^2_{\text{parapet}}}{2} = \frac{25 \text{ lb/ft} \cdot 3.33^2 \text{ ft}^2}{2} \times 12 \text{ in./ft} = 1663 \text{ lb-in.}$$

The maximum moment is close to that occurring at mid-height. The moment from wind load is the superposition of one-half moment at the upper support due to wind load on the parapet only, plus the midspan moment in a simply supported beam with that same wind load:

$$M_{midspan} = -\frac{1663}{2} + \frac{qL^2}{8} = -\frac{1663}{2} + \frac{25 \text{ lb/ft} \times 16.67^2 \text{ ft}^2}{8} \times 12 \text{ in./ft}$$

$$= 9589 \text{ lb-in.}$$

The unfactored moment due to eccentric axial load is

$$M_{gravity} = Pe = 1050 \text{ lb} \times 2.48 \text{ in.} = 2604 \text{ lb-in.}$$

Unfactored moment diagrams due to eccentric axial dead load and wind are shown in Fig. 8.28.

At each horizontal plane through the wall, the combinations of (P, M) must lie within the allowable-stress interaction diagram.

The critical point will be halfway up the wall.

As before, try 8 in. nominal units, and a specified compressive strength, f'_m, of 1500 lb/in.² This can be satisfied using units with a net-area compressive strength of 1900 lb/in.², and Type S PCL mortar. Work with a strip with a width of 1 ft (measured along the length of the wall in plan).

At mid-height,

$$P = 0.6 \times 700 \text{ lb} + 0.6 \times (3.33 + 8.33) \text{ ft} \times 48 \text{ lb/ft} = 755.8 \text{ lb}$$

$$M = P_{eccentric}\left(\frac{e}{2}\right) + \text{wind} = 0.6 \times 700 \text{ lb} \times \left(\frac{2.48 \text{ in.}}{2}\right) + 9589 \text{ lb-in.}$$

$$= 10,110 \text{ lb-in.}$$

In each foot of wall, the actions are P = 756 lb and M = 10,110 lb-in. This combination of actions lies just within the allowable-stress interaction diagram computed by spreadsheet, and the design is satisfactory.

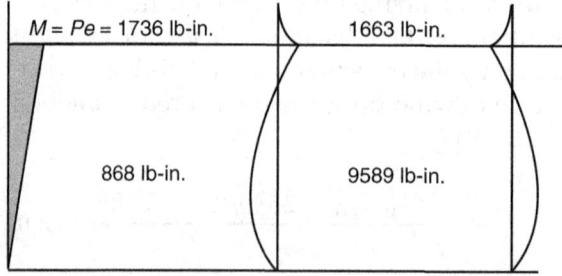

Figure 8.28 Unfactored moment diagrams due to eccentric axial dead load and wind.

8.5 Allowable-Stress Design of Reinforced Shear Walls

8.5.1 Introduction to Allowable-Stress Design of Reinforced Shear Walls

In this section, we shall study the behavior and design of reinforced masonry shear walls. The discussion follows the same approach used previously for unreinforced masonry shear walls.

8.5.2 Design Steps for Allowable-Stress Design of Reinforced Shear Walls

Reinforced masonry shear walls must be designed for the effects of

1. Gravity loads from self-weight, plus gravity loads from overlying roof or floor levels
2. Moments and shears from in-plane shear loads

Actions are shown in Fig. 8.29.

Flexural capacity of reinforced shear walls using allowable-stress procedures is calculated using moment-axial force interaction diagrams as discussed in the section on masonry walls loaded out-of-plane.

Shearing capacity is calculated using Sec. 2.3.5 of the 2008 MSJC *Code*. The approach is different from that of strength design.

1. Calculate an average shear stress, $f_v = V/bd$.
2. Check whether all shear can be resisted by masonry alone, using the relatively low allowable-shearing stresses from Sec. 2.3.5.2.2 of the 2008 MSJC *Code*.
   For $(M/Vd) < 1$,

$$F_v = \left(\frac{1}{3}\right)\left[4.0 - \left(\frac{M}{Vd}\right)\right]\sqrt{f'_m}$$

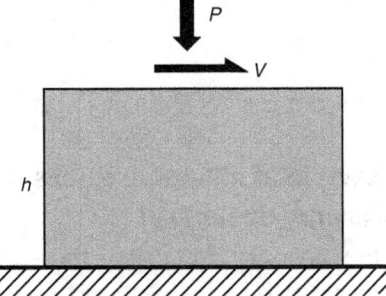

FIGURE 8.29 Design actions for reinforced masonry shear walls.

but not greater than

$$80 - 45(M/Vd) \text{ lb/in.}^2$$

and for $(M/Vd) \geq 1$,

$$F_v = \sqrt{f'_m}$$

but not greater than 35 lb/in.²

3. If (2) can be satisfied, no shear reinforcement is required. If (2) cannot be satisfied, then both the following conditions must be satisfied:

   a. Enough shear reinforcement must be provided to resist all shear (not just the difference between the applied shear and the capacity of the masonry). The minimum required shear reinforcement is calculated as

   $$A_v = \frac{Vs}{F_s d}$$

   b. The shear stress in the masonry must be checked against the relatively high allowable-shearing stresses from Sec. 2.3.5.2.3 of the 2008 MSJC *Code*:

   For $(M/Vd) < 1$,

   $$F_v = \left(\tfrac{1}{2}\right)\left[4.0 - \left(\frac{M}{Vd}\right)\right]\sqrt{f'_m}$$

   but not greater than

   $$120 - 45(M/Vd) \text{ lb/in.}^2$$

   and for $(M/Vd) \geq 1$,

   $$F_v = 1.5\sqrt{f'_m}$$

   but not greater than 75 lb/in.².

8.5.3 Example of Allowable-Stress Design of Reinforced Clay Masonry Shear Wall

Consider the masonry shear wall shown in Fig. 8.30.

Design the wall. Unfactored in-plane lateral loads at each floor level are due to earthquake, and are shown in Fig. 8.31, along with the corresponding shear and moment diagrams.

Allowable-Stress Design of Reinforced Masonry Elements

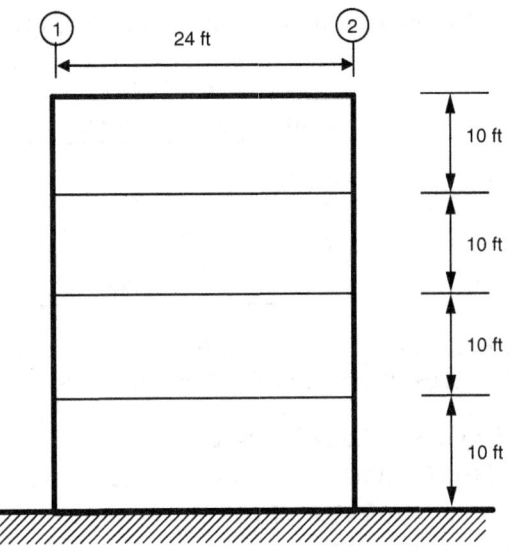

FIGURE 8.30 Reinforced masonry shear wall to be designed.

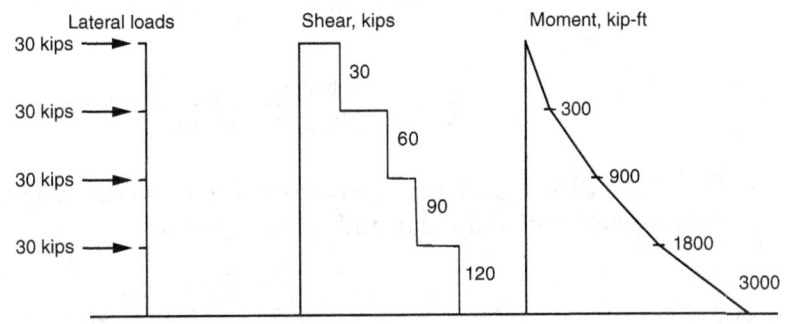

FIGURE 8.31 Unfactored in-plane lateral loads, shear and moment diagrams for reinforced masonry shear wall.

Assume an 8-in. nominal clay masonry wall, grouted solid, with Type S PCL mortar. The total plan length of the wall is 24 ft (288 in.), and its thickness is 7.5 in. Assume an effective depth d of 285 in.

	Clay Masonry
Unit Strength	6600
Mortar	Type S
f'_m	2500
E_m	1.75 × 10^6
n	16.6

Reinforcement = Grade 60; $E_s = 29 \times 10^6$ psi

Unfactored axial loads on the wall are given in the following table:

Level (Top of wall)	DL (kips)	LL (kips)
4	90	15
3	180	35
2	270	55
1	360	75

Check shear for the assumed wall thickness.
Use 2009 IBC ASD Load Combination 8: $0.6D + 0.7E$

$V = 0.7 \times 120,000 \text{ lb} = 84,000 \text{ lb}$
$M = 0.7 \times 3000 \text{ kip-ft} = 2100 \text{ kip-ft}$
$P = 0.6 \times 360 \text{ kips} = 216 \text{ kips}$

By MSJC *Code* Sec. 2.3.5.2.1,

$$f_v = \frac{V}{bd} = \frac{84,000 \text{ lb}}{7.5 \times 285 \text{ in.}^2} = 39.3 \text{ psi}$$

$$\frac{M}{Vd} = \frac{3000 \text{ kip-ft} \times 12 \text{ in./ft}}{120 \text{ kips} \times 285 \text{ in.}} = 1.05$$

By *Code* Sec. 2.3.5.2.2(b), where shear reinforcement is not provided to resist the calculated shear, and $(M/Vd) > 1$:

$$F_v = \sqrt{f'_m} = \sqrt{2500} = 50.0 \text{ psi}$$

but

$$F_v \leq 35 \text{ psi} \quad \Leftarrow \text{Governs}$$

$f_v = 39.3 \text{ psi} > F_v \quad \therefore \quad$ shear reinforcement is needed

Now check against the higher allowable shear stresses of *Code* Sec. 2.3.5.2.3, for $(M/Vd > 1)$:

$$F_v = 1.5\sqrt{f'_m} = 1.5\sqrt{2500} = 75.0 \text{ psi} \quad \Leftarrow \text{Governs}$$

$$F_v \leq 75 \text{ psi}$$

$f_v = 39.3 \text{ psi} \leq F_v \quad \therefore \quad$ OK

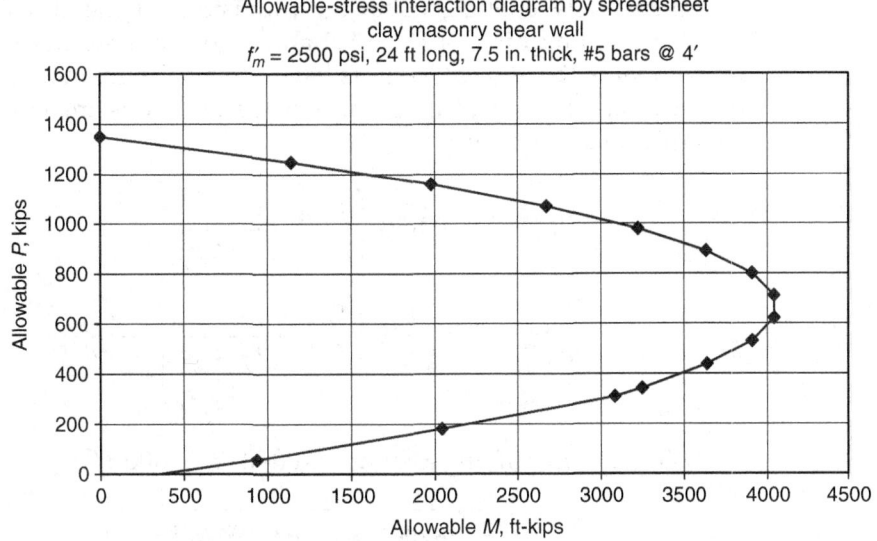

FIGURE 8.32 Plot of allowable-stress moment-axial force interaction diagram calculated by spreadsheet.

Design shear reinforcement using MSJC *Code* Sec. 2.3.2.1:

$$F_s = 24{,}000 \text{ psi}$$

$$A_v = \left(\frac{Vs}{F_s d}\right) = \frac{84{,}000 \text{ kips} \times 24 \text{ in.}}{24{,}000 \text{ psi} \times 285 \text{ in.}} = 0.29 \text{ in.}^2$$

Use #5 bars horizontally at 24 in. on center.

Now consider the flexural design. The spreadsheet for this problem illustrates how to calculate an allowable-stress moment-axial force interaction diagram for this wall (Fig. 8.32). That spreadsheet is first used to check the wall with reinforcement consisting of #5 bars @ 4 ft.

The interaction diagram shows that the wall, as designed in Fig. 8.32, can resist the combination of axial load and moment (216 kips, 2100 kip-ft).

8.5.4 Minimum and Maximum Reinforcement Ratios for Flexural Design by the Allowable-Stress Approach

The allowable-stress provisions of the 2008 MSJC *Code* have no requirements for minimum nor maximum flexural reinforcement, except for a maximum-reinforcement requirement for special reinforced masonry shear walls. The MSJC is actively studying this issue, however, and provisions will probably be added within 3 to 6 years.

Interestingly enough, using the provisions of the 2008 MSJC *Code* (E_m a constant multiple of f'_m, and F_b and F_s constant multiples of f'_m and f_y respectively), the allowable-stress balanced steel area, though having no physical significance of its own, is actually a constant fraction of the strength balanced area:

$$\rho_{bal}^{strength} = 0.85\beta_1 \left(\frac{f'_m}{f_y}\right)\left(\frac{\varepsilon_{mu}}{\varepsilon_{mu}+\varepsilon_y}\right)$$

$$\rho_{bal}^{allowable} = \left(\frac{1}{2}\right)k_b\left(\frac{F_b}{F_s}\right) = \left(\frac{1}{2}\right)\frac{n}{\left(F_s/F_b\right)+n}\left(\frac{F_b}{F_s}\right)$$

Using the simplifying assumption that $E_m \approx 800\, f'_m$,

f'_m, lb/in.²	$\sigma_{bal}^{allowable}/\sigma_{bal}^{strength}$
1500	0.347
2000	0.347
2500	0.347

Therefore, a design in which flexural reinforcement is kept below the allowable-stress balanced steel area, will coincidentally result in a design in which flexural reinforcement is below the balanced area for strength design as well.

8.5.5 Additional Comments on the Design of Reinforced Shear Walls

Reinforced masonry shear walls, like unreinforced ones, are relatively easy to design by either strength or allowable-stress approaches. Although shear capacities per unit area is small, the available area is large.

With either strength or allowable-stress approaches, it is rarely necessary to use shear reinforcement. In this sense, the best shear design strategy for shear walls is like that for shear design of beams—use enough cross-sectional area to eliminate the need for shear reinforcement. Seismic requirements may still dictate some shear reinforcement, however.

8.6 Required Details for Reinforced Bearing Walls and Shear Walls

Bearing walls that resist out-of-plane lateral loads, and shear walls, must be designed to transfer lateral loads to the floors above and below. Examples of such connections are shown in Figs. 8.33 through 8.37.

Allowable-Stress Design of Reinforced Masonry Elements

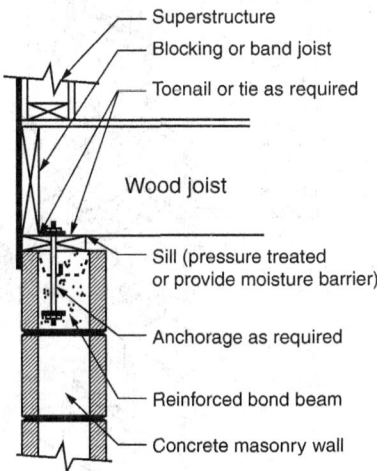

FIGURE 8.33 Example of wall-to-foundation connection. (*Source*: Figure 1 of National Concrete Masonry Association TEK 05-07A.)

These connections would have to be strengthened for regions subject to strong earthquakes or strong winds. Section 1604.8.2 of the 2009 IBC has additional requirements for anchorage of diaphragms to masonry walls. Section 12.11 of ASCE 7-05 has additional requirements for anchorage of structural walls for structures assigned to Seismic Design Categories C and higher.

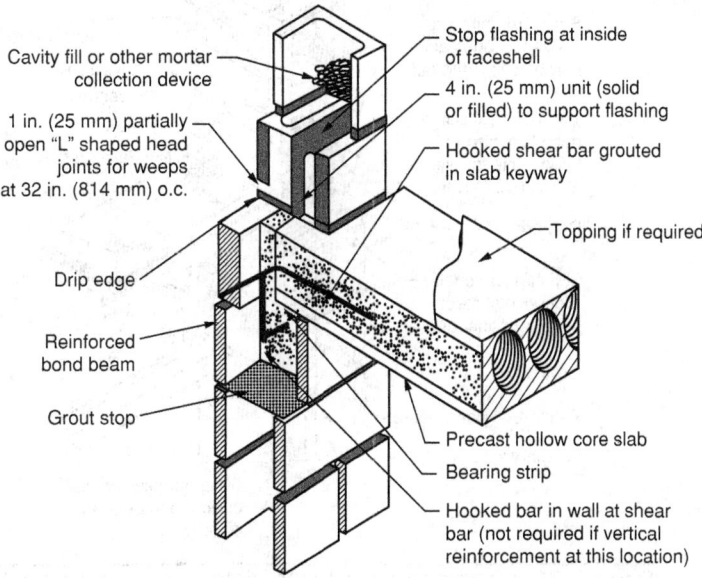

FIGURE 8.34 Example of wall-to-floor connection, planks perpendicular to wall. (*Source*: Figure 14 of National Concrete Masonry Association TEK 05-07A.)

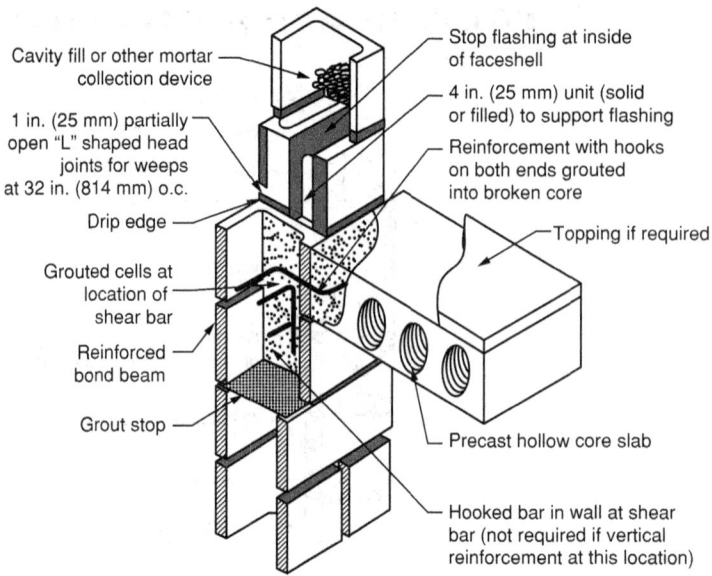

FIGURE 8.35 Example of wall-to-floor connection, planks parallel to wall. (*Source*: Figure 15 of National Concrete Masonry Association TEK 05-07A.)

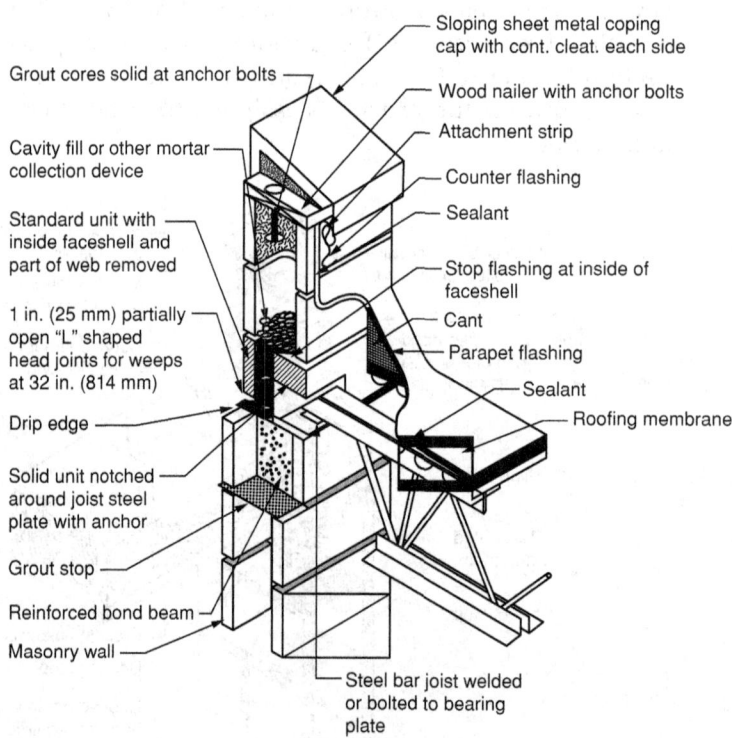

FIGURE 8.36 Example of wall-to-roof detail. (*Source*: Figure 11, "Steel Joist with Pocket," of National Concrete Masonry Association TEK 05-07A.)

Allowable-Stress Design of Reinforced Masonry Elements 327

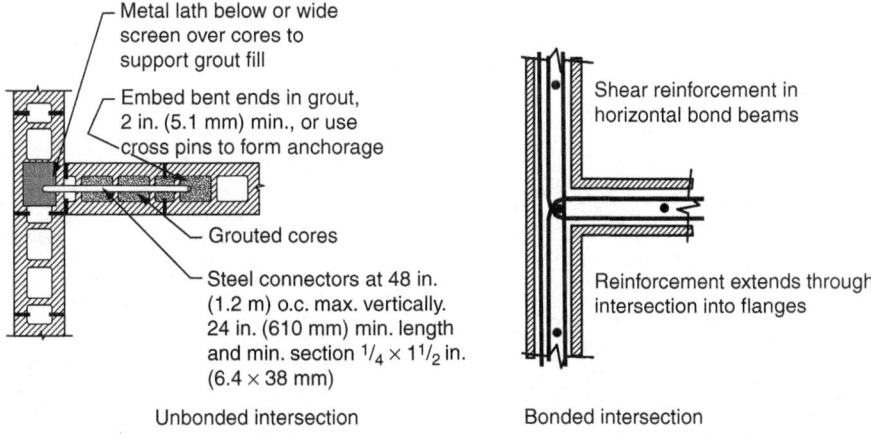

FIGURE 8.37 Examples of wall-to-wall connection details. (*Source*: Figure 2 of National Concrete Masonry Association TEK 14-08B.)

8.6.1 Wall-to-Foundation Connections

As shown in Fig. 8.33, CMU walls (or the inner CMU wythe of a drainage wall) must be connected to the concrete foundation. Bond breaker should be used only between the outer veneer wythe and the foundation.

8.6.2 Wall-to-Floor Details

Examples of a wall-to-floor detail shown in Figs. 8.34 and 8.35. In the latter detail (floor or roof planks oriented parallel to walls), the planks are actually cambered. They are shown on the outside of the walls so that this camber does not interfere with the coursing of the units. Some designers object to this detail because it could lead to spalling of the cover. If it is modified so that the planks rest on the face shells of the walls, then the thickness of the topping must vary to adjust for the camber, and form boards must be used against both sides of the wall underneath the planks, so that the concrete or grout that is cast into the bond beam does not run out underneath the cambered beam.

8.6.3 Wall-to-Roof Details

An example of a wall-to-roof detail is shown in Fig. 8.36.

8.6.4 Typical Details of Wall-to-Wall Connections

Typical details of wall-to-wall connections are shown in Fig. 8.37.

CHAPTER 9
Comparison of Design by the Allowable-Stress Approach versus the Strength Approach

In this chapter, designs carried out by the allowable-stress approach of the 2008 MSJC *Code* are compared with those carried out by the strength approach. In the past 10 years, the MSJC has devoted considerable effort to harmonize these two design approaches, so that they give quite similar results. Some differences still exist, however, and it is useful to be aware of them.

In this chapter, designs are compared in general terms by comparing the load side and the resistance side of each set of design equations—strength and allowable stress.

9.1 Comparison of Allowable-Stress and Strength Design of Unreinforced Panel Walls

Allowable-stress design and strength design for unreinforced panel walls loaded out-of-plane can be compared in tabular form, as shown in Table 9.1. In that table, for allowable-stress design, load effects are denoted by W and resistances by R. Using strength design, load effects are the unfactored load effects W, multiplied by the load factor for wind (1.6). Nominal resistances are proportional to the same section properties as for allowable-stress design, but the modulus of rupture is 2.5 times the

Allowable-stress design		Strength design	
Load effects	Resistances	Load effects	Resistances
W	R	1.6 W	ϕ (2.5 R)
			0.6 (2.5 R)
		1.6 W	1.5 R
		W	0.94 R

TABLE 9.1 Comparison of Allowable-Stress and Strength Design for Unreinforced Panel Walls

allowable stress. Design resistances are therefore proportional to 2.5R, multiplied by the strength-reduction factor of 0.6.

For a given load effect W, the strength design provisions require a slightly smaller resistance than the allowable-stress provisions. In future editions of the MSJC *Code*, this minor discrepancy may be eliminated. For the time being, the two sets of provisions can be regarded as essentially identical.

9.2 Comparison of Allowable-Stress Design and Strength Design of Unreinforced Bearing Walls

Allowable-stress design and strength design for unreinforced bearing walls loaded out-of-plane are not as easy to compare as they are for unreinforced panel walls. This is because bearing walls are subjected to combinations of axial load and moment from eccentric gravity load, out-of-plane wind, or both.

For the critical strength loading combination ($0.9D + 1.6L$), location (mid-height of the wall), and criterion (maximum tensile stress), the maximum tensile stress from the factored load combination is more than 1.6 times the maximum tensile stress from the allowable-load combination, because D is multiplied by 0.9 rather than 1.6. Therefore, strength design might routinely require grouting, while allowable-stress design would not.

9.3 Comparison of Allowable-Stress Design and Strength Design of Unreinforced Shear Walls

Assessing the comparative level of safety of strength design and allowable-stress design is relatively simple. Assume that the critical loading combination for strength design will be $0.9D + 1.6W$, because this gives the smallest shear capacity for the third shear strength criterion, which

Allowable-stress design		Strength design	
Load effects	Resistances	Load effects	Resistances
1.5 W	R	1.6 W	φ (2.50) R
			0.80 (2.50) R
W	0.67 R	W	1.25 R

TABLE 9.2 Comparison of Allowable-Stress and Strength Design for Unreinforced Shear Walls

usually controls. Allowable-stress design for shear uses a peak shear stress equal to (3/2) times the average stress, so the load effect is essentially multiplied by 1.5. Resistances are calculated based on an allowable resistance R. For strength design, load effects are multiplied by a load factor of 1.6. The equations for nominal shear capacity for strength design, divided by the equations for allowable-stress shear capacity, produces capacity ratios varying from 1.5 for the third shear criterion, to 2.5 for the first and second (300/120). This discrepancy will probably be addressed in future editions of the MSJC *Code*. Using a ratio of 2.50 for comparison purposes, the net result is shown in Table 9.2.

If allowable-stress design and strength design are compared in terms of equal load effects, and if the strength-design resistances are higher than their allowable-stress counterparts by a factor that is usually 2.5, then the final results for strength design require about half as much cross-sectional area as is required by allowable-stress design.

9.4 Comparison of Allowable-Stress and Strength Designs for Anchor Bolts

Allowable-stress design and strength design for anchor bolts are not simple to compare. The resistance (capacity) sides of allowable-stress equations for failure modes governed by masonry are about one-third of the corresponding strength equations. For pullout, the corresponding factor is 0.4 and for steel failure, the factor is 0.6.

Because anchor bolts can be subjected to many different loading combinations, a typical load factor for strength design is difficult to establish. For purposes of this book, it will be assumed that wind governs (which is often the case), and therefore a typical weighted load factor (representing gravity plus wind loads) is about 1.5.

Because safety factors are different for behavior governed by masonry and behavior governed by steel, they are presented separately in Table 9.3 and Table 9.4.

Allowable-stress design		Strength design	
Load effects	Resistances	Load effects	Resistances
W	R	1.5 W	φ (1/0.33) R
			0.5 (3.0) R
W	R	W	R

TABLE 9.3 Comparison of Allowable-Stress and Strength Design for Anchor Bolts, Masonry Controls

Allowable-stress design		Strength design	
Load effects	Resistances	Load effects	Resistances
W	R	1.5 W	φ (1/0.6) R
			0.9 (1.67) R
W	R	W	R

TABLE 9.4 Comparison of Allowable-Stress and Strength Design for Anchor Bolts, Steel Controls

In each case, we can see that the allowable-stress and strength equations for anchor bolt design have been harmonized so that they give essentially identical results.

9.5 Comparison of Allowable-Stress and Strength Designs for Reinforced Beams and Lintels

The design objective that the lintel be deep enough to preclude the need for transverse reinforcement means that the flexural design is governed by stresses in tensile reinforcement in each case. Allowable-stress design for flexure uses $D+L$. Resistances are based on an allowable stress of 24,000 lb/in.^{2}. Assume that the critical loading combination for strength design will be $1.2D + 1.6L$, or about $1.4(D + L)$. Resistances are based on specified yield stress (60,000 lb/in.^{2}). There is practically no difference between internal lever arms for the two design approaches. The net result is shown in Table 9.5.

If allowable-stress design and strength design are compared in terms of equal load effects, the resistance for strength design is higher than for allowable-stress design, by a factor of (38.6/24), or 1.61. While allowable-stress design is more conservative, the actual difference in required reinforcement is quite small.

The required depth of the lintel can also be compared for each design approach. Allowable-stress design for shear uses $D + L$. Resistances are based on an allowable stress of $\sqrt{f'_m}$. Assume that the critical loading

Allowable-stress design		Strength design	
Load effects	Resistances	Load effects	Resistances
D + L	24	1.4 (D + L)	ϕ 60
		1.4 (D + L)	0.90 (60)
		1.4 (D + L)	54
		D + L	38.6

*As governed by flexure.

TABLE 9.5 Comparison of Allowable-Stress and Strength Design for Reinforced Beams and Lintels*

Allowable-stress design		Strength design	
Load effects	Resistances	Load effects	Resistances
D + L	$\sqrt{f'_m}$	1.4 (D + L)	$\phi\,(2.25)\sqrt{f'_m}$
		1.4 (D + L)	$0.80\,(2.25)\sqrt{f'_m}$
		1.4 (D + L)	$1.8\sqrt{f'_m}$
		D + L	$1.29\sqrt{f'_m}$

*As governed by shear.

TABLE 9.6 Comparison of Allowable-Stress and Strength Design for Reinforced Beams and Lintels*

combination for strength design will be $1.2D + 1.6L$, or about $1.4(D + L)$. Resistances are based on $2.25\sqrt{f'_m}$. The net result is shown in Table 9.6.

If allowable-stress design and strength design are compared in terms of equal load effects, the required depth for allowable-stress design is greater than for strength design, by a factor of 1.29. While allowable-stress design is more conservative, the actual difference in required depth is probably not important in most cases.

9.6 Comparison of Allowable-Stress and Strength Designs for Reinforced Curtain Walls

Because the design of reinforced curtain walls is almost always governed by flexure, comparison of the results of allowable-stress and strength design approaches is similar to that of beams and lintels governed by flexure.

	Allowable-stress design		Strength design	
	Load effects	Resistances	Load effects	Resistances
	$D + L$	24	$1.4(D + L)$	$\phi\, 60$
			$1.4(D + L)$	$0.90(60)$
			$1.4(D + L)$	54
			$D + L$	38.6

*Governed by flexure.

TABLE 9.7 Comparison of Allowable-Stress and Strength Design for Reinforced Bearing Walls*

9.7 Comparison of Allowable-Stress and Strength Designs for Reinforced Bearing Walls

For most masonry walls of practical interest, design is controlled by stresses in tensile reinforcement, and the shape of the compressive stress block (triangular versus rectangular) does not make much difference. Allowable-stress design for flexure uses $D + L$. Resistances are based on an allowable stress of 24,000 lb/in.² Assume that the critical loading combination for strength design will be $1.2D + 1.6L$, or about $1.4(D + L)$. Resistances are based on specified yield stress (60,000 lb/in.²). There is practically no difference between internal lever arms for the two design approaches. The net result is shown in Table 9.7.

If allowable-stress design and strength design are compared in terms of equal load effects, the resistance for strength design is higher than for allowable-stress design, by a factor of (34.3/24), or 1.61. While allowable-stress design is more conservative, the actual difference in required reinforcement is quite small.

9.8 Comparison of Allowable-Stress and Strength Designs for Reinforced Shear Walls

For design governed by flexure, the comparison is basically similar to what we have seen for beams. Flexural design is governed by stresses in tensile reinforcement in each case. Allowable-stress design for flexure uses $D + L$. Resistances are based on an allowable stress of 24,000 lb/in.² Assume that the critical loading combination for strength design will be $1.6L$. Resistances are based on yield stress (60,000 lb/in.²). There is practically no difference between internal lever arms for the two design approaches. The net result is shown in Table 9.8.

Comparison of Design:Allowable-Stress vs. the Strength Approach

Allowable-stress design		Strength design	
Load effects	Resistances	Load effects	Resistances
D + L	24	1.6 (D + L)	ϕ 60
		1.6 (D + L)	0.90 (60)
		1.6 (D + L)	54
		D + L	33.8

*Governed by flexure.

TABLE 9.8 Comparison of Allowable-Stress and Strength Design for Reinforced Bearing Walls*

Allowable-stress design		Strength design	
Load effects	Resistances	Load effects	Resistances
D + L	$\sqrt{f'_m}$	1.6 (D + L)	$\phi(2.25)\sqrt{f'_m}$
		1.6 (D + L)	$0.80(2.25)\sqrt{f'_m}$
		1.6 (D + L)	$1.80\sqrt{f'_m}$
		D + L	$1.13\sqrt{f'_m}$

*As governed by shear.

TABLE 9.9 Comparison of Allowable-Stress and Strength Design for Reinforced Shear Walls*

If allowable-stress design and strength design are compared in terms of equal load effects, the resistance for strength design is considerably higher than that for allowable-stress design, by a factor of (33.8/24), or 1.41.

For design governed by shear, the comparison is basically what we have already seen for shear for beams. Allowable-stress design for shear uses $D + L$. Resistances are based on an allowable stress of about $\sqrt{f'_m}$. Assume that the critical loading combination for strength design is $1.6L$. Minimum resistances are based on $2.25\sqrt{f'_m}$. The net result is shown in Table 9.9.

If allowable-stress design and strength design are compared in terms of equal load effects, available resistance for strength design is about 1.13 times that for allowable-stress design. While strength design is slightly less conservative, the actual difference in required area of masonry is not significant.

CHAPTER 10
Lateral Load Analysis of Shear-Wall Structures

The preceding chapters contained an introduction to the behavior and design of masonry shear walls by the strength approach and by the allowable-stress approach. In the design examples of Secs. 5.3, 6.4, 7.3, and 8.5, it was clear that all walls would resist equal shears. In the design examples with a perforated wall (Secs. 5.3.6 and 7.3.6), it was clear that all wall segments would resist equal shears. This, however, is not generally the case.

Consider, for example, the building of Fig. 10.1 with a uniformly distributed wind pressure (wind from the south) of 35 lb/ft². The openings on the east wall introduce two problems:

- How are north-south shears distributed between the two walls which are oriented in that direction and what is the resulting building response?
- How is the shear on the east wall distributed among the three segments comprising that wall?

The classical design steps for this problem are as follows:

1. Classify the floor diaphragm as "rigid" or "flexible."
2. Based on that classification, solve for the distribution of shear to walls and to wall segments.

342 Chapter Ten

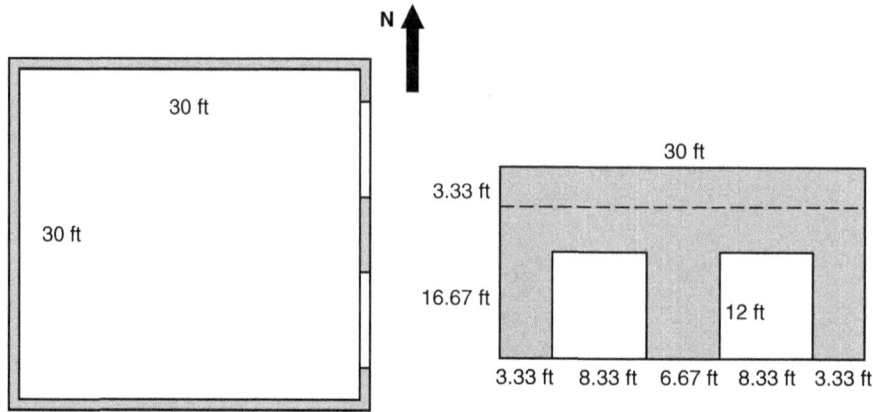

FIGURE 10.1 Example of building with perforated walls.

In the following sections, each of those steps is explained in more detail. The steps are then simplified, greatly reducing the required effort for most problems. Finally, readers are encouraged to use the simplified steps for rigid and for flexible diaphragms to bound the answer, eliminating the need to classify the diaphragms as rigid or flexible.

10.1 Classification of Horizontal Diaphragms as Rigid or Flexible

When a wall-type building is loaded laterally, its response, and the distribution of lateral load to its shear walls, depends on the in-plane flexibility of its horizontal diaphragms with respect to the in-plane flexibility of its walls. Horizontal diaphragms are classified as rigid, flexible, or semirigid in accordance with Sec. 1602 of the 2009 IBC, which also references Sec. 12.3.1 of ASCE7-05, as modified by Sec. 1613.6.1 of the 2009 IBC.

- Diaphragms whose in-plane diaphragm deformation is less than two times the in-plane wall deformation (average story drift), or which meet certain prescriptive requirements, are required to be considered as "rigid" (Sec. 1602 of the 2009 IBC).

- Diaphragms whose in-plane deformation is greater than two times the in-plane wall deformation (average story drift), or which meet certain prescriptive requirements, are permitted to be considered as "flexible."

- Diaphragms not classified as "rigid" or "flexible" according to the above criteria are considered "semirigid," and their flexibility must be explicitly considered in distributing wind or seismic forces (Sec. 12.3.1 of ASCE 7-05).

While some computer programs permit analysis of buildings including the effects of in-plane deformations of horizontal diaphragms, this is generally unnecessary. For almost all practical cases, it is sufficient to classify the diaphragm as either rigid or flexible, and then to analyze the building with the help of simplifying assumptions consistent with that classification. For preliminary design, it is even possible to first analyze the building assuming that the diaphragm is rigid, then analyze it assuming that the diaphragm is flexible, and takes the more critical of the two cases to arrive at design actions for each shear wall.

10.1.1 Prescriptive Characteristics of Rigid Diaphragms

In accordance with Sec. 12.3.1.2 of ASCE 7-05, horizontal diaphragms are permitted to be considered as rigid if they can be described as diaphragms of concrete labs or concrete filled metal deck with span-to-depth ratios of 3 or less in structures that have no horizontal irregularities.

10.1.2 General Characteristics of Flexible Diaphragms

In accordance with Sec. 12.3.1.1 of ASCE 7-05, horizontal diaphragms are permitted to be considered as flexible if they can be described as diaphragms constructed of untopped steel decking or wood structural panels in structures in which the vertical elements are masonry shear walls.

10.2 Lateral Load Analysis of Shear-Wall Structures with Rigid Floor Diaphragms

Many possible approaches are available for the lateral load analysis of shear-wall structures with rigid floor diaphragms. These are enumerated below; example analyses are conducted using each numbered approach; and the numbered approaches are then compared with respect to solution accuracy and time. Finally, based on those comparisons, recommendations are made for the best way to analyze such structures.

Method 1: The first method (referred to in this chapter as Method 1) is the linear elastic, finite element analysis of walls with openings (e.g., SAP 2000©), assuming rigid floor diaphragms. This is reasonably quick with modern programs, but cannot address cracking.

Most computer programs developed for the analysis of buildings consider floor diaphragms to be rigid in their own planes. Each floor level has only three horizontal degrees of freedom (two horizontal displacements and rotation about a vertical axis). This approach, while reasonable for frame structures, whose floors are much more rigid in their own planes than the vertical frames, is not correct for wall structures, whose horizontal diaphragms are usually about as rigid as their vertical diaphragms (walls).

Using this approach, because all but three horizontal degrees are condensed out at each floor level, in-plane actions in floor diaphragms cannot be calculated. For some loading conditions, in-plane actions can be inferred from differences in shears in vertical elements above and below each floor level. This process is always tedious, however, and is also inaccurate for cases involving multimodal response to dynamic loads.

Method 2: The second method (referred to in this chapter as Method 2), consists of the approximate analysis of panels with openings, followed by an analysis of the building. In order of increasing complexity, several variations of Method 2 are available:

Method 2a: Consider only shearing deformations of walls. Assume that shearing stiffness is proportional to length in plan. Ignore plan torsion due to eccentricity between center of stiffness of building and line of application of lateral load.

Method 2b: This is the same as Method 2a, but considers plan torsion.

Method 2c: This is the same as Method 2b, but considers flexural and shearing deformations of wall segments.

In the remainder of this section, evidence will be presented for the following observations:

- Method 1 is accurate, but requires a computer and considerable work.
- Method 2a gives reasonably accurate results with very little effort, and is therefore quite cost-effective for design.
- Method 2b gives almost the same accuracy as Method 2a, and requires significantly more effort. For most buildings, the effects of plan torsion are not significant, and Method 2b is not cost-effective.
- Method 2c generally requires much more effort than Method 2b, is not more accurate than Method 2a, and is not justified.

10.2.1 Example Application of Method 1

A reference answer for the example problem of this section is obtained using the finite element method. The roof diaphragm is assumed rigid. A masonry modulus of 1.5×10^6 lb/in.² is used. The applied lateral load of 35 lb/ft² results in a load of 12,600 lb applied through the plan center of

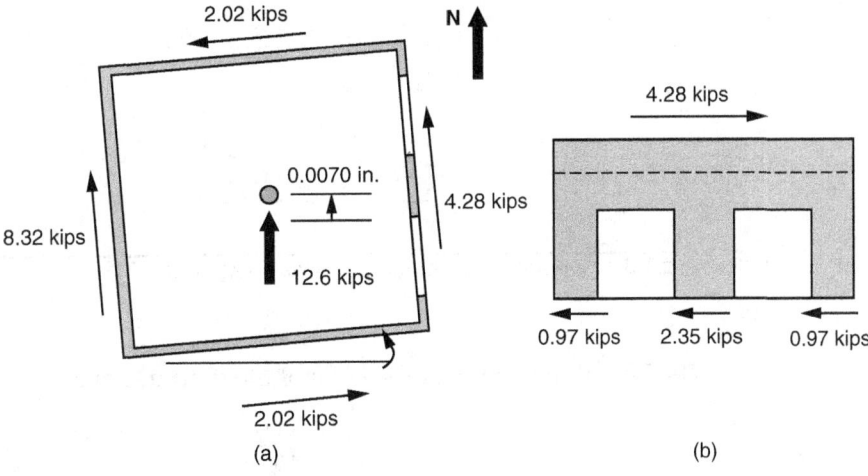

FIGURE 10.2 Solution to example problem using Method 1 (finite element method).

the building, at the level of the roof diaphragm. Using the example structure of Fig. 10.1, the solution of Fig. 10.2 is obtained. That solution is also summarized in Tables 10.1 and 10.2.

The structure rotates very slightly counter-clockwise. The plan center of the building displaces 0.0070 in. to the north. Shears applied to the tops of the walls, just below the roof, are shown in Fig. 10.2a. Figure 10.2b shows the distribution of shears in the segments of the east wall.

Modeling and analysis time: 30 min

10.2.2 Simplest Hand Method (Method 2a)

In the simplest hand method (Method 2a), consider only shearing deformations of the walls. Assume that shearing stiffness is proportional to length in plan. Ignore plan torsion due to eccentricity between center of stiffness of building, and line of application of lateral load.

Plan torsion can be ignored if the structure has reasonable plan length of walls in each principal plan direction. The deformation of wall segments can be idealized as due to shearing deformations only provided that the segments have an aspect ratio (ratio of height to plan length) of about 1.0 or less.

Shearing Stiffness of Each Wall Segment

The shearing deformation of a wall segment with height H and plan length L is shown in Fig. 10.3.

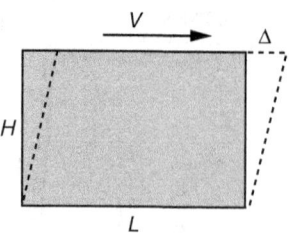

FIGURE 10.3 Shearing deformation of a wall segment.

The lateral deflection at the level of the diaphragm is

$$\Delta = \frac{VH}{A'G}$$

where G = shearing modulus of masonry
 H = wall height
 A' = effective shear area of the wall, taken for convenience as the product of the wall length in plan, L, and the wall thickness, t

Assume that G, t, and H are uniform. Therefore the deflection is proportional to V and inversely proportional to L:

$$\Delta \propto \frac{V}{L}$$

Finally, the stiffness of the segment is the applied shear divided by the deflection, and is simply proportional to the plan length of the wall segment.

$$\text{Stiffness} = \frac{V}{\Delta} \propto L$$

Distribution of Shears among Wall Segments
So the stiffness of each wall is proportional to its length in plan. In the elastic range, shears are distributed to the walls in proportion to their stiffnesses; that is, in proportion to their plan lengths.

10.2.3 Example of Simplest Hand Method (Method 2a)
To illustrate the simplest hand method (Method 2a), apply it to the example problem previously considered. The plan lengths of shear walls are shown in Fig. 10.4.

Given the total shear of 12.6 kips, the shear to wall 2 and wall 4 are easily computed using the proportions of wall lengths. The shear in each wall is the total shear multiplied by the ratio of the plan length of that wall divided by the total length of walls in the direction of applied load.

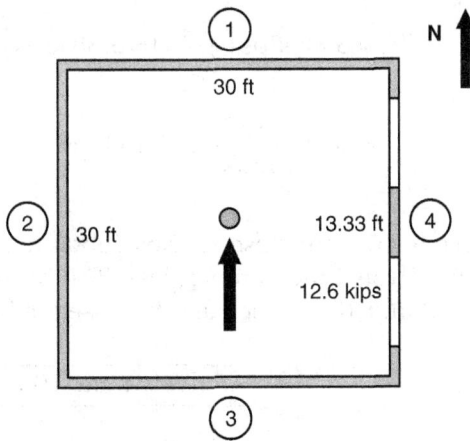

FIGURE 10.4 Plan lengths of wall segments for example problem using simplest hand method (Method 2a).

Because plan torsion is neglected, there are no shears in the walls oriented perpendicular to the applied load.

$$V_2 = 12.6 \text{ kips} \left(\frac{L_1}{L_{total}}\right) = 12.6 \text{ kips} \left(\frac{30 \text{ ft}}{30 + 3.33 + 6.67 + 3.33 \text{ ft}}\right)$$

$$= 12.6 \left(\frac{30}{43.33}\right) = 8.72 \text{ kips}$$

$$V_4 = 12.6 \text{ kips} \left(\frac{L_2}{L_{total}}\right) = 12.6 \text{ kips} \left(\frac{3.33 + 6.67 + 3.33 \text{ ft}}{30 + 3.33 + 6.67 + 3.33 \text{ ft}}\right)$$

$$= 12.6 \left(\frac{13.33}{43.33}\right) = 3.88 \text{ kips}$$

$$V_1 = V_3 = 0$$

The resulting shears in wall 2 and wall 4 are shown in Fig. 10.5.

The 3.88 kips applied to the east wall will be distributed to the three segments of that wall in proportion to their plan lengths. The equations for wall shears are provided below, and the calculated shears in each wall segment are shown in Fig. 10.6.

$$V_A = V_C = 3.88 \text{ kips} \left(\frac{L_A}{L_{total}}\right) = 3.88 \text{ kips} \left(\frac{3.33 \text{ ft}}{3.33 + 6.67 + 3.33 \text{ ft}}\right)$$

$$= 3.88 \left(\frac{3.33}{13.33}\right) = 0.97 \text{ kips}$$

$$V_B = 3.88 \text{ kips} \left(\frac{L_B}{L_{\text{total}}}\right) = 3.88 \text{ kips} \left(\frac{6.67 \text{ ft}}{3.33 + 6.67 + 3.33 \text{ ft}}\right)$$

$$= 3.88 \left(\frac{6.67}{13.33}\right) = 1.94 \text{ kips}$$

These results are close to those obtained by finite-element analysis, but the analysis time is much less. The results are also summarized in Tables 10.1 and 10.2, which are discussed in Sec. 10.2.6.

Analysis time: 10 min

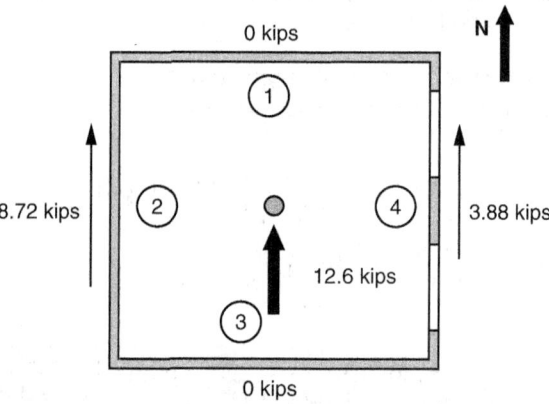

FIGURE 10.5 Shears in wall 2 and wall 4 of example using the simplest hand method (Method 2a).

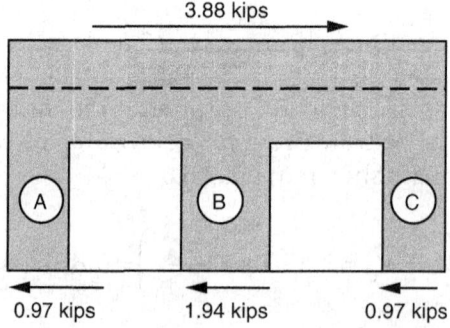

FIGURE 10.6 Shears in wall segments of wall 4 using the simplest hand method (Method 2a).

10.2.4 More Complex Hand Method (Method 2b)

Now modify Method 2a to consider plan torsion due to eccentricity between center of stiffness of building, and line of application of lateral load.

Concept of Center of Rigidity

To do this, we must introduce the concept of the center of rigidity, or shear center. The center of rigidity of a building is that plan location through which lateral load must be applied so as not to produce any twisting of the building in plan.

The center of rigidity of a symmetrical building is located at the geometric centroid of the building's plan area. Examples are shown in Fig. 10.7.

If the building is unsymmetrical, however, the center of rigidity is not located at the geometric centroid of the building's plan area. The classic example is the channel section, for which the shear center is actually located outside the plan area. Examples are shown in Fig. 10.8.

To include the effects of plan torsion, it is necessary to locate the center of rigidity of the building. Lateral loading applied through some arbitrary point is then treated as loading through the center of rigidity (which

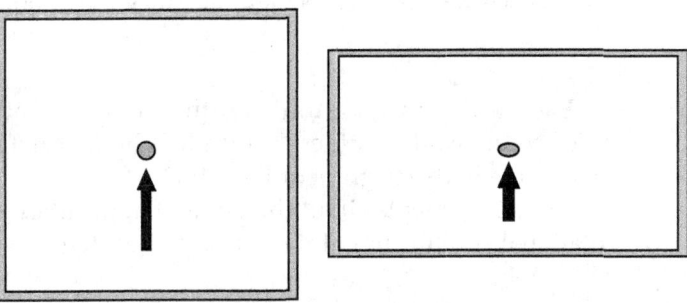

FIGURE 10.7 Examples of location of center of rigidity for symmetrical buildings.

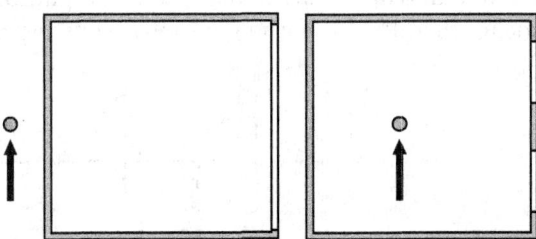

FIGURE 10.8 Examples of location of center of rigidity for unsymmetrical buildings.

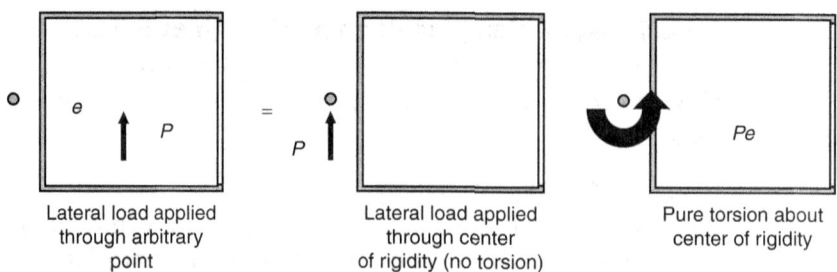

FIGURE 10.9 Decomposition of lateral load into a lateral load applied through the center of rigidity plus pure torsion about the center of rigidity.

causes no torsional response), plus a plan torsion about the center of rigidity (which produces pure torsional response). This is shown in Fig. 10.9.

Location of Center of Rigidity

The location of the center of rigidity along any building axis depends on the relative stiffnesses of the walls of the building that are oriented perpendicular to that axis. Refer to Fig. 10.10.

Define k_{y1} as the stiffness in the y direction of wall 1. If only shearing deformations are considered, the stiffness of each wall is proportional to its plan area. Then if the load P is applied through the center of rigidity (no twist), each wall will deflect laterally an equal amount in the y direction:

$$\Delta_{y1} = \Delta_{y2} = \Delta_y$$

Each wall applies a force on the underside of the roof diaphragm, equal to the wall's stiffness multiplied by that deflection. This is shown by the free-body diagram of Fig. 10.11.

Taking moments about the point of application of the external load P, rotational equilibrium of the roof requires that

$$k_{y1}\Delta_y \cdot x_1 = k_{y2}\Delta_y \cdot x_2$$

Canceling the common term Δ_y, and again noting that if only shearing deformations are considered, the forces applied by each wall are proportional to that wall's plan area, the above equation locates the line of action of the applied load P at the geometric centroid of the wall areas:

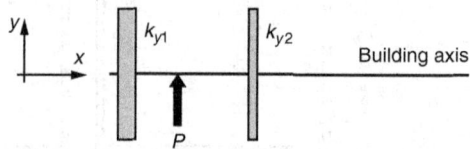

FIGURE 10.10 Location of the center of rigidity in one direction.

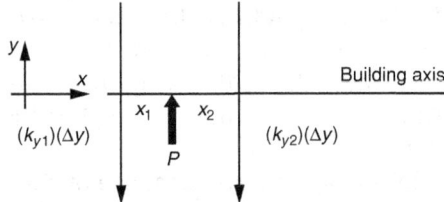

FIGURE 10.11 Free-body diagram of diaphragm showing applied loads and reactions from shear walls.

Along the x axis, the geometric centroid of the wall areas is given by

$$\bar{x} = \frac{\sum k_{yi} x_i}{\sum k_{yi}} = \frac{\sum \text{Area}_{yi} x_i}{\sum \text{Area}_{yi}}$$

so the location of the center of rigidity along the x axis is given by

$$\bar{x}_r = \frac{\sum k_{yi} x_i}{\sum k_{yi}} = \frac{\sum \text{Area}_{yi} x_i}{\sum \text{Area}_{yi}}$$

and similarly, if k_{yx} is the stiffness in the x direction of wall i, the location of the center of rigidity along the y axis is given by

$$\bar{y}_r = \frac{\sum k_{xi} y_i}{\sum k_{xi}} = \frac{\sum \text{Area}_{xi} y_i}{\sum \text{Area}_{xi}}$$

Response to Lateral Load Applied through an Arbitrary Point

As introduced in Sec. 10.2.4 above, once the center of rigidity is located in plan, any lateral load can be treated as the superposition of a load through the center of rigidity (producing no twist), plus a moment equal to the load times its eccentricity, acting about the center of rigidity (producing only twist) (Fig. 10.12).

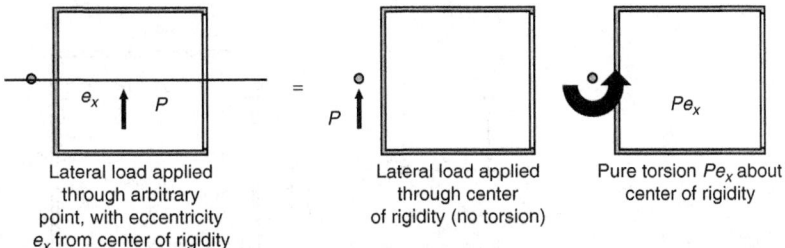

FIGURE 10.12 Decomposition of lateral load into lateral load through center of rigidity plus torsion about the center of rigidity.

In Fig. 10.12, a load P, acting in the y direction with an eccentricity e_x along the x axis, is decomposed into that same load P, acting in the y direction through the center of rigidity, plus a counter-clockwise moment Pe_x about the center of rigidity. Load in the x direction is handled in an analogous way.

Now let's examine the response of the structure under each of those load cases.

Response of the Structure due to Lateral Load Applied through Center of Rigidity

Due only to the lateral load applied through the center of rigidity, the structure does not twist, as shown in Fig. 10.13.

Shear forces are produced only in those walls oriented parallel to the direction of applied load. The shear forces are proportional to the shear stiffnesses (plan areas) of the walls

$$F_{xi} = 0$$

$$F_{yi} = \left(\frac{k_{yi}}{\sum k_{yi}}\right) P = \left(\frac{\text{Area}_{yi}}{\sum \text{Area}_{yi}}\right) P$$

The displacement of the structure in the y direction is

$$\Delta_y = \frac{P_y}{\sum k_{yi}}$$

Analogous expressions apply for load in the x direction.

Response of the Structure due to Torsional Moment Applied at Center of Rigidity

As shown in Fig. 10.14, due only to the torsional moment applied at the center of rigidity, the structure undergoes twist. The center of rigidity does not translate.

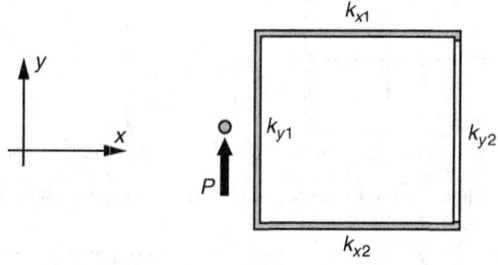

Figure 10.13 Lateral load applied through the center of rigidity.

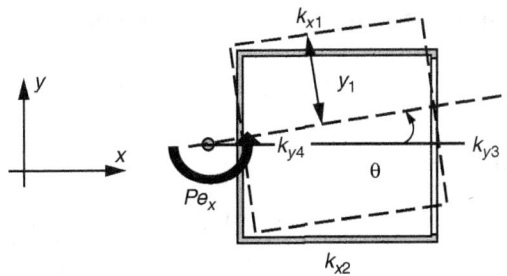

FIGURE 10.14 Pure rotation in plan of a structure with lateral load applied through the center of rigidity.

The force in each wall depends on that wall's stiffness and perpendicular distance from the center of rigidity.

Let x_i and y_i be the perpendicular distances from wall i to the center of rigidity. Then if the roof rotates in plan through some angle θ, the force exerted by each wall on the roof is proportional to that wall's shear stiffness times the displacement of the top of the wall in its own plane. The displacement of each wall in its own plane is the product of the angle θ and the perpendicular distance of the wall from the center of rigidity. To simplify the calculation, forces from each wall can be computed separately in the x and y directions.

For example, in Fig. 10.14, wall 1 is located at a perpendicular distance y_1 from the center of rigidity. As a result of a counter-clockwise rotation θ of the roof about the center of rigidity, the top of wall 1 moves to the left (parallel to the x axis) a distance $\theta \cdot y_1$. As a result of that movement, wall 1 applies a force on the underside of the roof, equal to

$$\text{Force}_{x1} = k_{k1} \cdot \theta \cdot y_i$$

Rotational equilibrium of the roof requires that the applied moment, Pe_x, be equilibrated by the summation of moments from each wall:

$$P \cdot e_x = \sum (\text{Force}_{xi} \cdot y_i + \text{Force}_{yi} \cdot x_i)$$

$$P \cdot e_x = \sum \left(k_{xi} y_i^2 + k_{yi} x_i^2 \right) \theta$$

This lets us solve for the plan rotation, θ:

$$\theta = \frac{Pe_x}{J}$$

The shear force applied to wall 1 is then

$$\text{Force}_{x1} = k_{k1} \cdot \theta \cdot y_i = \frac{Pe_x}{J} y_1 k_x$$

In general,

$$\text{Force}_{xi} = \frac{P_y e_x}{J} y_i k_{xi}$$

$$\text{Force}_{yi} = \frac{P_y e_x}{J} x_i k_{yi}$$

Response of the Structure due to Direct Shear Plus Plan Torsion

Finally, consider a structure loaded by a combination of direct shear plus the torsional moment applied at the center of rigidity (Fig. 10.15). The structure undergoes displacement and twist

$$\bar{x} = \frac{\sum k_{yi} x_i}{\sum k_{yi}}$$

$$\bar{y} = \frac{\sum k_{xi} y_i}{\sum k_{xi}}$$

$$\Delta_x = \frac{P_x}{\sum k_{xi}}$$

$$\Delta_y = \frac{P_y}{\sum k_{yi}}$$

$$\theta = \frac{P_y e_x + P_x e_y}{J}$$

$$\text{Force}_{xi} = \left(\frac{k_{xi}}{\sum k_{xi}}\right) P_x + \left(\frac{k_{xi} y_i}{J}\right)(P_x e_y + P_y e_x)$$

$$\text{Force}_{yi} = \left(\frac{k_{yi}}{\sum k_{yi}}\right) P_y + \left(\frac{k_{yi} x_i}{J}\right)(P_x e_y + P_y e_x)$$

Example of More Complex Hand Method

Now apply these principles to the structure considered in this section (Fig. 10.16).

Lateral Load Analysis of Shear-Wall Structures 355

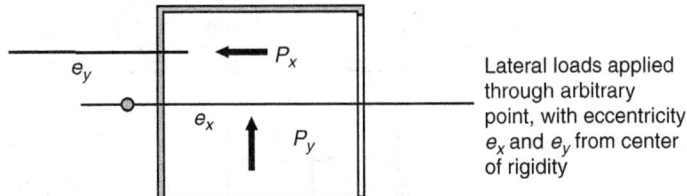

FIGURE 10.15 Structure loaded by a combination of load through the center of rigidity plus plan torsion about the center of rigidity.

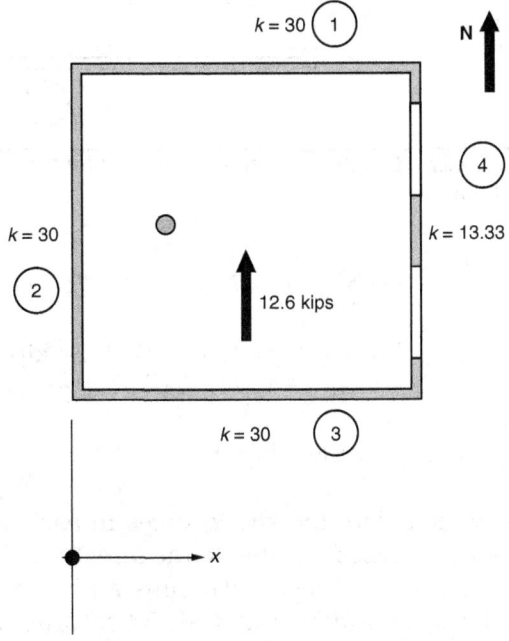

FIGURE 10.16 Application of rigid-diaphragm analysis to the structure considered in this section (Method 2b).

The center of rigidity is calculated using the equations below. As shown in Fig. 10.17, the center of rigidity is located at 5.77 ft from the line of action of the applied load.

$$\bar{x} = \frac{\sum k_{yi} x_i}{\sum k_{yi}} = \frac{30(0) + 13.33(30)}{30 + 13.33} = 9.23 \text{ ft}$$

$$e_x = 15 \text{ ft} - 9.23 \text{ ft} = 5.77 \text{ ft}$$

$$e_y = 0$$

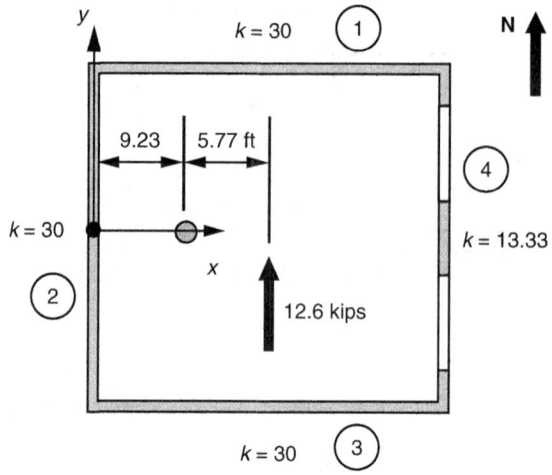

FIGURE 10.17 Location of center of rigidity for the example of this section (Method 2b).

$$J = \sum (k_{xi}y_i^2 + k_{yi}x_i^2)$$

$$J = 2(30 \text{ kip/ft}) \cdot (15 \text{ ft})^2 + (30 \text{ kip/ft}) \cdot (9.23 \text{ ft})^2$$
$$+ 13.33 \text{ kip/ft} \cdot (30 \text{ ft} - 9.23 \text{ ft})^2$$

$$J = 21,806 \text{ kip-ft}$$

Now calculate the shear forces in each wall. Because wall 2 and wall 4 are oriented parallel to the direction of the applied lateral force, they experience direct shear plus additional shear due to plan torsion. Because wall 1 and wall 3 are oriented perpendicular to the direction of the applied lateral force, they experience only shear due to plan torsion.

Wall 1:

$$\text{Force}_{xi} = \left(\frac{k_{xi}}{\sum k_{xi}}\right)P_x + \left(\frac{k_{xi}y_i}{J}\right)(P_x e_y + P_y e_x)$$

$$\text{Force}_{xi} = \left(\frac{k_{xi}y_i}{J}\right)(P_y e_x)$$

$$\text{Force}_{x1} = \left(\frac{k_{x1}y_1}{J}\right)(P_y e_x)$$

$$\text{Force}_1 = \left(\frac{30 \cdot 15 \text{ ft}}{21,806}\right)(P_y \cdot 5.77 \text{ ft}) = 0.12 P_y = 1.50 \text{ kips}$$

Wall 2:

$$\text{Force}_{yi} = \left(\frac{k_{yi}}{\sum k_{yi}}\right)P_y + \left(\frac{k_{yi}x_i}{J}\right)(P_xe_y + P_ye_x)$$

$$\text{Force}_{y2} = \left(\frac{k_{y2}}{\sum k_{yi}}\right)P_y + \left(\frac{k_{y2}x_2}{J}\right)(P_ye_x)$$

$$\text{Force}_{y2} = \left(\frac{30}{30+13.33}\right)P_y + \left(\frac{30 \cdot 9.23x_2}{21,806}\right)(P_y \cdot 5.77 \text{ ft})$$

$$\text{Force}_{y2} = 0.692P - 0.073P$$

$$\text{Force}_{y2} = 8.72 \text{ kips} - 0.92 \text{ kips} = 7.80 \text{ kips}$$

Wall 3:

$$\text{Force}_{x3} = \left(\frac{k_{x3}y_3}{J}\right)(P_ye_x)$$

$$\text{Force}_1 = \left(\frac{30 \cdot 15 \text{ ft}}{21,806}\right)(P_y \cdot 5.77 \text{ ft}) = 0.12P_y = 1.50 \text{ kips}$$

Wall 4:

$$\text{Force}_{yi} = \left(\frac{k_{yi}}{\sum k_{yi}}\right)P_y + \left(\frac{k_{yi}x_i}{J}\right)(P_xe_y + P_ye_x)$$

$$\text{Force}_{y4} = \left(\frac{k_{y4}}{\sum k_{yi}}\right)P_y + \left(\frac{k_{y4}x_4}{J}\right)(P_ye_x)$$

$$\text{Force}_{y4} = \left(\frac{13.33}{30+13.33}\right)P_y + \left(\frac{13.33 \cdot (30-9.23)}{21,806}\right)(P_y \cdot 5.77 \text{ ft})$$

$$\text{Force}_{y4} = 0.308P + 0.073P$$

$$\text{Force}_{y4} = 3.88 \text{ kips} + 0.92 \text{ kips} = 4.80 \text{ kips}$$

The final shear forces acting on the walls due to load applied through the center of rigidity and due to plan torsion about the center of rigidity, are shown separately in Fig. 10.18, and combined in Fig. 10.19.

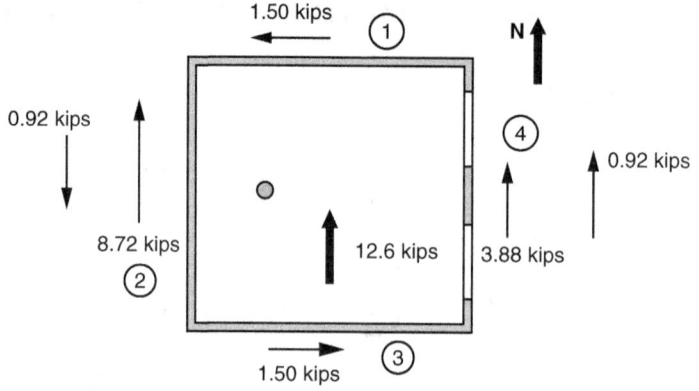

FIGURE 10.18 Shear forces acting on walls due to direct shear and due to torsion (Method 2b).

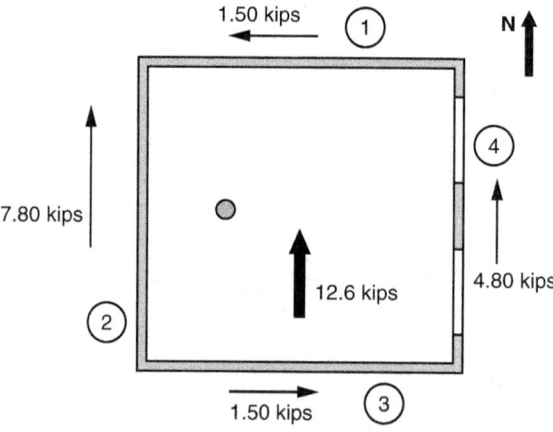

FIGURE 10.19 Combined shear forces acting on walls of example structure (Method 2b).

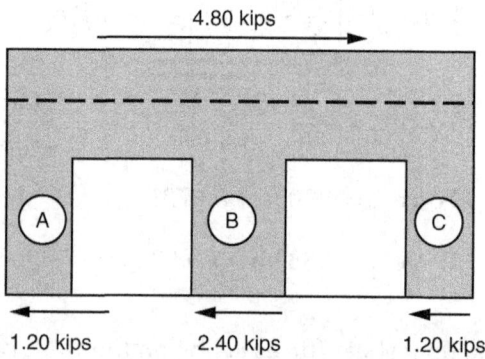

FIGURE 10.20 Distribution of shears to segments of the east wall of example structure (Method 2b).

Finally, distribute the shear of 4.80 kips to the segments of the east wall in proportion to their plan lengths, as shown in Fig. 10.20.

These results are quite close to those obtained by finite-element analysis. The results are also summarized in Tables 10.1 and 10.2, which are discussed in Sec. 10.2.6.

Analysis time: 60 min

10.2.5 Most Complex Hand Model (Method 2c)

The most complex hand model (Method 2c) is similar to the more complex hand method (Method 2b), with the additional complication that the stiffness of each wall segment includes flexural stiffness as well as shearing stiffness. In previously published versions of that method, this is often accomplished by modifying the shearing stiffness of each wall segment by a factor that depends on the aspect ratio of that segment.

Because Method 2c is inherently more complex than Method 2b, the time involved in it must be even greater. This increased time does not always result in increased accuracy, however, because any variant of Model 2 has problems in handling perforated walls in which the wall segments have unequal heights (such as that shown in Fig. 10.21). Such segments cannot simply be idealized as subjected to equal displacements at the diaphragm levels. Some previously published methods for applying Methods 2 to such systems even lead to the counter-intuitive and clearly suspect conclusion that the stiffness of walls is increased by perforating them. As a consequence, Method 2c is not pursued further here.

10.2.6 Comparison of Results from Each Method

In Table 10.1, results are compared for shear forces in the walls and wall segments of this example structure, using Method 1 (the finite element reference method), and variations on Method 2 (hand methods).

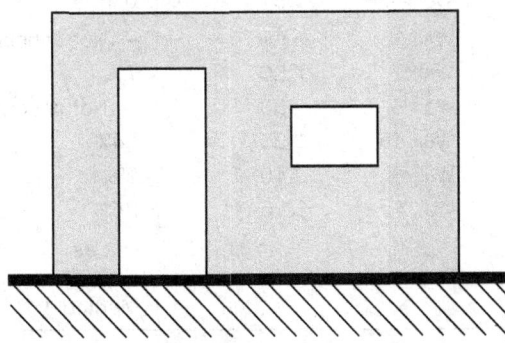

FIGURE 10.21 Example of perforated wall with segments of unequal height

Element	Shear force in each element, kips			
	Method 1 (Finite element method)	Method 2a (Shearing deformations only, neglect plan torsion)	Method 2b (Shearing deformations only, include plan torsion)	Method 2c (Shearing and flexural deformations)
Wall 1	2.02	0	1.50	?
Wall 2	8.32	8.72	7.80	?
Wall 3	2.02	0	1.50	?
Wall 4a	0.97	0.97	1.20	?
Wall 4b	2.35	1.94	2.40	?
Wall 4c	0.97	0.97	1.20	?

TABLE 10.1 Comparison of Results Obtained for Calculating Shear-Wall Forces by Each Method

Analysis method	Element	Shear, kips	% Error	Required time
Method 1 (finite element method)	Wall 1	2.02	Used as reference	30 min, computer required
	Wall 2	8.32		
	Wall 3	2.02		
	Wall 4a	0.97		
	Wall 4b	2.35		
	Wall 4c	0.97		
Method 2a (shearing stiffness only, neglect plan torsion)	Wall 1	0	—(Not critical)	10 min, no computer required
	Wall 2	8.72	5%	
	Wall 3	0	—(Not critical)	
	Wall 4a	0.97	0%	
	Wall 4b	1.94	17%	
	Wall 4c	0.97	0%	
Method 2b (shearing stiffness only, include plan torsion)	Wall 1	1.50	—(Not critical)	60 min, no computer required
	Wall 2	7.80	7%	
	Wall 3	1.50	—(Not critical)	
	Wall 4a	1.20	24%	
	Wall 4b	2.40	2%	
	Wall 4c	1.20	24%	
Method 2c (shearing and flexural stiffness)			Not as accurate as Method 1	Probably 120 min, no computer required

TABLE 10.2 Comparison of Results Obtained for Calculating Shear-Wall Forces by Each Method

In Table 10.2 the same results are presented, but now organized by analysis method, and including the percent error (compared to the reference solution of Method 1) and the required time.

10.2.7 Comments on Analysis of Shear-Wall Structures with Rigid Diaphragms

1. Using Method 1 (finite element analysis) as a reference, the simplest hand method (Method 2a), which considers wall shearing deformations only, and neglects plan torsion, gives acceptable results for this example, and is very efficient in terms of time.

2. Extending Method 2a to include plan torsion (Method 2b), requires much more time, and produces results that are not much more accurate.

3. Although Method 2b might appear more accurate because it correctly predicts shears in perpendicular walls (walls 1 and 3 in this example) due to plan torsion, those calculated shears are not critical for design. Critical shears in those walls occur under lateral loads acting parallel to the wall orientation. In this case, for example, shears in the perpendicular walls are 1.50 kips for a northward lateral load of 12.6 kips. If the same lateral load were applied in the EW direction, walls 1 and 3 would each have a shear of half that applied load, or 6.3 kips.

4. Further extending Method 2b to include flexural as well as shearing deformations (Method 2c) is not productive. It requires much more time, and produces results that are not necessarily any better. Hand methods following Method 2c present conceptual difficulties for openings of different heights. For example, they can sometimes predict that a wall with irregular openings will have greater in-plane stiffness than an otherwise identical wall with no openings.

10.2.8 Conclusions regarding the Lateral Load Analysis of Low-Rise Buildings with Rigid Diaphragms

1. For shear-wall buildings with rigid diaphragms, the distribution of shears to individual walls can generally be determined efficiently and with sufficient accuracy by considering only shearing deformations of the walls (i.e., by assuming that wall stiffness is proportional to plan area, or simply to plan length if walls are of equal thickness) and by neglecting plan torsion. This approach presumes that the building has a reasonable number of shear walls in each principal plan direction.

2. If more accuracy is desired, the distribution of shears to individual walls can be determined efficiently by linear elastic finite element analysis.

3. More sophisticated hand methods (e.g., including plan torsion, or considering flexural as well as shearing deformations) are generally not cost-effective.

10.3 Lateral Load Analysis and Design of Shear-Wall Structures with Flexible Floor Diaphragms

10.3.1 Introduction

In Sec. 10.1, it was noted that the classical approach to analyzing structures including the effect of horizontal diaphragm flexibility, is first to classify the horizontal diaphragm as rigid or as flexible compared to the vertically oriented lateral force-resisting system, and then to evaluate the actions on the shear walls comprising the lateral force-resisting system. The beginning of this section continues with that approach. It finishes, however, with what might be termed, "the simplest of all possible worlds." For preliminary design, it is even possible to first analyze the building assuming that the diaphragm is rigid, then analyze it assuming that the diaphragm is flexible, and take the more critical of the two cases to arrive at design actions for each shear wall.

10.3.2 Exact Approach to Flexible Diaphragms

General-purpose finite element programs and a few building analysis programs, permit floor diaphragms to be analyzed as elements with in-plane flexibility (e.g., as membrane elements). Effects of in-plane floor flexibility can be included in the analysis, and floor member actions can be evaluated. While this approach appears correct, its accuracy decreases greatly if the floor diaphragm is cracked.

10.3.3 Approximate Approach to Flexible Diaphragms

In cases involving flexible diaphragms, floor actions and structural response can usually be approximated by assuming that the diaphragm is completely flexible compared to the vertical elements that support it. In other words, the diaphragm can be designed as a series of beam elements acting in the horizontal plane, and supported by vertical elements. This approach is adopted in this section. Even more simply, diaphragm reactions (shears on shear walls), can be estimated based on horizontal tributary areas.

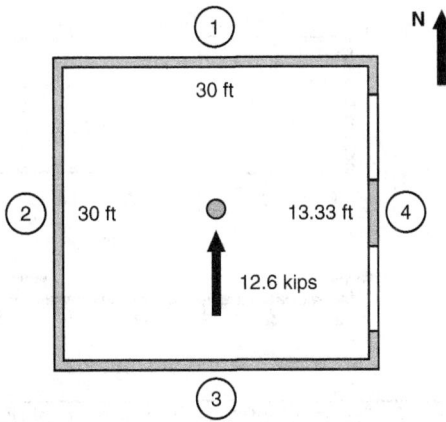

FIGURE 10.22 Plan view of example building with flexible roof diaphragm.

10.3.4 Example of Analysis of Shear-Wall Building with Flexible Diaphragms

Consider, for example, the building considered previously in this chapter, with a uniformly distributed wind pressure (wind from the south) of 35 lb/ft². The building is shown in Fig. 10.22.

Because the left and right supports (walls 2 and 4) are very stiff compared to the diaphragm, the diaphragm behaves like a simply supported beam, and the reactions on its supporting walls are simply $1/2\ V$ for each wall. This solution is shown in Fig. 10.23.

As before, the 6.3 kips applied to the east wall is distributed to the three segments of that wall in proportion to their plan lengths.

Analysis time: 5 min

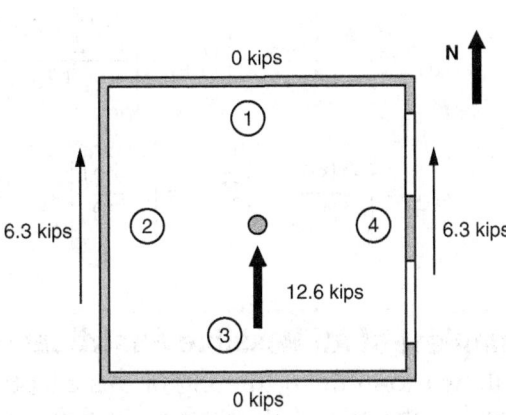

FIGURE 10.23 Results of example problem, assuming flexible diaphragm.

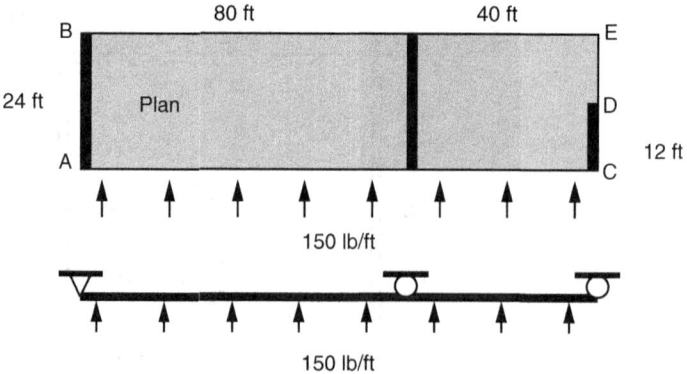

FIGURE 10.24 Example of a flexible horizontal diaphragm with more than two points of lateral support.

10.3.5 Example of Distribution of Shears with Flexible Diaphragm

A flexible horizontal diaphragm with more than two points of lateral support can be analyzed as a continuous beam, as shown in Fig. 10.24. Such a continuous beam could have different cross-sectional properties (in the horizontal plane) in different spans.

It is even simpler to analyze this continuous beam by tributary areas (i.e., according to the tributary length supported by each wall). The left-hand wall has a tributary length of 40 ft; the middle wall, 60 ft; and the right-hand wall, 20 ft.

$$V_{left} = \left(\frac{Area_{left}}{A_{total}}\right) V_{total} = \left[\frac{(80/2)}{120}\right] V_{total} = 0.333\, V_{total}$$

$$V_{middle} = \left(\frac{Area_{middle}}{A_{total}}\right) V_{total} = \left[\frac{(80/2 + 40/2)}{120}\right] V_{total} = 0.50\, V_{total}$$

$$V_{right} = \left(\frac{Area_{right}}{A_{total}}\right) V_{total} = \left[\frac{(40/2)}{120}\right] V_{total} = 0.167\, V_{total}$$

10.4 The Simplest of All Possible Analytical Worlds

As alluded to in the beginning of this chapter, it is rarely necessary to explicitly categorize a diaphragm as rigid or flexible. It is sufficient to estimate the design actions in each wall based first on the supposition of a rigid diaphragm, and then on the supposition of a flexible diaphragm. Finally, each wall is designed based on the more critical of those two assumptions.

For example, with the perforated wall discussed previously, the assumption of a rigid diaphragm leads to design shears of $\left(\dfrac{30}{43.33}\right)V$ and $\left(\dfrac{13.3}{43.3}\right)V$ for the left- and right-hand walls, respectively, while the assumption of a flexible diaphragm leads to design shears of $\dfrac{1}{2}V$ for each wall.

Taking the worse of the two results for each wall, we would have design shears of $\left(\dfrac{30}{43.33}\right)V$, or $0.69\ V$ for the left-hand wall, and $\dfrac{1}{2}V$, or $0.50\ V$, for the right-hand wall. In many practical cases, we would easily have sufficient capacity in each wall to resist these design shears, and we could easily finish the design without having to evaluate the flexibility of the horizontal diaphragm. For seismic design, when diaphragms are not flexible, then both inherent and accidental torsion must also be considered in accordance with Sec. 12.8.4 of ASCE 7-05.

CHAPTER 11
Design and Detailing of Floor and Roof Diaphragms

11.1 Introduction to Design of Diaphragms

The preceding chapter addresses the determination of forces in shear walls of structures subjected to lateral load. Those shear-wall forces can be used to calculate shear forces and moments in those diaphragms, and finally to design the diaphragms for those forces. While this procedure is in principle the same for every diaphragm (regardless of whether it is rigid, flexible, or semi-rigid), it is possible to apply it differently for rigid and for flexible diaphragms, because rigid diaphragms are often stronger than flexible ones.

In addition to the general principles set forth here, other code requirements may also apply. For example, for structures assigned to Seismic Design Categories C and higher, Sec. 12.11.2.2.1 of ASCE 7-05 imposes requirements for diaphragm connections, and implicitly introduces the concept of diaphragms as composed of sub-diaphragms, linked together by collectors.

11.1.1 Introduction to Design of Rigid Diaphragms

Rigid diaphragms usually have enough in-plane strength so that they do not have to be explicitly designed. They must, however, be connected to the walls that transfer their shear. The connections must be designed for that shear. Diaphragm chords and collectors also need to be checked, using the same procedures as for flexible diaphragms.

11.1.2 Introduction to Design of Flexible Diaphragms

Flexible diaphragms should be designed for shear and bending actions, and also to transfer those actions to walls. Of particular importance are

- Connections between precast planks (shear transfer)
- Average shear stress in thin topping
- Shear in nailed sheathing

Diaphragm chords and collectors also need to be checked, using the procedures demonstrated in Sec. 11.1.3.

11.1.3 Example of Design for Shears and Moments in Flexible Diaphragm

Calculate the distribution of shears to the two walls of the structure shown in Fig. 11.1, and design the roof diaphragm.

In spite of the difference in stiffness between wall AB and wall CD, and in contrast to the previous analysis of rigid diaphragm, we assume here that each wall resists equal shears of $(150 \text{ lb/ft} \times 120 \text{ ft}/2) = 9000$ lb. This result could be obtained either by visualizing this as a simply supported beam, or by distributing the wall load according to the tributary length of the diaphragm.

To make sure that shear acting within the roof diaphragm at a distance from wall CD be transferred to that wall, it is desirable to put a "drag strut" between points D and E.

If the roof diaphragm is considered simply supported in the horizontal plane (in view of the insignificant torsional stiffness of the walls), it is

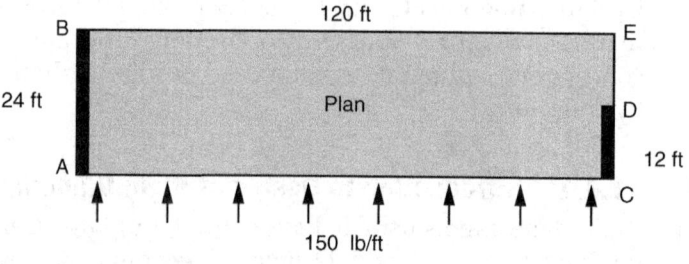

FIGURE 11.1 Example of a flexible diaphragm.

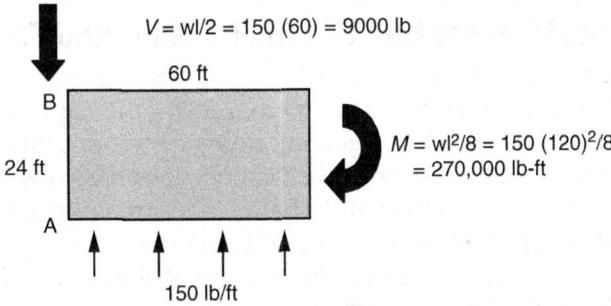

FIGURE 11.2 Example of design of a flexible diaphragm for shear and moment.

possible to calculate bending moments and shears in the roof diaphragm, as shown in Fig. 11.2.

Internal shears must be resisted by the roof diaphragm. The available net area is the area over which shear forces can be transmitted: the net cross-sectional area of the diaphragm for an integral diaphragm; or the cross-sectional area of the topping for a nonintegral diaphragm. Bending moments in the diaphragm are resisted by tensile and compressive forces in the chords of the diaphragm, calculated as illustrated in Fig. 11.3.

The compression chord is made up of a portion of the roof diaphragm itself, and need not be designed. The tension chord consists of deformed reinforcement, placed either in the slab topping, or directly in the bond beam at the level of the roof diaphragm. The required area can be calculated easily. Using strength design, for example,

$$A_s^{\text{required}} = \frac{T}{F_y} = \frac{11,250 \text{ lb}}{60,000 \text{ in.}^2} = 0.19 \text{ in.}^2$$

This is easily satisfied with one or two #4 bars.

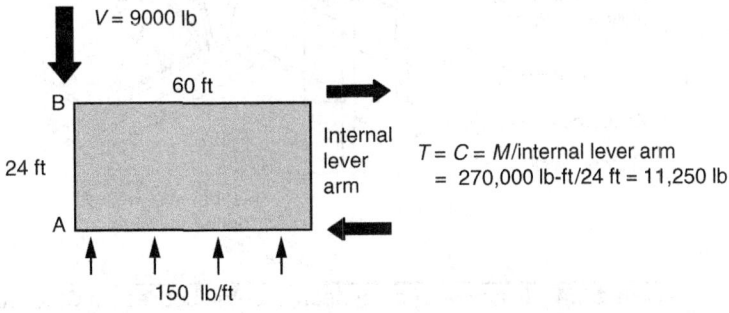

FIGURE 11.3 Example of computation of diaphragm chord forces.

11.2 Typical Connection Details for Roof and Floor Diaphragms

The connections between roof and floor diaphragms and their supporting walls must be designed to transfer the required design forces. In the remainder of this section, necessary connections to walls must be designed. Examples of such connections are shown in Figs. 11.4 and 11.5. These connections would have to be strengthened for regions subject to strong earthquakes or strong winds.

An example of a connection detail between a CMU wall and a roof or floor diaphragm composed of steel joists is shown in Fig. 11.4. Vertical reinforcement from the CMU wall is continous through or anchored in a horizontal bond beam, and the base plate for the joists is anchored to the bond beam.

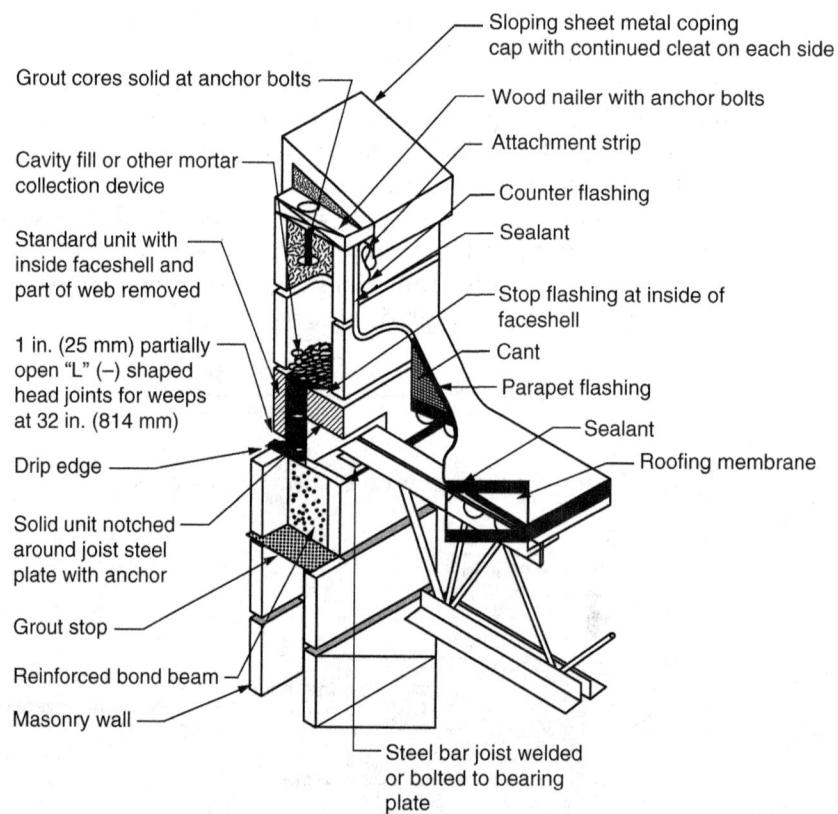

FIGURE 11.4 Example of a connection detail between a CMU wall and steel joists. (*Source*: Figure 11 of National Concrete Masonry Association TEK 05-07A.)

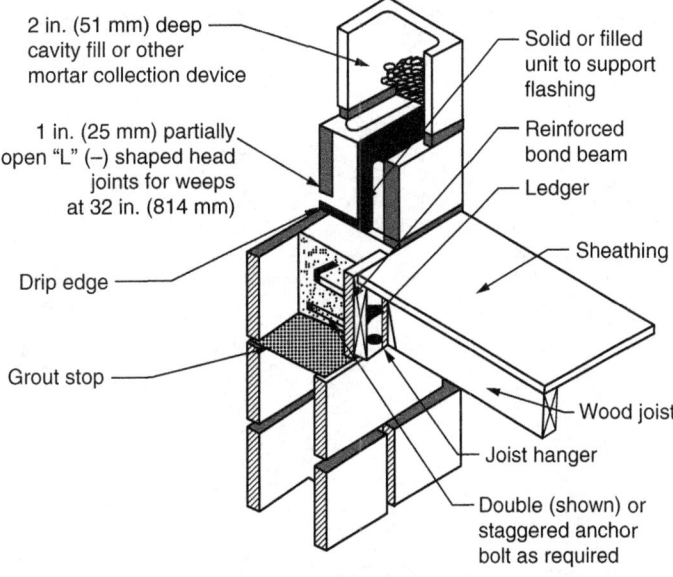

FIGURE 11.5 Example of a connection detail between a CMU wall and wooden joists. (*Source*: Figure 6 of National Concrete Masonry Association TEK 05-07A.)

An example of a connection detail between a CMU wall and a roof or floor diaphragm composed of wooden joists is shown in Fig. 11.5. Vertical reinforcement from the CMU wall is continuous through or anchored in a horizontal bond beam, and the base plate for the joists is anchored to the bond beam.

CHAPTER 12

Strength Design Example: One-Story Building with Reinforced Concrete Masonry

12.1 Introduction

Previous sections of this book have addressed the specification and detailing of masonry buildings requiring little or no structural calculation, and the structural calculation, by strength and allowable stress approaches, of individual masonry elements. Design of those individual masonry elements has included implicit consideration of how they work together to form a structural assemblage. For example, design calculations for masonry bearing walls were related to the required performance of horizontal diaphragms, and to the analysis and design of those diaphragms to obtain that performance.

It is now time to put this information together—to carry out the preliminary layout, specification, and structural design of a complete masonry building. The first such design example is a one-story commercial building (a warehouse), located in the outskirts of Austin, Texas. Design loads due to earthquake are neglected for this problem.

This combined example problem is carried out using the strength design approach. Previous comparisons are intended to facilitate estimation of the extent to which the design would change if the allowable-stress approach were used. This will also be commented on at the end of the example.

12.2 Design Steps for One-Story Building

The design steps for this one-story building are enumerated below:

1. Choose design criteria, calculate design loads, propose structural system:
   - Propose plan, elevation, materials, f_m'
   - Calculate D, L, W loads
   - Propose structural systems for gravity and lateral load
2. Design walls for gravity plus out-of-plane loads, using thickened wall sections at points of reaction of long-span joists (if necessary)
3. Design lintels
4. Conduct lateral force analysis, design roof diaphragm
5. Design wall segments for combined shear, flexure and axial loads
6. Design and detail connections
7. Design roof framing (this is not done here—a joist catalog would normally be used)
8. Design interior columns (this is not done here—structural steel tubes would be used)

12.3 Step 1: Choose Design Criteria

The plan and elevation of the building are shown in Figs. 12.1 and 12.2.

12.3.1 Design for Water-Penetration Resistance

A single-wythe barrier wall will be used. The wall will permit the passage of some water.

12.3.2 Locate Control Joints

On the north and south facades, space control joints at 20 ft, as shown in Fig. 12.3.

On the west facade, space control joints at 20 ft, as shown in Fig. 12.4.
On the east facade, space control joints at 20 ft, as shown in Fig. 12.5.

12.3.3 Design for Fire (2009 IBC)

Design for fire follows the 2009 IBC.

Use and occupancy: Group F, Division 1 (moderate hazard)
Use Type I or Type II construction (noncombustible material)
No area or height restrictions

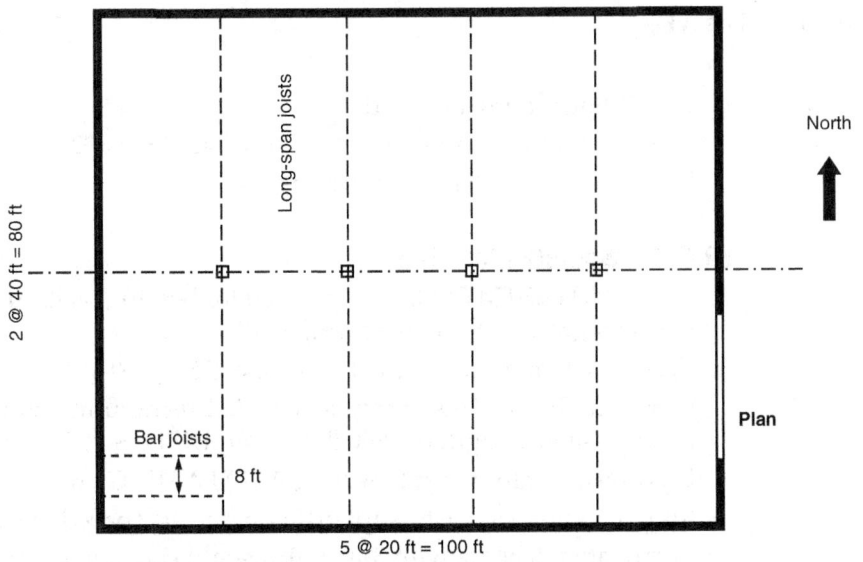

FIGURE 12.1 Plan of example one-story building.

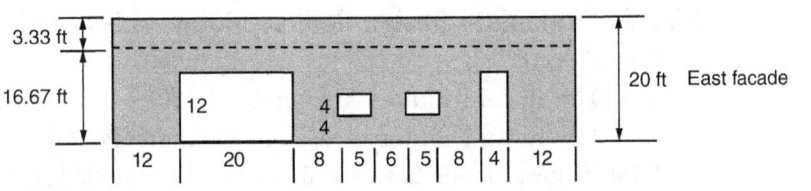

FIGURE 12.2 Elevation of example one-story building.

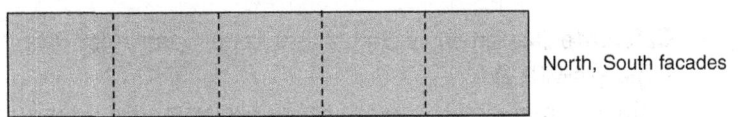

FIGURE 12.3 Locations of control joints on north and south facades.

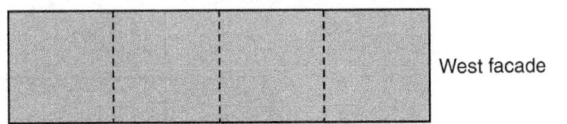

FIGURE 12.4 Locations of control joints on west facade.

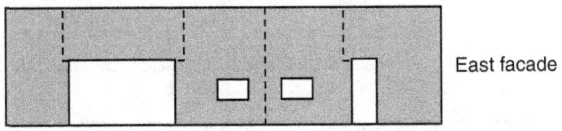

FIGURE 12.5 Spacing of control joints on east facade.

2- or 3-hour rating required
Meet separation requirements of 2009 IBC Table 602
Use grouted 8-in. CMU for bearing walls

12.3.4 Specify Materials

8-in. CMU (ASTM C90), fully grouted for bearing walls, ungrouted or partially grouted for non-bearing walls

Type S PCL mortar, specified by proportion (ASTM C270)

$f_m' = 1500$ lb/in.^{2} This is can be satisfied using units with a net-area compressive strength of 1900 lb/in.^{2}, and Type S PCL mortar

Deformed reinforcement meeting ASTM A615, Gr. 60

Roof of long-span joists, supporting bar joists spaced at 8 ft

Corrugated decking with 3-in. lightweight concrete topping

Roof supported by structural steel tube columns at midspan

12.3.5 Calculate Design Roof Load due to Gravity (2009 IBC)

$D = 60$ lb/ft^2

$L = 20$ lb/ft^2 for tributary area up to 200 ft^2

$L = 12$ lb/ft^2 for tributary areas greater than 600 ft^2

Linear interpolation between those two limits (2009 IBC, Sec. 1607.11.2)

12.3.6 Calculate Design Wind Load

Calculate Design Base Shear and Long-Span Joist Reactions due to Wind (MWFRS)

A three-dimensional view of the one-story building is shown in Fig. 12.6.

The critical direction for wind will be NS, because the area is greater on the north and south sides, and the area of shear walls is less in the NS direction.

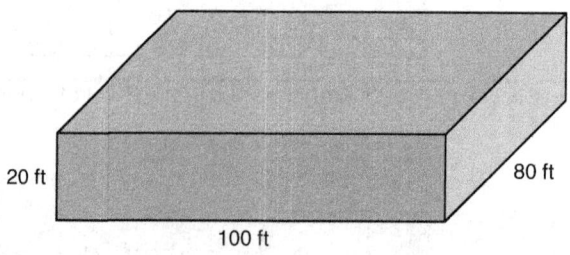

FIGURE 12.6 Three-dimensional view of one-story building.

Strength Design Example: One-Story Building

The following steps are the same as mentioned in Sec. 3.4 of the book. The sections and figures mentioned here are from ASCE 7-05:

1. Determine the *basic wind speed V* and *wind directionality factor* K_d in accordance with Sec. 6.5.4 of ASCE 7-05. The basic wind speed for Austin is 90 mi/h (ASCE 7-05, Fig. 6-1a). The wind directionality factor K_d is 0.85 (ASCE 7-05, Table 6-6, Buildings).

2. Determine the *importance factor I* in accordance with Sec. 6.5.5 of ASCE 7-05. Assume that the importance factor is 1.0.

3. Determine the exposure category or exposure categories and velocity pressure exposure coefficient K_z or K_h, as applicable, in accordance with Sec. 6.5.6 of ASCE 7-05. Assume Exposure C (open terrain with scattered obstructions). The velocity pressure exposure coefficients K_h and K_z are determined from Table 12.1, ASCE 7-05, Table 6-3, for Exposure C (Cases 1 & 2).

4. Determine a *topographic factor* K_{zt} in accordance with Sec. 6.5.7 of ASCE 7-05. Because the structure is not located on a hill, ridge, or escarpment, $K_{zt} = 1.0$.

5. Determine a *gust effect factor G* or G_f, as applicable, in accordance with Sec. 6.5.8 of ASCE 7-05. Assume a rigid structure; the gust effect factor, G, is 0.85.

6. Determine an *enclosure classification* in accordance with Sec. 6.5.9 of ASCE 7-05. Assume that the building is enclosed. This assumes that the openings on the east side are metal doors or armored glass that will not be broken by wind.

7. Determine an *internal pressure coefficient* GC_{pi} in accordance with Sec. 6.5.11.1 of ASCE 7-05. The internal pressure coefficient GC_{pi} is ±0.18.

8. Determine the *external pressure coefficients* C_p or GC_{pf}, or *force coefficients* C_f, as applicable, in accordance with Sec. 6.5.11.2 or 6.5.11.3 of ASCE 7-05, respectively.

Height above ground level, z	K_h, K_z
1–15	0.85
20	0.90

TABLE 12.1 Velocity Pressure Exposure Coefficients for One-Story Example Building

The external pressure coefficients for main wind force resisting systems GC_p are given in Fig. 6-6 of ASCE 7-05. From the plan views in Fig. 6-6 of ASCE, the windward pressure is $q_z GC_p$. The leeward pressure is $q_h GC_p$. The difference between the q_z and the q_h is that the former varies as a function of the height above ground level, while the latter is uniform over the building height, and is evaluated using the height of the building.

For wind blowing in the NS direction, $L/B = 0.8$. From Fig. 6-6 (cont'd) of ASCE, on the windward side of the building the external pressure coefficient C_p is 0.8. On the leeward side of the building, it is –0.5.

9. Determine the *velocity pressure* q_z or q_h, as applicable, in accordance with Sec. 6.5.10 of ASCE 7-05.

   The velocity pressure is

   $$q_z = 0.00256\, K_z K_{zt} K_d V^2.$$

   $$I = 1.0$$

   $$K_d = 0.85$$

   $$V = 90\ \text{ml/h}$$

   $$K_{zt} = 1.0$$

   $$q_z = 17.63 K_z\ \text{lb/ft}^2$$

10. Determine the *design wind load P* or *F* in accordance with Secs. 6.5.12 and 6.5.13 of ASCE 7-05, as applicable.

    For main force-resisting systems,

    $$p = qGC_p - q_i(GC_{pi})$$

    where $q = q_z$ for windward walls evaluated at height z above the ground
    $ = q_h$ for leeward walls, side walls, and roofs, evaluated at height h
    $q_i = q_h$ for windward walls, side walls, leeward walls, and roofs of enclosed buildings and for negative internal pressure evaluation in partially enclosed buildings
    $ = q_z$ for positive internal pressure evaluation in partially enclosed buildings where height z is defined as the level of the highest opening in the building that could affect the positive internal pressure. For buildings sited in wind-borne debris regions, glazing in the lower 60 ft

that is not impact-resistant or protected with an impact-resistant covering, the glazing shall be treated as an opening in accordance with Sec. 6.5.9.3 of ASCE 7-05. For positive internal pressure evaluation, q_i may conservatively be evaluated at height h ($q_i = q_h$)

G = gust effect factor from Sec. 6.5.8 of ASCE 7-05
C_p = external pressure coefficient from Fig. 6-6 or Fig. 6-8 of ASCE 7-05
C_{pi} = internal pressure coefficient from Fig. 6-5 of ASCE 7-05

Because the building is enclosed, the internal pressures on the windward and leeward sides are of equal magnitude and opposite direction, will produce zero net base shear, and therefore need not be considered.

On the windward side of the building,

$$p = q_z G C_p$$

$$p = (17.63\, K_z) G C_p$$

Because C_p is positive in sign, this pressure is positive in sign, indicating that the pressure acts inward against the windward wall. If the wind comes from the south, for example, the force on the windward wall acts toward the north. These values are shown in the "Windward side" columns of the spreadsheet in Table 12.2.

On the leeward side of the building,

$$p = q_h G C_p$$

$$p = (17.63\, K_h) G C_p$$

Because C_p is negative in sign, this pressure is negative in sign, indicating that the pressure acts outward against the leeward wall. If the wind comes from the south, for example, the force on

Building floor	Height above ground	Tributary area	Windward side						Leeward side					
			K_z	q_z	G	C_p	p	Force	K_h	q_h	G	C_p	p	Force
Roof	16.67	1166.5	0.867	15.28	0.85	0.8	10.39	12.12	0.867	15.28	0.85	−0.5	−6.49	−7.58
Ground	0	833.5	0.85	14.99	0.85	0.8	10.19	8.49	0.867	15.28	0.85	−0.5	−6.49	−5.41
Total force								20.61						−12.99

TABLE 12.2 Spreadsheet for Computation of Base Shear and Long-Span Joist Reactions for Example One-Story Building (MWFRS)

the leeward wall acts toward the north. These values are shown in the "Leeward side" columns of the spreadsheet of Table 12.2.

The design base shear due to wind load is the summation of 20.61 kips acting inward on the upwind wall and 12.99 kips acting outward on the downwind wall, for a total of 33.6 kips.

Calculate Design Pressure on Wall Elements due to Wind (Components and Cladding)

The critical region will be at the parapet, near a corner.

1. Determine the *basic wind speed V* and *wind directionality factor* K_d in accordance with Sec. 6.5.4. The basic wind speed for Austin is 90 mi/h (ASCE 7-05, Fig. 6-1a). The wind directionality factor K_d is 0.85 (ASCE 7-05, Table 6-6, buildings).

2. Determine the *importance factor I* in accordance with Sec. 6.5.5 of ASCE 7-05. Assume that the importance factor is 1.0.

3. Determine the exposure category or exposure categories and velocity pressure exposure coefficient K_z or K_h, as applicable, in accordance with Sec. 6.5.6 of ASCE 7-05. Assume Exposure B (urban and suburban areas). The velocity pressure exposure coefficients K_h and K_z are determined from Table 12.3 (Table 6-3 of ASCE 7-05), for Exposure B and Case 1 (components and cladding).

4. Determine a *topographic factor* K_{zt} in accordance with Section 6.5.7 of ASCE 7-05. Because the structure is not located on a hill, ridge, or escarpment, $K_{zt} = 1.0$.

5. Determine a *gust effect factor G* or G_f, as applicable, in accordance with Sec. 6.5.8 of ASCE 7-05. Assume a rigid structure; the gust effect factor, G, is 0.85.

6. Determine an *enclosure classification* in accordance with Sec. 6.5.9 of ASCE 7-05. Assume that the building is enclosed.

7. Determine an *internal pressure coefficient* GC_{pi} in accordance with Sec. 6.5.11.1 of ASCE 7-05. The internal pressure coefficient GC_{pi} is ±0.18.

Height above ground level, z	K_h, K_z
1–15	0.85
20	0.90

TABLE 12.3 Velocity Pressure Exposure Coefficients for One-Story Example Building

8. Determine the *external pressure coefficients* C_p or GC_{pf}, or *force coefficients* C_f, as applicable, in accordance with Sec. 6.5.11.2 or 6.5.11.3 of ASCE 7-05, respectively. The external pressure coefficients for components and cladding GC_p are given in Fig. 6-8 of ASCE 7-05.

   In computing the effective area of the cladding element, it is permitted to use an effective area equal to the product of the span and an effective width not less than one-third the span (ASCE 7-05, Sec. 6.2, "Effective Wind Area").

   Assume a panel with a span equal to the diaphragm height of 16.67 ft. Assume an effective width of one-third of that span, or 5.56 ft. The resulting effective area is 92.7 ft². From Fig. 6-17 of ASCE 7-05, a panel in Zone 5 has a positive pressure coefficient of 0.75, and a negative pressure coefficient of –1.4.

9. Determine the *velocity pressure* q_z or q_h, as applicable, in accordance with Sec. 6.5.10 of ASCE 7-05.

   The velocity pressure is

   $$q_z = 0.00256 \, K_z K_{zt} K_d V^2 I$$

   $$I = 1.0$$

   $$K_d = 0.85$$

   $$V = 90 \text{ mi/h}$$

   $$K_{zt} = 1.0$$

   $$q_z = 17.63 \, K_z \text{ lb/ft}^2$$

   Note that the above expression for q_z has K_z embedded in it.

10. Determine the *design wind load* P or F in accordance with Secs. 6.5.12 and 6.5.13 of ASCE 7-05, as applicable.

    For components and cladding of low-rise buildings and buildings with $h \leq 60$ ft:

    $$p = q_h[(GC_p) - (GC_{pi})]$$

    where q_h = velocity pressure evaluated at mean roof height h using exposure defined in Sec. 6.5.6.3.1
    GC_p = external pressure coefficients from Figs. 6-11 through 6-17 of ASCE 7-05
    GC_{pi} = internal pressure coefficients from Fig. 6-5 of ASCE 7-05

Building height, h	Height above ground, z	Maximum inward pressure (windward wall)								Total
		External pressure				Internal pressure				
		K_h	q_h	GC_p	$p_{outside}$	K_h	q_h	GC_{pi}	p_{inside}	p_{total}
16.67	16.67	0.867	15.28	0.75	11.46	0.867	15.28	−0.18	−2.75	14.21

TABLE 12.4 Spreadsheet for Calculation of Wind Pressure on Windward Side of One-Story Example Building (Components and Cladding)

Windward Side of Building On the windward side of the building, the maximum inward pressure will be produced on the cladding, due to the combination of GC_p acting inward (positive sign) and GC_{pi} also acting inward (negative sign).

$$p = q_h[(GC_p) - (GC_{pi})]$$

$$p = (17.63\, K_h)[(GC_p) - (GC_{pi})]$$

$q_h = q_z$, evaluated at 16.67 ft
$q_i = q_h$, evaluated at 16.67 ft
$GC_p = 0.75$ (Fig. 6-17 of ASCE 7-05)
$GC_{pi} = \pm 0.18$ (Fig. 6-5 of ASCE 7-05)

These values are shown in the spreadsheet in Table 12.4. The maximum inward pressure is the sum of 11.46 psf on the outside plus 2.75 psf on the inside, for a total of 14.21 psf acting inward.

Leeward Side of Building On the leeward side of the building, the maximum outward pressure will be produced on the cladding, due to the combination of GC_p acting outward (negative sign) and GC_{pi} also acting outward (positive sign).

$$p = q_h[(GC_p) - (GC_{pi})]$$

$$p = (17.63\, K_h)[(GC_p) - (GC_{pi})]$$

$q = q_h$, evaluated at 16.67 ft
$q_i = q_h$, evaluated at 16.67 ft
$GC_p = -1.4$ (Fig. 6-17)
$GC_{pi} = \pm 0.18$ (Fig. 6-5).

These values are shown in the spreadsheet in Table 12.5. The maximum outward pressure is the sum of −21.39 psf on the outside plus 2.75 psf on the inside, for a total of 24.14 psf acting outward.

| Building height, h | Height above ground, z | Maximum outward pressure (leeward wall) ||||||| Total |
| | | External pressure |||| Internal pressure ||| |
		K_h	q_h	GC_p	$p_{outside}$	K_h	q_h	GC_{pi}	p_{inside}	p_{total}
16.67	16.67	0.867	15.28	−1.4	−21.39	0.867	15.28	0.18	2.75	24.14

TABLE 12.5 Spreadsheet for Calculation of Wind Pressure on Leeward Side of One-Story Example Building (Components and Cladding)

Wall elements must therefore be designed for a pressure of 14.21 lb/ft² acting inward, and 24.14 lb/ft² acting outward. The latter governs.

Wind Pressures on Roof Now evaluate the wind pressures on the roof. Those pressures are calculated using components and cladding coefficients, which depend on the tributary area of the roof element under consideration. In this case we are interested in the reactions from the long-span joists. Repeat the above steps, starting with Step 8:

8. Determine the *external pressure coefficients* C_p or GC_{pf}, or *force coefficients* C_f, as applicable, in accordance with Sec. 6.5.11.2 or 6.5.11.3 of ASCE 7-05, respectively. The external pressure coefficients for components and cladding GC_p are given in Fig. 6-8 of ASCE 7-05.

    In computing the effective area of the cladding element, it is permitted to use an effective area equal to the product of the span and an effective width not less than one-third the span (ASCE 7-05, Sec. 6.2, "Effective Wind Area").

    The long-span joists have a span of 40 ft. Assume an effective width of one-third of that span, or 13.33 ft. The resulting effective area is 533 ft². For simplicity and continuity from the chapter dealing with calculation of wind loads, Fig. 6-17 of ASCE7-05 will be used, even though this building is less than 60 ft in height to the roof. Also for simplicity, values from Zone 1 will be used. From Fig. 6-17 of ASCE 7-05, a roof element in Zone 1 has a negative pressure coefficient of −0.9.

9. Determine the *velocity pressure* q_z or q_h, as applicable, in accordance with Sec. 6.5.10 of ASCE 7-05.

    The velocity pressure is

    $$q_z = 0.00256\, K_z K_{zt} K_d V^2 I$$

    $$I = 1.0$$

    $$K_d = 0.85$$

$$V = 90 \text{ mi/h}$$
$$K_{zt} = 1.0$$
$$q_z = 17.63 \, K_z \text{ lb/ft}^2$$

Note that the above expression for q_z has K_z embedded in it.

10. Determine the *design wind load P or F* in accordance with Secs. 6.5.12 and 6.5.13, as applicable. For components and cladding of one-story buildings and buildings with $h \leq 60$ ft:

$$p = q_h[(GC_p) - (GC_{pi})]$$

where q_h = velocity pressure evaluated at mean roof height h using exposure defined in Sec. 6.5.6.3.1
GC_p = external pressure coefficients from Figs. 6-11 through 6-17 of ASCE 7-05
GC_{pi} = internal pressure coefficients from Fig. 6-5 of ASCE 7-05

On the roof of the building, upward pressure will be critical. The maximum upward pressure will be produced due to the combination of GC_p acting outward (negative sign) and GC_{pi} also acting outward (positive sign).

$$p = q_h[(GC_p) - (GC_{pi})]$$

$$p = (17.63 \, K_h)[(GC_p) - (GC_{pi})]$$

$q_h = q_z$, evaluated at 16.67 ft
$q_i = q_h$, evaluated at 16.67 ft
$GC_p = 0.9$ (Fig. 6-17)
$GC_{pi} = \pm 0.18$ (Fig. 6-5)

These values are shown in the spreadsheet of Table 12.6. The maximum outward (uplift) pressure is the sum of -13.75 psf on

Building height, h	Height above ground, z	Maximum outward pressure (roof)								Total
		External pressure				Internal pressure				
		K_h	q_h	GC_p	$p_{outside}$	K_h	q_h	GC_{pi}	p_{inside}	p_{total}
16.67	16.67	0.867	15.28	−0.9	−13.75	0.867	15.28	0.18	2.75	16.50

TABLE 12.6 Spreadsheet for Calculation of Wind Pressure on Roof of One-Story Example Building (Components and Cladding)

the outside plus 2.75 psf on the inside, for a total of 16.50 psf acting outward. This is much less than the dead load of 60 psf, so net wind uplift does not exist.

12.3.7 Propose Structural Systems for Gravity and Lateral Load

Gravity loads are carried from the corrugated decking to the bar joists, from the bar joists to the long-span joists, and from the long-span joists to the north and south walls, and to the interior steel columns.

Lateral loads are resisted by perpendicular walls (vertical strips), which transfer their loads to the roof diaphragm and the foundation. Loads transferred to the roof diaphragm are carried to shear walls oriented parallel to the load.

12.4 Step 2: Design Walls for Gravity plus Out-of-Plane Loads

12.4.1 Design of West Wall for Gravity plus Out-of-Plane Loads

The west wall carries gravity load from a portion of the tributary area of the roof, plus wind loads.

Suppose that the load is applied over a 4-in. bearing plate, and assume that bearing stresses vary linearly under the bearing plate as shown in Fig. 12.7.

Then the eccentricity of the applied load with respect to the centerline of the wall is

$$e = \frac{t}{2} - \frac{\text{plate}}{3} = \frac{7.63 \text{ in.}}{2} - \frac{4 \text{ in.}}{3} = 2.48 \text{ in.}$$

Compute the load on the west wall from the roof, based on the tributary area shown in Fig. 12.8.

The tributary area of an entire bar joist is the product of the span (20 ft) and the distance between bar joists (8 ft). The tributary area of bar joist loading the west wall is one-half that, or 80 ft².

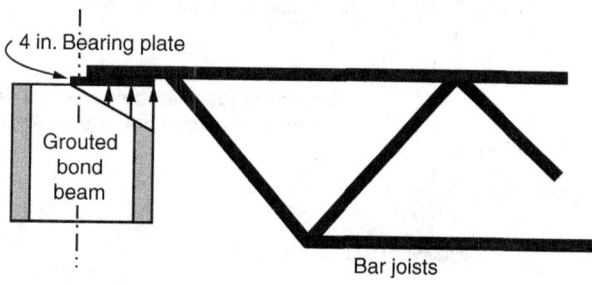

FIGURE 12.7 Assumed variation of bearing stresses under bearing plate.

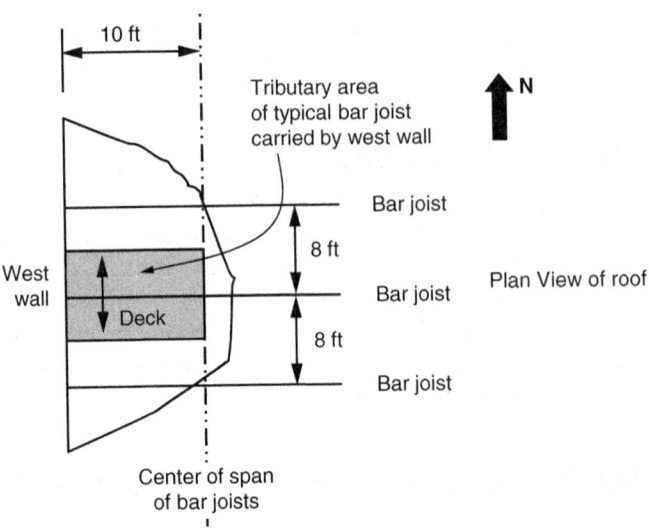

FIGURE 12.8 Tributary area of typical bar joist on west wall.

The roof dead load is 60 lb/ft². For tributary areas up to 200 ft², the roof live load is 20 lb/ft².

Assume as before that the critical loading combination is $0.9D + 1.6W$. Because wind loads are small and do not produce net uplift, they are neglected for simplicity. The factored gravity load acting on the wall per foot of length is therefore

$$w_u = 10 \text{ ft} \cdot 0.9(60 \text{ lb/ft}^2) = 540 \text{ lb/ft}$$

As in previous example problems, try an initial design of the wall as unreinforced (Fig. 12.9):

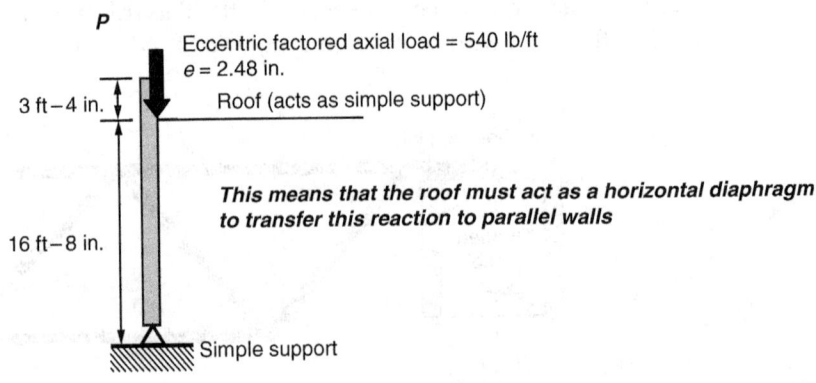

FIGURE 12.9 West bearing wall of example one-story building.

At each horizontal plane through the wall, the following conditions must be met:

- Maximum compressive stress from factored axial loads must not exceed the slenderness-dependent values in Eqs. (3-12) or (3-13) as appropriate, reduced by a ϕ factor of 0.60.
- Maximum compressive stress from factored loads (including a moment magnifiers) must not exceed 0.80 f_m' in the extreme compression fiber, reduced by a ϕ factor of 0.60.
- Maximum tension stress from factored loads (including a moment magnifier) must not exceed the modulus of rupture in the extreme tension fiber, reduced by the ϕ factor of 0.60.

For each condition, the more critical of the two possible loading combinations must be checked. Because there is wind load, and because previous examples showed little problem with the first two criteria, the third criterion (net tension) may well be critical. For this criterion, the critical loading condition could be either 1.2D + 1.6L or 0.9D + 1.6W. Both loading conditions must be checked.

We also must check various points on the wall. Critical points are just below the roof reaction (moment is high and axial load is low, so maximum tension may govern); and at the base of the wall (axial load is high, so maximum compression may govern).

To avoid having to check a large number of loading combinations and potentially critical locations, it is worthwhile to assess them first, and check only the ones that will probably govern.

Because the eccentric axial load places the outer fibers of the wall in tension, the critical wind condition will be suction, for which the design wind pressure is 24.44 lb/ft².

Due to wind only, the unfactored moment at the base of the parapet (roof level) is

$$M = \frac{qL_{parapet}^2}{2} = \frac{24.14 \text{ lb/ft} \cdot 3.33^2 \text{ ft}^2}{2} \cdot 12 \text{ in./ft} = 1606 \text{ lb-in.}$$

The maximum moment is close to that occurring at mid-height. The moment from wind load is the superposition of one-half moment at the upper support due to wind load on the parapet only, plus the midspan moment in a simply supported beam with that same wind load:

$$M_{midspan} = -\frac{1606}{2} + \frac{qL^2}{8} = -\frac{1606}{2} + \frac{24.14 \text{ lb/ft} \cdot 16.67^2 \text{ ft}^2}{8} \cdot 12 \text{ in./ft}$$

$$= 9259 \text{ lb-in.}$$

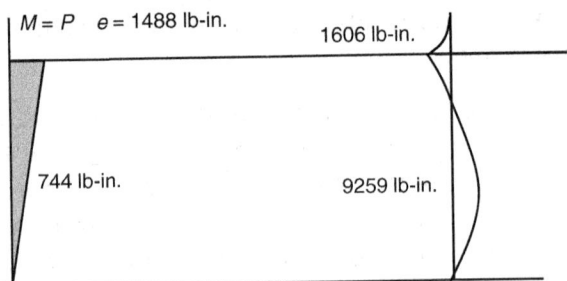

FIGURE 12.10 Unfactored moment diagrams due to eccentric dead load and wind.

Unfactored moment diagrams due to eccentric dead load and wind are shown in Fig. 12.10.

From previous examples with unreinforced bearing walls in Sec. 5.2.5, we know that loading combination 1.2D + 1.6L was not close to critical directly underneath the roof. Because the wind-load moments directly underneath the roof are not very large, they will probably not be critical either. The critical location will probably be at mid-height; the critical loading condition will probably be 0.9D + 1.6W; and the critical criterion will probably be net tension, because this masonry wall is unreinforced.

Work with a strip with a width of 1 ft (measured along the length of the wall in plan). Stresses are calculated using the critical section, consisting of the bedded area only (2008 MSJC *Code*, Sec. 1.9.1.1).

Check the net tensile stress. At the mid-height of the wall, the axial force due to 0.9D is

$$P_u = 0.9(600 \text{ lb}) + 0.9(3.33 \text{ ft} + 8.33 \text{ ft}) \cdot 48 \text{ lb/ft} = 1044 \text{ lb}$$

At the mid-height of the wall, the factored design moment, M_u, is given by

$$M_u = P_{u\,\text{eccentric}} \frac{e}{2} + M_{u\,\text{wind}} = \left(\frac{1}{2}\right) 0.9 \cdot 600 \text{ lb} \cdot 2.48 \text{ in.} + 1.6 \cdot 9259 \text{ lb-in.}$$

$$= 15{,}484 \text{ lb-in.}$$

$$f_{\text{tension}} = -\frac{P_u}{A} + \frac{M_u c}{I}$$

$$f_{\text{tension}} = -\frac{1044 \text{ lb}}{30 \text{ in.}^2} + \frac{15{,}484 \text{ lb-in.} \left(7.63/2\right) \text{ in.}}{309 \text{ in.}^4}$$

$$= -34.8 \text{ lb/in.}^2 + 191.17 \text{ lb/in.}^2 = 156.4 \text{ lb/in.}^2$$

$$0.60 f_r = 0.60 \cdot 63 \text{ lb/in.}^2 = 37.8 \text{ lb/in.}^2$$

The maximum tensile stress exceeds the prescribed value, and the design is not satisfactory. It will be necessary to grout or reinforce the wall. Recheck using a grouted wall.

Check the net tensile stress. At the mid-height of the wall, compute the axial force due to 0.9D. Use a self-weight of 120 lb/ft³ for a fully grouted wall.

$$P_u = 0.9 \,(600 \text{ lb}) + 0.9 \,(3.33 \text{ ft} + 8.33 \text{ ft}) \cdot 120 \text{ lb/ft} \cdot (7.63/12) = 1430 \text{ lb}$$

At the mid-height of the wall, the factored design moment, M_u, is given by

$$M_u = P_{u \text{ eccentric}} \frac{e}{2} + M_{u \text{ wind}} = \left(\frac{1}{2}\right) 0.9 \cdot 600 \text{ lb} \cdot 2.48 \text{ in.} + 1.6 \cdot 9259 \text{ lb-in.}$$

$$= 15{,}484 \text{ lb-in.}$$

For a solid 8-in. wall, $A = (7.63 \text{ in.} \cdot 12 \text{ in.}) = 91.56 \text{ in.}^2$, and $I = (12 \text{ in.} \cdot 7.63^3 \text{ in.}^3)/12 = 444 \text{ in.}^4$.

$$f_{\text{tension}} = -\frac{P_u}{A} + \frac{M_u c}{I}$$

$$f_{\text{tension}} = -\frac{1430 \text{ lb}}{91.56 \text{ in.}^2} + \frac{15{,}484 \text{ lb-in.} \left(7.63/2\right)}{444 \text{ in.}^4} = -15.62 + 133.04 \text{ lb/in.}^2$$

$$= 117.4 \text{ lb/in.}^2$$

$$0.60 f_r = 0.60 \cdot 170 \text{ lb/in.}^2 = 102 \text{ lb/in.}^2$$

The maximum tensile stress still exceeds the prescribed value. The wall will have to be reinforced.

Try #4 bars @ 48 in. Assuming a fully grouted wall, the factored axial load and moment per foot of length are 1430 lb and 15,484 in.-lb. The design moment-axial force interaction diagram is shown in Fig. 12.11.

Relevant cells from the spreadsheet are reproduced in Table 12.7.

The strength is just sufficient. Use #4 bars @ 48 in. Because the compressive stress block remains in the face shell for this axial load (refer to the spreadsheet in Table 12.7), the wall can be partially grouted, and the interaction diagram is still valid.

Because the axial load per foot of plan length is small, slenderness effects will not affect this answer significantly.

12.4.2 Design of East Wall for Gravity plus Out-of-Plane Loads

The loads on the east wall are identical to those on the west wall, except for the presence of the openings. Because the west wall had to be reinforced,

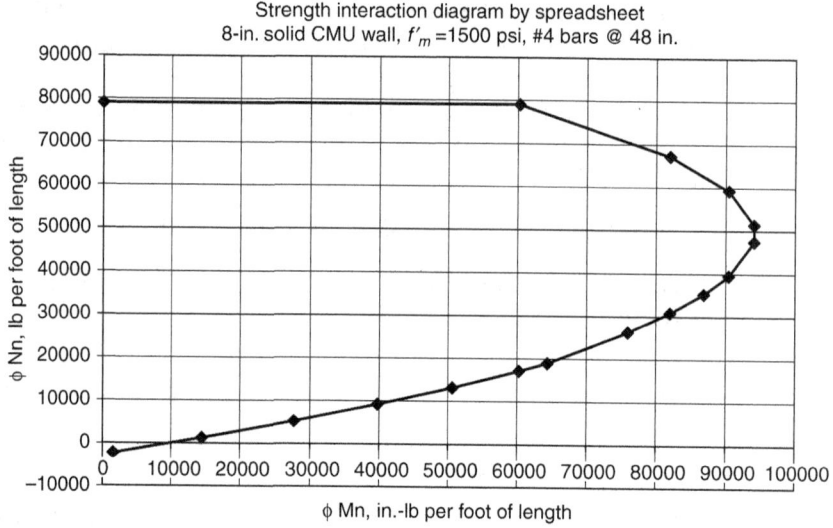

FIGURE 12.11 Design moment-axial force interaction diagram for west wall of example one-story building.

so will the east wall. Loads on each wall segment are increased by the ratio of the tributary width of the wall segment, to the actual width. The elevation of the east wall is shown in Fig. 12.12.

By inspection, Wall Segment B, with a ratio of tributary width to actual width of (20.5/8), is most critical. At the mid-height of the wall, the factored axial load and factored moment per foot of length, P_u and M_u, is given by:

$$P_u = 1430 \, lb$$

$$M_u = P_{u\,eccentric} \frac{e}{2} + M_{u\,wind} = \left(\frac{1}{2}\right) 0.9 \cdot 600 \, lb \cdot 2.48 \, in. + 1.6 \cdot 9259 \, lb\text{-}in.$$

$$= 15,484 \, lb\text{-}in.$$

The factored moment on Wall Segment B is obtained by multiplying that factored design moment per foot of length, times the tributary wall length of 20.5 ft, giving a total factored design moment on the critical Wall Segment B of 317,422 lb-in.

As before, the strength moment-axial force interaction diagram for a solidly grouted 8-in. masonry wall loaded out of plane is shown in Fig. 12.13. For vertical reinforcement consisting of #5 bars spaced at 48 in. and an axial load close to zero, out-of-plane design flexural capacity is about 15,000 in.-lb per foot of length. For Wall Segment B, with a plan length of 8 ft, the out-of-plane design flexural capacity is about

Spreadsheet for calculating strength moment-axial force interaction diagram for west wall of one-story building

Reinforcement at mid-depth						
Specified thickness	7.625					
emu	0.0025					
f'_m	1500					
f_y	60000					
E_s	29000000					
d	3.8125					
(c/d) balanced	0.54717					
Tensile reinforcement area	0.2					
Effective width	48					
phi	0.9					
Because compression reinforcement is not supported, it is not counted						
	c/d	c	C_{mas}	f_s	Moment	Axial force
Points controlled by steel	0.01	0.038125	1757	−60000	1501	−2305
	0.1	0.38125	17568	−60000	14467	1253
	0.2	0.7625	35136	−60000	27729	5206
	0.3	1.14375	52704	−60000	39785	9158
	0.4	1.525	70272	−60000	50635	13111
	0.5	1.90625	87840	−60000	60280	17064
	0.54717	2.086085	96127	−60000	64411	18929
Points controlled by masonry	0.54717	2.086085	96127	−60000	64411	18929
	0.7	2.66875	122976	−31071	75953	26271
	0.8	3.05	140544	−18125	81981	30807
	0.9	3.43125	158112	−8056	86803	35213
	1	3.8125	175680	0	90420	39528
	1.2	4.575	210816	0	94037	47434
	1.3	4.95625	228384	0	94037	51386
	1.5	5.71875	263520	0	90420	59292
	1.7	6.48125	298656	0	81981	67198
	2	7.625	351360	0	60280	79056
Pure axial load					0	79013

TABLE 12.7 Spreadsheet for Calculating Strength Moment-Axial Force Interaction for West Wall of One-Story Building

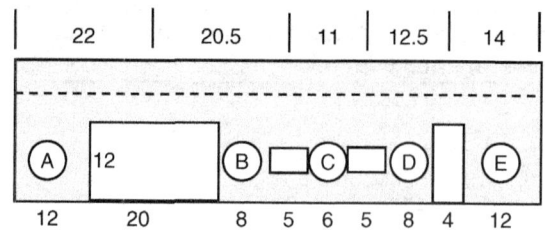

FIGURE 12.12 East wall of example one-story building.

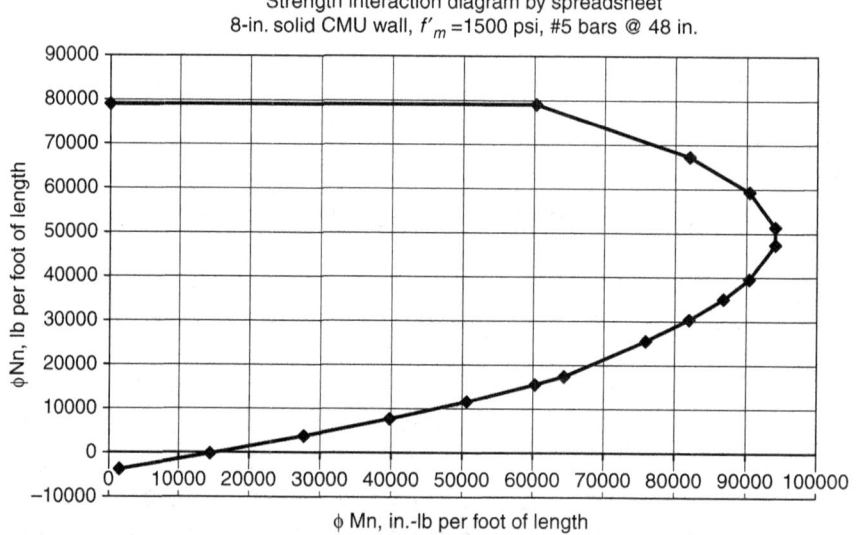

FIGURE 12.13 Strength moment-axial force interaction diagram for solidly grouted 8-in. CMU wall loaded out-of-plane (#5 bars @ 48 in.).

120,000 in.-lb. This corresponds to two #5 bars, or a total area of flexural reinforcement of 0.62 in.^{2}.

Flexural capacity at low axial loads is approximately proportional to the area of flexural reinforcement. To achieve a design capacity of 317,422 lb-in., the total steel area will need to be about that 0.62 in.^{2}, multiplied by the ratio of 317,422 kip-in. divided by 120,000 lb-in., or at least 1.64 in.^{2}.

Continue a trial design using 3 to #7 bars per wall segment (steel area equals 3 · 0.60 in.^{2}, or 1.80 in.^{2}). For the critical Wall Segment B, the effective width is 3t on each side of each bar. This is essentially equal to the total segment length of 8 ft. Capacity is insensitive to effective width at low axial loads. The moment-axial force interaction diagram for a uniform wall is given in Fig. 12.13, and the moment-axial force interaction diagram for Wall Segment B is given in Fig. 12.14.

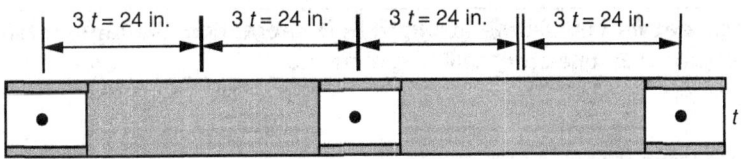

FIGURE 12.14 Trial design Wall Segment B of east wall as governed by out-of-plane wind load.

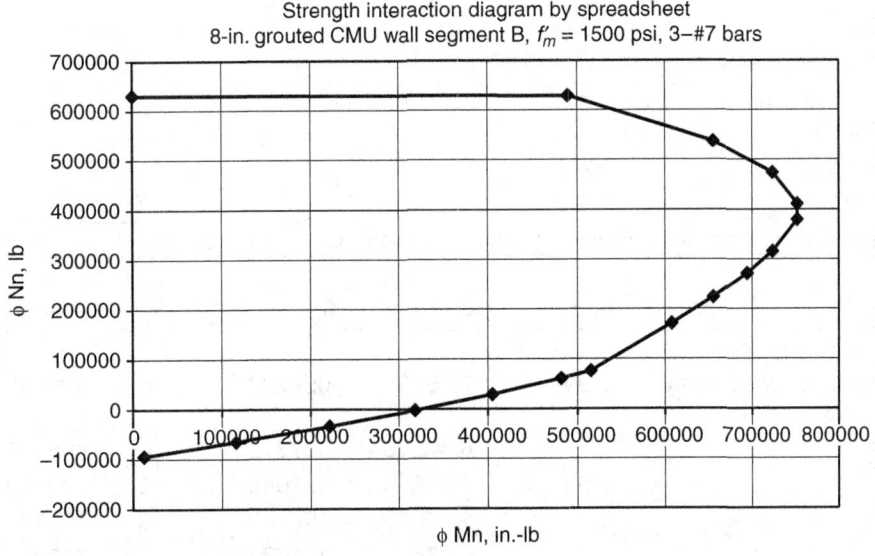

FIGURE 12.15 Design moment-axial force interaction diagram for Wall Segment B of one-story example building.

Using the trial reinforcement, we obtain a moment-axial force interaction diagram (strength) for the trial design of the critical Wall Segment B. This is shown in Fig. 12.15.

The cells of the spreadsheet are shown in Table 12.8.

At an axial load of close to zero, the design out-of-plane flexural capacity of the critical Wall Segment B is about 320,000 lb-in. The design is satisfactory.

For simplicity, use 3 to #7 bars vertically in every wall segment of the east wall. Again, slenderness effects will not change the answer significantly.

We have completed the design of the wall segments of the east wall. Now consider the design of the horizontally spanning lintel between Wall Segment A and Wall Segment B, against out-of-plane wind. This is shown in Fig. 12.16. The seismic provisions of the 2009 IBC would in

Spreadsheet for calculating strength M-N interaction diagram for fully grouted CMU Wall Segment B, one-story building example

Reinforcement at mid-depth						
Specified thickness	7.625					
emu	0.0025					
f'_m	1500					
f_y	60000					
E_s	29000000					
d	3.8125					
(c/d) balanced	0.54717					
Tensile reinforcement area	1.8					
Effective width	96					
phi	0.9					
Because compression reinforcement is not supported, it is not counted						
	c/d	c	C_{mas}	f_s	Moment	Axial force
Pure axial load					0	630893
Points controlled by masonry	1.99	7.586875	699206	0	489427	629286
	1.7	6.48125	597312	0	655849	537581
	1.5	5.71875	527040	0	723362	474336
	1.3	4.95625	456768	0	752297	411091
	1.2	4.575	421632	0	752297	379469
	1	3.8125	351360	0	723362	316224
	0.9	3.43125	316224	−8056	694428	271552
	0.8	3.05	281088	−18125	655849	223617
	0.7	2.66875	245952	−31071	607624	171021
	0.54717	2.086085	192254	−60000	515289	75828
Points controlled by steel	0.54717	2.086085	192254	−60000	515289	75828
	0.5	1.90625	175680	−60000	482242	60912
	0.4	1.525	140544	−60000	405083	29290
	0.3	1.14375	105408	−60000	318279	−2333
	0.2	0.7625	70272	−60000	221831	−33955
	0.1	0.38125	35136	−60000	115738	−65578
	0.01	0.038125	3514	−60000	12008	−94038

TABLE 12.8 Spreadsheet for Calculating Moment-Axial Force Interaction Diagram for Wall Segment B of One-Story Example Building

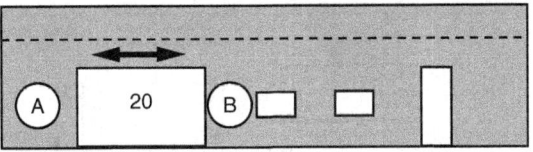

FIGURE 12.16 Design of lintel on east wall for out-of-plane loads.

many cases require that the roof diaphragm be connected to the walls at intervals of about 4 ft. For purposes of this example, the possible constraints imposed by such requirements are ignored in computing the span of the lintel out of plane. As will be seen later in this example, this has the additional advantage of showing that out-of-plane loads do not govern for the design of such lintels, even ignoring intermediate supports that will actually be there.

Again, suction will be critical. The span of the lintel out-of-plane is the distance between connectors to the diaphragm, assumed here as 4 ft. The depth of the lintel is (20 − 12) ft. The factored design out-of-plane moment on the lintel is

$$M_u^{\text{out-of-plane}} = (1.6)\frac{qL^2}{8} = \frac{(1.6)24.14 \text{ lb/ft}^2 \cdot (20-12) \text{ ft} \cdot 4^2 \text{ ft}^2 \cdot 12 \text{ in./ft}}{8}$$

$$M_u^{\text{out-of-plane}} = 7416 \text{ lb-in.}$$

The corresponding required nominal capacity is that value divided by 0.90, or 8240 lb-in.

Conservatively assume an effective depth equal to 90 percent of one-half the wall thickness.

$$A_s^{\text{required}} = \frac{M_u}{\phi d f_y} \approx \frac{7416 \text{ lb-in.}}{0.90 \times \left(0.9 \times \frac{7.63}{2} \text{ in.}\right) 60,000 \text{ lb/in.}^2} = 0.04 \text{ in.}^2$$

This will easily be satisfied using 2 to #4 bars in a bond beam at the level of the roof, plus 2 to #4 bars at the top of the parapet, plus 2 to #4 bars in the lowest course of the lintel. For out-of-plane loading, one-half the bars will work at a time.

12.4.3 Design North and South Walls for Gravity plus Out-of-Plane Wind Loads

The north and south walls span vertically between the foundation slab and the roof diaphragm. They support gravity loads from self-weight alone, because the long-span joists rest on pilasters (thickened wall sections)

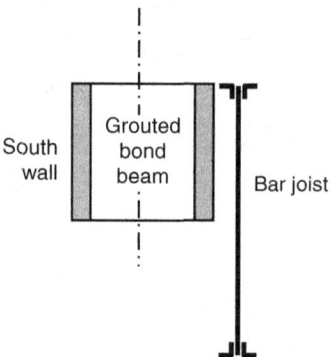

Figure 12.17 Placement of bar joists adjacent to north and south walls.

that are separated from the north and south walls by vertically oriented control joints. They also support out-of-plane wind loads.

Because the bar joists are oriented parallel to the north and south walls, a bar joist will be placed right next to those walls (Fig. 12.17). The north and south walls will therefore not support any gravity loads except their own weight.

At each horizontal plane through the wall, the following conditions must be met:

- Maximum compressive stress from factored axial loads must not exceed the slenderness-dependent values in Eqs. (3-12) or (3-13) as appropriate, reduced by a ϕ-factor of 0.60.
- Maximum compressive stress from factored loads (including a moment magnifiers) must not exceed $0.80\,f_m'$ in the extreme compression fiber, reduced by a ϕ-factor of 0.60.
- Maximum tension stress from factored loads (including a moment magnifier) must not exceed the modulus of rupture in the extreme tension fiber, reduced by the ϕ-factor of 0.60.

Check midspan moment in the wall. Wind pressures, and corresponding wind moments, will be the same as before.

Check the net tensile stress. At the mid-height of the wall, compute the axial force due to 0.9D. Assume a fully grouted wall, with a unit weight of 120 lb/ft³.

$$P_u = 0.9(3.33\text{ ft} + 8.33\text{ ft}) \cdot 120 \cdot (7.63/12)\text{ lb/ft} = 801\text{ lb}$$

At the mid-height of the wall, the factored design moment, M_u, is given by

$$M_u = M_{u\,\text{wind}} = 1.6 \cdot 9259 \text{ lb-in.} = 14{,}814 \text{ lb-in.}$$

For a solid-grouted 8-in. wall,

$A = (7.63 \text{ in.} \cdot 12 \text{ in.}) = 91.56 \text{ in.}^2$
$I = (12 \text{ in.} \cdot 7.63^3 \text{ in.}^3)/12 = 444 \text{ in.}^4$

$$f_{tension} = -\frac{P_u}{A} + \frac{M_u c}{I}$$

$$f_{tension} = -\frac{801 \text{ lb}}{91.56 \text{ in.}^2} + \frac{14{,}814 \text{ lb-in.} \left(7.63/2\right)}{444 \text{ in.}^4} = -8.75 + 127.29 \text{ lb/in.}^2$$

$$= 118.5 \text{ lb/in.}^2$$

$0.60 f_r = 0.60 \cdot 170 \text{ lb/in.}^2 = 102 \text{ lb/in.}^2$

The maximum tensile stress exceeds the prescribed value, and the wall will have to be reinforced. As with the west wall, use #4 bars @ 48 in.

12.4.4 Design Pilasters (Columns) in North and South Walls

Because the north and south walls have vertically oriented control joints at each pilaster, the pilasters really behave like 16- by 16-in. beam-columns. In the specific context of the MSJC Code, however, they do not have to meet the prescriptive requirements for columns (including transverse reinforcement), because they are not isolated elements. The cross section of a typical pilaster is shown in Fig. 12.18.

The load on the pilasters comes from eccentric gravity loads from the long-span joists. Their tributary area is shown in Fig. 12.19.

The tributary area is 400 ft², which corresponds to a reduced live load of 20 lb/ft² · 0.80 = 16 lb/ft², according to Sec. 1607.11.2 of the 2009 IBC. This is irrelevant, however, because as before, axial load significantly increases pilaster capacity below the balance point, and the governing loading combination is $0.9D + 1.6W$.

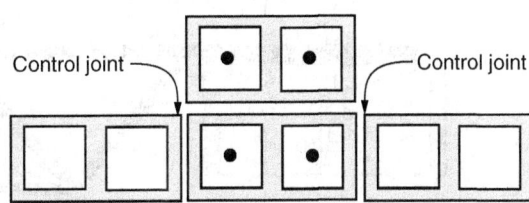

FIGURE 12.18 Cross section of typical pilaster in north and south walls of example one-story building.

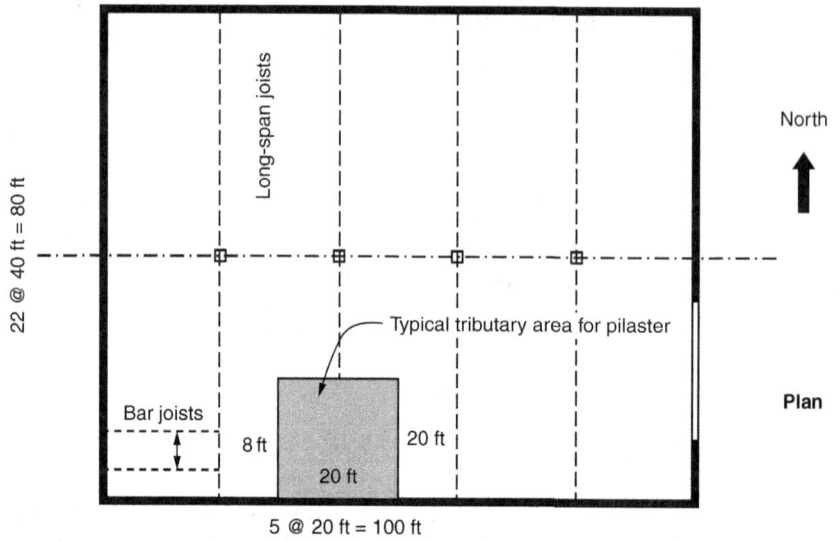

FIGURE 12.19 Tributary area supported by typical pilaster.

The factored axial load on each pilaster is therefore

$$P_u = 0.9 \cdot 60 \text{ lb/ft}^2 \cdot 400 \text{ ft}^2 = 21,600 \text{ lb}$$

The long-span joists rest on the pilasters through full-width base plates. Assuming a triangular stress distribution under the bearing plates, the eccentricity of gravity load can be calculated, as shown in Fig. 12.20.

The corresponding factored moment at the top of the pilaster due to eccentric gravity load is therefore

$$M_u = P_u e = 21,600 \text{ lb} \cdot 2.61 \text{ in.} = 56,376 \text{ lb-in.}$$

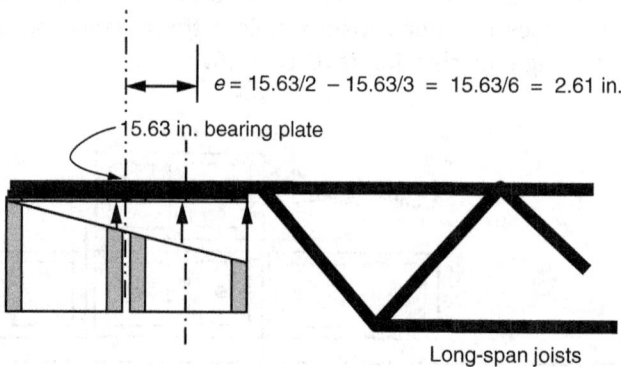

FIGURE 12.20 Distribution of bearing stresses under bearing plates of pilasters.

Due to wind only, the factored moment at the base of the parapet (roof level) is

$$M_u = \frac{qL^2}{2} = \frac{1.6 \cdot 24.14 \text{ lb/ft}^2 \cdot 1.33 \text{ ft} \cdot 3.33^2 \text{ ft}^2}{2} \cdot 12 \text{ in./ft} = 3418 \text{ lb-in.}$$

The maximum wind-load moment is close to that occurring at mid-height. The moment from wind load is the superposition of one-half moment at the upper support due to wind load on the parapet only, plus the midspan moment in a simply supported beam with that same wind load. Because the north and south walls are separated from the pilasters by control joints, the wind-load moment on the pilasters is due to their frontal area alone:

$$M_{u \text{ midspan}} = -\frac{3418}{2} + \frac{qL^2}{8}$$

$$M_{u \text{ midspan}} = -\frac{3418}{2} + \frac{1.6 \cdot 24.14 \text{ lb/ft}^2 \cdot 1.33 \text{ ft} \cdot 16.67^2 \text{ ft}^2}{8} \cdot 12 \text{ in./ft}$$

$$= 19,704 \text{ lb-in.}$$

Factored moment diagrams due to eccentric dead load and wind are as shown in Fig. 12.21.

Maximum factored design moment is the summation of that due to eccentric gravity load and that due to wind:

$$M_u = 28,188 + 19,704 = 47,892 \text{ lb-in.}$$

Now design the pilaster to have sufficient capacity.

The effective depth, d, of the pilaster is computed based on the specified dimensions of nominal 8-in. CMU (Fig. 12.22).

Using a spreadsheet as before, a strength-based, moment-axial force interaction diagram can be generated. For simplicity, the contribution

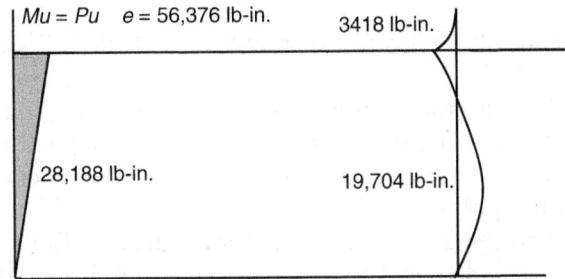

FIGURE 12.21 Factored moment diagrams due to eccentric dead load and wind on pilasters.

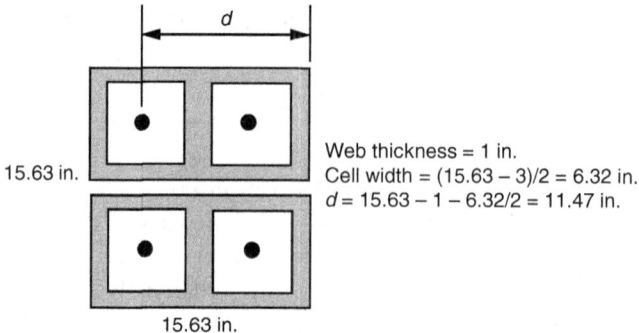

FIGURE 12.22 Effective depth, d, of pilasters.

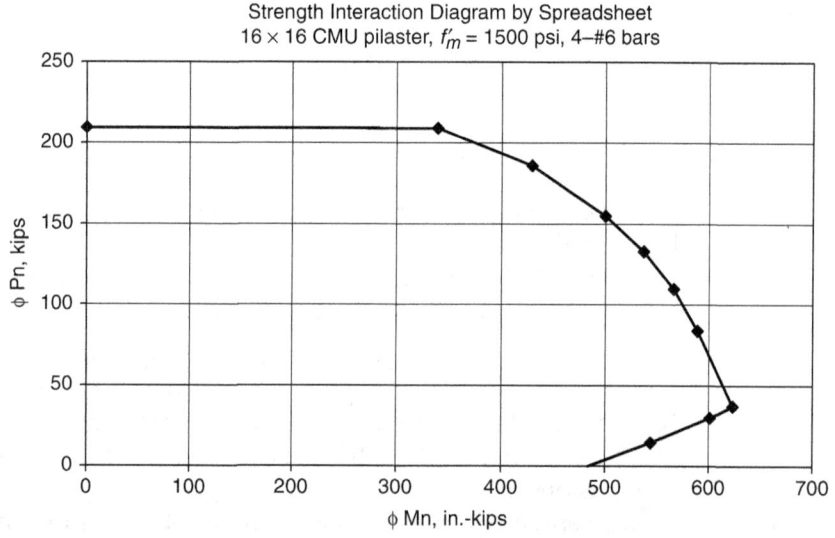

FIGURE 12.23 Strength moment-axial force interaction diagram for typical pilaster.

of compressive reinforcement is neglected. This diagram is shown in Fig. 12.23.

Because the reinforcement is located at a good distance from the geometric centroid of the cross section, the shape of the interaction diagram is familiar, with the maximum moment capacity corresponding to the balance point.

The spreadsheet cells are reproduced in Table 12.9.

Using 4 to #6 bars, the capacity of the pilaster is much greater than required. Wind uplift will not govern. Shear is very small, and will not govern.

Strength Design Example: One-Story Building

Spreadsheet for calculating strength M-N interaction diagram for 16-in. CMU pilaster, one-story building	
Depth	15.63
emu	0.0025
f'_m	1.5
f_y	60
E_s	29000
d	11.47
(c/d) Balanced	0.54717
Width	15.63
phi	0.9

Steel layers are counted from the extreme compression fiber to the extreme tension fiber
Distances are measured from the extreme compression fiber
Compression in masonry and reinforcement is taken as positive
Stress in compressive reinforcement is set to zero, because the reinforcement may not be laterally supported

Row of reinforcement	Distance	Area
1	4.16	0.88
2	11.47	0.88

	c/d	c	C_{mas}	$f_s(1)$	$f_s(2)$	Moment	Axial force
Pure axial load						0	210
Points controlled by masonry	1.35	15.48	232	0.00	0.00	339	209
	1.2	13.76	207	0.00	0.00	429	186
	1	11.47	172	0.00	0.00	500	155
	0.9	10.32	155	0.00	−8.06	537	133
	0.8	9.18	138	0.00	−18.13	566	110
	0.7	8.03	120	0.00	−31.07	589	84
	0.54717	6.28	94	0.00	−60.00	623	37
Points controlled by steel	0.54717	6.28	94	0.00	−60.00	623	37
	0.5	5.74	86	0.00	−60.00	601	30
	0.4	4.59	69	0.00	−60.00	544	14
	0.3	3.44	52	−15.15	−60.00	429	−13
	0.2	2.29	34	−58.97	−60.00	217	−63
	0.1	1.15	17	−60.00	−60.00	114	−80
	0.01	0.11	2	−60.00	−60.00	12	−93

TABLE 12.9 Spreadsheet for Calculating Strength Moment-Axial Force Interaction Diagram for Typical Pilaster

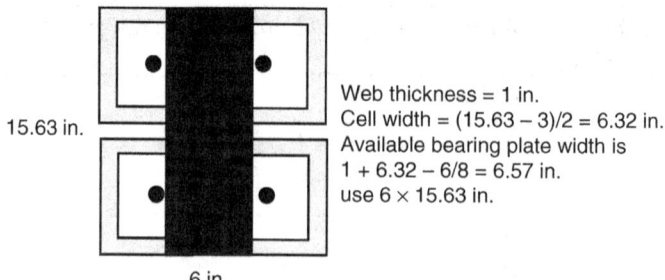

FIGURE 12.24 Bearing plate under long-span joists.

12.4.5 Bearing Plate under Long-Span Joists

The 2008 MSJC *Code* specifies an strength-reduction factor of 0.60 for bearing (Sec. 3.1.4.6), and provides formulas for nominal bearing capacity (Sec. 3.1.7):

$$\phi P_{n\,\text{bearing}} = \phi\, 0.6\, A_n f'_m = 0.6 \cdot 0.6 \cdot 15.63^2 \text{ in.}^2 \cdot 1500 \cdot \text{lb/in.}^2 = 131{,}920 \text{ lb}$$

This design capacity is far in excess of the factored axial load. The required area of the bearing plate is about 10 percent of the cross-sectional area of the pilaster. Use a bearing plate measuring about 6 in. × 15 in (Fig. 12.24).

12.5 Step 3: Design Lintels

Only the lintel over the 20-ft opening will be critical (Fig. 12.25).

As with the previous design of the area over the lintel for out-of-plane flexure in the horizontal plane, the structural span of the lintel is 20 ft, plus one-half of one-half unit on each side, or 20.67 ft. Conservatively, arching action will be neglected.

12.5.1 Calculate Gravity Load on Lintel

The lintel supports some direct gravity load from the roof decking, based on the tributary area shown in Fig. 12.26.

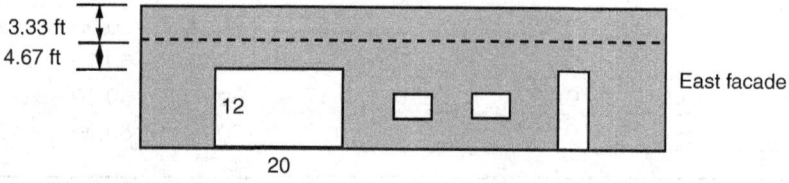

FIGURE 12.25 East facade of one-story building, showing critical 20-ft lintel.

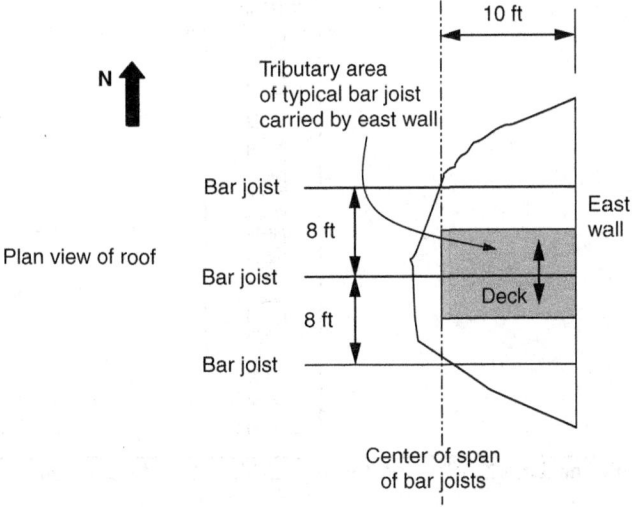

FIGURE 12.26 Tributary area supported by bar joists bearing on lintel of east wall.

The tributary area supported by the entire lintel is 10 ft (tributary width from the Fig. 12.26), multiplied by the span of 20 ft, or 200 ft². Because this is exactly equal to the upper limit of the tributary area at which live load reduction starts, no live load reduction is applied.

Factored gravity loads per foot of length on the lintel are itemized in Table 12.10. Because the lintel is uncoupled from the wall system by the control joints at each end, gravity loads alone constitute the critical loading case for it, and the governing loading combination is $1.2D + 1.6L$. In calculating the self-weight of the parapet and wall above the opening, the masonry is assumed to be fully grouted, with a unit weight of 76 lb/ft².

As in previous lintel-design examples, the bars in the lintel will probably be placed in the lower part of an inverted bottom course. As shown in Fig. 12.27, the effective depth d is calculated using the minimum cover

Description	Calculation	Load factor	Factored load, lb/ft
Roof DL	60 lb/ft² × 10 ft	1.2	720.0
Roof LL	20 lb/ft² × 10 ft	1.6	320.0
Parapet + wall	8 ft × 76 lb/ft²	1.2	729.6
Total			1769.6

TABLE 12.10 Factored Gravity Loads Acting on 20-ft Lintel of East Wall

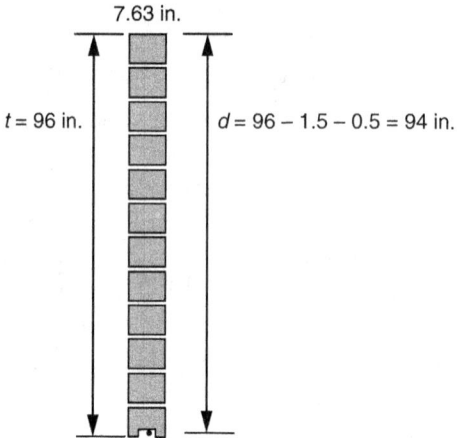

FIGURE 12.27 Section through 20-ft lintel of east wall.

of 1.5 in. (Sec. 1.15.4.1 of the 2008 MSJC *Code*), plus one-half the diameter of an assumed #8 bar.

Calculate the factored design moment and shear for the lintel:

$$V_u = \frac{w_u L}{2} = \frac{1770 \text{ lb/ft} \cdot 20.67 \text{ ft}}{2} = 36,586 \text{ lb}$$

$$M_u = \frac{w_u L^2}{8} = \frac{1770 \text{ lb/ft} \cdot 20.67^2 \text{ ft}^2 \cdot 12 \text{ in./ft}}{8} = 1134 \text{ kip-in.}$$

Because this is a reinforced element, shearing capacity is calculated using Sec. 3.3.4.1.2.1 of the 2008 MSJC *Code*:

$$V_{nm} = \left[4.0 - 1.75\left(\frac{M_u}{V_u d_v}\right)\right] A_n \sqrt{f'_m} + 0.25 P_u$$

As $(M_u/V_u d_v)$ increases, V_{nm} decreases. Because $(M_u/V_u d_v)$ need not be taken greater than 1.0 (2008 MSJC *Code*, Sec. 3.3.4.1.2.1), the most conservative (lowest) value of V_{mm} is obtained with $(M_u/V_u d_v)$ equal to 1.0. Also, axial load, P_u, is zero:

$$V_{nm} = [4.0 - 1.75(1.0)] A_n \sqrt{f'_m}$$

$$V_{nm} = 2.25 A_n \sqrt{f'_m}$$

$$V_u = 36,586 \text{ lb} \le \phi V_n = 0.8 \cdot 2.25 \cdot 7.63 \text{ in.} \cdot 94 \text{ in.} \sqrt{1500} \text{ lb/in.}^2 = 50,000 \text{ lb}$$

and the design is acceptable for shear.

Now check the required flexural reinforcement:

$$M_n \approx A_s f_y 0.9d$$

In our case,

$$M_n^{\text{required}} = \frac{M_u}{\phi} = \frac{M_u}{0.9} = \frac{1,134,000 \text{ lb-in.}}{0.9} = 1,260,000 \text{ lb-in.}$$

Solve for the required steel area:

$$A_s^{\text{required}} = \frac{M_u}{\phi f_y 0.9d} = \frac{1,134,000 \text{ lb-in.}}{0.9 \times 60,000 \text{ psi} \times 0.9 \times 94 \text{ in.}} = 0.25 \text{ in.}^2$$

Because of the depth of the beam, this can easily be satisfied with 2 to #4 bars in the lowest course. Also include 2 to #4 bars at the level of the roof (bond beam reinforcement). This will be consistent with the requirements of the design of the lintel for out-of-plane bending in a horizontal plane. Finally, to guard against possible cracking of the lintel near the top, use two more #4 bars at the top course of the parapet.

12.6 Summary So Far

Thus far in the design, all walls are fully grouted. The north, west, and south walls have #4 bars vertically, at a horizontal spacing of 48 in.

As a result of the design of the wall segments of the east wall for out-of-plane bending, the lintel of the east wall for out-of-plane bending, and the design of the lintel of the east wall as a beam, reinforcement in the east wall is as shown in Fig. 12.28. Each wall segment has three #7 vertical bars; and three sets of horizontal reinforcement are provided, in the form of two #4 bars at the top of the parapet and the bottom of the lintel, and 2 to #7 bars at the level of the roof. The #4 and #7 bars will be continued around the perimeter of the entire building.

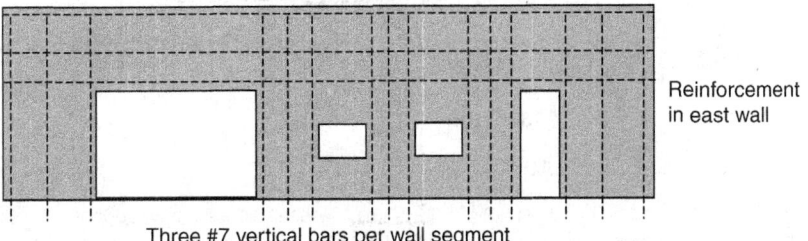

FIGURE 12.28 Reinforcement in east wall of one-story building.

12.7 Step 4: Conduct Lateral Force Analysis, Design Roof Diaphragm

Lateral force analysis will be critical in the north-south direction, because the area normal to the wind is greater, and the area of shear walls is less.

12.7.1 Check Roof Diaphragm

Compute Moment and Shear in Roof Diaphragm

From the wind-load analysis of Step 1, the unfactored design base shear on the building, due to wind from the north or south (the critical directions), is 33.6 kips.

As shown in Fig. 12.29, some of this is transmitted to the roof diaphragm; the rest is transmitted to the foundation slab. To compute the amount transferred to the roof diaphragm, idealize the 33.6 kips as applied uniformly over either the north or south wall of the building. In reality, it is applied to both, but the simplifying assumption can be used to calculate the diaphragm actions. As before, a simple support is assumed at the base of the wall.

This roof reaction is distributed over the roof length of 100 ft, giving a horizontal load on the diaphragm of 201.6 lb/ft (Fig. 12.30).

Design Roof Chords

Next, we need to design the roof chords. The load factor for W is 1.6:

$$M_{u\,roof} = \frac{q_u L^2}{8} = \frac{1.6 \cdot 201.6 \text{ lb/ft} \cdot 100^2 \text{ ft}^2}{8} = 403,000 \text{ lb-ft}$$

The required chord force is this factored moment, divided by the distance between chords (80 ft), and divided by the ϕ factor for axial

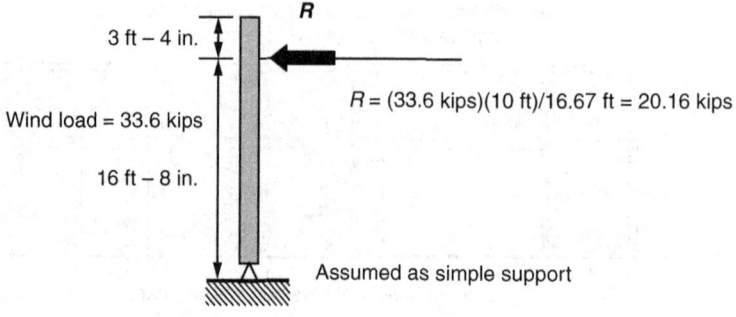

FIGURE 12.29 Wind load transmitted to roof diaphragm.

Strength Design Example: One-Story Building

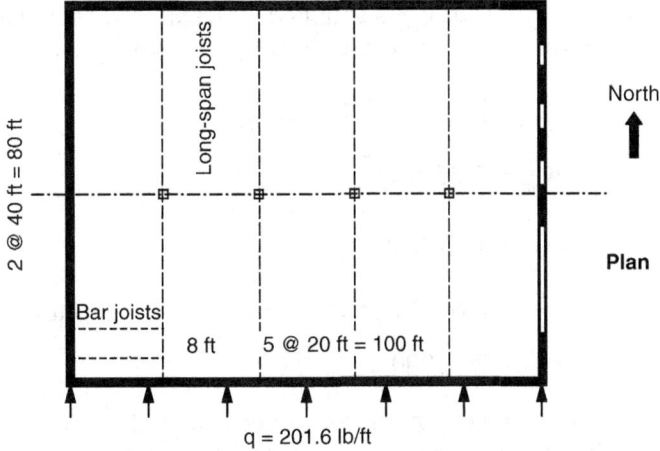

FIGURE 12.30 Plan view of one-story building showing wind loads transferred to roof diaphragm.

tension (0.9). The required steel area is this chord force, divided by the specified yield strength of the reinforcement (60,000 lb/in.²):

$$T_{u\,chord} = \frac{M_u}{\phi H} = \frac{403{,}000 \text{ lb-ft}}{0.90 \cdot 80 \text{ ft}} = 5600 \text{ lb}$$

$$A_s^{required} = \frac{T_{u\,chord}}{f_y} = \frac{5600 \text{ lb}}{60{,}000 \text{ lb/in.}^2} = 0.09 \text{ in.}^2$$

We have already specified two #4 bars around the perimeter of the roof, so that will be fine.

Check Shear Capacity of Roof Diaphragm

Next, we need to check the shear capacity of the roof diaphragm. The load factor for W is 1.6:

$$V_{u\,roof} = \frac{q_u L}{2} = \frac{1.6 \cdot 201.6 \text{ lb/ft} \cdot 100 \text{ ft}}{2} = 16{,}128 \text{ lb}$$

Because this is a reinforced element, shearing capacity is calculated using Sec. 3.3.4.1.2.1 of the 2008 MSJC *Code*:

$$V_{nm} = \left[4.0 - 1.75\left(\frac{M_u}{V_u d_v}\right)\right] A_n \sqrt{f'_m} + 0.25 P_u$$

As $(M_u/V_u d_v)$ increases, V_{nm} decreases. Because $(M_u/V_u d_v)$ need not be taken greater than 1.0 (2008 MSJC *Code* Sec. 3.3.4.1.2.1), the most conservative (lowest) value of V_{nm} is obtained with $(M_u/V_u d_v)$ equal to 1.0. Also, axial load, P_u, is zero. In our case, the shearing capacity is given by

$$V_{nm} = [4.0 - 1.75(1.0)] A_n \sqrt{f'_m}$$

$$V_{nm} = 2.25 A_n \sqrt{f'_m}$$

$$V_u = 16{,}128 \text{ lb} \le \phi V_n$$

$$= 0.8 \times 2.25 \times 3.00 \text{ in.} \times 80 \text{ ft} \times 12 \text{ in./ft} \times \sqrt{1500} \text{ lb/in.}^2$$

$$= 201{,}000 \text{ lb}$$

For convenience, in the above calculation the specified strength of the lightweight concrete topping has been taken as the same 1500 psi used for masonry, and any reduction in shear capacity due to lightweight aggregate has been neglected. These simplifications are believed justified in view of the large excess shear capacity of the roof diaphragm.

Also, according to Eq. (3-20)

$$V_n \le 4\sqrt{f'_m} A_n$$

This does not govern, and the shear design is acceptable.

12.7.2 Compute Design Shears on Walls

The total factored design shear applied to the west and east walls due to north-south load is the factored load on the roof diaphragm. This is 1.6 × 20.16 kips, or 32.26 kips.

Neglect plan torsion. Assume that shear is distributed to walls and wall segments in proportion to the segment lengths on each side. This is consistent with a rigid roof diaphragm, because of the topping on the roof.

On the west side, the total length is 80 ft. On the east side, the total wall segment length is (12 + 8 + 6 + 8 + 12) ft, or 46 ft.

Calculate the shear in the west and east walls:

$$V_{u\,west} = 32.26 \text{ kips} \left(\frac{80}{80 + 46}\right) = 20.48 \text{ kips}$$

$$V_{u\,east} = 32.26 \text{ kips} \left(\frac{46}{80 + 46}\right) = 11.78 \text{ kips}$$

Distribute the shear to the wall segments of the east wall in proportion to their plan length. The results are shown in Table 12.11.

Wall segment	Plan length, ft	Design shear, kips
A	12	3.07
B	8	2.05
C	6	1.54
D	8	2.05
E	12	3.07
Total		11.78

TABLE 12.11 Design Shear in Each Segment of East Wall due to Design Wind Load

12.8 Step 5: Design Wall Segments

12.8.1 Design of West Wall In-Plane

The capacity of the west will obviously be governed by shear. This will be no problem. The factored design shear in the west wall, 20.48 kips, is far less than the shear capacity, reduced by the strength-reduction factor for shear:

Conservatively neglect the beneficial effects of axial load, and conservatively take $M_u/V_u d_v = 1$.

$$V_{nm} = \left[4.0 - 1.75\left(\frac{M_u}{V_u d_v}\right)\right] A_n \sqrt{f'_m} + 0.25 P_u$$

$$V_{nm} = [4.0 - 1.75(1.0)]\, 7.63 \text{ in.} \times 80 \text{ ft} \times 12 \text{ in./ft}\, (\sqrt{1500} \text{ lb/in.}^2) + 0$$

$$V_{nm} = 638.3 \text{ kips}$$

The corresponding design shear capacity is

$$\phi V_n = 0.80 \times 638.3 \text{ kips} = 510.6 \text{ kips}$$

The design shear capacity far exceeds the factored design shear of 20.48 kips, and the west wall is satisfactory for shear.

12.8.2 Design of East Wall In-Plane

Because shear has been distributed in proportion to plan length, the nominal in-plane shear stress in each wall segment is equal. Check any wall segment, for example, Wall Segment A. The factored design shear in the wall segment is 3.07 kips.

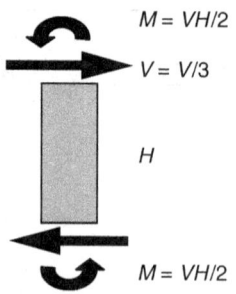

FIGURE 12.31 Assumed variation of shear and moment in each segment of east wall.

Assuming a point of inflection at mid-height (Fig. 12.31), the corresponding moment is $(VH/2)$, or $(3.07 \text{ kips} \times 12 \text{ ft}/2) = 18.42$ kip-ft = 221 kip-in.

12.8.3 Capacity of Wall Segment A as Governed by Flexure

Conservatively neglect the effects of axial load, and assume an internal lever arm of 90 percent of the total depth of the wall segment. Compute the flexural capacity of the wall segment with three #7 bars. Neglect the contribution of the compressive reinforcement, and the middle layer of reinforcement:

$$M_n \approx A_s f_y \cdot 0.9 t$$

$$M_n \approx 0.60 \text{ in.}^2 \times 60 \text{ kip/in.}^2 \times 0.9 \times 144 \text{ in.}$$

$$M_n \approx 4666 \text{ kip-in.}$$

Using a ϕ factor of 0.90 for flexure (*Code* Sec. 3.1.4.1), the design flexural capacity of 4666 kip-in. far exceeds the factored design moment of 221 kip-in.

12.8.4 Capacity of Wall Segment A as Governed by Shear

Now check the capacity of a typical wall segment as governed by shear. All segments have the same nominal shear stress. Check Segment A, with a plan length of 12 ft.

For a single wall segment,

$$V_n = V_{nm} = \left[4.0 - 1.75 \left(\frac{M_u}{V_u d_v} \right) \right] A_n \sqrt{f'_m} + 0.25 P_u$$

Conservatively neglect the effects of axial load. Then

$$V_n = \left[4.0 - 1.75\left(\frac{V_u L/2}{V_u d_v}\right)\right] \cdot (12\text{ft} \times 12 \text{ in.}/\text{ft} \times 7.63 \text{ in.}^2)\sqrt{1500 \text{ lb}/\text{in.}^2}$$

$$V_n = \left[4.0 - 1.75\left(\frac{12 \cdot 12 \text{ in.}}{2 \cdot 144 \cdot 0.9 \text{ in.}}\right)\right] \cdot 1099 \text{ in.}^2 \cdot 38.73 \text{ lb}/\text{in.}^2$$

$$V_n = \left[4.0 - 1.75(0.555)\right] \cdot 1099 \text{ in.}^2 \cdot 38.73 \text{ lb}/\text{in.}^2$$

$$V_n = 3.03 \cdot 1099 \text{ in.}^2 \cdot 38.73 \text{ lb}/\text{in.}^2 = 128{,}875 \text{ lb}$$

Using a ϕ factor of 0.80 for shear (MSJC *Code* Sec. 3.1.4.3), the design shear capacity far exceeds the factored design shear.

12.9 Step 6: Design and Detail Connections

12.9.1 Wall-Slab Connections for North and South Walls
Use #4 foundation dowels @ 48 in. Use #6 foundation dowels connected to longitudinal reinforcement in pilasters.

12.9.2 Wall-Slab Connections for West Wall
Use #4 foundation dowels @ 48 in.

12.9.3 Wall-Slab Connections for East Wall
Use #7 foundation dowels connected to wall segment reinforcement.

12.9.4 Connections between Walls and Roof Diaphragm
Walls will be solid grouted. Bar joists will be embedded into bond beams at roof level. Long-span joists will rest on bearing plates embedded into column (pilaster) sections. Angles at the edge of roof diaphragm will be connected to walls using 1/2-in. anchor bolts spaced at 48 in.

CHAPTER 13
Strength Design Example: Four-Story Building with Clay Masonry

13.1 Introduction

The second example extends the synthesized design principles of Chap. 12, to a multi-story, hotel-type structure. To illustrate the calculation and application of seismic loads from the 2009 IBC, seismic loading is included in this example. To emphasize the feasibility of masonry in a zone of significant seismic risk, the building will be located in Charleston, South Carolina. The principal lateral force-resisting elements of the structure are transverse shear walls.

This combined example problem is carried out using strength design. Previous comparisons are intended to facilitate estimation of the extent to which the design would change if the allowable-stress design were used.

13.2 Design Steps for Four-Story Example

1. Choose design criteria, specify materials:
    - Propose plan, elevation, materials, f'_m
    - Calculate D, L, W, E loads
    - Propose structural systems for gravity and lateral load
2. Design transverse shear walls for gravity and earthquake loads

3. Design exterior walls for gravity and wind loads
   - Earthquake loads will be carried by longitudinal walls in-plane
   - Out-of-plane wind loads will be carried by longitudinal walls out-of-plane using vertical and horizontal strips

13.3 Step 1: Choose Design Criteria, Specify Materials

The plan and elevation of the building are shown Figs. 13.1 and 13.2.

13.3.1 Architectural Constraints for Four-Story Example Building

Water-penetration resistance: A single-wythe, fully grouted clay masonry wall with through-wall units will be used. The wall will resist water penetration.

Movement joints: Expansion joints will probably not be needed. The building can move. Expansion of clay walls will not be restrained. If needed, use horizontal expansion joints every two bays.

13.3.2 Design for Fire

Design for fire is carried out in accordance with the 2009 IBC.

Use and occupancy: Group B

Use Type I or Type II construction (noncombustible material)

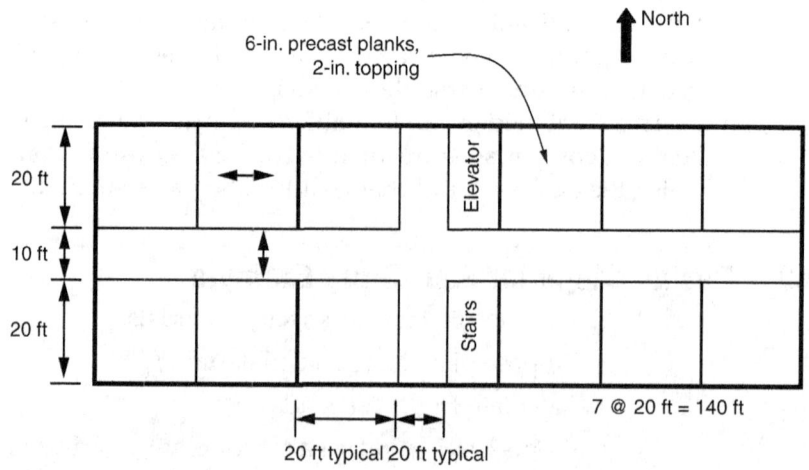

Figure 13.1 Plan view of typical floor of four-story example building.

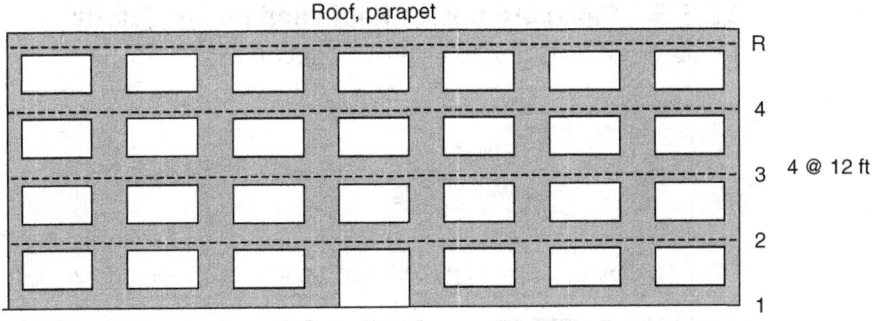

FIGURE 13.2 Plan view of typical floor facade of four-story example building.

No area or height restrictions
2- or 3-hour rating required
Must meet separation requirements of Table 602 of the 2009 IBC
Bearing walls: 4-h rating   (8-in. nominal grouted masonry OK)
Shafts: 2-h rating   (8-in. nominal grouted masonry OK)
Floors: 2-h rating   (planks and topping OK)

13.3.3 Specify Materials

8-in. through-wall clay units (ASTM C652), fully grouted.

Type S PCL mortar, specified by proportion (ASTM C270).

$f_m' = 2500$ lb/in.^{2}, use clay units with a net-area compressive strength of 6600 lb/in.^{2}, and Type S PCL mortar.

Deformed reinforcement meeting ASTM A615, Gr. 60.

Cover floors and roof of hollow-core planks with 2-in. topping, reinforced with welded-wire reinforcement.

13.3.4 Structural Systems

Gravity load: Gravity load on roof and floors will be transferred to transverse walls. Gravity load on corridor will be transferred to spine walls.

Lateral load: Lateral load (earthquake will govern) will be transferred by floor and roof diaphragms to the transverse shear walls, which will act as statically determinate cantilevers.

13.3.5 Calculate Design Roof Load due to Gravity

Design roof load due to gravity is calculated below.

Dead load	Planks	60 lb/ft^2
	Topping	25 lb/ft^2
	HVAC, roofing	30 lb/ft^2
		115 lb/ft^2 total
Live load	Ignore reduction of live load based on tributary area	20 lb/ft^2

13.3.6 Calculate Design Floor Load due to Gravity

Design floor load due to gravity is calculated below.

Dead load	Planks	60 lb/ft^2
	Topping	25 lb/ft^2
	HVAC, floor finish, partitions	30 lb/ft^2
		115 lb/ft^2 total
Live load	Use weighted average of corridor and guest rooms. Ignore reduction of live load based on tributary area	60 lb/ft^2

13.3.7 Calculate Design Lateral Load from Earthquake

Design earthquake loads are calculated according to Sec. 1613 of the 2009 IBC. That section essentially references ASCE 7-05 *(Supplement)*. Seismic design criteria are given in Chap. 11. The seismic design provisions of ASCE 7-05 *(Supplement)* begin in Chap. 12 of ASCE 7-05, which prescribes basic requirements (including the requirement for continuous load paths) (Sec. 12.1); selection of structural systems (Sec. 12.2); diaphragm characteristics and other possible irregularities (Sec. 12.3); seismic load effects and combinations (Sec. 12.4); direction of loading (Sec. 12.5); analysis procedures (Sec. 12.6); modeling procedures (Sec. 12.7); and specific design approaches. Four procedures are prescribed: an equivalent lateral force procedure (Sec. 12.8); a modal response-spectrum analysis (Sec. 12.9); a simplified alternative procedure (Sec. 12.14); and a seismic response history procedure (Chap. 16 of ASCE 7-05). The equivalent lateral force procedure is described here, because it is relatively simple, and is permitted in most situations. The simplified alternative procedure is permitted only in a few situations. The other procedures are permitted in all situations, and are required only in a few situations.

Now discuss each step in more detail, following the example of Sec. 3.5 for a building in Charleston, South Carolina. Section references refer to ASCE 7-05.

Step 1: Determine S_s, the mapped MCE (maximum considered earthquake), 5 percent damped, spectral response acceleration parameter at short periods as defined in Sec. 11.4.1 of ASCE 7-05.

Step 2: Determine S_1, the mapped MCE, 5 percent damped, spectral response acceleration parameter at a period of 1 s as defined in Sec. 11.4.1.

Determine the parameters S_S and S_1 from the 0.2-s and 1-s spectral response maps shown in Figs. 22-1 through 22-7 of ASCE 7-05.

For Charleston, South Carolina, $S_S = 2.00\ g$ and $S_1 = 0.50\ g$.

Step 3: Determine the *Site Class* (A through F, a measure of soil response characteristics and soil stability) in accordance with Sec. 20.3 and Table 20.3-1 of ASCE 7-05.

Assume Site Class D (stiff soil).

Step 4: Determine the MCE spectral response acceleration for short periods (S_{MS}) and at 1 s (S_{M1}), adjusted for Site Class effects, using Eqs. (11.4-1) and (11.4-2) of ASCE 7-05 respectively.

The acceleration-dependent site coefficient, F_a, is 1.0 (Table 11.4-1). The velocity-dependent site coefficient, F_v, is 1.5 (Table 11.4-2 of ASCE 7-05).

Then the maximum considered short-period response acceleration is

$$S_{MS} = F_a \cdot S_s = 1.0 \cdot 2.00g = 2.00g$$

and the maximum considered 1-s response acceleration is

$$S_{M1} = F_v \cdot S_1 = 1.5 \cdot 0.50g = 0.75g$$

Step 5: Determine the design response acceleration parameter for short periods, S_{DS}, and for a 1-s period, S_{D1}, using Eqs. (11.4-3) and (11.4-4) of ASCE 7-05 respectively.

The design response acceleration is two-thirds of the maximum considered acceleration. Continuing with our example for Charleston, South Carolina, the design response acceleration for short periods is

$$S_{DS} = \frac{2}{3} \cdot S_{MS} = \frac{2}{3} \cdot 2.00g = 1.33g$$

and the design response acceleration for a 1-s period is

$$S_{D1} = \frac{2}{3} \cdot S_{M1} = \frac{2}{3} \cdot 0.75g = 0.50g$$

Step 6: If required, determine the design response spectrum curve as prescribed by Sec. 11.4.5, and shown in Fig. 13.3.

Because the equivalent lateral force procedure is being used, the response spectrum curve is not required. Nevertheless, for pedagogical completeness, it was developed in Sec. 3.5.

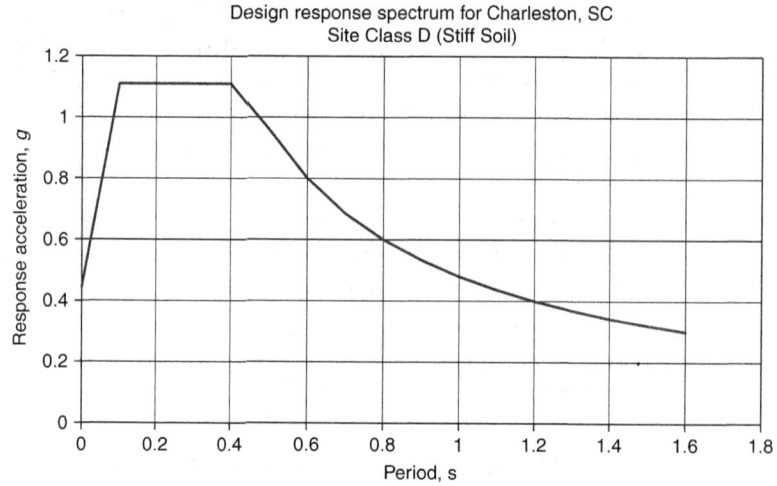

FIGURE 13.3 Design response spectrum for Charleston, South Carolina.

Step 7: Determine the structure's importance factor, I, and Occupancy Category using Sec. 11.5 of ASCE 7-05 and Table 13.1.

Assume that the structure is assigned an Occupancy Category II. This corresponds to an Importance Factor of 1.0.

Step 8: Determine the structure's Seismic Design Category using Sec. 11.6 and Table 13.2.

Occupancy category	I
I or II	1.0
III	1.25
IV	1.5

Source: Table 11.5-1 of ASCE 7-05.

TABLE 13.1 Importance Factors

	Occupancy category		
Value of S_{D1}	I or II	III	IV
$S_{D1} < 0.067$	A	A	A
$0.067 \leq S_{D1} < 0.133$	B	B	C
$0.133 \leq S_{D1} < 0.20$	C	C	D
$0.20 \leq S_{D1}$	D	D	D

Source: Table 11.6-1 of ASCE 7-05.

TABLE 13.2 Seismic Design Category Based on 1-s Period Response Acceleration Parameter

Because S_{D1} exceeds 0.20, the structure is assigned to Seismic Design Category D.

Step 9: Calculate the structure's seismic base shear using Secs. 12.8.1 and 12.8.2 of ASCE 7-05.

In accordance with ASCE 7-05 *(Supplement)*, Sec. 12.8.1.1,

$$C_s = \frac{S_{DS}}{\left(\frac{R}{I}\right)} \qquad (12.8\text{-}2)$$

In our case,

$S_{DS} = 1.33\ g$
$R = 5$ (meet detailing provisions for special reinforced masonry shear wall)
$I = 1.00$ (ASCE 7-05, Table 11.5-1)

$$C_s = \frac{S_{DS}}{\left(\frac{R}{I}\right)} = \frac{1.33}{\left(\frac{5}{1}\right)} = 0.267$$

The value of C_s computed in accordance with Eq. (12.8-2) need not exceed the following:

$$C_s = \frac{S_{D1}}{T\left(\frac{R}{I}\right)} \qquad \text{for} \qquad T \leq T_L \qquad (12.8\text{-}3)$$

The corresponding equation for $T > T_L$ does not apply. In addition, C_s shall not be less than 0.01. In our case, the value of C_s given by Eq. (12.8-3) is

$$C_s = \frac{S_{D1}}{T\left(\frac{R}{I}\right)} = \frac{0.50}{T\left(\frac{R}{I}\right)} = \frac{0.50}{5\,T} = \frac{0.1}{T}$$

and in our case, from Sec. 12.8.1.1 of ASCE 7-05 *(Supplement 2)*,

$$C_s = \frac{S_{D1}}{T\left(\frac{R}{I}\right)} \qquad \text{for} \qquad T \leq T_L$$

$$C_s = \frac{S_{D1} T_L}{T^2 \left(\frac{R}{I}\right)} \qquad \text{for} \qquad T > T_L$$

On the left end of the plateau in the design response spectrum, at a period $T = T_0 = 0.08$ s, $C_s = 1.25$, and Eq. (12.8-3) doesn't govern. Near the right end of the plateau, at $T = 0.40$ s, $C_s = 0.25$, and Eq. (12.8-3) might barely govern. Conservatively assume that the structure is stiff enough that Eq. (12.8-3) doesn't govern.

Because the structure is assigned to SDC D, the redundancy factor, ρ, is required to be taken as 1.3 (Sec. 12.3.4.1) unless certain conditions are met.

Finally, in accordance with ASCE 7-05, Sec. 12.4.2, the design horizontal seismic load effect E_h is

$$E_h = \rho Q_E \qquad (12.4\text{-}1)$$

Now compute the seismic base shear. In accordance with ASCE 7-05, Sec. 12.8.1, the effects of horizontal seismic forces Q_E come from V. The design seismic base shear is given by

$$V = C_s W$$
$$V = 0.267\, W$$

This is multiplied by the redundancy factor of 1.3, giving a product of 0.347. In other words, the building must be designed for 34.7 percent of its weight, applied as a lateral force.

Step 10: Distribute seismic base shear vertically using Sec. 12.8.3 of ASCE 7-05.

This force is distributed triangularly over the height of the building. The weight of a typical floor is its area, times the dead load per square foot, plus the interior transverse wall weight, plus the spine wall weight, plus the weight of the exterior walls. For simplicity, assume that the roof weighs the same as a typical floor, and ignore the parapet.

Floor weight: 115 lb/ft² × 50 × 140 ft² = 805 kips
Transverse wall weight: 7 × 20 × 12 ft² × 80 lb/ft² = 134.4 kips
Spine wall weight: 2 × 130 × 12 ft² × 80 lb/ft² = 249.6 kips
Perimeter wall weight: 2 × (140 + 50) × 12 ft² × 80 lb/ft²
= 364.8 kips

Total weight of a typical floor is 1553.8 kips.

The design base shear is calculated assuming a linear distribution of forces over the height of the structure.

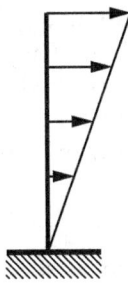

Level	W	H	WH	WH/SUM
R	1553.8	48	74,582	0.40
4	1553.8	36	55,937	0.30
3	1553.8	24	37,291	0.20
2	1553.8	12	18,646	0.10
	6215.2		186,456	

Total design base shear is 6215.2 kips × 0.347 = 2154.6 kips.

At the roof level, the factored design lateral force is the design base shear (2154.6 kips), multiplied by 0.40 (the quotient of WH/SUM) for the triangular distribution, or 861.8 kips. At the next level down, the factored design lateral force is 2154.6 kips, multiplied by 0.30, and so forth.

At each level, the factored design moment is the summation of the products of the factored design lateral forces above that level, each multiplied by its respective height above that level. The load factor for seismic loads is 1.0.

Factored design shear and moment diagrams for the four-story example building are shown in Table 13.3 and Fig. 13.4.

Level	F_u, k	H, ft	V_u, k	M_u, k-ft
R	861.8	48	861.8	0
4	646.4	36	1508.2	10,342
3	430.9	24	1939.1	28,440
2	215.5	12	2154.6	51,709
				77,564

TABLE 13.3 Factored Design Lateral Forces for Four-Story Example Building

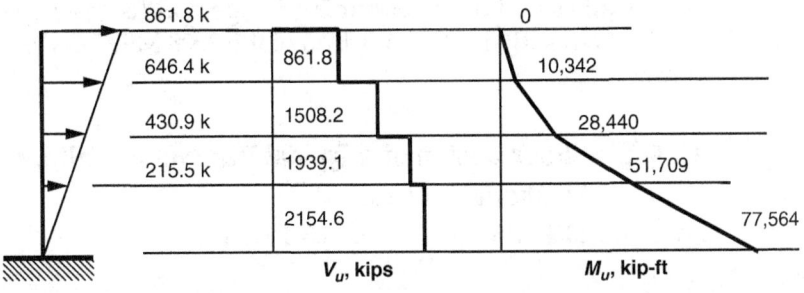

FIGURE 13.4 Factored design shears and moments for four-story example building.

Step 11: Distribute seismic base shear horizontally using Sec. 12.8.4.

These last three steps are structure-dependent. They depend on the seismic response modification coefficient assigned to the structural system, on the structure's plan structural irregularities, on the structure's vertical structural irregularities, and on the structure's redundancy.

Plan structural irregularities include

- Plan eccentricities between the center of mass and the center of stiffness
- Re-entrant corners
- Out-of-plane offsets
- Nonparallel systems

These can increase seismic response.

Vertical structural irregularities include

- Stiffness irregularity
- Mass irregularity
- Vertical geometric irregularity
- In-plane discontinuity in vertical lateral-force-resisting elements
- Discontinuity in capacity—weak story

These can also increase seismic response. Structures with low redundancy have a higher probability of failure, which is compensated for by increasing design seismic forces.

13.4 Step 2: Design Transverse Shear Walls for Gravity plus Earthquake Loads

The transverse direction is critical for this building. The 16 transverse walls are conservatively assumed to be uncoupled, so that each functions as an independent cantilever. As shown in Fig. 13.5, design each transverse wall as an I beam, assuming flange widths of 4 ft. This is less than the limits specified in Sec. 1.9.4.2.3 of the 2008 MSJC *Code*, and is therefore conservative.

13.4.1 Shear Design of a Typical Transverse Wall for Earthquake Loads

From the 2008 MSJC *Code*, Sec. 3.3.4.1.2.1,

$$V_n = V_{nm} = \left[4.0 - 1.75\left(\frac{M_u}{V_u d_v}\right)\right] A_n \sqrt{f'_m} + 0.25 P_u$$

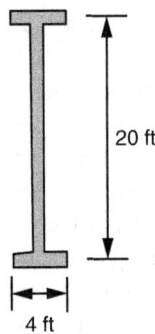

FIGURE 13.5 Effective flange width used for each transverse shear wall.

Include the effects of axial load, assuming that the typical transverse wall carries its self-weight plus the distributed floor weight on a tributary width of 20 ft:

Self-weight of wall: 19.2 kips/floor
Floor weight: 115 lb/ft² × 20 × 20 = 46 kips/floor

Total unfactored axial dead load at base is 4 × (19.2 + 46) = 260.8 kips.
Total unfactored axial live load at base is (20 + 3 × 60 psf) × 20 × 20 = 80 kips.
Then

$$V_n = V_{nm} = \left[4.0 - 1.75\left(\frac{M_u}{V_u d_v}\right)\right] A_n \sqrt{f'_m} + 0.25 P_u$$

$$V_n = \left[4.0 - 1.75\left(\frac{74,504 \cdot 12 \text{ kip-in.}}{2069.6 \cdot 20 \cdot 12 \text{ in.}}\right)\right] \cdot (20 \cdot 12 \cdot 7.50 \text{ in.}^2)\sqrt{2500 \text{ lb/in.}^2}$$

$$+ 0.25 \cdot 0.9 \cdot 260,800$$

$$V_n = [4.0 - 1.75(1.80)] \cdot 1800 \text{ in.}^2 \cdot 50.0 \text{ lb/in.}^2 + 58,680$$

But the ratio $(M_u/V_u d_v)$ need not be taken greater than 1.0. Take it equal to that value.

$$V_n = [4.0 - 1.75(1.0)] \cdot 1800 \text{ in.}^2 \cdot 50.0 \text{ lb/in.}^2 + 58,680$$

$$V_n = 2.25 \cdot 1800 \text{ in.}^2 \cdot 50.0 \text{ lb/in.}^2 + 58,680$$

$$V_n = 202,500 + 58,680 = 261,180 \text{ lb} = 261 \text{ kips}$$

The φ factor for shear is 0.80 (*Code* Sec. 3.1.4.3).

$$\phi V_n = 0.80 \cdot 261 \text{ kips} = 209 \text{ kips}$$

This considerably exceeds (1/16 walls) times the factored design base shear (1/16 × 2154.6 kips = 134.7 kips). The wall must also meet the prescriptive reinforcement requirements and the capacity design requirements corresponding to a "special reinforced shear wall."

In accordance with the 2008 MSJC *Code*, Sec. 1.17.3.2.6, the total reinforcement percentage (horizontal and vertical) shall be at least 0.002, with at least one-third of this placed in each direction.

The corresponding steel area per foot is 0.002 × 8 in. × 12 in. = 0.2 in.² per foot. If we put two-thirds of this vertically, that is equivalent to #4 bars at 18 in. If we put one-third of it horizontally, that is equivalent to #4 bars at 36 in. Meet minimum reinforcement requirements using #4 bars at 16 in. vertically, and #4 bars at 32 in. horizontally.

Capacity design requirements of Sec. 1.17.3.2.6.1 will be checked later.

13.4.2 Flexural Design of Transverse Shear Walls for Earthquake Loads

Each transverse shear wall has a plan length of 20 ft. The factored base moment per wall is (1/16) × 77,564 ft-kips, or 4848 ft-kips. The critical load case is $0.9D + 1.0E$. The factored axial load (see Sec. 13.4.1) is 0.9 × 260 kips, or 234 kips. Using a spreadsheet, the interaction diagram for the wall (with #5 bars spaced at 16 in. vertically in the web and flanges) is shown in Fig. 13.6. The spreadsheet is identical to that used previously for shear walls with rectangular cross-section. It is valid only for low

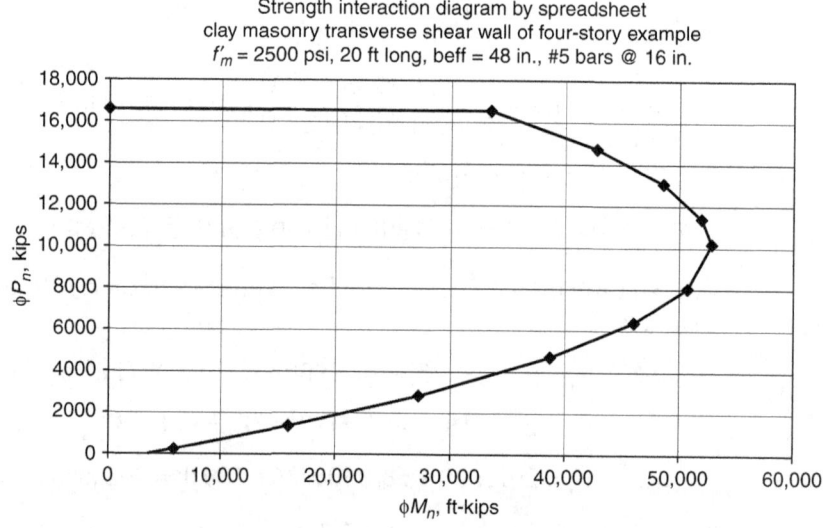

Figure 13.6 Strength moment-axial force interaction diagram for transverse masonry shear wall.

axial loads (i.e., axial loads that are low enough to keep the neutral axis within the compression flange).

Selected cells from the spreadsheet are reproduced in Table 13.4. Examination of the values in the spreadsheet shows that at a factored axial load of 234 kips, the design moment capacity of the wall is 5871 kip-ft, is greater than the required capacity of 4848 kip-ft. The position of the neutral axis is 7.50 in. from the extreme compression fiber, just within the flange, so the interaction diagram is still valid.

Flexural reinforcement consisting of #5 bars @ 16 in. is required. Each 4-ft flange has 5 bars, and the web has an additional 12 bars. The total area of reinforcement is 0.31 in.² (5 + 5 + 12) = 6.82 in.².

In the web, this is equivalent to a steel percentage of

$$\rho_{vertical} = \frac{A_s}{bt} = \frac{0.31 \text{ in.}^2}{7.50 \times 16 \text{ in.}^2} = 0.00258$$

This exceeds the required minimum of 0.0007 each way, and is satisfactory so far.

Now check ρ_{max}, continuing to consider the wall as a "special" reinforced masonry shear wall (R = 5, α = 4).

$$\rho_{max} = \frac{0.64 f'_m \left(\frac{\varepsilon_{mu}}{\alpha \varepsilon_y + \varepsilon_{mu}} \right) - \frac{N_u}{bd\phi}}{f_y \left(\frac{\alpha \varepsilon_y - \varepsilon_{mu}}{\alpha \varepsilon_y + \varepsilon_{mu}} \right)}$$

$$\rho_{max} = \frac{0.64 \times 2500 \text{ psi} \left(\frac{0.0035}{4 \times 0.00207 + 0.0035} \right) - \frac{320,800 \text{ lb}}{7.50 \text{ in.} \times 237 \text{ in.} \times 0.9}}{60000 \text{ psi} \left(\frac{4 \times 0.00207 - 0.0035}{4 \times 0.00207 + 0.0035} \right)}$$

$$\rho_{max} = \frac{1600 \text{ psi}(0.2971) - 200.53}{60000 \text{ psi}(0.4058)} = \frac{475.36 - 200.53}{24348}$$

$$\rho_{max} = 0.0113$$

Check the maximum permitted area of flexural reinforcement. Because of the flange, the wall is equivalent to rectangular wall 48 in. wide.

$$A_{smax} = \rho_{max} b \times 48 \text{ in.} = 0.0113 (48 \text{ in.}) \, 237 \text{ in.} = 128.4 \text{ in.}^2$$

We have much less than this (6.82 in.²), and the design is satisfactory.

Spreadsheet for calculating strength moment-axial force interaction diagram for transverse shear wall of four-story building

Depth	240
emu	0.0035
f'_m	2.5
F_y	60
E_s	29000
d	237
(c/d) balanced	0.628483
Width	48
Phi	0.9

Steel layers are counted from the extreme compression fiber to the extreme tension fiber

Distances are measured from the extreme compression fiber

Reinforcement consists of #5 bars at 16-in. intervals, assumed lumped at 32 in. for this spreadsheet

Compression in masonry and reinforcement is taken as positive

Stress in compressive reinforcement is set to zero, because the reinforcement is not laterally supported

Row of reinforcement	Distance	Area
1	3.00	1.55
2	35.00	0.62
3	67.00	0.62
4	99.00	0.62
5	141.00	0.62
6	173.00	0.62
7	205.00	0.62
8	237.00	1.55

	c/d	c	C_{mas}	$f_s(1)$	$f_s(2)$	$f_s(3)$	$f_s(4)$	$f_s(5)$	$f_s(6)$	$f_s(7)$	$f_s(8)$	Moment	Axial force
Pure axial load												0	16579
Points controlled by masonry	1.01	239.37	18384	0.00	0.00	0.00	0.00	0.00	0.00	0.00	0.00	33438	16545
	0.9	213.30	16381	0.00	0.00	0.00	0.00	0.00	0.00	0.00	−11.28	42762	14728
	0.8	189.60	14561	0.00	0.00	0.00	0.00	0.00	0.00	−8.24	−25.38	48605	13065
	0.7	165.90	12741	0.00	0.00	0.00	0.00	0.00	−4.34	−23.92	−43.50	51954	11391
	0.628483	148.95	11439	0.00	0.00	0.00	0.00	0.00	−16.39	−38.19	−60.00	52845	10181
Points controlled by steel	0.628483	148.95	11439	0.00	0.00	0.00	0.00	0.00	−16.39	−38.19	−60.00	52845	10181
	0.5	118.50	9101	0.00	0.00	0.00	0.00	−19.27	−46.68	−60.00	−60.00	50741	8037
	0.4	94.80	7281	0.00	0.00	0.00	−4.50	−49.47	−60.00	−60.00	−60.00	46065	6372
	0.3	71.10	5460	0.00	0.00	0.00	−39.83	−60.00	−60.00	−60.00	−60.00	38718	4708
	0.1875	44.44	3413	0.00	0.00	−51.54	−60.00	−60.00	−60.00	−60.00	−60.00	27240	2825
	0.1	23.70	1820	0.00	−48.39	−60.00	−60.00	−60.00	−60.00	−60.00	−60.00	15949	1360
	0.03165	7.50	576	0.00	−60.00	−60.00	−60.00	−60.00	−60.00	−60.00	−60.00	5871	234
	0.01	2.37	182	−26.98	−60.00	−60.00	−60.00	−60.00	−60.00	−60.00	−60.00	2074	−158

TABLE 13.4 Spreadsheet for Calculating Strength Moment-Axial Force Interaction Diagram for Transverse Shear Wall of Four-Story Building Example

Now check *Code* Sec. 3.1.3 (capacity design for shear). We have designed the wall for the calculated design shear, which is normally sufficient. The wall is a special reinforced masonry shear wall, however, as required in areas of high seismic risk, so that the capacity design requirements of *Code* Sec. 1.17.3.2.6.1.1 of the 2008 MSJC *Code* apply.

First try to meet the capacity design provisions of 1.17.32.6.1 of the 2008 MSJC *Code*. At an axial load of 234 kips, the nominal flexural capacity of this wall is the design capacity of 5871 ft-kips, divided by the strength reduction factor of 0.9, or 6523 ft-kips. The ratio of this nominal flexural capacity to the factored design moment is 5736 divided by 4848, or 1.35. Including the additional factor of 1.25, that gives a ratio of 1.68.

$$\phi V_n \geq 1.68 V_u$$

$$V_n \geq \frac{1.68}{\phi} V_u = \frac{1.68}{0.8} V_u = 2.10 V_u = 2.10 \times 134.7 = 283.2 \text{ kips}$$

Because $V_n = 261$ kips, the wall will require a small amount of shear reinforcement.

This can probably be met by prescriptive seismic requirements. We need a total steel percentage of 0.002 (summation of horizontal and vertical reinforcement), with at least 0.0007 horizontally and vertically. Vertical reinforcement is 0.00258, greater than the required sum, so horizontal reinforcement must meet only the minimum of 0.0007.

Use #4 bars @ 32 in. horizontally.

$$\rho_{horizontal} = \frac{A_s}{bt} = \frac{0.20 \text{ in.}^2}{7.50 \times 32 \text{ in.}^2} = 0.000833$$

From the 2008 MSJC *Code*, Sec. 3.3.4.1.2.2,

$$V_{ns} = 0.5 \left(\frac{A_v}{s}\right) f_y d_v$$

$$V_{ns} = 0.5 \left(\frac{0.20 \text{ in.}^2}{16 \text{ in.}}\right) 60 \text{ kips/in.}^2 \times 240 \text{ in.}$$

$$V_{ns} = 90 \text{ kips}$$

$$V_n = V_{nm} + V_{ns} = 261 \text{ kips} + 90 \text{ kips} = 351 \text{ kips}$$

This exceeds the required nominal shear capacity of 283.2 kips, and the design is satisfactory for shear.

Summary: Use #5 vertical bars @ 16 in.
                         Use #4 horizontal bars @ 32 in.

13.4.3 Comments on Design of Transverse Shear Walls

The most laborious part of this design is calculation of the design lateral force for earthquake loads. Once that calculation is done, design of the lateral-force-resisting system is straightforward, even for a region of high seismic risk such as Charleston.

This structural system would have continued to be feasible up to about six stories.

13.5 Step 3: Design Exterior Walls for Gravity plus Out-of-Plane Wind

This design follows the same steps as in the low-rise building example of Chap. 12. The critical panel will be at the top of the building, where the wind load is highest.

1. The panel must be designed for out-of-plane wind. Load effects in vertical jamb strips will be increased by the ratio of the plan length of openings to the total plan length.

2. Since the windows occupy at least half the plan length of the perimeter frame, it is possible that the panels will have to be designed as combinations of vertical strips spanning between floor slabs, and horizontal strips spanning between transverse walls.

3. The lintels above the windows and door must be designed for in-plane bending and for out-of-plane bending as in the low-rise building example of Chap. 12.

13.6 Overall Comments on Four-Story Building Example

1. Although it is located in a region of high seismic risk, this building needs comparatively little reinforcement, because of the large plan area of its bearing walls.

2. Considerable simplicity in design and analysis was achieved by letting transverse shear walls resist lateral loads as statically determinate cantilevers.

3. Masonry bearing wall construction is inexpensive and straightforward for this type of building.

CHAPTER 14
Structural Design of AAC Masonry

14.1 Introduction to Autoclaved Aerated Concrete (AAC)

Autoclaved aerated concrete (AAC) is a concrete-like material with very light weight, obtained by uniformly distributed, closed air bubbles (Fig. 14.1). Material specifications for this product are prescribed in ASTM C1386.

Because AAC typically has one-sixth to one-third the density of conventional concrete, and about the same ratio of compressive strength, it is useful for cladding and infills, and for bearing-wall components of low- to medium-rise structures. Because its thermal conductivity is one-sixth or less that of concrete, it is energy-efficient. Because its fire rating is slightly longer than that of conventional concrete of the same thickness, it is very fire-resistant. It is not susceptible to mold. Because of its internal porosity, it has very low sound transmission, and is acoustically very effective.

14.1.1 Historical Background of AAC

AAC was first produced commercially in Sweden, in 1923. Since that time, its production and use have spread to more than 40 countries on all continents, including North America, Central and South America, Europe, the Middle East, the Far East, and Australia. This wide experience has produced many case studies of use in different climates and under different

Figure 14.1 Close-up view of AAC.

building codes. Background material on experience with AAC in Europe, is given in RILEM (1993).

In the United States, modern uses of AAC began in 1990, for residential and commercial projects in the southeastern states. U.S. production of plain and reinforced AAC started in 1995 in the southeast, and has since spread to other parts of the country. A nationwide group of AAC manufacturers was formed in 1998 as the Autoclaved Aerated Concrete Products Association (http://www.aacpa.org/). Design provisions for AAC are provided in the *Code* and *Specification* of the Masonry Standards Joint Committee (MSJC), and in the technical manuals available on the web site of the AACPA. The AACPA includes one manufacturer in Monterrey, Mexico, and many technical materials are available in Spanish as well as English.

14.1.2 AAC Elements

AAC can be used to make unreinforced, masonry-type units, and also factory-reinforced floor panels, roof panels, wall panels, lintels, beams, and other special shapes (Fig. 14.2). These elements can be used in a variety of applications including residential, commercial, and industrial construction. Reinforced wall panels can be used as cladding systems as well as load bearing and nonload-bearing exterior and interior wall systems. Reinforced floor and roof panels can be efficiently used to provide the horizontal diaphragm system while supporting the necessary gravity loads.

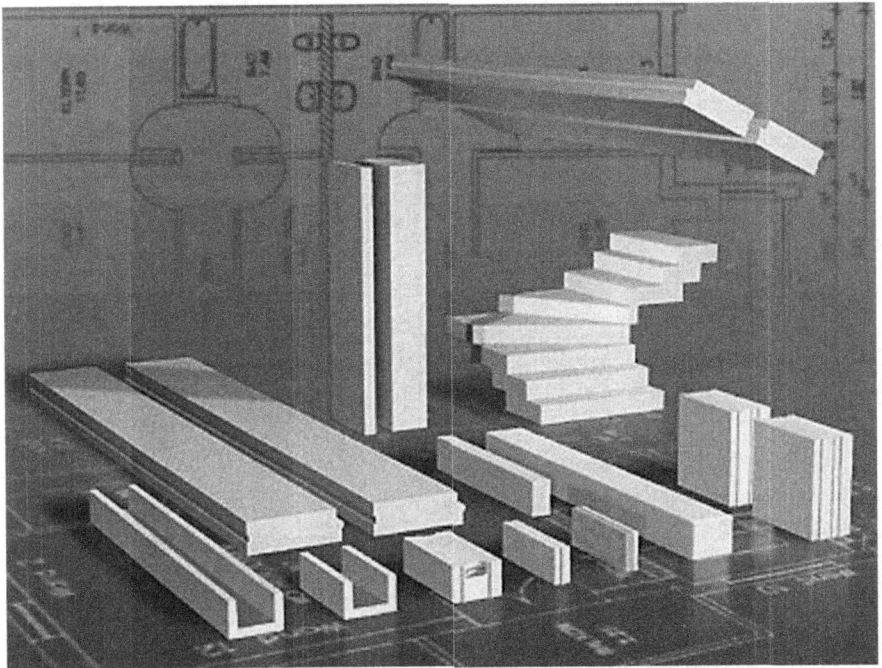

FIGURE 14.2 Examples of AAC elements. (*Courtesy of Ytong International.*)

14.1.3 Materials Used in AAC

Materials for AAC vary with manufacture and location, and are specified in ASTM C1386. They include some or all of the following: fine silica sand; Class F fly ash; hydraulic cements; calcined lime; gypsum; expansive agents such as finely ground aluminum powder or paste, and mixing water. Details of the mixture designs used by each producer depend on the available materials and the precise manufacturing process, and are not publicly available. The finely ground aluminum powder or paste produces expansion by combining with the alkaline slurry to produce hydrogen gas. AAC can be reinforced internally in the manufacturing process with welded wire cages, and also at the job site with conventional reinforcement.

14.1.4 How AAC Is Made

Overall steps in the manufacture of AAC are shown in Fig. 14.3, and described below:

Sand is ground to the required fineness in a ball mill, if necessary, and is stored along with other raw materials. The raw materials are then batched by weight and delivered to the mixer. Measured amounts of water and expansive agent are added to the mixer, and the cementitious slurry is mixed.

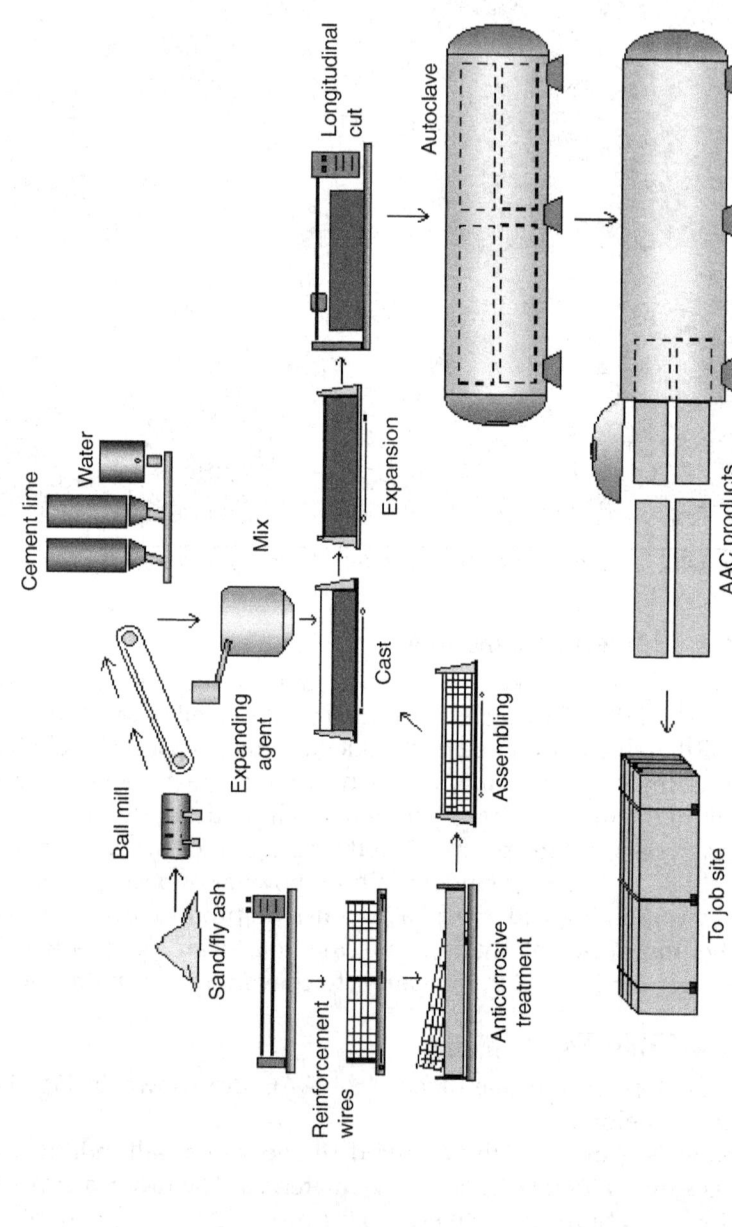

FIGURE 14.3 Overall steps in manufacture of AAC.

Steel molds are prepared to receive the fresh AAC. If reinforced AAC panels are to be produced, steel reinforcing cages are secured within the molds. After mixing, the slurry is poured into the molds. The expansive agent creates small, finely dispersed voids in the fresh mixture, which increases the volume by approximately 50 percent in the molds within 3 h.

Within a few hours after casting, the initial hydration of cementitious compounds in the AAC gives it sufficient strength to hold its shape and support its own weight.

After cutting, the aerated concrete product is transported to a large autoclave, where the curing process is completed. Autoclaving is required to achieve the desired structural properties and dimensional stability. The process takes about 8 to 12 h under a pressure of about 174 psi (12 bars) and a temperature of about 360°F (180°C) depending on the grade of material produced. During autoclaving, the wire-cut units remain in their original positions in the AAC block. After autoclaving, they are separated for packaging.

AAC units are normally placed on pallets for shipping. Unreinforced units are typically shrink-wrapped, while reinforced elements are banded only, using corner guards to minimize potential localized damage that might be caused by the banding.

14.1.5 AAC Strength Classes

AAC is produced in different densities and corresponding compressive strengths, in accordance with ASTM C1386 (Precast Autoclaved Aerated Concrete Wall Construction Units). Densities and corresponding strengths are described in terms of "strength classes" (Table 14.1).

Strength class	Specified compressive strength lb/in² (MPa)	Nominal dry bulk density lb/ft³ (kg/m³)	Density limits lb/ft³ (kg/m³)
AAC 2	290 (2)	25 (400)	22–28 (350–450)
		31 (500)	28–34 (450–550)
AAC 4	580 (4)	31 (500)	28–34 (450–550)
		37 (600)	34–41 (550–650)
		44 (700)	41–47 (650–750)
		50 (800)	47–53 (750–850)
AAC 6	870 (6)	44 (700)	41–47 (650–750)
		50 (800)	47–53 (750–850)

*Other strength classes within these ranges and densities may be produced depending on specific design requirements.

TABLE 14.1 Typical Material Characteristics of AAC in Different Strength Classes*

AAC unit type	Width, in. (mm)	Height, in. (mm)	Length, in. (mm)
Standard block	2–15 (50–375)	8 (200)	24 (610)
Jumbo block	4–15 (100–375)	16–24 (400–610)	24–40 (610–1050)

TABLE 14.2 Dimensions of Plain AAC Wall Units

Product type	Thickness, in. (mm)	Height or width, in. (mm)	Typical length, ft (mm)
Wall panel	2–15 (50–375)	24 (610)	20 (6090)
Floor panel	4–15 (100–375)	24 (610)	20 (6090)
Lintel/beam	4–15 (100–375)	8–24 (200–610)	20 (6090)

TABLE 14.3 Dimensions of Reinforced AAC Wall Units

14.1.6 Typical Dimensions of AAC Units

Typical dimensions for plain AAC wall units (masonry-type units) are shown in Table 14.2.

Typical dimensions for reinforced AAC wall units (panels) are shown in Table 14.3.

14.2 Applications of AAC

AAC can be used in a wide variety of structural and nonstructural applications (Barnett et al. 2005), examples of which are shown in Figs. 14.4 to 14.6. Figure 14.4 shows an AAC residence in Arizona, in which the AAC is used as structure and envelope.

Figure 14.5 shows an AAC hotel in Tampico, Mexico, in which the AAC is again used as structure and envelope.

Figure 14.6 shows an AAC cladding application on a high-rise building in Monterrey, Mexico.

14.3 Structural Design of AAC Elements

14.3.1 Integrated U.S. Design Context for AAC Elements and Structures

Prior to October 2003, proposed AAC masonry buildings in the United States had to be approved on a case-by-case basis. Since that date, project approvals can be obtained under the general evaluation-service reports

Structural Design of AAC Masonry 445

FIGURE 14.4 AAC residence in Monterrey, Mexico. (*Courtesy of Xella Mexicana.*)

FIGURE 14.5 AAC hotel in Tampico, Mexico. (*Courtesy of Xella Mexicana.*)

Figure 14.6 AAC cladding, Monterrey, Mexico. (*Courtesy of Xella Mexicana.*)

ICC AC 15 (2003) and ICC ESR-1371 (2004). Since early 2005, project approvals for AAC masonry structures can be obtained through the inclusion of design provisions for AAC masonry in the mandatory-language Appendix A of the 2005 MSJC *Code* and *Specification*. It is also expected that reinforced AAC panels will be analogously addressed through ACI 318. This design context is shown schematically in Fig. 14.7, and is applied in the rest of this chapter. Because the basic behavior of structural

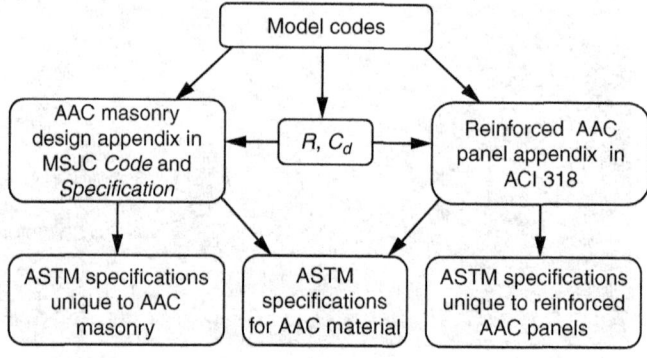
Figure 14.7 Integrated U.S. design background for AAC elements and structures.

elements of AAC masonry is the same as that of structural elements of clay or concrete masonry, previous sections on basic behavior are not repeated. Design provisions are slightly different, however, and their use is illustrated in detail.

Loads for structural design of AAC should be taken from appropriate load codes, such as ASCE 7-05. AAC masonry elements are designed using the provisions of Appendix A of the 2008 MSJC *Code* and *Specification*. Reinforced AAC panels are designed using manufacturers' recommendations.

14.3.2 ASTM Specifications for AAC Construction

ASTM traditionally deals with specifications for materials and methods of test. For the past several years, standards-development work regarding AAC has been going on in two ASTM committees:

- In 1998, ASTM Subcommittee C-27.60 (Precast Concrete Elements of AAC) developed a material standard for AAC: C1386-98 (Standard Specification for Precast Autoclaved Aerated Concrete Wall Units). Subcommittee C27-60 has also developed a standard for reinforced AAC panels: C1452-00 (Standard Specification for Reinforced Autoclaved Aerated Concrete Units). That subcommittee has also developed a standard method of test for determining the modulus of AAC.

- In 2003, ASTM Subcommittee C-15.10 (Autoclaved Aerated Concrete Masonry) developed a standard for AAC masonry: C 1555-03a (Standard Practice for Autoclaved Aerated Concrete Masonry). That standard references the AAC material provisions of ASTM C1386-98, and also contains construction provisions. It has been incorporated into the 2005 MSJC *Specification*.

14.3.3 U.S. Design and Construction Provisions for Elements and Structures of AAC Masonry

In the United States, development of masonry design provisions by an ANSI consensus process is the responsibility of the Masonry Standards Joint Committee (MSJC), sponsored by the American Concrete Institute (ACI), the American Society of Civil Engineers (ASCE), and The Masonry Society (TMS). The MSJC *Code* and *Specification* is essentially referenced directly by U.S. model codes (*International Building Code* and *NFPA Code*).

The MSJC design provisions cover a wide variety of design approaches (strength, allowable-stress, empirical) and materials (clay, concrete, glass block). Based on the combination of test results from The University of Texas at Austin, the University of Alabama at Birmingham, and elsewhere, a strength design approach was developed for AAC masonry, with provisions that are generally similar to current strength-design provisions for other types of masonry, and for reinforced concrete. The proposed design provisions, commentary, and "super-commentary" were introduced, refined by, and approved by MSJC in 2004, in the form of a mandatory-language Appendix to the 2005 MSJC *Code* and *Specification*. They produce final designs similar to those produced by guidelines recently published by the American Concrete Institute for reinforced AAC panels. Those guidelines are not discussed further here.

Design of AAC masonry elements is based on the specified compressive strength of the AAC material, f'_{AAC}. Conformance with this specified compressive strength is verified by testing of 4-in. cubes of the AAC material only. In contrast to concrete or clay masonry, prism tests are not used. The reason for this is that the compressive strength of AAC masonry elements is close to the compressive strength of the material, because the thin-bed mortar is stronger than the AAC material itself, and the volume of the thin-bed mortar joints is small compared to the total volume of the AAC masonry element. The design equations of the 2008 MSJC *Code* have been calibrated against this value for f'_{AAC}.

Flexural resistance of AAC masonry elements is computed assuming yielded flexural reinforcement and an appropriate equivalent rectangular stress block. Maximum reinforcement is limited to ensure tension-controlled behavior. Deformed reinforcement must be used, and must be surrounded by grout. Development and splice requirements are the same as for conventional masonry; only the grout is considered, and bond failure and splitting are addressed.

In-plane shear resistance of AAC masonry elements is computed as the sum of resistance from masonry plus deformed reinforcement in intermediate bond beams only. In-plane shear resistance from AAC masonry is checked with respect to web shear, crushing of the diagonal strut, and sliding shear. Out-of-plane resistance of AAC masonry elements is computed using beam shear equations similar to those used for conventional masonry. Capacity design for shear is required.

These design requirements are accompanied by corresponding construction requirements in the MSJC *Specification*, which is mandated by the MSJC *Code*. Construction requirements address quality assurance, materials, and execution.

14.3.4 Handling, Erection, and Construction with AAC Elements

AAC masonry units are laid with a polymer-modified, thin-bed mortar. AAC panels are lifted and placed using specially designed clamps, and are aligned using alignment bars.

When AAC elements are used as a load-bearing wall system, the floor and roof systems are usually designed and detailed as horizontal diaphragms to transfer lateral loads to shear walls. The tops of the panels are connected to the floor or roof diaphragms using a cast-in-place reinforced concrete ring beam.

AAC floor and roof panels can be erected on concrete, steel, or masonry construction. All bearing surfaces should be in level and minimum required bearing areas (to prevent local crushing) should be maintained. Most floor and roof panels are connected by keyed joints that are reinforced and filled with grout to lock the panels together and provide diaphragm action to resist lateral loads. A cast-in-place reinforced concrete ring beam is normally placed along the perimeter of the diaphragm, completing the system.

14.4 Design of Unreinforced Panel Walls of AAC Masonry

14.4.1 Steps in Flexural Design of Panel Walls of AAC Masonry

Nominal flexural capacity corresponds to a maximum flexural compressive stress of $0.85 f'_{AAC}$, or a maximum flexural tensile stress equal to the modulus of rupture. Because the modulus of rupture is much lower than $0.85 f'_{AAC}$, it governs. Design actions are factored, and design capacities are computed using those nominal capacities and the appropriate strength-reduction factor.

Load Factors

Load factors are as discussed earlier, in Sec. 3.6.1. As prescribed in Sec. 1605.2 of the 2009 IBC, the two loading combinations involving wind are

4. $1.2D + 1.6W + f_1 L + 0.5 (L_r$ or S or $R)$
6. $0.9D + 1.6W + 1.6H$

Of these, the second will usually govern. Both combinations have a load factor for W of 1.6.

Modulus of Rupture

According to Sec. A.1.8.3 of the 2008 MSJC *Code*, nominal flexural capacity of unreinforced AAC masonry is computed using a modulus of rupture, f_{rAAC}, equal to twice the splitting tensile strength, f_{tAAC}. According to Sec. A.1.8.2 of the 2008 MSJC *Code*, that splitting tensile strength is given as

$$f_{t\,AAC} = 2.4 \sqrt{f'_{AAC}}$$

Strength-Reduction Factors

For combinations of flexure and axial load in unreinforced masonry, $\phi = 0.60$ (Sec. A.1.5.2 of the 2008 MSJC *Code*).

14.4.2 Example of Design of a Single-Wythe Panel Wall of AAC Masonry (Solid Units)

Check the design of the panel wall shown in Fig. 14.8, for a wind load w of 20 lb/ft², using Class 4 AAC units with a nominal thickness of 8 in., laid using thin-bed mortar.

The panel wall will be designed as unreinforced AAC masonry. The design follows the steps, using a nominal thickness of 8 in. The panel could be designed as a two-way panel. Nevertheless, because of its aspect ratio, the vertical strips will carry practically all the load. Therefore, design it as a one-way panel, consisting of a series of vertically spanning, simply supported strips. AAC masonry units are solid, and are fully bedded.

The specified compressive strength, f'_{AAC}, for Class 4 AAC is 580 psi (Table 14.1). The corresponding splitting tensile strength is

$$f_{t\,AAC} = 2.4\sqrt{f'_{AAC}}$$

$$f_{t\,AAC} = 2.4\sqrt{580\;\text{lb/in.}^2}$$

$$f_{t\,AAC} = 57.8\;\text{lb/in.}^2$$

The modulus of rupture is twice this value, or 115.6 psi.

Calculate the maximum factored design bending moment and corresponding factored design flexural tensile stress in a strip, 1-ft wide,

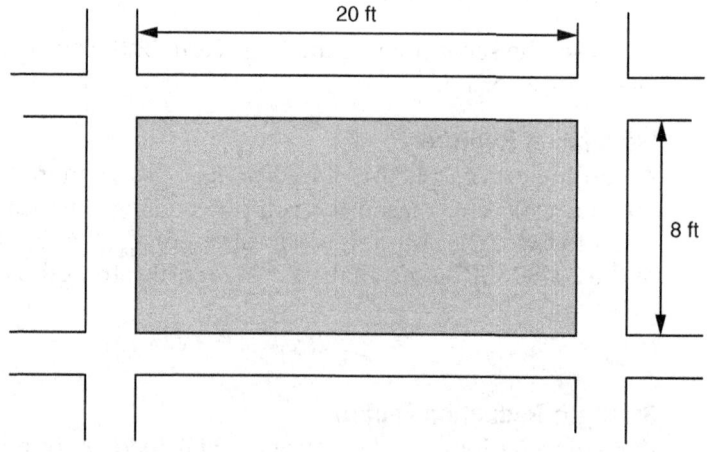

FIGURE 14.8 Example panel wall to be designed using AAC masonry.

with a nominal thickness of 8 in. The specified thickness of the wall is 7.9 in.

$$M_{u\max} = \frac{w_u \lambda^2}{8} = \frac{1.6 \cdot 20 \text{ lb/ft } (8 \text{ ft})^2}{8} \times 12 \text{ in./ft} = 3072 \text{ lb-in.}$$

$$f_t = \frac{Mc}{I} = \frac{3072 \text{ lb-in.} \cdot \left(7.9/2\right) \text{ in.}}{\left[12 \text{ in.} \cdot (7.9 \text{ in.})^3 / 12\right]} = 24.6 \text{ lb/in.}^2$$

The factored flexural tensile stress, 26.4 lb/in.², is less than the modulus of rupture (115.6 psi), reduced by a strength-reduction factor of 0.6, or 69.4 lb/in.². The design is therefore satisfactory. We should also check one-way (beam) shear. An example of this is given below.

14.4.3 Example of Check of Shear Capacity for an Unreinforced Panel Wall of AAC Masonry

Check the effect of shear in the example of Sec. 14.4.2. Although Sec. A.2.5 of the 2008 MSJC *Code* is not clear on this point, it is most logical to use the out-of-plane shear capacity from Sec. A.3.4.1.2.5. Compute the out-of-plane shear capacity on a 1-ft wide strip:

$$V_{nAAC} = 0.8 \sqrt{f'_{AAC}} \, bd$$

$$V_{nAAC} = 0.8 \sqrt{580 \text{ lb/in.}^2} \times 12 \text{ in.} \times 7.9 \text{ in.}$$

$$V_{nAAC} = 1826 \text{ lb}$$

On a 1-ft wide strip, the factored wind load of 1.6 times 20 lb/ft² produces a factored design shear of

$$V_u = \frac{q_u L}{2} = \frac{1.6 \cdot 20 \text{ lb/ft} \cdot (8 \text{ ft})}{2} = 128 \text{ lb}$$

This is far less than the nominal capacity, reduced by the strength-reduction factor for shear in AAC masonry (0.8), and one-way shear does not govern the design.

14.4.4 Overall Comments on Design of Unreinforced Panel Walls of AAC Masonry

- Nonload-bearing masonry, without calculated reinforcement, can easily resist wind loads.
- If noncalculated reinforcement is included, it will not act until the masonry has cracked.

- Elements such as the ones we have calculated in this section can be designed in many cases by prescription.
- The previous justifications (based on the strip method) for assuming that all load is carried by vertically spanning strips continue to be valid for AAC masonry panel walls.

14.5 Design of Unreinforced Bearing Walls of AAC Masonry

14.5.1 Steps in Design of Unreinforced Bearing Walls of AAC Masonry

In the 2008 MSJC *Code*, design of unreinforced bearing walls of AAC masonry is similar to the design of panel walls, except that axial load must be considered. In Sec. A.2.2 of the 2008 MSJC *Code*, no explicit equations are given for computing flexural strength. The usual assumption of plane sections is invoked, and tensile and compressive stresses in masonry are to be assumed proportional to strain.

Nominal capacities in masonry are reached at an extreme fiber tension equal to the modulus of rupture (Sec. A.1.8.3 of the 2008 MSJC *Code*), and at a compressive stress of $0.85 f'_{AAC}$. Compressive capacity is given by Eq. (A-3) and Eq. (A-4) of the 2008 MSJC *Code*.

For

$$\frac{kh}{r} = \frac{h}{r} \leq 99$$

$$P_n = 0.80 \left\{ 0.85 A_n f'_{AAC} \left[1 - \left(\frac{h}{140\, r} \right)^2 \right] \right\}$$

and for

$$\frac{kh}{r} = \frac{h}{r} > 99$$

$$P_n = 0.80 \left[0.85 A_n f'_{AAC} \left(\frac{70\, r}{h} \right)^2 \right]$$

The strength reduction factor, ϕ, is equal to 0.60 (Sec. A.1.5.2 of the 2008 MSJC *Code*).

Unlike the strength design of unreinforced bearing walls of concrete or clay masonry, second-order effects are not directly addressed.

14.5.2 Example of Design of Unreinforced AAC Masonry Bearing Wall with Concentric Axial Load

The bearing wall shown in Fig. 14.9 has an unfactored, concentric axial load of 1050 lb/ft. Using AAC masonry, design the wall.

According to the 2009 IBC, and in the context of these example problems (dead load, wind load, and roof live load), the following loading combinations must be checked for strength design:

4. $1.2D + 1.6W + f_1 L + 0.5 (L_r$ or S or $R)$
6. $0.9D + 1.6W + 1.6H$

The second of these is usually critical, because roof live load must be considered off as well as on.

To apply those loading combinations, let us assume that the total unfactored wall load of 1050 lb/ft represents 700 lb/ft of dead load and 350 lb/ft of live load.

At each horizontal plane through the wall, the following conditions must be met:

- Maximum compressive stress from factored axial loads must not exceed the slenderness-dependent values in Eqs. (A-3) or (A-4) as appropriate, reduced by a ϕ factor of 0.60.
- Maximum compressive stress from factored loads (including a moment magnifiers) must not exceed $0.85 f'_{AAC}$ in the extreme compression fiber, reduced by a ϕ factor of 0.60.
- Maximum tension stress from factored loads must not exceed the modulus of rupture in the extreme tension fiber, reduced by the ϕ factor of 0.60.

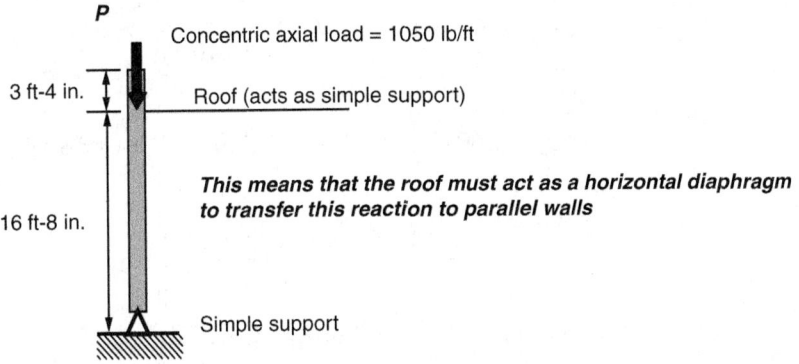

FIGURE 14.9 Unreinforced AAC masonry bearing wall with concentric axial load.

For each condition, the more critical of the two possible loading combinations must be checked. Because there is no wind load, this example will be worked using the loading combination $1.2D + 1.6L$.

In theory, we must check various points on the wall. In this problem, however, the wall has only axial load, which increases from top to bottom due to the wall's self-weight. Therefore we need to check only at the base of the wall.

Try 8-in. nominal units and Class 4 AAC, with a specified compressive strength, f'_{AAC}, of 580 lb/in.^{2} and a unit weight of 40 lb/ft^3. Using the specified thickness of 7.9 in., that corresponds to a unit weight of 26.3 lb/ft^2. Work with a strip with a width of 1 ft (measured along the length of the wall in plan). Stresses are calculated using the critical section, consisting of the entire cross-sectional area (2008 MSJC *Code*, Sec. 1.9.1.1).

At the base of the wall, the factored axial force is

$$P_u = 1.2(700 \text{ lb}) + 1.6(350 \text{ lb}) + 1.2\,(20 \text{ ft} \times 26.3 \text{ lb/ft}) = 2031 \text{ lb}$$

To calculate stiffness-related parameters for the wall, we use the average cross section, corresponding to the fully bedded gross cross section (2008 MSJC *Code*, Sec. 1.9.3).

$$r = \sqrt{\frac{I}{A}} = \sqrt{\frac{bh^3/12}{bh}} = \frac{h}{\sqrt{12}} = \frac{7.9 \text{ in.}}{\sqrt{12}} = 2.28 \text{ in.}$$

$$\frac{kh}{r} = \frac{16.67 \text{ ft} \times 12 \text{ in./ft}}{2.28 \text{ in.}} = 87.8$$

This is less than the transition slenderness of 99, so the nominal axial capacity is based on the curve that is an approximation to inelastic buckling:

$$\phi P_n = \phi\, 0.80 \left\{ 0.85\, A_n f'_m \left[1 - \left(\frac{h}{140\,r} \right)^2 \right] \right\}$$

$$\phi P_n = 0.60 \cdot 0.80 \cdot \left\{ 0.85 \times 7.9 \text{ in.} \times 12 \text{ in.} \times 580 \text{ lb/in.}^2 \right.$$

$$\left. \times \left[1 - \left(\frac{16.67 \text{ ft} \times 12 \text{ in./ft}}{140 \times 2.28 \text{ in.}^2} \right)^2 \right] \right\}$$

$$\phi P_n = 22{,}433 \text{ lb} \times 0.607 = 13{,}627 \text{ lb}$$

The factored axial load, P_u, 2031 lb, is far less than this, and this part of the design is satisfactory.

Now check the net compressive stress. Because the load is concentric, there is no bending stress. At the base of the wall,

$$f_a = \frac{P_u}{A} = \frac{1.2(700 \text{ lb}) + 1.6(350 \text{ lb}) + 1.2(20 \text{ ft} \times 26.3 \text{ lb/ft})}{7.9 \times 12 \text{ in.}^2}$$

$$= \frac{2031 \text{ lb}}{30 \text{ in.}^2} = 21.4 \text{ lb/in.}^2$$

$$f_a = 21.4 \text{ lb/in.}^2$$

The maximum permitted compressive stress is

$$0.60 \cdot 0.85 f'_{AAC} = 0.60 \times 0.85 \times 580 \text{ lb/in.}^2 = 296 \text{ lb/in.}^2$$

The maximum compressive stress is much less than this, and the design is satisfactory for this also.

Clearly, because this example involves concentric axial loads only, the first criterion (axial capacity reduced by slenderness effects) is more severe than the second (maximum compressive stress from axial loads and bending moments). Because there is no moment, there is no tensile stress, and the third criterion is automatically satisfied. The design is satisfactory.

It would probably be possible to achieve a satisfactory design with a smaller nominal wall thickness. To maintain continuity in the example problems that follow, however, the design will stop at this point.

Although the 2008 MSJC *Code* has no explicit minimum eccentricity requirements for walls, the leading coefficient of 0.80 for nominal axial compressive capacity effectively imposes a minimum eccentricity of about $0.1t$.

14.5.3 Example of Design of Unreinforced AAC Masonry Bearing Wall with Eccentric Axial Load

Now consider the same bearing wall of the previous example, but make the gravity load eccentric. As before, suppose that the load is applied over a 4-in. bearing plate, and assume that bearing stresses vary linearly under the bearing plate as shown in Fig. 14.10.

Then the eccentricity of the applied load with respect to the centerline of the wall is

$$e = \frac{t}{2} - \frac{\text{Plate}}{3} = \frac{7.9 \text{ in.}}{2} - \frac{4 \text{ in.}}{3} = 2.62 \text{ in.}$$

The wall is as shown in Fig. 14.11.

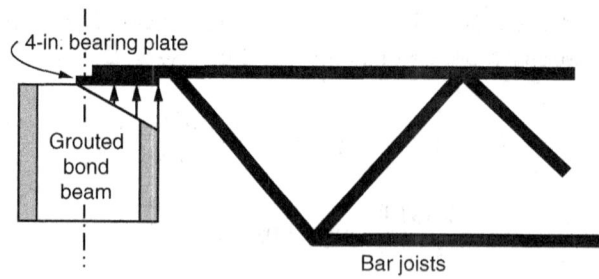

FIGURE 14.10 Assumed linear variation of bearing stresses under bearing plate of AAC masonry wall.

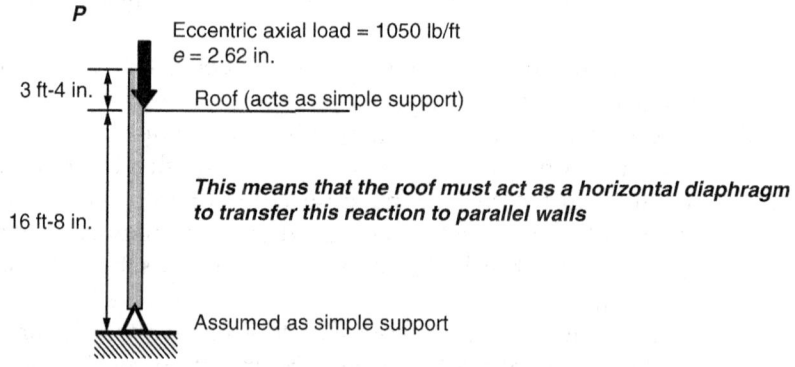

FIGURE 14.11 Unreinforced AAC masonry bearing wall with eccentric axial load.

At each horizontal plane through the wall, the following conditions must be met:

- Maximum compressive stress from factored axial loads must not exceed the slenderness-dependent values in Eqs. (A-3) or (A-4) as appropriate, reduced by a ϕ-factor of 0.60.
- Maximum compressive stress from factored loads (including a moment magnifiers) must not exceed $0.85 f'_{AAC}$ in the extreme compression fiber, reduced by a ϕ-factor of 0.60.
- Maximum tension stress from factored loads must not exceed the modulus of rupture in the extreme tension fiber, reduced by the ϕ-factor of 0.60.

For each condition, the more critical of the two possible loading combinations must be checked. Because there is no wind load, this example will be worked using the loading combination $1.2D + 1.6L$.

We must check various points on the wall. Critical points are just below the roof reaction (moment is high and axial load is low, so maximum tension may govern) and at the base of the wall (axial load is high, so maximum compression may govern). Check each of these locations.

As before, try 8-in. nominal units, and a specified compressive strength, f'_{AAC}, of 580 lb/in.^{2}. Work with a strip with a width of 1 ft (measured along the length of the wall in plan). Stresses are calculated using the gross section, because the units are solid and are fully bedded using thin-bed mortar (2008 MSJC *Code*, Sec. 1.9.1.1).

Just below the roof reaction, the axial force is

$$P_u = 1.2(700 \text{ lb}) + 1.6(350 \text{ lb}) + 1.2(3.33 \text{ ft} \times 26.3 \text{ lb/ft}) = 1505 \text{ lb}$$

To calculate stiffness-related parameters for the wall, we use the average cross section, corresponding to the fully bedded gross cross section (2008 MSJC *Code*, Sec. 1.9.3).

$$r = \sqrt{\frac{I}{A}} = \sqrt{\frac{bh^3/12}{bh}} = \frac{h}{\sqrt{12}} = \frac{7.9 \text{ in.}}{\sqrt{12}} = 2.28 \text{ in.}$$

$$\frac{kh}{r} = \frac{16.67 \text{ ft} \times 12 \text{ in./ft}}{2.28 \text{ in.}} = 87.8$$

This is less than the transition slenderness of 99, so the nominal axial capacity is based on the curve that is an approximation to inelastic buckling:

$$\phi P_n = \phi \, 0.80 \left\{ 0.85 \, A_n f'_m \left[1 - \left(\frac{h}{140 \, r} \right)^2 \right] \right\}$$

$$\phi P_n = 0.60 \cdot 0.80 \cdot \left\{ 0.85 \times 7.9 \text{ in.} \times 12 \text{ in.} \times 580 \text{ lb/in.}^2 \right.$$

$$\left. \times \left[1 - \left(\frac{16.67 \text{ ft} \times 12 \text{ in./ft}}{140 \times 2.28 \text{ in.}} \right)^2 \right] \right\}$$

$$\phi P_n = 22,433 \text{ lb} \times 0.607 = 13,627 \text{ lb}$$

Because the factored axial force is much less than slenderness-dependent nominal capacity, reduced by the appropriate ϕ factor, the axial force check is satisfied.

Now check the net compressive stress. Because the loading is eccentric, there is bending stress:

$$f_{compression} = \frac{P_u}{A} + \frac{M_u c}{I}$$

The factored design axial load, P_u, is computed in the preceding equations. The factored design moment, $M_{u'}$ is given by:

$$M_u = P_u e = (1.2 \times 700 + 1.6 \times 350) \text{ lb} \times 2.62 \text{ in.} = 3668 \text{ lb-in.}$$

$$f_{compression} = \frac{P_u}{A} + \frac{M_u c}{I}$$

$$f_{compression} = \frac{1505 \text{ lb}}{7.9 \times 12 \text{ in.}^2} + \frac{3668 \text{ lb-in.} \left(7.9/2\right)}{12 \times 7.9^3/12 \text{ in.}^4} = 15.9 + 29.4 \text{ lb/in.}^2$$

$$= 45.3 \text{ lb/in.}^2$$

$$0.60 \times 0.85 f'_{AAC} = 0.60 \times 0.85 \times 580 \text{ lb/in.}^2 = 296 \text{ lb/in.}^2$$

The net compressive stress does not exceed the prescribed value. Clearly, because this example involves eccentric axial loads, the first criterion (axial load reduced by slenderness effects) is less severe than the second (maximum compressive stress from axial loads and bending moments).

Now check the net tensile stress. At the mid-height of the wall, the axial force due to $0.9D$ is

$$P_u = 0.9 (700 \text{ lb}) + 0.9 (3.33 \text{ ft} + 8.33 \text{ ft}) \times 26.3 \text{ lb/ft} = 906 \text{ lb}$$

At the mid-height of the wall, the factored design moment, $M_{u'}$ is given by:

$$M_u = P_{u \text{ eccentric}} \frac{e}{2} = \left(\frac{1}{2}\right) 0.9 \cdot 700 \text{ lb} \times 2.62 \text{ in.} = 825 \text{ lb-in.}$$

$$f_{tension} = -\frac{P_u}{A} + \frac{M_u c}{I}$$

$$f_{tension} = -\frac{906 \text{ lb}}{7.9 \times 12 \text{ in.}^2} + \frac{825 \text{ lb-in.} \left(7.9/2\right)}{12 \times 7.9^3/12 \text{ in.}^4} = -9.56 + 6.61 \text{ lb/in.}^2$$

$$= -2.95 \text{ lb/in.}^2$$

The maximum tensile stress is actually negative, indicating net compression, and the design is satisfactory.

The other critical section could be at the base of the wall, where the checks of all three criteria are identical to those of the example of Sec. 14.5.2. All are satisfied, and the design is therefore satisfactory.

14.5.4 Example of Design of Unreinforced AAC Masonry Bearing Wall with Eccentric Axial Load plus Wind

Now consider the same AAC masonry bearing wall of the example of Sec. 14.5.3, but add a uniformly distributed wind load of 25 lb/ft². The wall is as shown in Fig. 14.12.

At each horizontal plane through the wall, the following conditions must be met:

- Maximum compressive stress from factored axial loads must not exceed the slenderness-dependent values in Eqs. (A-3) or (A-4) as appropriate, reduced by a ϕ factor of 0.60.
- Maximum compressive stress from factored loads (including a moment magnifiers) must not exceed $0.85 f'_{AAC}$ in the extreme compression fiber, reduced by a ϕ factor of 0.60.
- Maximum tension stress from factored loads must not exceed the modulus of rupture in the extreme tension fiber, reduced by the ϕ factor of 0.60.

For each condition, the more critical of the two possible loading combinations must be checked. Because there is wind load, and because the previous two examples showed little problem with the first two criteria, the third criterion (net tension) may well be critical. For this criterion, the critical loading condition could be either $1.2D + 1.6L$ or $0.9D + 1.6W$. Both loading conditions must be checked.

We must check various points on the wall. Critical points are just below the roof reaction (moment is high and axial load is low, so net tension may govern); at the mid-height of the wall, where moment from eccentric gravity load and wind load are highest; and at the base of the

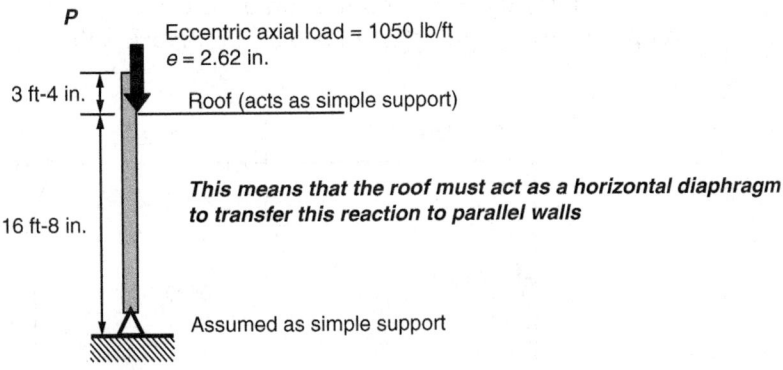

Figure 14.12 Unreinforced masonry bearing wall with eccentric axial load and wind load.

wall (axial load is high, so the maximum compressive stress may govern). Check each of these locations.

To avoid having to check a large number of loading combinations and potentially critical locations, it is worthwhile to assess them first, and check only the ones that will probably govern.

Due to wind only, the unfactored moment at the base of the parapet (roof level) is

$$M = \frac{qL_{parapet}^2}{2} = \frac{25 \text{ lb/ft} \times 3.33^2 \text{ ft}^2}{2} \times 12 \text{ in./ft} = 1663 \text{ lb-in.}$$

The maximum moment is close to that occurring at mid-height. The moment from wind load is the superposition of one-half moment at the upper support due to wind load on the parapet only, plus the midspan moment in a simply supported beam with that same wind load:

$$M_{midspan} = -\frac{1663}{2} + \frac{qL^2}{8} = -\frac{1663 \text{ lb-in.}}{2} + \frac{25 \text{ lb/ft} \times 16.67^2 \text{ ft}^2}{8} \times 12 \text{ in./ft}$$

$$= 9589 \text{ lb-in.}$$

Unfactored moment diagrams due to eccentric axial load and wind are as shown in Fig. 14.13.

From the example of Sec. 14.5.2, we know that loading combination $1.2D + 1.6L$ was not close to critical directly underneath the roof. Because the wind-load moments directly underneath the roof are not very large, they will probably not be critical either. The critical location will probably be at mid-height; the critical loading condition will probably be $0.9D + 1.6W$; and the critical criterion will probably be net tension, because this AAC masonry wall is unreinforced.

As before, try 8-in. nominal units of Class 4 AAC, with a specified compressive strength, f'_{AAC}, of 580 lb/in.². Work with a strip with a width

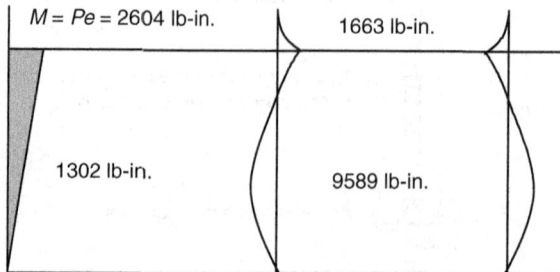

Figure 14.13 Unfactored moment diagrams due to eccentric axial load and wind.

of 1 ft (measured along the length of the wall in plan). Stresses are calculated using the critical section, consisting of the full bedded area (gross area) (2008 MSJC Code, Sec. 1.9.1.1).

Now check the net tensile stress. At the mid-height of the wall, the axial force due to $0.9D$ is

$$P_u = 0.9\,(700\text{ lb}) + 0.9\,(3.33\text{ ft} + 8.33\text{ ft}) \times 26.3\text{ lb/ft} = 906\text{ lb}$$

At the mid-height of the wall, the factored design moment, M_u, is given by:

$$M_u = P_{u\,\text{eccentric}}\frac{e}{2} + M_{u\,\text{wind}} = \left(\frac{1}{2}\right)0.9\cdot 700\text{ lb} \times 2.62\text{ in.} + 1.6 \times 9589\text{ lb-in.}$$

$$= 16{,}167\text{ lb-in.}$$

$$f_{\text{tension}} = -\frac{P_u}{A} + \frac{M_u c}{I}$$

$$f_{\text{tension}} = -\frac{906\text{ lb}}{7.9 \times 12\text{ in.}^2} + \frac{16{,}167\text{ lb-in.}\left(7.9/2\right)}{12 \times 7.9^3/12\text{ in.}^4}$$

$$= -9.56 + 129.5\text{ lb/in.}^2 = 120.0\text{ lb/in.}^2$$

The specified compressive strength, f'_{AAC}, for Class 4 AAC is 580 psi (Table 14.1). The corresponding splitting tensile strength is

$$f_{t\,\text{AAC}} = 2.4\sqrt{f'_{\text{AAC}}}$$

$$f_{t\,\text{AAC}} = 2.4\sqrt{580\text{ lb/in.}^2}$$

$$f_{t\,\text{AAC}} = 57.8\text{ lb/in.}^2$$

The modulus of rupture is twice this value, or 115.6 psi. The maximum permissible stress is this value, multiplied by the strength-reduction factor of 0.6.

$$0.60 f_r = 0.60 \times 115.6\text{ lb/in.}^2 = 69.4\text{ lb/in.}^2$$

The maximum tensile stress exceeds the prescribed value, and the design is not satisfactory. It will be necessary to reinforce the wall, as illustrated in the design example of Sec. 14.9. Thickening the wall, though possible, is probably not a cost-effective option.

14.5.5 Comments on the above Examples for Design of Unreinforced AAC Masonry Bearing Walls

1. In retrospect, it probably would not have been necessary to check all three criteria at all locations. With experience, a designer could realize that the location with highest wind moment would govern, and could therefore check only the mid-height of the wall.

2. The addition of wind load to the second example, to produce the third example, changes the critical location from just under the roof, to the mid-height of the simply supported section of the wall. The wind load of 25 lb/ft^2 in the third example produces maximum tensile stresses above the allowable values for AAC masonry, and makes it necessary to thicken the wall or reinforce it.

14.5.6 Extension of the above Concepts to AAC Masonry Walls with Openings

AAC masonry bearing walls with openings are handled as in Sec. 5.2.7. That material is not repeated here.

14.5.7 Final Comment on the Effect of Openings in Unreinforced AAC Masonry Bearing Walls

As the summation of the plan lengths of openings in a bearing wall exceeds about one-half the plan length of the wall, even the higher allowable stresses (or moduli of rupture) corresponding to fully grouted walls will be exceeded, and it will generally become necessary to use reinforcement. Design of reinforced AAC masonry bearing walls is addressed later in this chapter.

14.6 Design of Unreinforced Shear Walls of AAC Masonry

Unreinforced masonry shear walls must be designed for the effects of:

1. Gravity loads from self-weight plus gravity loads from overlying roof or floor levels
2. Moments and shears from in-plane shear loads

Actions are shown in Fig. 14.14.

For unreinforced AAC masonry, the 2008 MSJC *Code* requires that maximum tensile stresses from in-plane flexure, alone or in combination with axial loads, not exceed the in-plane modulus of rupture from Sec. A.1.8.3 of the 2008 MSJC *Code*.

Shear must also be checked. According to Sec. A.2.5 of the 2008 MSJC *Code*, the nominal shear capacity of AAC masonry is the least of the

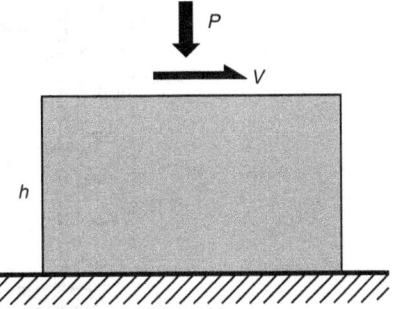

FIGURE 14.14 Design actions for unreinforced shear walls.

following three equations, related respectively to web-shear cracking, crushing of the diagonal strut, and sliding.

$$V_{nAAC} = \min \begin{cases} 0.95\, \lambda_w t \sqrt{f'_{AAC}} \sqrt{1 + \dfrac{P_u}{2.4\sqrt{f'_{AAC}}\, \lambda_w t}} \\[6pt] 0.17 f'_{AAC} t\, \dfrac{h \cdot \lambda_w^2}{h^2 + (\tfrac{3}{4}\lambda_w)^2} \\[6pt] \mu_{AAC} P_u \end{cases}$$

The strength-reduction factor for shear is 0.80 (2008 MSJC *Code*, Sec. A.1.5.3).

14.6.1 Example of Design of Unreinforced Shear Wall of AAC Masonry

Consider the simple structure of Fig. 14.15, the same one whose bearing walls have been designed previously in this book. Use nominal 8-in. AAC masonry units, Class 4, $f'_{AAC} = 580$ lb/in.², laid with thin-bed mortar and

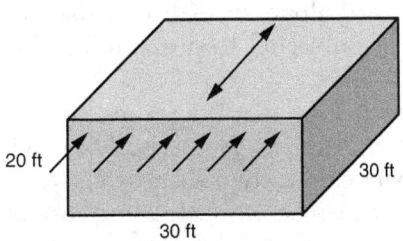

FIGURE 14.15 Example problem for strength design of unreinforced shear wall.

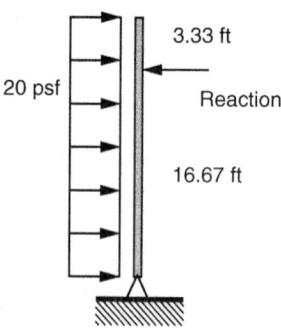

FIGURE 14.16 Calculation of reaction on roof diaphragm, strength design of unreinforced AAC masonry shear wall.

fully bedded. The roof applies a gravity load of 1050 lb/ft to the walls; the walls measure 16 ft, 8 in. height to the roof, and have an additional 3 ft, 4 in. parapet. The walls are loaded with a wind load of 20 lb/ft.² The roof acts as a one-way system, transmitting gravity loads to the front and back walls. At this stage, all loads are unfactored; load factors will be applied later.

Now design the shear wall. The critical section for shear is just under the roof, where axial load in the shear walls is least, coming from the parapet only. As a result of the wind loading, the reaction transmitted to the roof diaphragm is calculated using Fig. 14.16.

$$\text{Reaction} = \frac{20 \text{ lb/ft}^2 \cdot \left(\dfrac{20^2 \text{ ft}^2}{2}\right)}{16.67 \text{ ft}} = 240 \text{ lb/ft}$$

Total roof reaction acting on one side of the roof is

$$\text{Reaction} = 240 \text{ lb/ft} \cdot 30 \text{ ft} = 7200 \text{ lb}$$

This is divided evenly between the two shear walls, so the shear per wall is 3600 lb.

In Fig. 14.17, for simplicity, the lateral load is shown as if it acted on the front wall alone. In reality, it also acts on the back wall, so that the structure is subjected to pressure on the front wall and suction on the back wall.

The horizontal diaphragm reaction transferred to each shear wall is 240 lb/ft, multiplied by the building width of 30 ft, and then divided equally between the two shear walls, for a total of 3600 lb per shear wall.

Using the conservative loading case of $0.9D + 1.6W$,

$$V_u = 1.6 V_{\text{unfactored}} = 1.6 \cdot 3600 \text{ lb} = 5760 \text{ lb}$$

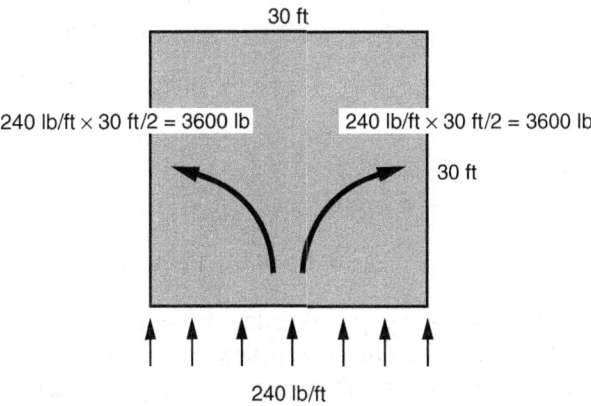

FIGURE 14.17 Transmission of forces from roof diaphragm to shear walls.

Compute the axial force in the wall at that level. To be conservative, use the loading combination $0.9D + 1.6W$.

$$P_u = 0.9 \times 3.33 \text{ ft} \times 26.3 \text{ lb/ft}^2 \times 30 \text{ ft} = 2365 \text{ lb}$$

The nominal shear capacity at that level, as governed by web-shear cracking, crushing of the diagonal strut, and sliding, respectively, is given below. In accordance with Sec. A.1.8.5 of the 2008 MSJC *Code*, the coefficient of friction between AAC and leveling bed mortar is 1.0.

$$V_{nAAC} = \min \begin{cases} 0.95 \lambda_w t \sqrt{f'_{AAC}} \sqrt{1 + \dfrac{P_u}{2.4\sqrt{f'_{AAC}}\, \lambda_w t}} \\ 0.17 f'_{AAC} t \dfrac{h \cdot \lambda_w^2}{h^2 + (\tfrac{3}{4}\lambda_w)^2} \\ \mu_{AAC} P_u \end{cases}$$

$$V_{nAAC} = \min \begin{cases} 0.95 \times 30 \text{ ft} \times 12 \text{ in./ft} \times 7.9 \text{ in.} \sqrt{580 \text{ lb/in.}^2} \sqrt{1 + \dfrac{2365 \text{ lb}}{2.4\sqrt{580 \text{ lb/in.}^2} \times 30 \text{ ft} \times 12 \text{ in./ft} \times 7.9 \text{ in.}}} \\ 0.17 \times 580 \text{ lb/in.}^2 \times 7.9 \text{ in.} \times \dfrac{16.67 \text{ ft} \times 12 \text{ in./ft} \times (30 \text{ ft} \times 12 \text{ in/ft})^2}{\left[(16.67 \text{ ft} \times 12 \text{ in./ft})^2 + \left(\dfrac{3}{4} \times 30 \text{ ft} \times 12 \text{ in./ft}\right)^2\right]} \\ 1.0 \times 2365 \text{ lb} \end{cases}$$

$$V_{nAAC} = \min \begin{cases} 65{,}534 \text{ lb} \\ 178{,}842 \text{ lb} \\ 2365 \text{ lb} \end{cases}$$

The design shear capacity is

$$\phi V_n = 0.80 \times 2365 \text{ lb} = 1892 \text{ lb}$$

The design shear capacity is less than the factored design shear of 5760 lb.

This design issue is complex. Some designers might not use the third equation, reasoning that this wall does not have an unbonded interface, because the assumption of unreinforced masonry is consistent with an uncracked condition. If a designer considers the possibility of an unbonded interface and opts to include the third equation for shear capacity, this design issue would normally be addressed using shear friction or dowel action across this interface. Such provisions do not exist in the 2008 MSJC *Code*, and are being developed at this writing. For the time being, the interface is regarded as uncracked, and the third equation is not included. The design capacities associated with the other two limit states (web-shear cracking and crushing of the diagonal strut) greatly exceed the factored design shear, and do not govern.

Now check for the net flexural tensile stress. The critical section is at the base of the wall, where in-plane moment is maximum. Because the roof spans between the font and back walls, the distributed gravity load on the roof does not act on the side walls, and their axial load comes from self-weight only. Again, use the conservative loading combination of $0.9D + 1.6W$:

$$f_{tension} = \frac{M_u c}{I} - \frac{P_u}{A} = \frac{V_u h c}{I} - \frac{P_u}{A} \leq \phi f_r$$

$$f_{tension} = \frac{1.6 \times 3600 \text{ lb} \cdot 16.67 \text{ ft} \times 12 \text{ in./ft} \left(\frac{30 \text{ ft} \times 12 \text{ in./ft}}{2}\right)}{\left[\frac{7.9 \text{ in.} \times (30 \text{ ft} \times 12 \text{ in./ft})^3}{12}\right]}$$

$$- \frac{0.9 \times 20 \text{ ft} \times 26.3 \text{ lb/ft}}{7.9 \times 12 \text{ in.}^2}$$

$$f_{tension} = 1.6 \times 4.22 \text{ lb/in.}^2 - 0.9 \times 5.55 \text{ lb/in.}^2$$

$$f_{tension} = 6.75 \text{ lb/in.}^2 - 4.00 \text{ lb/in.}^2$$

$$f_{tension} = 1.76 \text{ lb/in.}^2$$

The specified compressive strength, f'_{AAC}, for Class 4 AAC is 580 psi (Table 14.1). The corresponding splitting tensile strength is

$$f_{t\,AAC} = 2.4\sqrt{f'_{AAC}}$$

$$f_{t\,AAC} = 2.4\sqrt{580\text{ lb/in.}^2}$$

$$f_{t\,AAC} = 57.8\text{ lb/in.}^2$$

The modulus of rupture is twice this value, or 115.6 psi. The maximum permissible stress is this value, multiplied by the strength-reduction factor of 0.6.

$$0.60 f_r = 0.60 \times 115.6\text{ lb/in.}^2 = 69.4\text{ lb/in.}^2$$

The net tension in the wall is less than this value, and the design is satisfactory.

When the wind blows against the side walls, these walls transfer their loads to the roof diaphragm, and the front and back walls act as shear walls. The side walls must be checked for this loading direction also, following the procedures of previous examples in this book.

In-plane, the (h/r) value for this shear wall is much less than the triggering value of 45, and the moment magnifier can be taken as 1.0 (2008 MSJC *Code*, Sec. 3.2.2.4).

14.6.2 Comments on Example Problem with Design of Unreinforced AAC Masonry Shear Walls

Clearly, unreinforced AAC masonry shear walls have large shear capacity because of their large cross-sectional area. Sliding needs to be addressed by the addition of shear-friction provisions in the MSJC *Code*. If this area is reduced by openings, then shear capacities will decrease, and in-plane flexural capacities as governed by net flexural tension may decrease even faster.

14.7 Design of Reinforced Beams and Lintels of AAC Masonry

The most common reinforced masonry beam is a lintel. Lintels are beams that support masonry over openings. Strength design of reinforced beams and lintels follows the steps given below:

1. *Shear design*: Calculate the design shear, and compare it with the corresponding resistance. Revise the lintel depth if necessary.

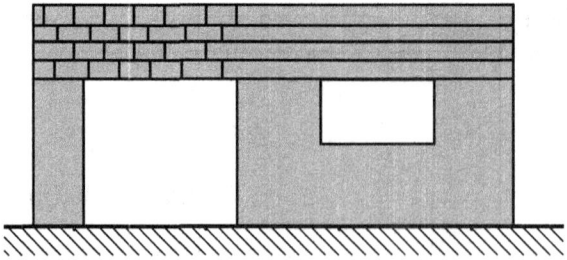

Figure 14.18 Example of masonry lintel.

2. *Flexural design:*
   a. Calculate the design moment.
   b. Calculate the required flexural reinforcement. Check that it fits within minimum and maximum reinforcement limitations. Because deformed reinforcement is required to be surrounded by grout, and because AAC masonry units are manufactured solid, it is usually more cost-effective to place deformed horizontal reinforcement for lintels in a bond beam of concrete masonry units.

In many cases, the depth of the lintel is determined by architectural considerations. In other cases, it is necessary to determine the number of courses of masonry that will work as a beam. For example, consider the lintel in Fig. 14.18.

The depth of the beam, and hence the area that is effective in resisting shear, is determined by the number of courses that we consider to comprise it. Because it is not very practical to put shear reinforcement in masonry beams, the depth of the beam may be determined by this. In other words, the beam design may start with the number of courses that are needed to that shear can be resisted by masonry alone.

14.7.1 Physical Properties of Steel Reinforcing Bars

Physical properties of steel reinforcing bars are given in Table 14.4.
Cover requirements are given in Sec. 1.15.4 of the 2008 MSJC *Code*.

14.7.2 Example of Lintel Design Using AAC Masonry

Suppose that we have a uniformly distributed load of 1050 lb/ft, applied at the level of the roof of the structure shown in Fig. 14.19. Design the lintel. Assume AAC masonry with Class 4 AAC units having a nominal thickness of 8 in., a weight of 26.3 lb/ft^2, and a specified compressive strength of 580 lb/in.^{2}. Use thin-bed mortar. The lintel has a span of 10 ft, and a total depth (height of parapet plus distance between the roof and

Designation	Diameter, in.	Area, in.²
Bars		
#3	0.375	0.11
#4	0.500	0.20
#5	0.625	0.31
#6	0.750	0.44
#7	0.875	0.60
#8	1.000	0.79
#9	1.128	1.00
#10	1.270	1.27
#11	1.410	1.56

TABLE 14.4 Physical Properties of Steel Reinforcing Bars

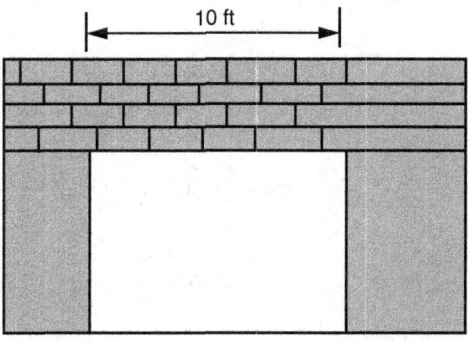

FIGURE 14.19 Example for design of an AAC masonry lintel.

the lintel) of 4 ft. These are shown in the schematic of Fig. 14.19. Assume that 700 lb/ft of the roof load is D and the remaining 350 lb/ft is L. The governing loading combination is $1.2D + 1.6L$.

Again, first check whether the depth of the lintel is sufficient to avoid the use of shear reinforcement. Because the opening may have a movement joint on either side, again use a span equal to the clear distance, plus one-half of a half-unit on each side. So the span is 10 ft plus 8 in., or 10.67 ft.

$$M_u = \frac{w_u l^2}{8}$$

$$= \frac{[(700 + 4 \text{ ft} \times 26.3 \text{ lb/ft}) \times 1.2 + 350 \text{ lb/ft} \times 1.6] \times 10.67^2 \text{ ft}^2 \times 12 \text{ in./ft}}{8}$$

$$= 260{,}641 \text{ in.-lb}$$

$$V_u = \frac{w_u l}{2} = \frac{[(700 + 4\text{ ft} \times 26.3\text{ lb/ft}) \times 1.2 + 350\text{ lb/ft} \times 1.6] \times 10.67\text{ ft}}{2}$$

$$= 8142\text{ lb}$$

The bars in the lintel will probably be placed in the lower part of an inverted bottom course of concrete masonry as shown in Fig. 14.20.

The effective depth d is calculated using the minimum cover of 1.5 in. (Sec. 1.15.4.1 of the 2008 MSJC *Code*), plus one-half the diameter of an assumed #8 bar.

The nominal shear capacity, as governed by web-shear cracking, crushing of the diagonal strut, and sliding, respectively, is given below. In accordance with Sec. A.1.8.5 of the 2008 MSJC *Code*, the coefficient of friction between AAC and leveling bed mortar is 1.0. The MSJC shear equations were originally developed for walls. To apply them to beams, the dimension of the beam in the direction of the applied shear (the depth) is used as the wall length. Only the first equation is relevant.

$$V_{nAAC} = \min \begin{cases} 0.95 \lambda_w t \sqrt{f'_{AAC}} \sqrt{1 + \dfrac{P_u}{2.4 \sqrt{f'_{AAC}} \lambda_w t}} \\ 0.17 f'_{AAC} t \dfrac{h \cdot \lambda_w^2}{h^2 + (\frac{3}{4} \lambda_w)^2} \\ \mu_{AAC} P_u \end{cases}$$

$$V_{nAAC} = 0.95 \times 48\text{ in.} \times 7.9\text{ in.} \sqrt{580\text{ lb/in.}^2} \sqrt{1 + 0}$$

$$V_{nAAC} = 8676\text{ lb}$$

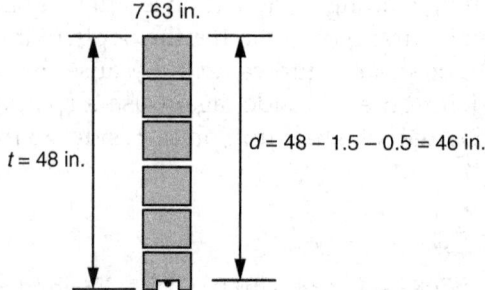

FIGURE 14.20 Example showing placement of bottom reinforcement in lowest course of lintel.

The design shear capacity is

$$\phi V_n = 0.80 \times 8676 \text{ lb} = 6941 \text{ lb}$$

The design shear capacity is less than the factored design shear of 8142 lb. The specified strength of the AAC in the lintel will have to be increased to Class 6, with a specified compressive strength of 870 psi.

$$V_{nAAC} = 0.95 \lambda_w t \sqrt{f'_{AAC}} \sqrt{1 + \frac{P_u}{2.4\sqrt{f'_{AAC}} \lambda_w t}}$$

$$V_{nAAC} = 0.95 \times 48 \text{ in.} \times 7.9 \text{ in.} \sqrt{870 \text{ lb/in.}^2} \sqrt{1+0}$$

$$V_{nAAC} = 10{,}626 \text{ lb}$$

The design shear capacity is

$$\phi V_n = 0.80 \times 10{,}626 \text{ lb} = 8501 \text{ lb}$$

This exceeds the factored design shear of 8142 lb, and the design is satisfactory so far.

Also, according to Eq. (A-11),

$$V_n \leq 4\sqrt{f'_{AAC}} A_n$$

$$V_n \leq 4\sqrt{870 \text{ lb/in.}^2} \; 48 \text{ in.} \times 7.9 \text{ in.}$$

$$V_n \leq 44{,}739 \text{ lb}$$

This does not govern and the shear design is acceptable. This shear design contains a subtle complexity. Because this design involves reinforced masonry, which is assumed to be cracked in flexure, this lintel could have vertical cracks. This design issue would normally be addressed using shear friction or dowel action across this interface. Such provisions do not exist in the 2008 MSJC *Code*, and are being developed at this writing. For the time being, this interface is regarded as bonded.

Now check the required flexural reinforcement:

$$M_n = A_s f_y \text{ (lever arm)}$$

$$M_n \approx A_s f_y \cdot 0.9 d$$

In our case,

$$M_n^{required} = \frac{M_u}{\phi} = \frac{M_u}{0.9} = \frac{260{,}641 \text{ lb-in.}}{0.9} = 289{,}601 \text{ lb-in.}$$

$$A_s^{\text{required}} \approx \frac{M_n^{\text{required}}}{0.9\,df_y} = \frac{289{,}601 \text{ lb-in.}}{0.9 \times 46 \text{ in.} \times 60{,}000 \text{ lb/in.}^2} = 0.12 \text{ in.}^2$$

Because of the depth of the beam, this can easily be satisfied with a #4 bar in the lowest course (of concrete masonry units). The corresponding nominal flexural capacity is approximately

$$M_n \approx A_s f_y (0.9\,d)$$

$$M_n \approx 0.20 \text{ in.}^2 \times 60{,}000 \text{ lb/in.}^2 \times 0.9 \times 46 \text{ in.}$$

$$M_n \approx 496{,}800 \text{ lb-in.}$$

Also include two #4 bars at the level of the roof (bond beam reinforcement, again using concrete masonry units). The flexural design is quite simple.

Sec. A.3.4.2.2.2 of the 2008 MSJC Code does require that the nominal flexural strength of a beam not be less than 1.3 times the nominal cracking capacity, calculated using the modulus of rupture from Code A.1.8.3. The specified compressive strength, f'_{AAC}, for Class 6 AAC is 870 psi (Table 14.1). The corresponding splitting tensile strength is

$$f_{t\,AAC} = 2.4\sqrt{f'_{AAC}}$$

$$f_{t\,AAC} = 2.4\sqrt{870 \text{ lb/in.}^2}$$

$$f_{t\,AAC} = 70.8 \text{ lb/in.}^2$$

The modulus of rupture is twice this value or 141.6 psi.

In our case, the nominal cracking moment for the 4-ft deep section is

$$M_{cr} = Sf_r = \frac{bt^2}{6}f_r = \frac{7.9 \text{ in.} \times 48^2 \text{ in.}^2}{6} \times 141.6 \text{ lb/in.}^2 = 429{,}496 \text{ lb-in.}$$

This value, multiplied by 1.3, is 558,345 lb-in., which exceeds the nominal capacity of this lintel with the provided #4 bar. Flexural reinforcement must be increased to

$$A_s \approx 0.20 \text{ in.}^2 \left(\frac{558{,}345 \text{ lb-in.}}{496{,}800 \text{ lb-in.}}\right) = 0.22 \text{ in.}^2$$

Use two #4 bars. Finally, Sec. A.3.3.5 of the 2008 MSJC Code imposes maximum flexural reinforcement limitations that are based on a series of critical strain gradients. These generally do not govern for members with little or no axial load, like this lintel. They may govern for members with significant axial load, such as tall shear walls.

14.8 Design of Reinforced Curtain Walls of AAC Masonry

Although reinforced curtain walls of AAC masonry are theoretically possible, it is much more cost-effective to use factory-reinforced panels spanning horizontally between columns, rather than field-reinforced AAC masonry. For this reason, the strength design of reinforced curtain walls of AAC masonry is not discussed further here. It is discussed in ACI 523.4R-09 (2009).

14.9 Design of Reinforced Bearing Walls of AAC Masonry

14.9.1 Example of Moment-Axial Force Interaction Diagram for AAC Masonry (Spreadsheet Calculation)

Construct the moment-axial force interaction diagram by the strength approach for a nominal 8-in. AAC masonry wall with Class 4 AAC (f'_{AAC} = 580 lb/in.²) and reinforcement consisting of #4 bars at 48 in., placed in the center of the wall.

The effective width of the wall is $6t$, or 48 in. The spreadsheet and corresponding interaction diagram are shown in Fig. 14.21 and Table 14.5. As noted in Sec. 6.3, because the reinforcement is located at the geometric centroid of the section, the balance-point axial load (about 100,000 lb)

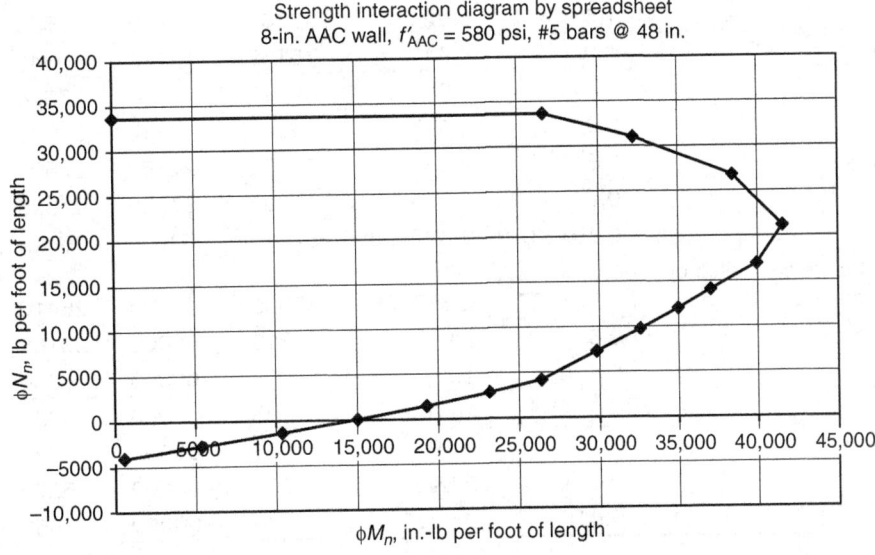

FIGURE 14.21 Moment-axial force interaction diagram (strength approach), spreadsheet calculation.

Example of spreadsheet for calculating moment-axial force interaction diagram for reinforced AAC bearing wall

Reinforcement at mid-depth						
Specified thickness	7.9					
emu	0.003					
f_m'	580					
f_y	60000					
E_s	29000000					
d	3.95					
(c/d) balanced	0.591837					
Tensile reinforcement area	0.31					
Effective width	48					
Phi	0.9					
Because compression reinforcement is not supported, it is not counted						
	c/d	c	C_{mas}	f_s	Moment	Axial force
Pure axial load					0	33623
Points controlled by masonry	2.387	9.42865	149490	0	26619	33635
	2.2	8.69	137779	0	32205	31000
	1.9	7.505	118991	0	38441	26773
	1.5	5.925	93940	0	41536	21137
	1.2	4.74	75152	0	39941	16909
	1	3.95	62627	0	37014	14091
	0.9	3.555	56364	−9667	34990	12008
	0.8	3.16	50101	−21750	32594	9756
	0.7	2.765	43839	−37286	29825	7263
	0.591837	2.337755	37065	−60000	26410	4155
Points controlled by steel	0.591837	2.337755	37065	−60000	26410	4155
	0.5	1.975	31313	−60000	23168	2861
	0.4	1.58	25051	−60000	19280	1451
	0.3	1.185	18788	−60000	15020	42
	0.2	0.79	12525	−60000	10386	−1367
	0.1	0.395	6263	−60000	5379	−2776
	0.01	0.0395	626	−60000	555	−4044

TABLE 14.5 Spreadsheet for Computing Moment-Axial Force Interaction Diagram for AAC Bearing Wall

does not correspond to the maximum moment capacity. As required by Sec. A.3.2 of the 2008 MSJC *Code*, the spreadsheet is like that of Sec. 6.3 (strength design of masonry bearing walls), except that the maximum useful compressive strain in the masonry is 0.003 (rather than 0.0025 or 0.0035); the equivalent rectangular compressive stress block has a height $0.85 f'_{AAC}$ (rather than $0.8 f'_m$), and β_1 is to 0.67 (rather than 0.8).

Plot of Interaction Diagram for AAC Masonry Bearing Wall by Spreadsheet

The moment-axial force interaction diagram for this AAC masonry bearing wall, plotted by spreadsheet is shown in Fig. 14.21. Slenderness effects are neglected.

Relevant cells from the spreadsheet are reproduced in Table 14.5.

14.9.2 Example of Design of AAC Masonry Walls Loaded Out-of-Plane

Once we have developed the moment-axial force interaction diagram, the actual design simply consists of verifying that the combination of factored design axial force and moment lies within the diagram of nominal axial and flexural capacity, reduced by strength-reduction factors. Consider the bearing wall designed previously as unreinforced, shown in Fig. 14.22. It has an eccentric axial load plus out-of-plane wind load of 25 lb/ft².

At each horizontal plane through the wall, the following condition must be met:

- Combinations of factored axial load and moment must lie within the moment-axial force interaction diagram, reduced by strength-reduction factors.

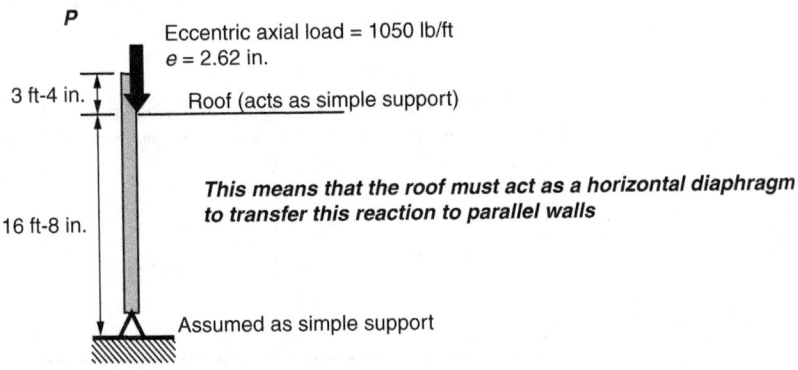

FIGURE 14.22 Reinforced masonry wall loaded by eccentric gravity axial load plus out-of-plane wind load.

Because flexural capacity increases with increasing axial load, the critical loading combination is probably $0.9D + 1.6W$.

From our previous experience, we know that the critical point on the wall is at the midspan of the lower portion. Due to wind only, the unfactored moment at the base of the parapet (roof level) is

$$M = \frac{qL_{parapet}^2}{2} = \frac{25 \text{ lb/ft} \times 3.33^2 \text{ ft}^2}{2} \times 12 \text{ in./ft} = 1663 \text{ lb-in.}$$

The maximum moment is close to that occurring at mid-height. The moment from wind load is the superposition of one-half moment at the upper support due to wind load on the parapet only, plus the midspan moment in a simply supported beam with that same wind load:

$$M_{midspan} = -\frac{1663}{2} + \frac{qL^2}{8} = -\frac{1663}{2} + \frac{25 \text{ lb/ft} \cdot 16.67^2 \text{ ft}^2}{8} \times 12 \text{ in./ft}$$

$$= 9589 \text{ lb-in.}$$

The unfactored moment due to eccentric axial load is

$$M_{gravity} = Pe = 1050 \text{ lb} \times 2.62 \text{ in.} = 2751 \text{ lb-in.}$$

Unfactored moment diagrams due to eccentric axial load and wind are as shown in Fig. 14.23.

Check the adequacy of the wall with 8-in. nominal AAC units, Class 4 AAC (specified compressive strength, $f'_{AAC} = 580 \text{ lb/in.}^2$), unit weight = 26.3 lb/ft², and #4 bars spaced at 48 in. All design actions are calculated per foot of width of the wall.

At the mid-height of the wall, the axial force due to $0.9D$ is

$$P_u = 0.9 \,(700 \text{ lb}) + 0.9(3.33 \text{ ft} + 8.33 \text{ ft}) \times 26.3 \text{ lb/ft} = 906 \text{ lb}$$

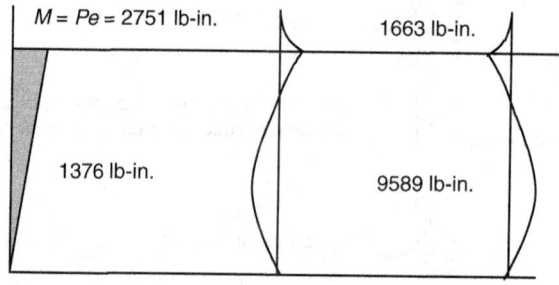

Figure 14.23 Unfactored moment diagrams due to eccentric axial load plus wind load.

At the mid-height of the wall, the factored design moment, M_u, is given by:

$$M_u = P_u \frac{e}{2} + M_{u\,wind} = \left(\frac{1}{2}\right) 0.9 \times 700 \text{ lb} \times 2.62 \text{ in.} + 1.6 \times 9589 \text{ lb-in.}$$

$$= 16{,}167 \text{ lb-in.}$$

In each foot of wall, the design actions are $P_u = 906$ lb and $M_u = 16{,}167$ lb-in. That combination lies within the interaction diagram of design capacities (Fig. 14.21), and the design is satisfactory.

Because this out-of-plane wall is checked for magnified moments in accordance with Sec. A.3.5 of the MSJC *Code*, the slenderness-dependent reduction factor is not applied to the moment-axial force interaction. The design is satisfactory.

Outside of a plastic hinge zone, Sec. A.3.3.1 of the 2008 MSJC *Code* imposes a maximum bar area of 4.5 percent of the cell. Using a 3-in. grouted core, the area ratio is $(0.625/3)^2$, or 0.043, satisfying the requirement. This bar size will easily satisfy the maximum reinforcement limitations of Sec. A.3.3.5 for out-of-plane flexure, and the design is satisfactory for flexure.

The provisions of the 2008 MSJC *Code* also require a check of the possible effects of secondary moments for reinforced walls loaded out-of-plane (MSJC *Code*, Sec. A.3.5.4).

In accordance with these sections, MSJC *Code* Eq. (A-18) is used to calculate the maximum moment, including possible secondary moments. That maximum moment is then compared with the interaction diagram. MSJC *Code* Eq. (A-18) is based on a member simply supported at top and bottom, which is the case here:

$$M_u = \frac{w_u h^2}{8} + P_{uf}\left(\frac{e_u}{2}\right) + P_u \delta_u$$

As calculated above, for each feet of wall length, the first two terms in this equation total 16,167 lb-in., and P_u equals 906 lb. In accordance with MSJC *Code* Secs. 3.3.5.3, δ_u is to be calculated using *Code* Eqs. (3-31) and (3-32), replacing M_{ser} with M_u. Because the cracking moment used in these equations is calculated without strength-reduction factors, it might exceed the factored design moment. Nevertheless, it is believed prudent to assume that reinforced masonry is cracked at bed joints.

For this problem, the cracked moment of inertia I_{cr} for use in MSJC *Code* Sec, A.3.5.4 is approximately and conservatively be taken as 40 percent of the gross moment of inertia. This relationship between cracked and gross inertia is commonly used for lightly reinforced concrete or masonry sections. Its use here is consistent with the assumption that the entire wall is initially cracked on the bed joints before any load is applied. Section properties are per foot of plan length.

$$\delta_u = \frac{5 M_u h^2}{48 E_m I_{cr}}$$

$$I_{cr} = 0.40 I_g = 0.40 \left(\frac{bt^3}{12}\right) = 0.40 \left[\frac{12 \text{ in.} \times (7.9 \text{ in.})^3}{12}\right]$$

$$= 0.40 \times 493.0 \text{ in.}^4 = 197.2 \text{ in.}^4$$

$$\delta_{u1} = \frac{5 \times 16{,}167 \text{ lb-in.} \times (16.67 \text{ ft} \times 12 \text{ in./ft.})^2}{48 \left[6500 \times (580 \text{ lb/in.}^2)^{0.6}\right](197.2 \text{ in.}^4)} = 1.16 \text{ in.}$$

$$M_{u2} = 16{,}167 \text{ lb-in.} + 906(1.16) \text{ lb-in.} = 17{,}213 \text{ lb-in.}$$

Check convergence:

$$\delta_{u2} = \frac{5 \times 17{,}213 \text{ lb-in.} \times (16.67 \text{ ft} \times 12 \text{ in./ft.})^2}{48 \left[6500 \times (580 \text{ lb/in.}^2)^{0.6}\right](197.2 \text{ in.}^4)} = 1.23 \text{ in.}$$

$$M_{u3} = 16{,}167 \text{ lb-in.} + 906(1.23) \text{ lb-in.} = 17{,}281 \text{ lb-in.}$$

Because the moment is changing by less than 0.4 percent, it can be assumed to have converged. The combination of factored axial force and factored moment (including secondary moments) remains within the moment-axial force interaction diagram, and the design is still satisfactory.

Finally, the provisions of the 2008 MSJC *Code* also require a check of out-of-plane deflections for reinforced AAC masonry walls loaded out-of-plane (MSJC *Code* Sec. A.3.5.5).

In accordance with these sections, MSJC *Code* Eq. (A-24) or (A-25) is used to calculate the mid-height deflection. These equations are based on a member simply supported at top and bottom, which is the case here. Conservatively assuming the section to be cracked, the out-of-plane deflection is given by the converged δ_{u2} from above (1.23 in.). That deflection is less than 0.007 h (equal to 0.007 times 16.67 ft, or 1.40 in.). The out-of-plane deflection requirement is satisfied, even though the AAC wall has a lower modulus of elasticity and is therefore more flexible than a CMU wall of comparable thickness.

14.9.3 Minimum and Maximum Reinforcement Ratios for Out-of-Plane Flexural Design of AAC Masonry Walls

The design provisions of the 2008 MSJC *Code* include requirements for minimum and maximum flexural reinforcement. In this section, the implications of those requirements for the out-of-plane flexural design of AAC masonry walls are addressed.

Minimum Flexural Reinforcement by 2008 MSJC *Code*

The 2008 MSJC *Code* has no requirements for minimum flexural reinforcement for out-of-plane design of masonry walls.

Maximum Flexural Reinforcement for AAC Walls by 2008 MSJC *Code*

The 2008 MSJC *Code* has a maximum reinforcement requirement (Sec. A.3.3.5) that is intended to ensure ductile behavior over a range of axial loads. As compressive axial load increases, the maximum permissible reinforcement percentage decreases. For compressive axial loads above a critical value, the maximum permissible reinforcement percentage drops to zero, and design is impossible unless the cross-sectional area of the element is increased.

For walls subjected to out-of-plane forces, for columns and for beams, the provisions of the 2008 MSJC *Code* set the maximum permissible reinforcement based on a critical strain condition in which the masonry is at its maximum useful strain, and the extreme tension reinforcement is set at a 1.5 times the yield strain.

The critical strain condition for walls with a single layer of concentric reinforcement and loaded out-of-plane is shown in Fig. 14.24, along with the corresponding stress state. The parameters for the equivalent rectangular

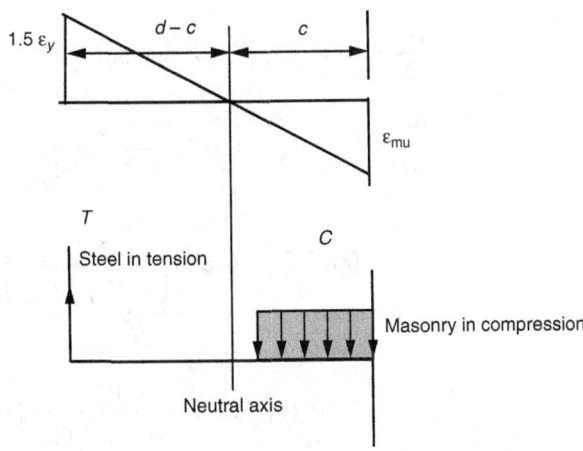

FIGURE 14.24 Critical strain condition for an AAC masonry wall loaded out-of-plane.

stress block are the same as those used for conventional flexural design. The height of the equivalent rectangular stress block is $0.80 f'_m$, and the depth is $0.80 c$. The tensile reinforcement is assumed to be at f_y.

Locate the neutral axis using the critical strain condition:

$$\frac{\varepsilon_{mu}}{1.5\varepsilon_y} = \frac{c}{d-c}$$

$$c = d\left(\frac{\varepsilon_{mu}}{1.5\varepsilon_y + \varepsilon_{mu}}\right)$$

Compute the tensile and compressive forces acting on the section, assuming concentric reinforcement with a percentage of reinforcement $\rho = \frac{A_s}{bd}$, where $d = \frac{t}{2}$.

The compressive force in the masonry is given by:

$$C_{masonry} = 0.85 f'_{AAC}\, 0.67\, cb$$

The tensile force in the reinforcement is given by:

$$T_{steel} = \rho b d f_y$$

Equilibrium of axial forces requires:

$$N_n = C - T$$

$$\frac{N_u}{\phi} = C - T$$

$$\frac{N_u}{\phi} = 0.85 f'_{AAC}\, 0.67\, cb - \rho d b f_y$$

$$\frac{N_u}{\phi} = 0.85 f'_{AAC}\, 0.67 d \left(\frac{\varepsilon_{mu}}{1.5\varepsilon_y + \varepsilon_{mu}}\right) b - \rho d b f_y$$

$$\rho = \frac{0.85 f'_{AAC}\, 0.67 d \left(\frac{\varepsilon_{mu}}{1.5\varepsilon_y + \varepsilon_{mu}}\right) b - \frac{N_u}{\phi}}{b d f_y}$$

$$\rho = \frac{0.85 f'_{AAC}\, 0.67 \left(\frac{\varepsilon_{mu}}{1.5\varepsilon_y + \varepsilon_{mu}}\right) - \frac{N_u}{bd\phi}}{f_y}$$

so

$$\rho_{max} = \frac{0.57 f'_{AAC} \left(\dfrac{\varepsilon_{mu}}{1.5\varepsilon_y + \varepsilon_{mu}} \right) - \dfrac{N_u}{bd\phi}}{f_y}$$

14.10 Design of Reinforced Shear Walls of AAC Masonry

Reinforced shear walls of AAC masonry must be designed for the effects of: (1) gravity loads from self-weight plus gravity loads from overlying roof or floor levels and (2) moments and shears from in-plane shear loads.

Actions are shown in Fig. 14.25.

Flexural capacity of reinforced AAC shear walls is calculated using moment-axial force interaction diagrams as discussed in the section on AAC masonry walls loaded out-of-plane. In contrast to the elements addressed in that section, a shear wall is subjected to flexure in its own plane rather than out-of-plane. It therefore usually has multiple layers of flexural reinforcement. Computation of moment-axial force interaction diagrams for shear walls is much easier using a spreadsheet.

From the 2008 MSJC *Code*, Sec. A.3.4.1.2, nominal shear strength is the summation of shear strength from AAC masonry and shear strength from shear reinforcement:

$$V_n = V_{nAAC} + V_{ns}$$

According to Sec. A.3.4.1.2 of the 2008 MSJC *Code*, the nominal shear capacity of AAC masonry is the least of the following three equations,

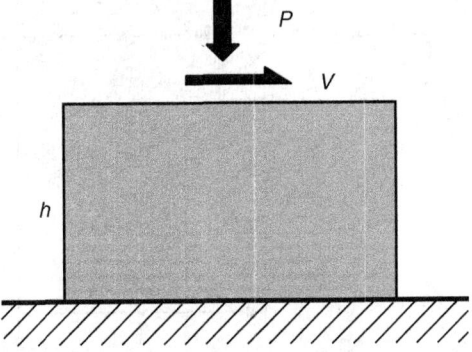

FIGURE 14.25 Design actions for reinforced AAC masonry shear walls.

related respectively to web-shear cracking, crushing of the diagonal strut, and sliding.

$$V_{nAAC} = \min \begin{cases} 0.95\,\lambda_w\,t\sqrt{f'_{AAC}}\sqrt{1 + \dfrac{P_u}{2.4\sqrt{f'_{AAC}}\,\lambda_w t}} \\[2ex] 0.17\,f'_{AAC}\,t\,\dfrac{h\cdot\lambda_w^2}{h^2 + (\tfrac{3}{4}\lambda_w)^2} \\[2ex] \mu_{AAC} P_u \end{cases}$$

The strength-reduction factor for shear is 0.80 (2008 MSJC *Code*, Sec. A.1.5.3).

Just as in reinforced concrete design, this model assumes that shear is resisted by reinforcement crossing a hypothetical failure surface oriented at 45 degrees, as shown in Fig. 14.26.

The nominal resistance from reinforcement is taken as the area associated with each set of shear reinforcement, multiplied by the number of sets of shear reinforcement crossing the hypothetical failure surface. Because the hypothetical failure surface is assumed to be inclined at 45 degrees, its projection along the length of the member is approximately equal to d_v, and the number of sets of shear reinforcement crossing the hypothetical failure surface can be approximated by (d_v/s):

$$V_{ns} = A_v f_y n$$

$$V_{ns} = A_v f_y \left(\dfrac{d_v}{s}\right)$$

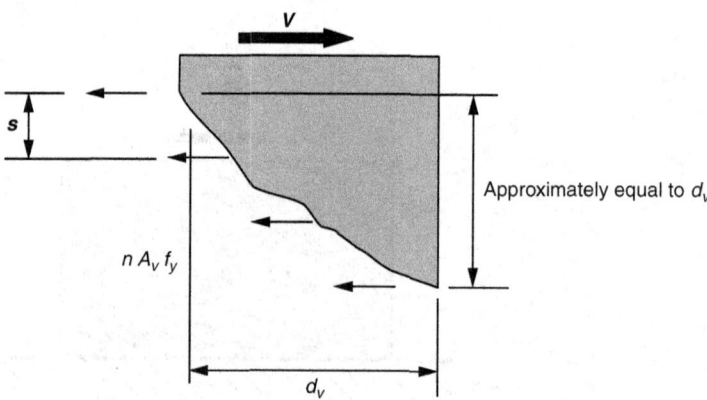

Figure 14.26 Idealized model used in evaluating the resistance due to shear reinforcement.

In contrast to Sec. 3.3.4.1.2.2 of the 2008 MSJC *Code* (strength design of clay or concrete masonry), an efficiency factor of 0.5 is not used. From the 2008 MSJC *Code*, Sec. A.3.4.1.2.4,

$$V_{ns} = \left(\frac{A_v}{s}\right) f_y d_v$$

Again in contrast to Sec. 3.3.4.1.2.2 of the 2008 MSJC *Code* (strength design of clay or concrete masonry), joint reinforcement is not permitted to be used, because it produces local bearing failures of the AAC. Only deformed reinforcement in grouted bond beams is permitted to be included in computing V_{ns} (2008 MSJC *Code*, Sec. A.3.4.1.1).

Finally, because shear resistance really comes from a truss mechanism in which horizontal reinforcement is in tension, and diagonal struts in the masonry are in compression, crushing of the diagonal compressive struts is controlled by limiting the total shear resistance V_n, regardless of the amount of shear reinforcement.

For $(M_u/V_u d_v) < 0.25$,

$$V_n = 6 A_n \sqrt{f'_{AAC}}$$

and for $(M_u/V_u d_v) > 1.00$,

$$V_n = 4 A_n \sqrt{f'_{AAC}}$$

As shown in Fig. 14.27, interpolation is permitted between these limits.

If these upper limits on V_n are not satisfied, the cross-sectional area of the section must be increased.

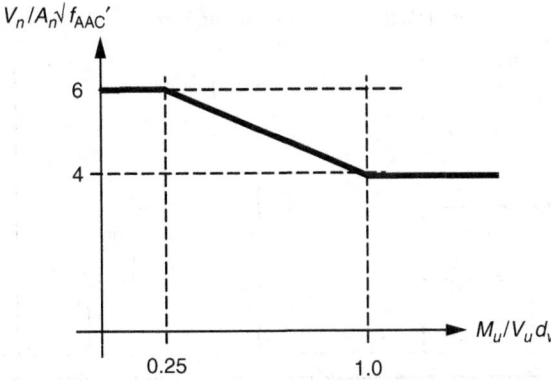

FIGURE 14.27 Maximum permitted nominal shear capacity of AAC masonry as a function of $(M_u/V_u d_v)$.

484 Chapter Fourteen

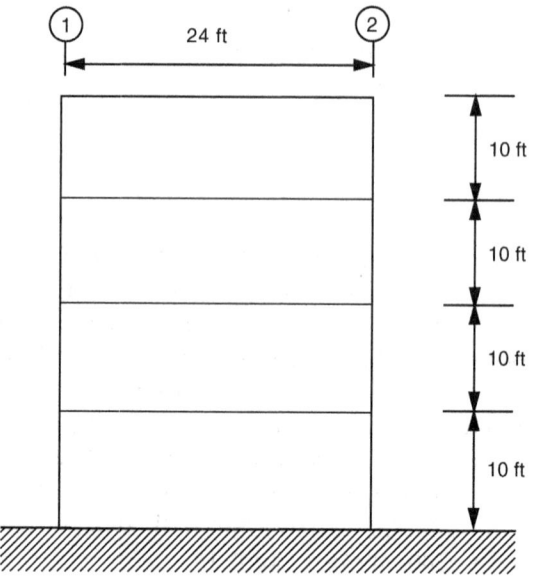

FIGURE 14.28 Reinforced AAC masonry shear wall to be designed.

14.10.1 Example of Design of Reinforced AAC Masonry Shear Wall

Consider the masonry shear wall shown in Fig. 14.28.

Design the wall. Unfactored in-plane lateral loads at each floor level are due to earthquake, and are shown in Fig. 14.29, along with the corresponding shear and moment diagrams.

Assume a 12-in. AAC masonry wall, Class 6 AAC (f'_{AAC} = 870 lb/in.²), with thin-bed mortar. The total plan length of the wall is 24 ft (288 in.),

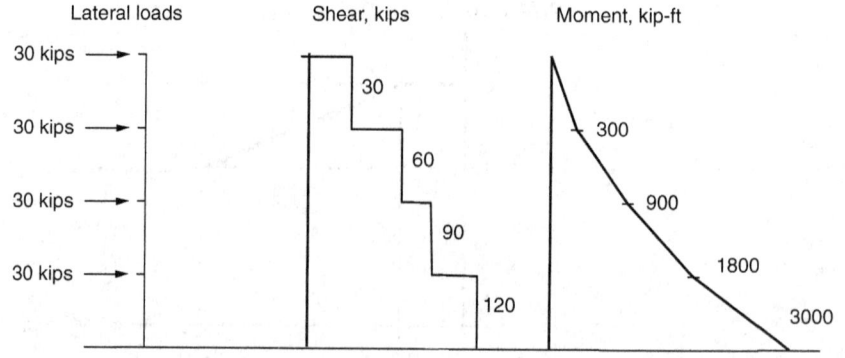

FIGURE 14.29 Unfactored in-plane lateral loads, shear, and moment diagrams for reinforced AAC masonry shear wall.

and its specified thickness is 11.9 in. Assume an effective depth d of 285 in. As is shown later, the reason for the higher strength class of AAC and the greater wall thickness is to increase the shear capacity of the AAC masonry. Shear design and capacity design requirements for shear are critical for this wall.

Unfactored axial loads on the wall are given in the table below.

Level (top of wall)	DL (kips)	LL (kips)
4	90	15
3	180	35
2	270	55
1	360	75

Use 2009 IBC SD Load Combination 7: $0.9D + 1.0E$. At the base of the wall, the factored axial load for the critical loading combination is $0.9D$, or 0.9×360 kips $= 324$ kips.

Check shear for assumed wall thickness. By Sec. A.3.4.1.2 of the 2008 MSJC Code,

$$V_n = V_{nAAC} + V_{ns}$$

$$V_{nAAC} = \min \begin{cases} 0.95 \lambda_w t \sqrt{f'_{AAC}} \sqrt{1 + \dfrac{P_u}{2.4\sqrt{f'_{AAC}}\, \lambda_w t}} \\[6pt] 0.17 f'_{AAC} t \dfrac{h \cdot \lambda_w^2}{h^2 + (\frac{3}{4}\lambda_w)^2} \\[6pt] \mu_{AAC} P_u \end{cases}$$

Take the coefficient of friction for the third equation as 1.0 (AAC against mortar).

$$V_{nAAC} = \min \begin{cases} 0.95 \times 288 \text{ in.} \times 11.9 \text{ in.} \sqrt{870 \text{ lb/in.}^2} \sqrt{1 + \dfrac{324{,}000 \text{ lb}}{2.4\sqrt{870 \text{ lb/in.}^2} \times 288 \text{ in.} \times 11.9 \text{ in.}}} \\[6pt] 0.17 \times 870 \text{ lb/in.}^2 \times 11.9 \text{ in.} \dfrac{40 \text{ ft} \times 24^2 \text{ ft}^2}{40^2 \text{ ft}^2 + (\frac{3}{4} \times 24)^2 \text{ ft}^2} \times 12 \text{ in./ft} \\[6pt] 1.0 \times 324 \text{ kips} \end{cases}$$

$$V_{nAAC} = \min \begin{cases} 146{,}761 \text{ lb} \\ 252{,}914 \text{ lb} \\ 324{,}000 \text{ lb} \end{cases}$$

$$\phi V_n > V_u \qquad \phi = 0.80 \qquad V_n = V_{nAAC} = 146.8 \text{ kips}$$

$$0.80(146.8 \text{ kips}) = 117.4 \text{ kips} < V_u = 120 \text{ kips}$$

Shear design is not satisfactory. Reinforcement will be required. Use horizontal reinforcement consisting of two #4 bars in bond beams at each story level, corresponding to a spacing of 10 ft. The additional nominal capacity due to reinforcement is

$$V_{ns} = \left(\frac{A_v}{s}\right) f_y d_v$$

$$V_{ns} = \left(\frac{2 \times 0.20 \text{ in.}^2}{10 \text{ ft}}\right) 60 \text{ kip/in.}^2 \times 24 \text{ ft}$$

$$V_{ns} = 57.6 \text{ kip}$$

$$V_n = V_{nAAC} + V_{ns}$$

$$V_n = 146.8 \text{ kips} + 57.6 \text{ kips}$$

$$V_n = 204.4 \text{ kips}$$

$$\phi V_n = 0.8\, V_n = 163.5 \text{ kips}$$

$$M_u = 3000 \times 12 \times 1000 \text{ in.-lb} = 36.0 \times 10^6 \text{ in.-lb}$$

$$V_u d_v = 120{,}000 \text{ lb} \qquad d_v = 285 \text{ in.}$$

$$M_u/V_u d_v = \frac{36 \times 10^6 \text{ in.-lb}}{120{,}000 \text{ lb}\,(285 \text{ in.})} = 1.05$$

For $(M_u/V_u d_v) > 1.00$,

$$V_n \leq 4 A_n \sqrt{f'_{AAC}}$$

$$V_n \leq 4 \times 11.9 \text{ in.} \times 288 \text{ in.} \sqrt{870 \text{ lb/in.}^2}$$

$$V_n \leq 404{,}351 \text{ lb}$$

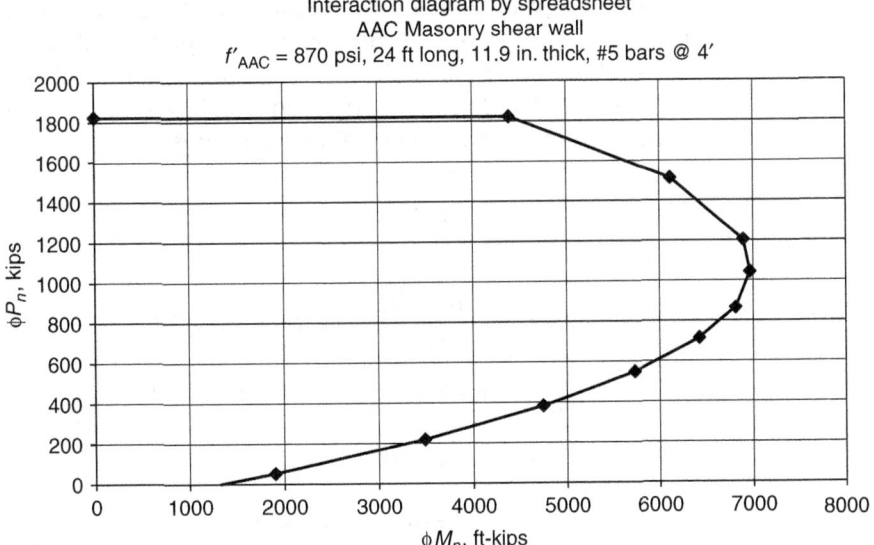

FIGURE 14.30 Moment-axial force interaction for reinforced AAC shear wall, neglecting slenderness effects.

This is satisfied, and design for shear is OK so far. Code Sec. 1.17.3.2.6.1 will be checked later.

Now check flexural capacity using a spreadsheet-generated moment-axial force interaction diagram. Try #5 bars @ 4 ft. Neglecting slenderness effects, the diagram is shown in Fig. 14.30.

At a factored axial load of $0.9D$, or 0.9×360 kips $= 324$ kips, the design flexural capacity of this wall is about 4000 ft-kips, and the design is satisfactory for flexure.

We have designed the wall for the calculated design shear. AAC masonry shear walls are required to be designed to meet the capacity design requirements of *Code* Sec. 1.17.3.2.6.1.1.

At an axial load of 324 kips, the nominal flexural capacity of this wall is the design capacity of 4000 ft-kips, divided by the strength reduction factor of 0.9, or 4000 ft-kips. The ratio of this nominal flexural capacity to the factored design moment is 4444 divided by 3000, or 1.48. Including the additional factor of 1.25, that gives a ratio of 1.85.

$$\phi V_n \geq 1.85 \, V_u$$

$$V_n \geq \frac{1.85}{\phi} V_u = \frac{1.85}{0.8} V_u = 2.31 \, V_u = 2.31 \times 120 = 277.8 \text{ kips}$$

Shear design is not satisfactory. Additional shear reinforcement will be required. We are still under the maximum upper limit for V_n. Use horizontal reinforcement consisting of two #5 bars in bond beams at each story level and at midheight, corresponding to a spacing of 5 ft. The additional nominal capacity due to reinforcement is

$$V_{ns} = \left(\frac{A_v}{s}\right) f_y d_v$$

$$V_{ns} = \left(\frac{2 \times 0.31 \text{ in.}^2}{5 \text{ ft}}\right) 60 \text{ kip/in.}^2 \times 24 \text{ ft}$$

$$V_{ns} = 178.6 \text{ kip}$$

$$V_n = V_{nAAC} + V_{ns}$$

$$V_n = 146.8 \text{ kips} + 178.6 \text{ kips}$$

$$V_n = 325.4 \text{ kips}$$

$$M_u = 3000 \times 12 \times 1000 \text{ in.-lb} = 36.0 \times 10^6 \text{ in.-lb}$$

$$V_u d_v = 120,000 \text{ lb} \qquad d_v = 285 \text{ in.}$$

$$M_u/V_u d_v = \frac{36 \times 10^6 \text{ in.-lb}}{120,000 \text{ lb}(285 \text{ in.})} = 1.05$$

For $(M_u/V_u d_v) > 1.00$,

$$V_n \le 4 A_n \sqrt{f'_{AAC}}$$

$$V_n \le 4 \times 7.9 \text{ in.} \times 288 \text{ in.} \sqrt{870 \text{ lb/in.}^2}$$

$$V_n \le 268,435 \text{ lb}$$

This is satisfied, and design for shear is OK so far.

Check ρ_{max}, assuming that the wall is classified as an ordinary reinforced AAC masonry shear wall ($\alpha = 1.5$). See derivation and discussion at the end of this section.

$$\rho_{max} = \frac{0.64 f'_m \left(\dfrac{\varepsilon_{mu}}{\alpha \varepsilon_y + \varepsilon_{mu}}\right) - \dfrac{N_u}{bd\phi}}{f_y \left(\dfrac{\alpha \varepsilon_y - \varepsilon_{mu}}{\alpha \varepsilon_y + \varepsilon_{mu}}\right)}$$

Structural Design of AAC Masonry

$$\rho_{max} = \frac{0.64 f'_m \left(\dfrac{\varepsilon_{mu}}{4\varepsilon_y + \varepsilon_{mu}}\right) - \dfrac{N_u}{bd\phi}}{f_y \left(\dfrac{4\varepsilon_y - \varepsilon_{mu}}{4\varepsilon_y + \varepsilon_{mu}}\right)}$$

In accordance with MSJC *Code* Sec. 3.3.3.5.1(d), the governing axial load combination is $D + 0.75 L + 0.525 Q_E$, and the axial load is (360,000 + 0.75 × 75,000 lb), or 416,250 lb.

$$\rho_{max} = \frac{0.57 f'_m \left(\dfrac{\varepsilon_{mu}}{1.5\varepsilon_y + \varepsilon_{mu}}\right) - \dfrac{N_u}{bd\phi}}{f_y \left(\dfrac{1.5\varepsilon_y - \varepsilon_{mu}}{1.5\varepsilon_y + \varepsilon_{mu}}\right)}$$

$$\rho_{max} = \frac{0.57\,(870\text{ psi})\left[\dfrac{0.003}{1.5\,(0.00207)+0.003}\right] - \dfrac{416,250\text{ lb}}{(11.9\text{ in.})(285\text{ in.})(0.9)}}{(60,000\text{ psi})\left[\dfrac{1.5\,(0.00207)-0.003}{1.5\,(0.00207)+0.003}\right]}$$

$$\rho_{max} = 0.103$$

Check maximum area of flexural reinforcement per 48 in. of wall length

$$A_{s\,max} = \rho_{max} b \times 48\text{ in.} = 0.103\,(11.9\text{ in.})\,48\text{ in.} = 58.8\text{ in.}^2$$

We have 0.31 in.² every 48 in., and the design is satisfactory.

Summary: Use #5 @ 4 ft vertically, grouted bond beams with two #5 bars @ 5 ft.

14.10.2 Minimum and Maximum Reinforcement Ratios for Flexural Design of AAC Masonry Shear Walls

Minimum Flexural Reinforcement by 2008 MSJC *Code*
The 2008 MSJC *Code* has no global requirements for minimum flexural reinforcement for AAC masonry shear walls.

Maximum Flexural Reinforcement by 2008 MSJC *Code*
The 2008 MSJC *Code* has a maximum reinforcement requirement (Sec. A.3.3.5) that is intended to ensure ductile behavior over a range of axial loads. As compressive axial load increases, the maximum permissible

reinforcement percentage decreases. For compressive axial loads above a critical value, the maximum permissible reinforcement percentage drops to zero, and design is impossible unless the cross-sectional area of the element is increased.

For walls subjected to in-plane forces, for columns, and for beams, the provisions of the 2008 MSJC *Code* set the maximum permissible reinforcement based on a critical strain condition in which the masonry is at its maximum useful strain, and the extreme tension reinforcement is set at a multiple of the yield strain, where the multiple depends on the expected curvature ductility demand on the wall. For "special" reinforced masonry shear walls, the multiple is 4; for "intermediate" walls, it is 3. For walls not required to undergo inelastic deformations, no upper limit is imposed.

The critical strain condition for walls loaded in-plane, and for columns and beams, is shown in Fig. 14.31 below, along with the corresponding stress state. The multiple is termed "α." The parameters for the equivalent rectangular stress block are the same as those used for conventional flexural design. The height of the stress block is $0.85 f'_{AAC}$, and the depth is $0.67 c$. The stress in yielded tensile reinforcement is assumed to be f_y. Compression reinforcement is included in the calculation, based on the assumption that protecting the compression toe will permit the masonry there to provide lateral support to the compression reinforcement. This assumption, while perhaps reasonable, is not consistent with that used for calculation of moment-axial force interaction diagrams.

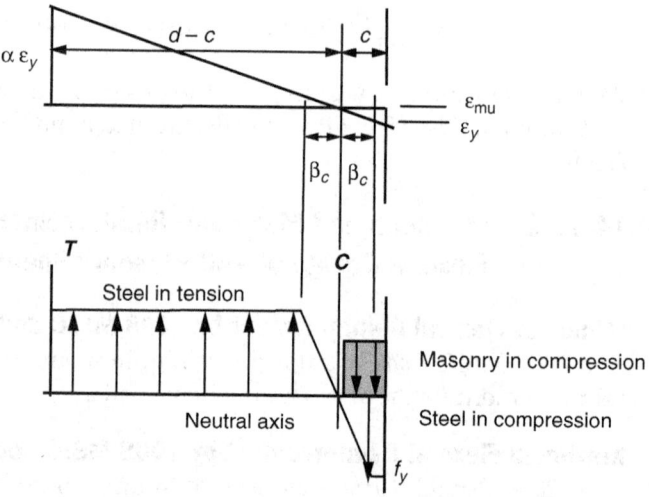

FIGURE 14.31 Critical strain condition for design of AAC masonry walls loaded in-plane, and for columns and beams.

Locate the neutral axis using the critical strain condition:

$$\frac{\varepsilon_{mu}}{\alpha \varepsilon_y} = \frac{c}{d-c}$$

$$c = d\left(\frac{\varepsilon_{mu}}{\alpha \varepsilon_y + \varepsilon_{mu}}\right)$$

Compute the tensile and compressive forces acting on the section, assuming uniformly distributed flexural reinforcement, with a percentage of reinforcement $\rho = \frac{A_s}{bd}$. On each side of the neutral axis, the distance over which the reinforcement is in the elastic range is βc, where β is given by proportion as $\beta = \frac{\varepsilon_y}{\varepsilon_{mu}}$.

The compressive force in the AAC masonry is given by:

$$C_{masonry} = 0.85 \, f'_{AAC} \, 0.67 \, cb$$

The compressive force in the reinforcement is given by:

$$C_{steel} = \rho \beta \, cbf_y \left(\frac{1}{2}\right) + \rho(1-\beta)cbf_y$$

$$C_{steel} = \rho\left(\frac{\varepsilon_y}{\varepsilon_{mu}}\right) cbf_y \left(\frac{1}{2}\right) + \rho\left[1 - \left(\frac{\varepsilon_y}{\varepsilon_{mu}}\right)\right]cbf_y$$

The tensile force in the reinforcement is given by:

$$T_{steel} = \rho \beta \, cbf_y \left(\frac{1}{2}\right) + \rho\,(d - c - \beta c)\,bf_y$$

$$T_{steel} = \rho\left(\frac{\varepsilon_y}{\varepsilon_{mu}}\right) cbf_y \left(\frac{1}{2}\right) + \rho\left[d - c - \left(\frac{\varepsilon_y}{\varepsilon_{mu}}\right)c\right]bf_y$$

Equilibrium of axial forces requires:

$$N_n = C - T$$

$$\frac{N_u}{\phi} = C - T$$

$$\frac{N_u}{\phi} = 0.85 f'_{AAC}\, 0.67\, cb$$

$$+ \rho\left(\frac{\varepsilon_y}{\varepsilon_{mu}}\right) cb f_y \left(\frac{1}{2}\right) + \rho\left[1-\left(\frac{\varepsilon_y}{\varepsilon_{mu}}\right)\right] cb f_y$$

$$- \rho\left(\frac{\varepsilon_y}{\varepsilon_{mu}}\right) cb f_y \left(\frac{1}{2}\right) - \rho\left[d - c - \left(\frac{\varepsilon_y}{\varepsilon_{mu}}\right)c\right] b f_y$$

$$\frac{N_u}{\phi} = 0.85 f'_{AAC}\, 0.67\, cb + \rho\left[1-\left(\frac{\varepsilon_y}{\varepsilon_{mu}}\right)\right] cb f_y - \rho\left[d - c - \left(\frac{\varepsilon_y}{\varepsilon_{mu}}\right)c\right] b f_y$$

$$\frac{N_u}{\phi} = 0.85 f'_{AAC}\, 0.67\, cb + \rho(2-d) cb f_y$$

$$\frac{N_u}{\phi} = 0.85 f'_{AAC}\, 0.67\, d\left(\frac{\varepsilon_{mu}}{\alpha \varepsilon_y + \varepsilon_{mu}}\right) b + 2\rho cb f_y - \rho db f_y$$

$$\frac{N_u}{\phi} = 0.85 f'_{AAC}\, 0.67\, d\left(\frac{\varepsilon_{mu}}{\alpha \varepsilon_y + \varepsilon_{mu}}\right) b + 2\rho d\left(\frac{\varepsilon_{mu}}{\alpha \varepsilon_y + \varepsilon_{mu}}\right) b f_y - \rho db f_y$$

$$\frac{N_u}{\phi} = 0.85 f'_{AAC}\, 0.67\, d\left(\frac{\varepsilon_{mu}}{\alpha \varepsilon_y + \varepsilon_{mu}}\right) b + \rho bd f_y\left[\left(\frac{2\varepsilon_{mu}}{\alpha \varepsilon_y + \varepsilon_{mu}}\right) - 1\right]$$

$$\rho bd f_y\left[1 - \left(\frac{2\varepsilon_{mu}}{\alpha \varepsilon_y + \varepsilon_{mu}}\right)\right] = 0.85 f'_{AAC}\, 0.67\, d\left(\frac{\varepsilon_{mu}}{\alpha \varepsilon_y + \varepsilon_{mu}}\right) b - \frac{N_u}{\phi}$$

$$\rho = \frac{0.85 f'_{AAC}\, 0.67\, d\left(\dfrac{\varepsilon_{mu}}{\alpha \varepsilon_y + \varepsilon_{mu}}\right) b - \dfrac{N_u}{\phi}}{bd f_y\left[1 - \left(\dfrac{2\varepsilon_{mu}}{\alpha \varepsilon_y + \varepsilon_{mu}}\right)\right]} = \frac{0.85 f'_{AAC}\, 0.67\left(\dfrac{\varepsilon_{mu}}{\alpha \varepsilon_y + \varepsilon_{mu}}\right) - \dfrac{N_u}{bd\phi}}{f_y\left[1 - \left(\dfrac{2\varepsilon_{mu}}{\alpha \varepsilon_y + \varepsilon_{mu}}\right)\right]}$$

$$\rho = \frac{0.85 f'_{AAC}\, 0.67\left(\dfrac{\varepsilon_{mu}}{\alpha \varepsilon_y + \varepsilon_{mu}}\right) - \dfrac{N_u}{bd\phi}}{f_y\left[\left(\dfrac{\alpha \varepsilon_y + \varepsilon_{mu}}{\alpha \varepsilon_y + \varepsilon_{mu}}\right) - \left(\dfrac{2\varepsilon_{mu}}{\alpha \varepsilon_y + \varepsilon_{mu}}\right)\right]} = \frac{0.85 f'_{AAC}\, 0.67\left(\dfrac{\varepsilon_{mu}}{\alpha \varepsilon_y + \varepsilon_{mu}}\right) - \dfrac{N_u}{bd\phi}}{f_y\left(\dfrac{\alpha \varepsilon_y - \varepsilon_{mu}}{\alpha \varepsilon_y + \varepsilon_{mu}}\right)}$$

so

$$\rho_{max} = \frac{0.57 f'_{AAC} \left(\dfrac{\varepsilon_{mu}}{\alpha \varepsilon_y + \varepsilon_{mu}} \right) - \dfrac{N_u}{bd\phi}}{f_y \left(\dfrac{\alpha \varepsilon_y - \varepsilon_{mu}}{\alpha \varepsilon_y + \varepsilon_{mu}} \right)}$$

14.10.3 Additional Comments of the Design of Reinforced AAC Shear Walls

Reinforced AAC masonry shear walls have lower shear capacity than otherwise similar shear walls of concrete or clay masonry. They will probably need to be thicker than clay or concrete masonry shear walls, and will probably need Class 6 AAC (the highest strength). Capacity design may govern the design for shear, because these walls are required to be ductile.

14.11 Seismic Design of AAC Structures

Because it has been used extensively in Europe for more than 70 years, AAC has been extensively researched there (RILEM 1993). Outside of the U.S., seismic qualification of AAC components and structures is based on experience in the Middle East and Japan. In the United States, it is based indirectly on that experience, and directly on an extensive experimental and analytical research program conducted at The University of Texas at Austin, and described further here and in Tanner et al. (2005a,b), Varela et al. (2006), and Klingner et al. (2005a,b). That research program developed design models, draft design provisions, and seismic design factors (R and C_d). In the rest of this chapter, the U.S. approach to seismic design of AAC structures is summarized, a design example is presented, and the research background for the design procedure is reviewed.

14.11.1 Basic Earthquake Resistance Mechanism of AAC Structures

Structures whose basic earthquake resistance depends on AAC elements are generally shear-wall structures. Lateral earthquake loads are carried by horizontal diaphragms to AAC shear walls, which transfer those loads to the ground. General response of shear wall structures to lateral loads is discussed in the *Masonry Designers' Guide* (MDG 2006), and is not repeated here.

Earthquake design of AAC shear-wall structures is similar to earthquake design of conventional masonry shear-wall structures. A complete design example is given later in this document.

14.11.2 Seismic Design Factors (R and C_d) for Ductile AAC Shear-Wall Structures in the United States

Because AAC structures (whether of masonry units or reinforced panels) in practically all parts of the United States must be designed for earthquake loads, it is necessary to develop seismic design factors (R and C_d) for use with ASCE 7, the seismic load document referenced by model codes such as the IBC.

The seismic force-reduction factor (R) is intended to account for ductility, and for structural overstrength. It is based on observation of the performance of different structural systems in previous strong earthquakes, on technical justification, and on tradition. Because AAC is a new material in the United States, its seismic design factors (R and C_d) must be based on laboratory test results and numerical simulation of the response of AAC structures to earthquake ground motions. The proposed factors must then be verified against the observed response of AAC structures in strong earthquakes.

Values of R and C_d for ductile AAC shear-wall structures have been proposed in two code-development forums.

- In October 2002, seismic design factors were proposed to and approved by ICC ES (a model-code evaluation service), as part of a proposed ICC ES listing for AAC structural components and systems produced by members of the Autoclaved Aerated Concrete Products Association (AACPA). That listing is intended to make it easier to use such systems throughout the United States, until consensus design provisions are incorporated in MSJC and ACI documents, and are referenced by model codes.

- In September 2006, the ICC Structural Committee approved the following R and C_d values shown in Table 14.6 for ordinary reinforced AAC masonry shear wall systems, for inclusion in the 2007 IBC Supplement. The values were approved in ICC

Response modification coefficient, R	System overstrength factor, Ω_0	Deflection amplification factor, C_D	System limitations and building height limitations (feet) by seismic design category as determined in Sec. 1613.5.6				
			A or B	C	D	E	F
2	2.5	2	NL	35	NP	NP	NP

TABLE 14.6 Seismic Design Factors for Ordinary Reinforced AAC Masonry Shear Walls

public comment hearings in May 2007, and were included in the 2007 Supplement to the 2006 IBC. They are also included in the 2009 IBC.

- Starting in 2005, the same seismic design factors were considered by the Building Seismic Safety Council. They were approved in 2008, and are currently proposed for approval by ASCE7.

14.12 Design Example: Three-Story AAC Shear-Wall Hotel

This example illustrates the preliminary design of a three-story AAC shear-wall hotel in Asheville, North Carolina, a zone of moderate seismic risk, using the loading provisions of the 2009 IBC and the AAC masonry design and detailing provisions of the 2008 MSJC *Code* and *Specification*. The principal lateral force-resisting elements of the structure are transverse shear walls.

The design proceeds using the following steps:

1. Choose design criteria:
   a. Propose plan, elevation, materials, f'_{AAC}
   b. Calculate D, L, W, E loads
   c. Propose structural systems for gravity and lateral load
2. Design transverse shear walls for gravity and earthquake loads
3. Design exterior walls for gravity and wind loads
   a. Earthquake loads will be carried by longitudinal walls in-plane
   b. Out-of-plane wind loads will be carried by longitudinal walls out-of-plane using vertical and horizontal strips

14.12.1 Step 1: Choose Design Criteria

The plan and elevation of the building are shown Figs. 14.32 and 14.33. The structure has a story height of 11 ft and a 2-ft parapet, making a total height of 35 ft.

Architectural Constraints

Water-penetration resistance: A single-wythe AAC masonry wall will be used. Exterior protection will be provided by low-modulus acrylic stucco.

Movement joints: To control crack widths from shrinkage of AAC walls, use vertical control joints every bay.

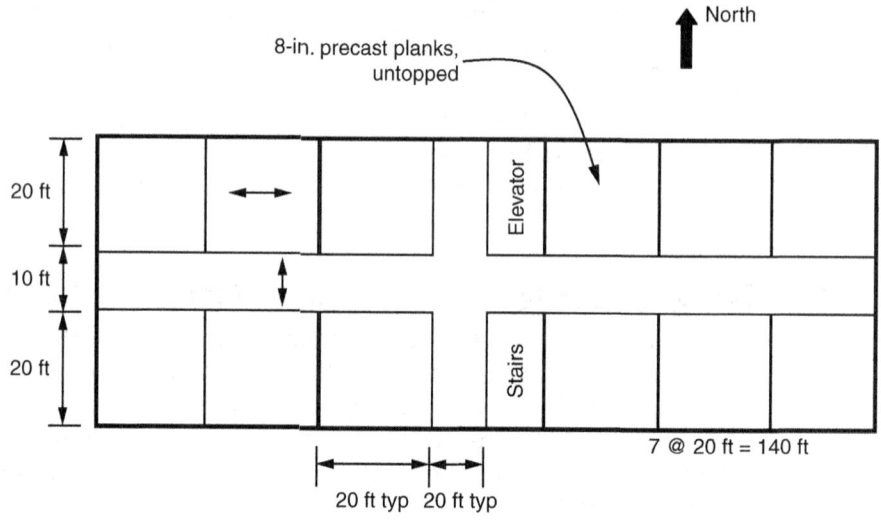

FIGURE 14.32 Plan of 3-story hotel example using AAC masonry.

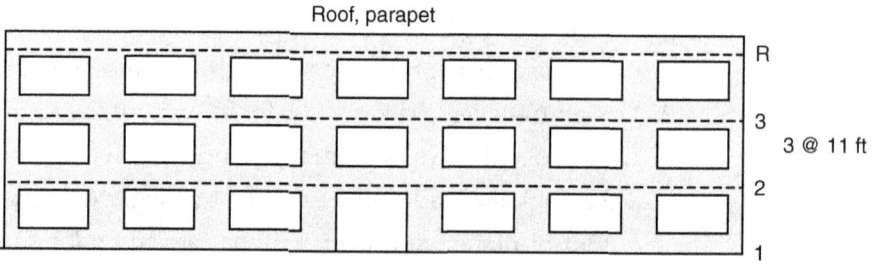

FIGURE 14.33 Elevation of 3-story hotel, typical facade example using AAC masonry.

Design for Fire

Use and occupancy: Group R-1
Use Type I or Type II construction (noncombustible material)
The building meets the area or height restrictions of Table 503
2- or 3-hours rating required
Must meet separation requirements of Table 602 of the 2009 IBC
Bearing walls: 4-h rating   (8-in. nominal AAC masonry OK)
Shafts: 2-h rating   (8-in. nominal AAC masonry OK)
Floors: 2-h rating   (planks and topping OK)

Specify Materials

12-in. AAC masonry units (ASTM C1555), fully mortared

Thin-bed mortar (ASTM C1555)

Class 6 AAC (f'_{AAC} = 6 MPa or 870 psi), assumed unit weight 45 pcf

Deformed reinforcement meeting ASTM A615, Gr. 60

Floors and roof of untopped AAC planks with diaphragm reinforcement

Structural Systems

Gravity load: Gravity load on roof and floors will be transferred to transverse walls. Gravity load on corridor will be transferred to spine walls.

Lateral load: Lateral load (earthquake will govern) will be transferred by floor and roof diaphragms to the transverse shear walls, which will act as statically determinate cantilevers.

Calculate Design Roof Load due to Gravity

Dead load	Planks	30 lb/ft^2
	EPDM membrane, gravel	20 lb/ft^2
	HVAC, roofing	30 lb/ft^2
		80 lb/ft^2 total
Live load	Ignore reduction of live load based on tributary area.	20 lb/ft^2

Calculate Design Floor Load due to Gravity

Dead load	Planks	30 lb/ft^2
	HVAC, floor finish, partitions	20 lb/ft^2
		50 lb/ft^2 total
Live load	Use weighted average of corridor and guest rooms. Ignore reduction of live load based on tributary area.	60 lb/ft^2

Calculate Design Lateral Load from Earthquake

Design earthquake loads are calculated according to Sec. 1613 of the 2009 IBC. That section essentially references ASCE 7-05 (*Supplement*). Seismic design criteria are given in Chap. 11. The seismic design provisions of ASCE 7-05 (*Supplement*) begin in Chap. 12, which prescribes basic requirements (including the requirement for continuous load paths) (Sec. 12.1);

selection of structural systems (Sec. 12.2); diaphragm characteristics and other possible irregularities (Sec. 12.3); seismic load effects and combinations (Sec. 12.4); direction of loading (Sec. 12.5); analysis procedures (Sec. 12.6); modeling procedures (Sec. 12.7); and specific design approaches. Four procedures are prescribed: an equivalent lateral force procedure (Sec. 12.8); a modal response-spectrum analysis (Sec. 12.9); a simplified alternative procedure (Sec. 12.14); and a seismic response history procedure (Chap. 16). The equivalent lateral-force procedure is described here, because it is relatively simple, and is permitted in most situations. The simplified alternative procedure is permitted in only a few situations. The other procedures are permitted in all situations, and are required in only a few situations.

The required seismic design steps are summarized below. Section references are to ASCE 7-05.

Determine Seismic Ground Motion Values
1. Determine S_S, the mapped MCE (maximum considered earthquake), 5 percent damped, spectral response acceleration parameter at short periods as defined in Sec. 11.4.1 of ASCE 7-05.
2. Determine S_1, the mapped MCE, 5 percent damped, spectral response acceleration parameter at a period of 1 s as defined in Sec. 11.4.1 of ASCE 7-05.
3. Determine the *site class* (A through F, a measure of soil response characteristics and soil stability) in accordance with Sec. 20.3 and Table 20.3-1.
4. Determine the MCE spectral response acceleration for short periods (S_{MS}) and at 1 s (S_{M1}), adjusted for Site Class effects, using Eqs. (11.4-1) and (11.4-2), respectively.
5. Determine the design response acceleration parameter for short periods, S_{DS}, and for a 1-s period, S_{D1}, using Eqs. (11.4-3) and (11.4-4), respectively.
6. If required, determine the design response spectrum curve as prescribed by Sec. 11.4.5.

Determine Seismic Base Shear Using the Equivalent Lateral Force Procedure
1. Determine the structure's importance factor, I, and occupancy category using Sec. 11.5.
2. Determine the structure's Seismic Design Category using Sec. 11.6.
3. Calculate the structure's seismic base shear using Secs. 12.8.1 and 12.8.2.

Distribute Seismic Base Shear Vertically and Horizontally
1. Distribute seismic base shear vertically using Sec. 12.8.3.
2. Distribute seismic base shear horizontally using Sec. 12.8.4.

Now apply these steps to our example in Asheville, North Carolina:

Step 1: Determine S_S, the mapped MCE (maximum considered earthquake), 5 percent damped, spectral response acceleration parameter at short periods as defined in Sec. 11.4.1 of ASCE 7-05.

Step 2: Determine S_1, the mapped MCE, 5 percent damped, spectral response acceleration parameter at a period of 1 s as defined in Sec. 11.4.1 of ASCE 7-05.

Determine the parameters S_S and S_1 from the 0.2- and 1-s spectral response maps shown in Figs. 22-1 through 22-7 of ASCE 7-05.

With the exception of some parts of the western United States (where maximum considered earthquakes have a deterministic basis), those maps correspond to accelerations with a 2 percent probability of exceedance within a 50-year period. Such an earthquake is sometimes described as a "2500-year earthquake." To see why, let p be the unknown annual probability of exceedance of that level of acceleration:

The probability of exceedance in a particular year is	p
The probability of non-exceedance in a particular year is	$(1-p)$
The probability of non-exceedance in 50 consecutive years is	$(1-p)^{50}$
The probability of exceedance within a 50-year period is	$[1-(1-p)^{50}]$
Solve for p, the annual probability of exceedance. Set the probability of exceedance within the 50-year period equal to the given 2%	$[1-(1-p)^{50}] = 0.02$ $(1-p)^{50} = 0.98$ $p = 1 - 0.98^{1/50}$ $p = 4.04 \times 10^{-4}$
The return period is the reciprocal of the annual probability of exceedance	$\dfrac{1}{p} = 2475$
The approximate return period is	2500 years

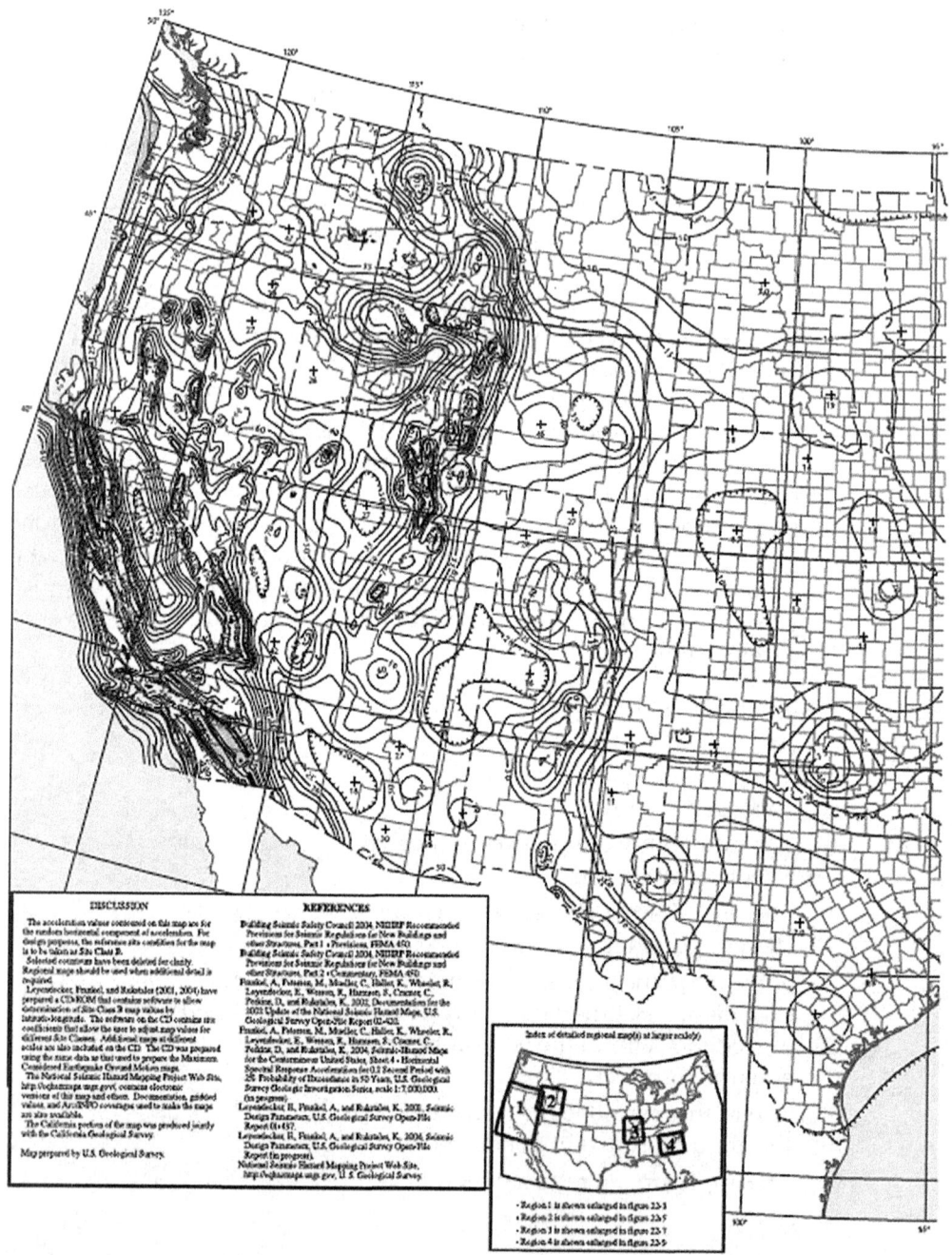

FIGURE 14.34 Figures 22-1 of ASCE 7-05. Maximum considered earthquake ground motion for the conterminous United States of 0.2 sec spectral response acceleration (5% of critical damping), site class B.

Structural Design of AAC Masonry 501

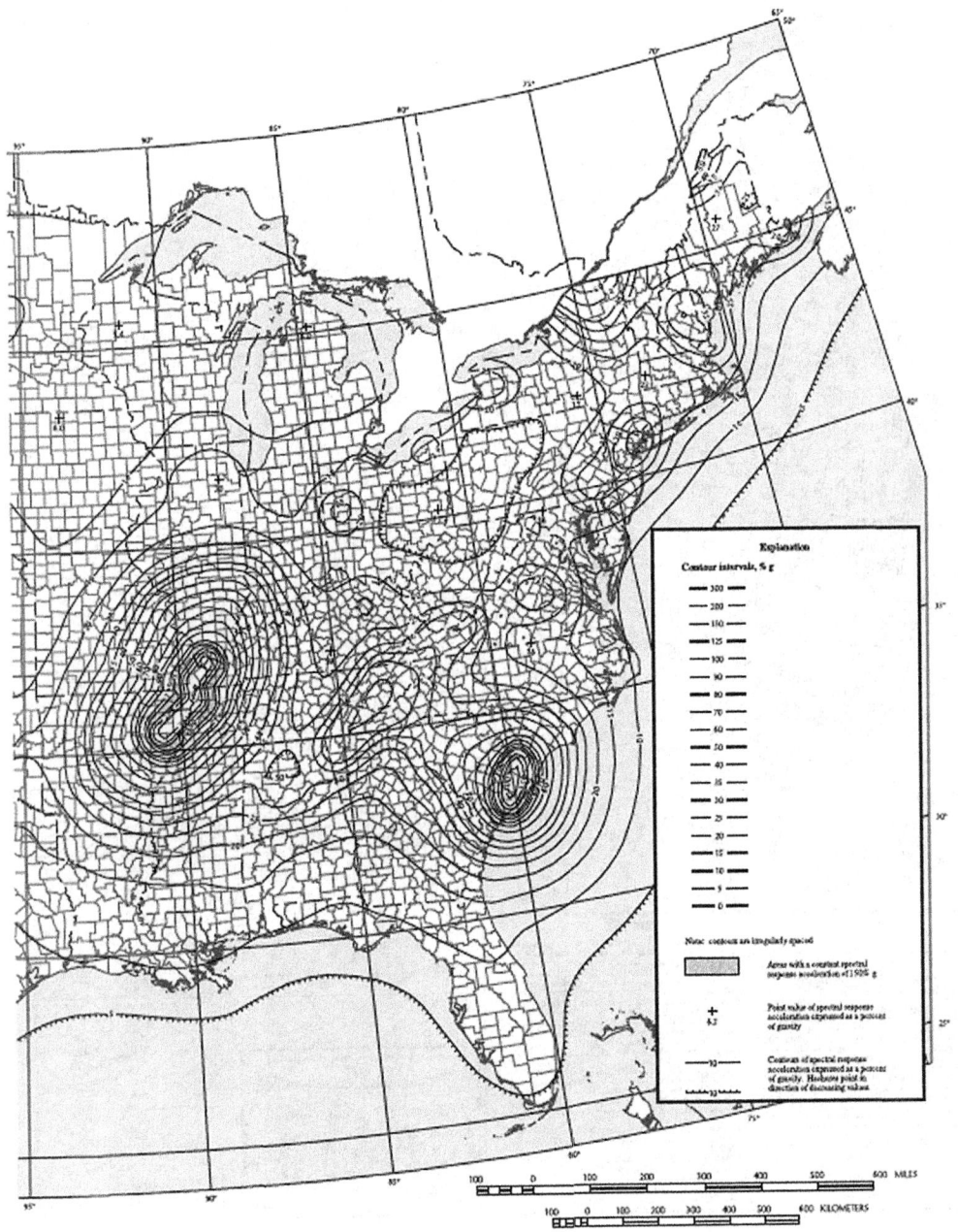

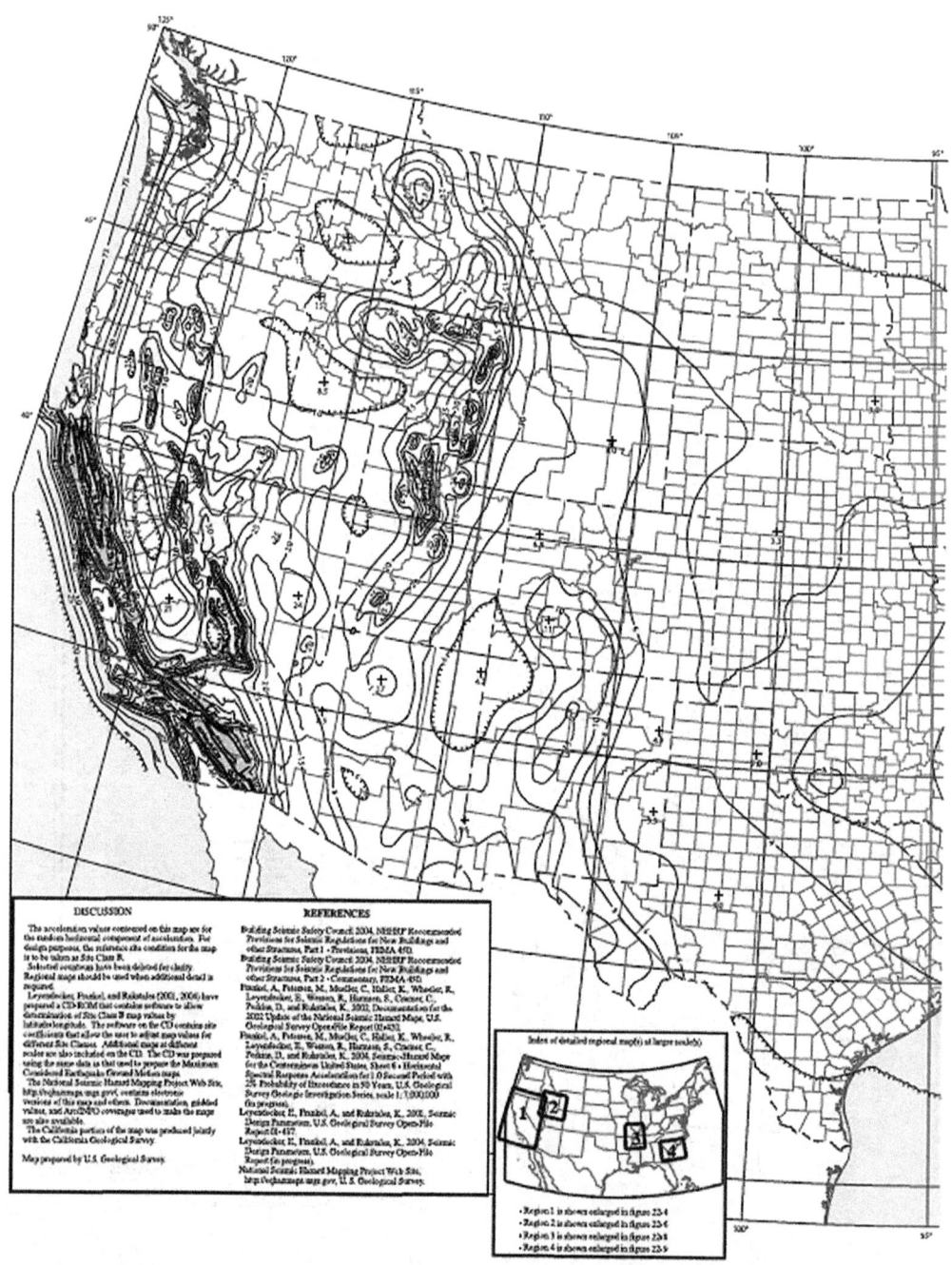

FIGURE 14.35 Figures 22-2 of ASCE 7-05. Maximum considered earthquake ground motion for the conterminous United States of 1.0 sec spectral response acceleration (5% of critical damping), site class B.

Structural Design of AAC Masonry   503

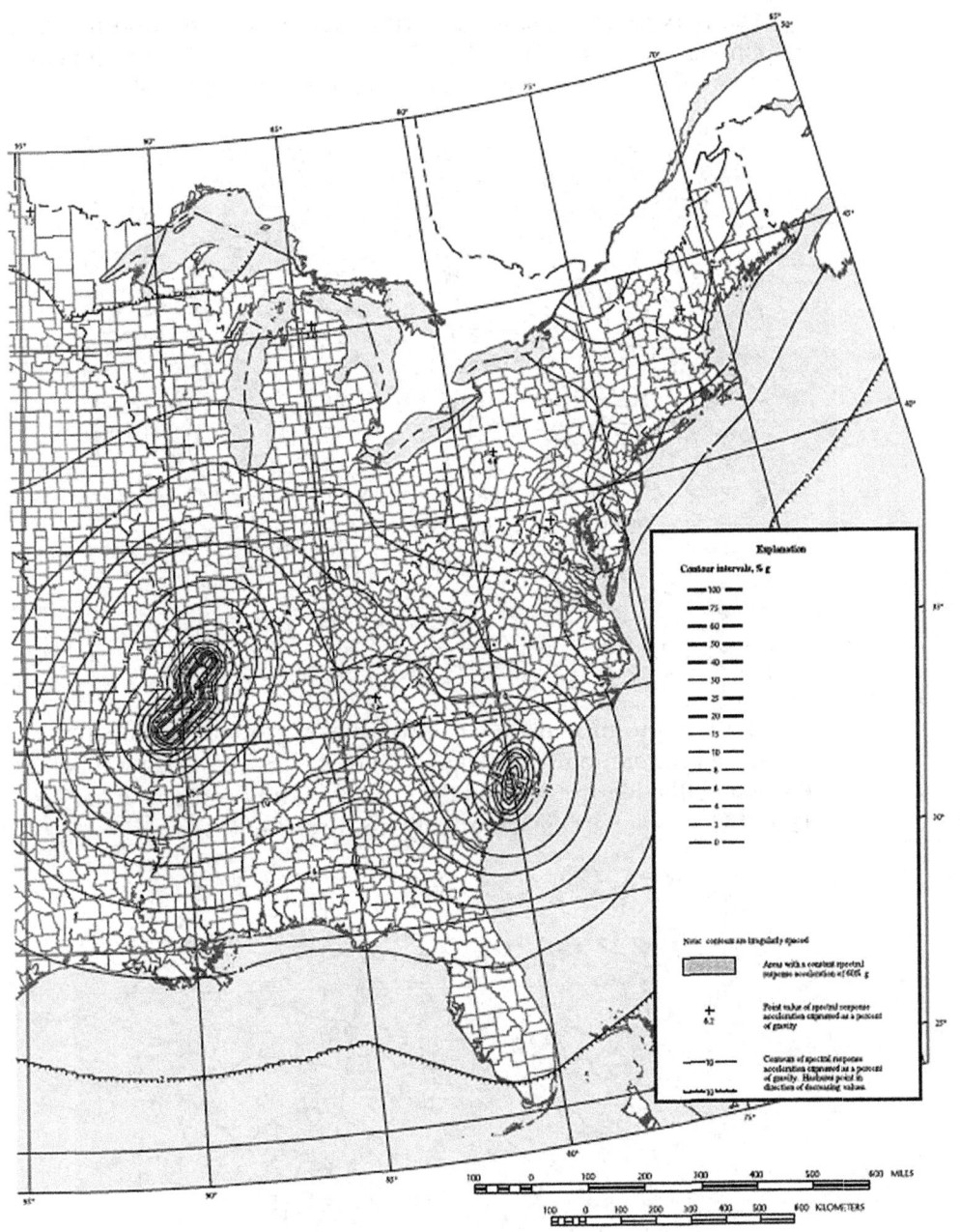

The maximum considered earthquake ground motion for 0.2 s response acceleration (from Fig. 22-1 of ASCE 7-05) is shown below. For Asheville, NC the corresponding contour is 40 percent g.

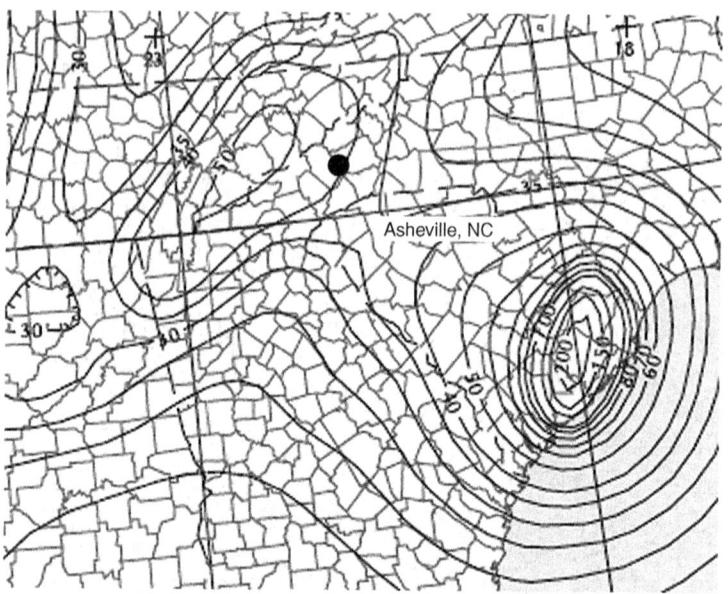

The maximum considered earthquake ground motion for 1 s response acceleration (from Fig. 22-2) of ASCE 7-05 is shown below. For Asheville, North Carolina the corresponding contour is between 11 and 12 percent g. Conservatively use 12 percent g.

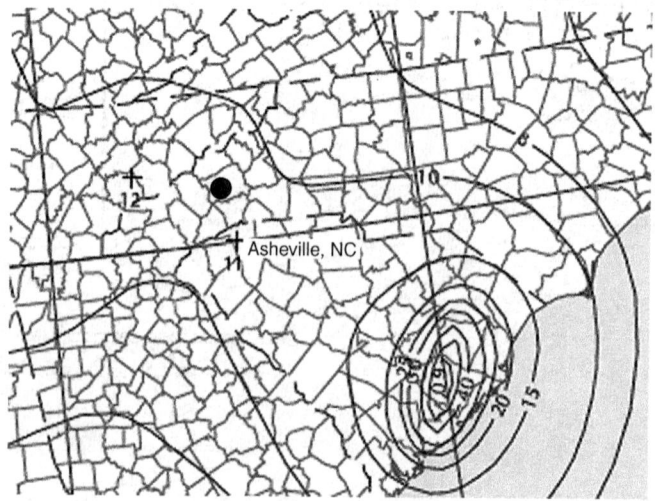

For Asheville, North Carolina, therefore, $S_S = 0.40\,g$ and $S_1 = 0.12\,g$.

Structural Design of AAC Masonry

Site class	$\bar{v}_s$	$\bar{N}$ or $\bar{N}_{ch}$	$\bar{S}_u$
A. Hard rock	>5000 ft/s	NA	NA
B. Rock	2500 to 5000 ft/s	NA	NA
C. Very dense soil and soft rock	1200 to 2500 ft/s	>50	>2000 psf
D. Stiff soil	600 to 1200 ft/s	15 to 50	1000 to 2000 psf
E. Soft clay soil	<600 ft/s	<15	<1000 psf
	Any profile with more than 10 ft of soil having the following characteristics: – Plasticity index PI > 20, – Moisture content w ≥ 40%, and – Undrained shear strength $\bar{S}_u$ < 500 psf		
F. Soils requiring site response analysis in accordance with Section 21.1	See Section 20.3.1		

For SI: 1 ft/s = 0.3048 m/s, 1 lb/ft² = 0.0479 kN/m²
Source: Table 20.3-1 of ASCE 7-05.

TABLE 14.7 Site Classification

Step 3: Determine the *site class* (A through F, a measure of soil response characteristics and soil stability) in accordance with Sec. 20.3 and Table 20.3-1.

In accordance with Table 20.3-1, site classes are assigned as shown in Table 14.7.

Assume Site Class D (stiff soil) in accordance with Sec. 11.4.2.

Step 4: Determine the MCE spectral response acceleration for short periods (S_{MS}) and at 1 s (S_{M1}), adjusted for Site Class effects, using Eqs. (11.4-1) and (11.4-2), respectively.

$$S_{MS} = F_a \cdot S_S \quad (11.4\text{-}1)$$

$$S_{M1} = F_v \cdot S_1 \quad (11.4\text{-}2)$$

For Site Class D, linear interpolation between the tabular values of 1.6 for $S_S \leq 0.25$ and 1.4 for $S_S = 0.5$, and using $S_S = 0.40$, gives an acceleration-dependent site coefficient, F_a, of 1.48 (Table 14.8).

For Site Class D, linear interpolation between the tabular values of 2.4 for $S_1 \leq 0.1$ and 2.0 for $S_1 = 0.2$, and using $S_1 = 0.12$, gives a velocity-dependent site coefficient, F_v, of 2.32 (Table 14.9).

	Mapped maximum considered earthquake spectral response acceleration parameter at short period				
Site class	$S_s \leq 0.25$	$S_s = 0.5$	$S_s = 0.75$	$S_s = 1.0$	$S_s \geq 1.25$
A	0.8	0.8	0.8	0.8	0.8
B	1.0	1.0	1.0	1.0	1.0
C	1.2	1.2	1.1	1.0	1.0
D	1.6	1.4	1.2	1.1	1.0
E	2.5	1.7	1.2	0.9	0.9
F	See Section 11.4.7				

Note: Use straight-line interpolation for intermediate values of S_S.
Source: Table 11.4-1 of ASCE 7-05.

TABLE 14.8 Site Coefficient, F_a

	Mapped maximum considered earthquake spectral response acceleration parameter at 1-s period				
Site class	$S_1 \leq 0.1$	$S_1 = 0.2$	$S_1 = 0.3$	$S_1 = 0.4$	$S_1 \geq 0.5$
A	0.8	0.8	0.8	0.8	0.8
B	1.0	1.0	1.0	1.0	1.0
C	1.7	1.6	1.5	1.4	1.3
D	2.4	2.0	1.8	1.6	1.5
E	3.5	3.2	2.8	2.4	2.4
F	See Section 11.4.7				

Note: Use straight-line interpolation for intermediate values of S_1.
Source: Table 11.4-2 of ASCE 7-05.

TABLE 14.9 Site Coefficient, F_v

Then the maximum considered short-period response acceleration is

$$S_{MS} = F_a \cdot S_S = 1.48 \cdot 0.40g = 0.59g$$

and the maximum considered 1 s response acceleration is

$$S_{M1} = F_v \cdot S_1 = 2.32 \cdot 0.12g = 0.28g$$

Step 5: Determine the design response acceleration parameter for short periods, S_{DS}, and for a 1 s period, S_{D1}, using Eqs. (11.4-3) and (11.4-4), respectively.

The design response acceleration is two-thirds of the maximum considered acceleration:

$$S_{DS} = \frac{2}{3} \cdot S_{MS} \qquad (11.4\text{-}3)$$

$$S_{D1} = \frac{2}{3} \cdot S_{M1} \qquad (11.4\text{-}4)$$

With the exception of some parts of the western United States (where design earthquakes have a deterministic basis), these design spectral ordinates correspond to an earthquake with a 10 percent probability of exceedance within a 50-year period. Such an earthquake is sometimes described as a "500-year earthquake." To see why, let p be the unknown annual probability of exceedance of that level of acceleration:

The probability of exceedance in a particular year is	p
The probability of non-exceedance in a particular year is	$(1-p)$
The probability of non-exceedance in 50 consecutive years is	$(1-p)^{50}$
The probability of exceedance within a 50-year period is	$[1-(1-p)^{50}]$
Solve for p, the annual probability of exceedance. Set the probability of exceedance within the 50-year period equal to the given 10%	$[1-(1-p)^{50}] = 0.10$ $(1-p)^{50} = 0.90$ $p = 1 - 0.90^{(1/50)}$ $p = 2.10 \times 10^{-3}$
The return period is the reciprocal of the annual probability of exceedance	$\dfrac{1}{p} = 475$
The approximate return period is	500 years

Continuing with our example for Asheville, North Carolina, the design response acceleration for short periods is

$$S_{DS} = \frac{2}{3} \cdot S_{MS} = \frac{2}{3} \cdot 0.59g = 0.39g$$

and the design response acceleration for a 1 s period is

$$S_{D1} = \frac{2}{3} \cdot S_{M1} = \frac{2}{3} \cdot 0.28g = 0.19g$$

Step 6: If required, determine the design response spectrum curve as prescribed by Sec. 11.4.5.

Because the equivalent lateral force procedure is being used, the response spectrum curve is not required. Nevertheless, for pedagogical completeness, it will be developed.

First, define

$$T_0 \equiv 0.2 \frac{S_{D1}}{S_{DS}} \quad \text{and} \quad T_S \equiv \frac{S_{D1}}{S_{DS}}$$

Then for our case,

$$T_0 \equiv 0.2 \frac{S_{D1}}{S_{DS}} = 0.2 \left(\frac{0.19g}{0.39g} \right) = 0.10 \text{ s}$$

$$T_S \equiv \frac{S_{D1}}{S_{DS}} = \left(\frac{0.19g}{0.39g} \right) = 0.49 \text{ s}$$

- For periods less than or equal to T_0, the design spectral response acceleration, S_a, is given by Eq. (11.4-5):

$$S_a = S_{DS} \left(0.4 + 0.6 \frac{T}{T_0} \right) \tag{11.4-5}$$

- For periods greater than T_0 and less than or equal to T_S, the design spectral response acceleration, S_a, is equal to S_{DS}.
- For periods greater than T_S and less than or equal to T_L (from Figs. 22-15 through 22-20), the design spectral response acceleration, S_a, is given by Eq. (11.4-6). In our case, $T_L = 8$ s (on the border between 8 and 12 s).

$$S_a = \frac{S_{D1}}{T} \tag{11.4-6}$$

- For periods greater than T_L, the design spectral response acceleration, S_a, is given by Eq. (11.4-7):

$$S_a = \frac{S_{D1} T_L}{T^2} \tag{11.4-7}$$

The resulting design acceleration response spectrum is given in Fig. 14.36.

Step 7: Determine the structure's importance factor, I, and occupancy category using Sec. 11.5.

Importance factors are given in Table 14.10.

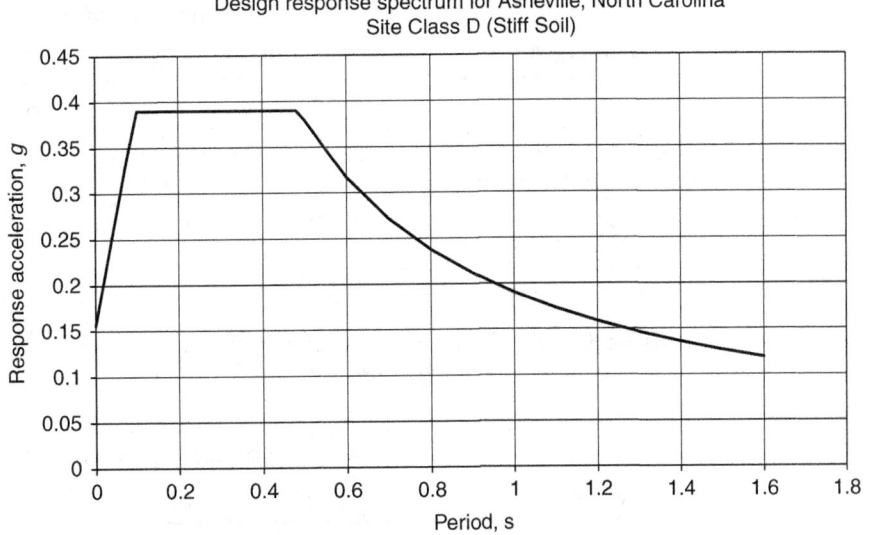

FIGURE 14.36 Design response spectrum for Asheville, North Carolina.

Occupancy category	I
I or II	1.0
III	1.25
IV	1.5

Source: Table 11.5-1 of ASCE 7-05.

TABLE 14.10 Importance Factors

Because the building is a hotel, it is assigned to Occupancy Category II in accordance with Table 1604.5 of the 2009 IBC. This corresponds to an Importance Factor of 1.0.

Step 8: Determine the structure's Seismic Design Category using Sec. 11.6.

Tables 14.11 and 14.12 must be checked, and the higher seismic design category from those two tables applies.

In our case, S_{DS} is 0.39, and S_{D1} is 0.19. Because S_{DS} is between 0.133 and 0.20 (Table 14.11), and S_{D1} is between 0.133 and 0.20 (Table 14.12), the structure is assigned to Seismic Design Category (SDC) C. Because the structure is less than or equal to 35 ft in height, and is assigned to SDC C, it can be designed as ordinary reinforced AAC masonry according to the provisions of the 2009 IBC.

	Occupancy category		
Value of S_{DS}	I or II	III	IV
$S_{DS} < 0.167$	A	A	A
$0.167 \leq S_{DS} < 0.33$	B	B	C
$0.33 \leq S_{DS} < 0.50$	C	C	D
$0.50 \leq S_{DS}$	D	D	D

Source: Table 11.6-1 of ASCE 7-05.

TABLE 14.11 Seismic Design Category Based on Short Period Response Acceleration Parameter

	Occupancy category		
Value of S_{D1}	I or II	III	IV
$S_{D1} < 0.067$	A	A	A
$0.067 \leq S_{D1} < 1.33$	B	B	C
$0.133 \leq S_{D1} < 0.20$	C	C	D
$0.20 \leq S_{D1}$	D	D	D

Source: Table 11.6-2 of ASCE 7-05.

TABLE 14.12 Seismic Design Category Based on 1-s Period Response Acceleration Parameter

Step 9: Calculate the structure's seismic base shear using Secs. 12.8.1 and 12.8.2.

In accordance with ASCE 7-05 (*Supplement*), Sec. 12.8.1.1,

$$C_s = \frac{S_{DS}}{\left(\dfrac{R}{I}\right)} \qquad \text{(Equation 12.8-2)}$$

In our case,

$S_{DS} = 0.39\,g$

$R = 2$ (ordinary AAC masonry shear wall systems)
$I = 1.00$ (ASCE 7-05, Table 11.5-1)

$$C_s = \frac{S_{DS}}{\left(\dfrac{R}{I}\right)} = \frac{0.39}{\left(\dfrac{2}{1}\right)} = 0.20$$

The value of C_s computed in accordance with Eq. 12.8-2 need not exceed the following:

$$C_s = \frac{S_{D1}}{T\left(\frac{R}{I}\right)} \quad \text{for} \quad T \leq T_L \quad \text{(Equation 12.8-3)}$$

The corresponding equation for $T > T_L$ does not apply. In addition, C_s shall not be less than $0.044\,S_{DS}I$, nor less than 0.01.

In our case, the value of C_s given by Eq. 12.8-3 is

$$C_s = \frac{S_{D1}}{T\left(\frac{R}{I}\right)} = \frac{0.19}{T\left(\frac{R}{I}\right)} = \frac{0.19}{2T}$$

On the left end of the plateau in the design response spectrum, at a period $T = T_0 = 0.10$ s, $C_s = 0.95$, and Eq. 12.8-3 doesn't govern. Near the right end of the plateau, at $T = 0.40$ s, $C_s = 0.24$, and Eq. 12.8-3 still doesn't govern.

The approximate period of the building as given by Eq. 12.8-7 of ASCE 7-05 is

$$T_a = C_t h_n^x$$

In our case, h_n (the height above the base to the highest level of the structure) is 35 ft. The values of C_t and x are given by Table 12.8-2 of ASCE 7-05 as 0.02 and 0.75, respectively.

$$T_a = 0.02(35)^{0.75}$$

$$T_a = 0.29 \text{ s}$$

This is less than the transition period $T_s = 0.49$ s, and the structure is on the plateau of the design spectrum. Equation 12.8-2 governs, and $C_s = 0.20$. Because the structure is assigned to SDC C, the redundancy factor, ρ, is permitted to be taken as 1.0 (Sec. 12.3.4.1). The structure has no vertical or horizontal irregularities, and the diaphragms are not flexible.

Finally, in accordance with ASCE 7-05, Sec. 12.4.2, the design horizontal seismic load effect E_h is

$$E_h = \rho Q_E \quad \text{(Equation 12.4-1)}$$

Now compute the seismic base shear. In accordance with ASCE 7-05, Sec. 12.8.1, the effects of horizontal seismic forces Q_E come from V. The design seismic base shear is given by

$$V = C_s W$$
$$V = 0.20\ W$$

This is multiplied by the redundancy factor of 1.0, giving a product of 0.20. In other words, the building must be designed for 20 percent of its weight, applied as a lateral force.

Step 10: Distribute seismic base shear vertically using Sec. 12.8.3.

This force is distributed triangularly over the height of the building. The weight of a typical floor is its area, times the dead load per square foot, plus the interior transverse wall weight, plus the spine wall weight, plus the weight of the exterior walls. For simplicity, assume that the roof weighs the same as a typical floor, and ignore the parapet.

Floor weight: 50 lb/ft² × 50 × 140 ft² = 350 kips
Transverse wall weight: 7 × 20 × 11 ft² × 45 lb/ft² = 69.3 kips
Spine wall weight: 2 × 130 × 11 ft² × 45 lb/ft² = 128.7 kips
Perimeter wall weight: 2 × (140 + 50) × 12 ft² × 45 lb/ft² = 188.1 kips

Total weight of a typical floor is 736.1 kips.

The design base shear is calculated assuming a linear distribution of forces over the height of the structure, because the fundamental period of the structure is less than 0.5 s (ASCE 7-05, Sec. 12.8.3). Total design base shear is 2208 kips × 0.20 = 441.7 kips.

Level	W	H	WH	WH/SUM
R	736.1	33	24,291	0.50
3	736.1	22	16,194	0.333
2	736.1	11	8,097	0.167
	2208.3		48,582	

At the roof level, the factored design lateral force is the factored design base shear (441.7 kips), multiplied by 0.50 (the quotient of WH/SUM) for the triangular distribution, or 220.9 kips. At the next level down, the factored design lateral force is 441.7 kips, multiplied by 0.333, and so forth.

At each level, the factored design moment is the summation of the products of the factored design lateral forces above that level, each

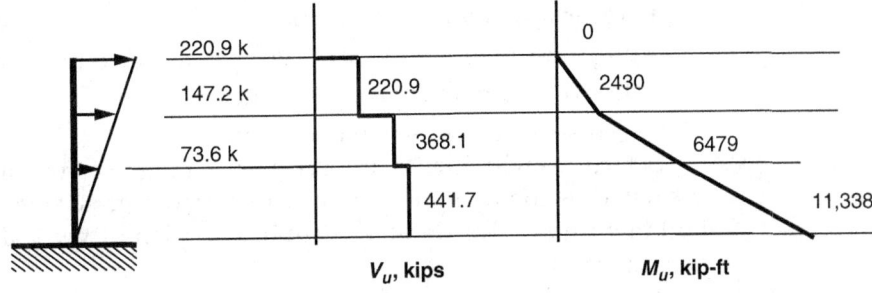

FIGURE 14.37 Graphs of factored design shears and moments for 3-story hotel example using AAC masonry.

multiplied by its respective height above that level. Because seismic loads based on ASCE 7-05 are at strength level, the load factor for seismic loads is 1.0. Factored design shear and moment diagrams for the building are shown in the Fig. 14.37 and Table 14.13.

Step 11: Distribute seismic base shear horizontally using Sec. 12.8.4.

The last three steps are structure-dependent. They depend on the seismic response modification coefficient assigned to the structural system, on the structure's plan structural irregularities, on the structure's vertical structural irregularities, and on the structure's redundancy.

Plan structural irregularities include:

- Plan eccentricities between the center of mass and the center of stiffness
- Re-entrant corners
- Out-of-plane offsets
- Nonparallel systems

These can increase seismic response.

Vertical structural irregularities include:

- Stiffness irregularity
- Mass irregularity

Level	F_u, k	H, ft	V_u, k	M_u, k-ft
R	220.9	33	220.9	0
3	147.2	22	368.1	5923
2	73.6	11	441.7	29,616
				44,424

TABLE 14.13 Factored Design Shears and Moments for 3-Story Hotel Example Using AAC Masonry

- Vertical geometric irregularity
- In-plane discontinuity in vertical lateral-force-resisting elements
- Discontinuity in capacity—weak story

These can also increase seismic response.

Structures with low redundancy have a higher probability of failure, which is compensated for by increasing design seismic forces. The building under consideration here has no plan or vertical structural irregularities.

In general, the effects of torsion must be included. Inherent torsion is a function of the walls, and is negligible in this case, because the walls are symmetrically arranged in plan. Accidental torsion is independent of the walls, and is prescribed by Secs. 12.8.4.1 and 12.8.4.2 of ASCE 7-05. For the sake of simplicity, torsion is ignored in this example.

14.12.2 Design Transverse Shear Walls for Gravity plus Earthquake Loads

(All references are according to 2008 MSJC *Code* and *Specification*)

The transverse direction is critical for this building. The 16 transverse walls are conservatively assumed to be uncoupled, so that each functions as an independent cantilever. As shown in Fig. 14.38, design each transverse wall as an I beam, assuming flange widths of 4 ft. This is less than the limits specified in Sec. 1.9.4.2.3 of the 2008 MSJC *Code*, and is therefore conservative.

Shear Design of a Typical Transverse Wall for Earthquake Loads

Check shear for assumed wall thickness. Include the effects of axial load, assuming that a typical transverse wall carries its self-weight plus the distributed floor weight on a tributary width of 20 ft:

Self-weight of wall: 9.9 kips/floor
Floor weight: 60 lb/ft^2 × 20 × 20 = 24 kips/floor

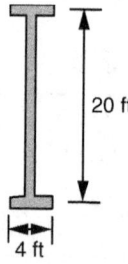

Figure 14.38 Typical transverse shear wall of 3-story hotel example with AAC masonry.

Total unfactored axial load at base, P, is $3 \times (9.9 + 24) = 101.7$ kips. Assume that the critical load case is $P_u = 0.9D = 91.5$ kips.

By Sec. A.3.4.1.2 of the 2008 MSJC *Code*,

$$V_n = V_{nAAC} + V_{ns}$$

$$V_{nAAC} = \min \begin{cases} 0.95 \lambda_w t \sqrt{f'_{AAC}} \sqrt{1 + \dfrac{P_u}{2.4\sqrt{f'_{AAC}} \lambda_w t}} \\ 0.17 f'_{AAC} t \dfrac{h \cdot \lambda_w^2}{h^2 + (\tfrac{3}{4}\lambda_w)^2} \\ \mu_{AAC} P_u \end{cases}$$

Take the coefficient of friction for the third equation as 1.0 (AAC against mortar).

$$V_{nAAC} = \min \begin{cases} 0.95 \times 240 \text{ in.} \times 11.9 \text{ in.} \sqrt{870 \text{ lb/in.}^2} \\ \quad \times \sqrt{1 + \dfrac{91,530 \text{ lb}}{2.4\sqrt{870 \text{ lb/in.}^2} \times 240 \text{ in.} \times 11.9 \text{ in.}}} \\ 0.17 \times 870 \text{ lb/in.}^2 \times 11.9 \text{ in.} \dfrac{33 \text{ ft} \times 20^2 \text{ ft}^2}{33^2 \text{ ft}^2 + (\tfrac{3}{4} \times 20)^2 \text{ ft}^2} \times 12 \text{ in./ft} \\ 1.0 \times 91.5 \text{ kips} \end{cases}$$

$$V_{nAAC} = \min \begin{cases} 96,457 \text{ lb} \\ 212,166 \text{ lb} \\ 91,500 \text{ lb} \end{cases}$$

$V_{nAAC} = 91,500$ lb

$\phi V_{nAAC} = 0.8 \times 91,500$ lb $= 73,200$ lb

This exceeds (1/16 walls) times the factored design base shear (1/16 × 441.7 kips = 27.6 kips), and the transverse walls are be satisfactory for shear thus far. While floor-level bond beams are required, no shear reinforcement is required.

In-Plane Flexural Design of Transverse Shear Walls for Earthquake Loads

Each transverse shear wall has a plan length of 20 ft. The factored design base moment per wall is (1/16) × 11,338 ft-kips, or 708.63 ft-kips. The critical load case is 0.9D + 1.0E. The factored axial load (see above) is 0.9 × 101.7 kips, or 91.5 kips. Because of the flanges, the effective width of the wall is taken as 48 in. This is valid provided that the compressive stress block does not leave the flange. This will be checked later.

Using a spreadsheet, the interaction diagram for the wall is shown in Fig. 14.39. At a factored axial load of 91.5 kips, the factored moment capacity is 1100 kip-ft, more than satisfactory. The neutral axis is located 4.7 in. from the extreme compression fiber, still in the flange, so the interaction diagram calculated using a 48-in. effective width is valid.

Flexural reinforcement consisting of one #4 bar at each end is required. The bars should be placed in grouted cores at least 12-in. square (at intersections of web and flanges).

Check splice requirements and percent area requirements. Assume a 2000-psi grout strength by the proportion specification of ASTM C476. By Sec. A.3.3.3.1,

$$l_d = \frac{0.13\, d_b^2 f_y \gamma}{K_{AAC}\sqrt{f_g'}} = \frac{0.13 \times 0.5^2 \times 60{,}000 \times 1.0}{\left(\dfrac{12-0.5}{2}\right)\sqrt{2000}} = 7.58 \text{ in.}$$

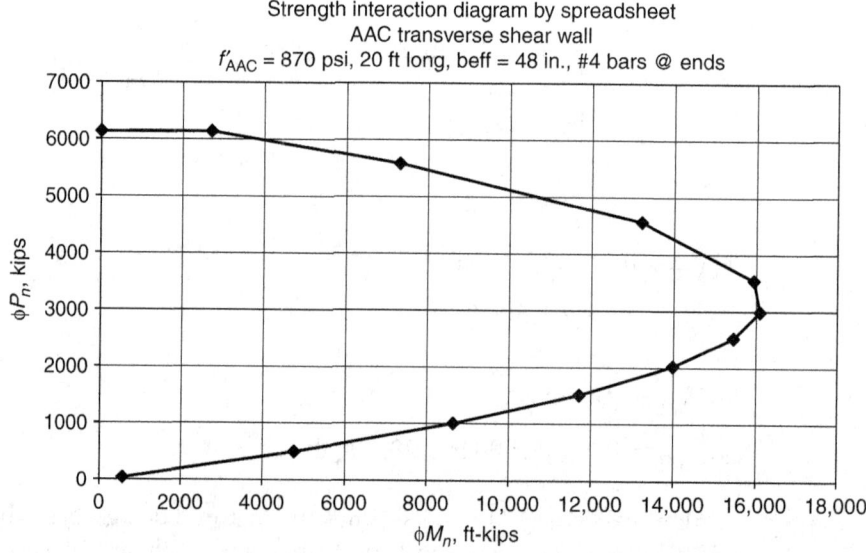

FIGURE 14.39 Strength interaction diagram by spreadsheet, AAC transverse shear wall.

Twelve inches governs. By *Code* Sec. A.3.3.1, the maximum percent area in a plastic hinge zone is 3 percent. For a 12-in. square core, a #4 bar easily satisfies this requirement. Because the wall is symmetrically reinforced, maximum reinforcement limitations (*Code* Sec. A.3.3.5) are satisfied.

Now check capacity design for shear (Sec. 1.17.3.2.6.1.1 of the 2008 MSJC *Code*). First try to meet the capacity design provisions of that section. At an axial load of 91.5 kips, the nominal flexural capacity of this wall is the design flexural capacity of 1100 ft-kips, divided by the strength reduction factor of 0.9, or 1222 ft-kips. The ratio of this nominal flexural capacity to the factored design moment is 1222 divided by 708.6, or 1.72. Including the additional factor of 1.25, that gives a ratio of 2.16.

$$\phi V_n \geq 2.16 V_u$$

$$V_n \geq \frac{2.16}{\phi} V_u = \frac{2.16}{0.8} V_u = 2.70 V_u = 2.70 \times 27.6 = 74.4 \text{ kips}$$

V_n (governed by sliding shear) is 91.5 kips, considerably greater than this. The wall is satisfactory without shear reinforcement. A nominal 8-in. wall could probably be used instead of 12 in. Horizontal reinforcement will be needed at diaphragm-level bond beams (Sec. 1.17.3.2.7.1 of the 2008 MSJC *Code*).

14.12.3 Comments of Design of Transverse Shear Walls for Seismic Loads

- The most laborious part of this design is the calculation of the design lateral force for earthquake loads. Once that calculation is done, design of the lateral force-resisting system is straightforward, even for a region of moderate seismic risk such as Asheville.
- This structural system could be designed for increased capacity. Increased flexural capacity would be quite easy to achieve, but increased shear capacity (to meet capacity design requirements) would probably require intermediate bond beams.

14.12.4 Design Exterior Walls for Gravity plus Out-of-Plane Wind

The critical panel will be at the top of the building, where the wind load is highest. The panel must be designed for out-of-plane wind. Load effects in vertical jamb strips will be increased by the ratio of the plan length of openings to the total plan length.

Use factored wind load (components and cladding) on wall ρ_u = 50 lbs/ft².

Reinforcement will be placed in 3-in. grouted cells at 4 ft. on center.

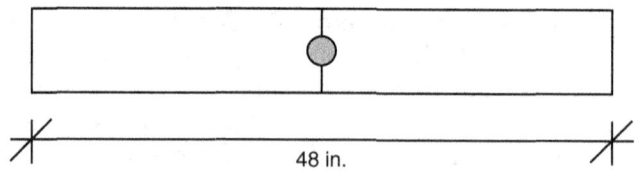

FIGURE 14.40 Plan view of section of exterior wall, 3-story example with AAC masonry.

The span is 11 ft, and the panel thickness is 12 in.
The plan view of a 4-ft section of wall is shown in Fig. 14.40.

The wall is considered simply supported at each floor diaphragm.

Flexural Capacity of Out-of-Plane Walls

1. Determine design moment

$$w_u = p \cdot \text{width} = 50 \text{ lb/ft}^2 \cdot 4\text{ft} = 200 \text{ lb/ft}$$

$$M_u = \frac{wl^2}{8} = \frac{200 \cdot (11)^2}{8} = 3025 \text{ lb-ft} = 36,300 \text{ lb-in.} \qquad \text{Sec. A.3.2}$$

2. Try a #4 bar, and neglect axial load

$$T = A_s f_y = 0.20 \text{ in.}^2 \times 60,000 \text{ lb/in.}^2 = 12,000 \text{ lb}$$

$$a = \frac{T}{0.85 f'_{AAC} b} = \frac{12,000 \text{ lb}}{0.85 \times 870 \text{ lb/in.}^2 \times 48 \text{ in.}} = 0.34 \text{ in.}$$

$$M_n = A_s f_y (d - \frac{a}{2}) = 12.0 \text{ kips} \times (6 - \frac{0.34}{2}) \text{ in.} = 70,000 \text{ lb-in.}$$

$$M_u = \phi M_n = 0.9 \times 70,000 \text{ lb-in.} = 63,000 \text{ lb-in.} > M_u \qquad \text{OK}$$

Use a #4 bar. Outside of a plastic hinge zone, *Code* Sec. A.3.3.1 imposes a maximum bar area of 4.5 percent of the cell. Using a 3-in. grouted core, the area ratio is $(0.5/3)^2$, or 0.028, easily satisfying the requirement. This bar size will easily satisfy the maximum reinforcement limitations of Sec. A.3.3.5 for out-of-plane flexure, and the design is satisfactory for flexure.

Shear Capacity

1. Determine factored loads and maximum shear force for a single panel.

$$w_u = 50 \text{ psf, and the panel is 4 ft wide.}$$

$$V_u = \frac{200 \text{ lb/ft} \times 11 \text{ ft}}{2} = 1100 \text{ lb}$$

2. Determine shear capacity of panel.

$$V_n = V_{nAAC} = 0.8\sqrt{f'_{AAC}}\,A_n = 0.8\sqrt{870} \times 48 \text{ in.} \times 6 \text{ in.} = 6796 \text{ lb}$$

$$\phi V_n = 0.8\,V_n = 0.8 \times 6796 = 5437 \text{ lb} > V_u = 1100 \text{ lb} \quad \text{OK}$$

14.12.5 Design Floor Diaphragms for In-Plane Actions

Design requirements for AAC floor diaphragms are not given in the 2008 MSJC *Code* and *Specification*, because that can be applied to many different types of floor systems. The design procedure given here based on the requirements of ICC Acceptance Criteria AC 215, which was developed based on research at The University of Texas at Austin. The procedure is also given at the AACPA web site (www.aacpa.org).

f'_{AAC} = 870 psi
f'_{grout} = 2000 psi
f_y = 60,000 psi
Ring beam reinforcement 2 #5
Grouted key reinforcement 1 #5

Factored transverse lateral load in each bay, F_u = 220.9 kips/16 bays = 13.81 kips. The plan view and sectional view of the diaphragm are shown in Fig. 14.41 and 14.42 respectively.

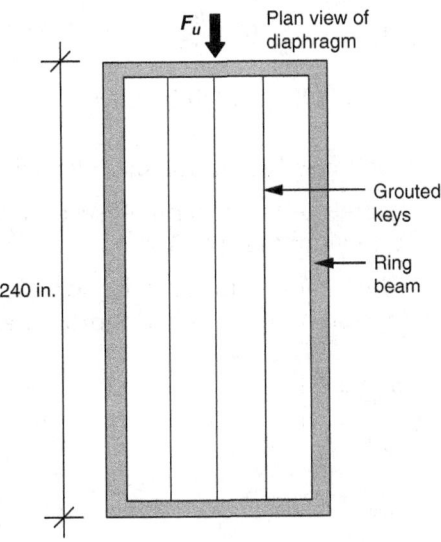

FIGURE 14.41 Plan view of AAC floor diaphragm, 3-story hotel example with AAC masonry.

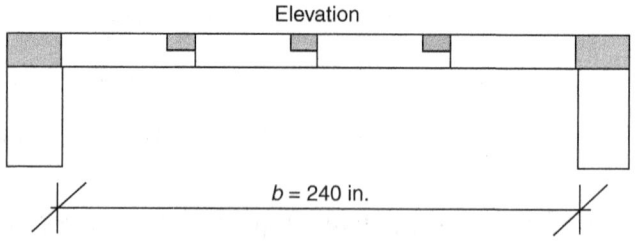

FIGURE 14.42 Sectional view of AAC floor diaphragm, 3-story hotel example with AAC masonry.

1. Design diaphragm for flexure, assuming that load is uniformly distributed along span

$$M = \frac{w_u l^2}{8} = \frac{F_u \times l}{8} = \frac{13{,}810 \text{ lb} \times 240 \text{ in.}}{8} = 414{,}188 \text{ lb-in.}$$

$$T = A_s f_y = 2 \times 0.31 \text{ in.}^2 \times 60{,}000 \text{ lb/in.}^2 = 37{,}200 \text{ lb}$$

$$a = \frac{C}{0.85 f'_{\text{grout}} b} = \frac{37{,}200 \text{ lb}}{0.85 \, (2000 \text{ lb/in.}^2) \, (240 \text{ in.})} = 0.09 \text{ in.}$$

d = length of key − ring beam/2 − 2 * U-block thickness = 240 in. − 4 in. − 4 in. = 238 in.

$$\phi M_n = \phi A_s f_y \cdot (d - \frac{a}{2}) = 0.9 \times 37{,}200 \text{ lb} \times (238 \text{ in.} - 0.09 \text{ in.})$$

$$= 7{,}970{,}000 \text{ lb-in.} \geq M_u \qquad \text{OK}$$

2. Design diaphragm for shear based on adhesion

   a. Panel-to-panel joint: A section of the panel-to-panel joint is shown in Fig. 14.43.

   The total resistance is the adhesion of the grouted area plus the adhesion of the thin-bed mortar area.

$$b_{\text{grout}} = 5 \text{ in.}$$
$$b_{\text{thin-bed}} = 3 \text{ in.}$$

$$V_{\text{grout}} = \tau_{\text{grout}} \cdot b_{\text{grout}} \cdot l = 36 \cdot 5 \cdot 240 = 43{,}200 \text{ lb}$$

$$V_{\text{thin-bed}} = \tau_{\text{thin-bed}} \cdot b_{\text{thin-bed}} \cdot l = 18 \cdot 3 \cdot 240 = 13{,}000 \text{ lb}$$

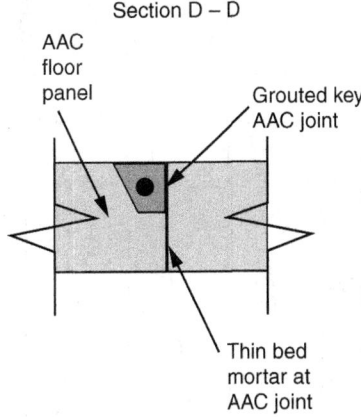

FIGURE 14.43 Section of panel-to-panel joint, AAC floor diaphragm.

$$V_{total} = V_{grout} + V_{thin\text{-}bed} = 55,200 \text{ lb}$$

$$\phi V_{total} = 0.67 \cdot 55,200 \text{ lb} = 36,980 \text{ lb} > V_u = \frac{F_u}{2} = 6,900 \text{ lb} \quad \text{OK}$$

    b. Panel-to-bond beam joint: A section of the panel-to-bond beam joint is shown in Fig. 14.44.

$$b_{grout} = 8 \text{ in.}$$

$$V_{grout} = \tau_{grout} \cdot b_{grout} \cdot l = 36 \text{ lb/in.}^2 \times 8 \text{ in.} \times 240 \text{ in.} = 69,100 \text{ lb}$$

$$\phi V_{total} = 0.67 \cdot 69,100 \text{ lb} = 46,300 \text{ lb} > V_u = \frac{F_u}{2} = 6,900 \text{ lb} \quad \text{OK}$$

3. Design diaphragm for shear based on truss model

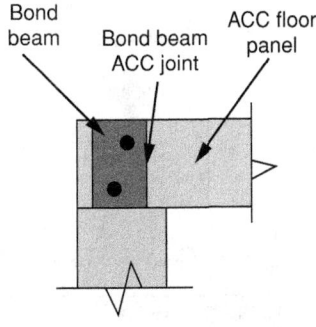

FIGURE 14.44 Section of panel-to-bond beam joint, AAC floor diaphragm.

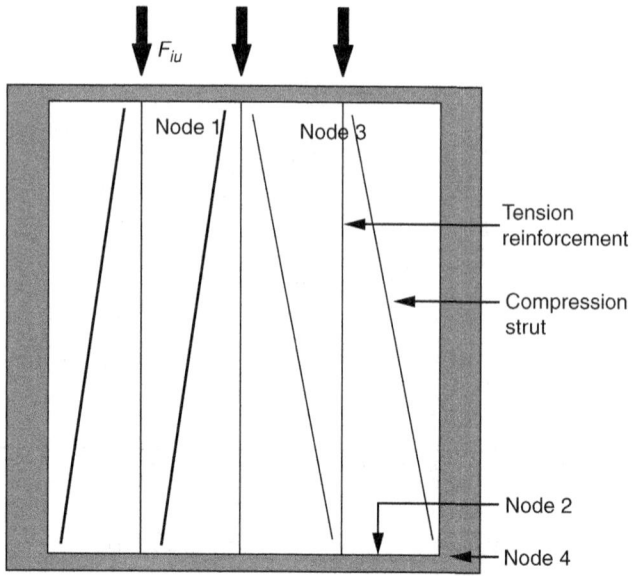

FIGURE 14.45 Truss model for design of AAC diaphragm.

Use one # 5 bar in each grouted key. Each plank is 2 ft wide, so there are 10 planks. The load applied to each node is 1/10 of the total load, or 1.38 kips. Refer to Fig. 14.45.

In this model the compression chords act as diagonal compression members. There are two types of nodes: loaded nodes (on the upper side of the Fig. 14.46) and unloaded nodes (on the lower side of Fig. 14.47).

The critical diagonal compression occurs in the panels next to the support. The component of that compression parallel to the transverse walls is one-half the total factored load on the panel, or one-half of 13.81 kips, or 6.91 kips. The total compressive force in the diagonal, and also the

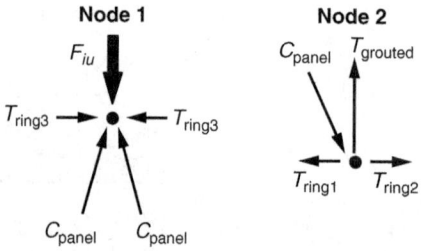

FIGURE 14.46 Loaded nodes for design of AAC diaphragm.

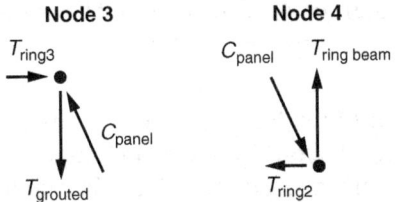

FIGURE 14.47 Unloaded notes for design of AAC diaphragm.

tension force in the associated tension tie, is essentially that shear, because of the aspect ratio of the panels.

$$C_{panel} = \frac{6.91 \text{ kips}}{\cos\left(\tan^{-1}\frac{2}{20}\right)} = \frac{6.91 \text{ kips} \times \sqrt{20^2 + 2^2}}{20} = 6.94 \text{ kips} = T_{grouted}$$

Check the capacity of compression strut:

$W_{strut} = 6$ in.

$T_{panel} = 8$ in.

$F_{strut} = 6.94 \text{ kips}/48 = 145 \text{ psi} < 0.75\,(0.85\,f'_{AAC}) = 0.75(0.85)(870)$

$= 555 \text{ psi}$ OK

Check the capacity of the tension tie in the grouted key:

$T_{grouted\,key} = 6.94 \text{ kips} < \Phi A_s f_y = 0.75 \times 0.31 \times 60{,}000 = 14{,}000 \text{ lb}$ OK

Tension ties in ring beams have already been checked, and are satisfactory.

14.12.6 Overall Comments on Seismic Design Example with AAC Masonry

- Although it is located in a region of moderate seismic risk, this building needs comparatively little reinforcement, because of the large plan area of its bearing walls.
- Considerable simplicity in design and analysis was achieved by letting transverse shear walls resist lateral loads as statically determinate cantilevers.

- Design of AAC bearing walls is inexpensive and straightforward for this type of building.
- Many types of floor and roof systems are possible with AAC. To adapt this design to other types of floor or roof elements, the unit weight would have to be changed appropriately; the connection details would have to be changed appropriately; and the diaphragm actions would have to be checked appropriately. For example, if hollow-core prestressed concrete planks were used, the unit weight would increase, and so would the seismic base shear and overturning moment. Shear design of the transverse shear walls would still govern. Details of the connections between walls and floor or roof would be similar to those used with the AAC planks. Shears in the horizontal diaphragms would be transferred in topping only.

14.13 References on AAC

Because the US code basis for structural design of AAC is relatively new, the references used to develop these code basis are included here in the final sections of this chapter.

ASTM C1386, *Standard Specification for Precast Autoclaved Aerated Concrete (PAAC) Wall Construction Units*, ASTM International, West Conshohocken, Pennsylvania, 1998.

ASTM C1452, *Standard Specification for Reinforced Autoclaved Aerated Concrete Units*, ASTM International, West Conshohocken, Pennsylvania, 2000.

ASTM C1555, *Standard Practice for Autoclaved Aerated Concrete Masonry*, ASTM International, West Conshohocken, Pennsylvania, (2003a).

ASTM C1591, *Standard Test Method for Determining the Modulus of Elasticity of AAC*, ASTM International, West Conshohocken, Pennsylvania, 2004.

Barnett, R. E., Tanner, J. E., Klingner, R. E., and Fouad, F. H., "Guide for Using Autoclaved Aerated Concrete Panels: I—Structural Design," *ACI Special Publication SP 226*, Caijun Shi and Fouad H. Fouad (eds.), American Concrete Institute, Farmington Hills, Michigan, April 2005, pp. 17–28.

IBC 2003, *International Building Code*, International Code Council, Falls Church, Virginia, 2003.

ICC AC 215, "Acceptance Criteria for Seismic Design Factors and Coefficients for Seismic-Force-Resisting Systems of Autoclaved Aerated Concrete (AAC)," *Evaluation Report AC215*, ICC Evaluation Service, Inc., Whittier, California, November 1, 2003.

ICC ESR-1371, "Autoclaved Aerated Concrete (AAC) Block Masonry Units," *Evaluation Report ESR-1371*, ICC Evaluation Service, Inc., Whittier, California, October 1, 2004.

Klingner, R. E., Tanner, J. E., Varela, J. L., Brightman, M., Argudo, J., and Cancino, U., "Technical Justification for Proposed Design Provisions for AAC Structures: Introduction and Shear Wall Tests," *ACI Special Publication SP 226*, Caijun Shi and Fouad H. Fouad (eds.), American Concrete Institute, Farmington Hills, Michigan, April 2005a, pp. 45–66.

Klingner, R. E., Tanner, J. E., and Varela, J. L., "Technical Justification for Proposed Design Provisions for AAC Structures: Assemblage Test and Development of R and C_d Factors," *ACI Special Publication SP 226*, Caijun Shi and Fouad H. Fouad (eds.), American Concrete Institute, Farmington Hills, Michigan, April 2005b, pp. 67–90.

MDG, *Masonry Designers' Guide*, 5th ed., Phillip J. Samblanet (ed.), The Masonry Society, Boulder, Colorado, 2006.

RILEM, "Autoclaved Aerated Concrete: Properties, Testing and Design," RILEM Recommended Practice, RILEM Technical Committees 78-MCA and 51-ALC, E & FN Spon, London, 1993.

Tanner, J. E., Varela, J. L., and Klingner, R. E., "Design and Seismic Testing of a Two-Story Full-scale Autoclaved Aerated Concrete (AAC) Assemblage Specimen," *Structures Journal*, American Concrete Institute, Farmington Hills, Michigan, vol. 102, no. 1, January–February 2005a, pp. 114–119.

Tanner, J. E., Varela, J. L., Klingner, R. E., Brightman M. J., and Cancino, U., "Seismic Testing of Autoclaved Aerated Concrete (AAC) Shear Walls: A Comprehensive Review," *Structures Journal*, American Concrete Institute, Farmington Hills, Michigan, vol. 102, no. 3, May–June 2005b, pp. 374–382.

Varela, J. L., Tanner, J. E., and Klingner, R. E., "Development of Seismic Force-Reduction and Displacement Amplification Factors for AAC Structures," *EERI Spectra*, vol. 22, no. 1, February 2006, pp. 267–286.

14.14 Additional References on AAC

14.14.1 MS Theses (The University of Texas at Austin)

Argudo, J., "Evaluation and Synthesis of Experimental Data for Autoclaved Aerated Concrete," August 2003.

Brightman, M., "AAC Shear Wall Specimens: Development of Test Setup and Preliminary Results," May 2000.

Cancino, U., "Behavior of Low-Strength Shear Walls of Autoclaved Aerated Concrete," December 2003.

14.14.2 PhD Dissertations (The University of Texas at Austin)

Tanner, J. E., "Design Provisions for Autoclaved Aerated Concrete (AAC) Structural Systems," PhD dissertation, Department of Civil Engineering, The University of Texas, Austin, May 2003.

Varela, J., "Development of R and C_d Factors for the Seismic Design of Autoclaved Aerated Concrete Structures," PhD dissertation, Department of Civil Engineering, The University of Texas, Austin, May 2003.

14.14.3 Referred Journal Publications

Barnett, R. E., Tanner, J. E., Klingner, R. E., and Fouad, F. H. "Guide for Using Autoclaved Aerated Concrete Panels: I—Structural Design," *ACI Special Publication SP 226*, Caijun Shi and Fouad H. Fouad (eds.), American Concrete Institute, Farmington Hills, Michigan, April 2005, pp. 17–28.

Klingner, R. E., Tanner, J. E., Varela, J. L., Brightman, M., Argudo, J., and Cancino, U., "Technical Justification for Proposed Design Provisions for AAC Structures: Introduction and Shear Wall Tests," *ACI Special Publication SP 226*, Caijun Shi and Fouad H. Fouad (eds.), American Concrete Institute, Farmington Hills, Michigan, April 2005a, pp. 45–66.

Klingner, R. E., Tanner, J. E., and Varela, J. L., "Technical Justification for Proposed Design Provisions for AAC Structures: Assemblage Test and Development of R and C_d Factors," *ACI Special Publication SP 226*, Caijun Shi and Fouad H. Fouad (eds.), American Concrete Institute, Farmington Hills, Michigan, April 2005b, pp. 67–90.

Tanner, J. E., Varela, J. L., and Klingner, R. E., "Design and Seismic Testing of a Two-Story Full-Scale Autoclaved Aerated Concrete (AAC) Assemblage Specimen," *Structures Journal*, American Concrete Institute, Farmington Hills, Michigan, vol. 102, no. 1, January–February 2005a, pp. 114–119.

Tanner, J. E., Varela, J. L., Klingner, R. E., Brightman M. J., and Cancino, U., "Seismic Testing of Autoclaved Aerated Concrete (AAC) Shear Walls: A Comprehensive Review," *Structures Journal*, American Concrete Institute, Farmington Hills, Michigan, vol. 102, no. 3, May–June 2005b, pp. 374–382.

Varela, J. L., Tanner, J. E., and Klingner, R. E., "Development of Seismic Force-Reduction and Displacement Amplification Factors for AAC Structures," *EERI Spectra*, vol. 22, no. 1, February 2006, pp. 267–286.

14.14.4 Referred Conference Proceedings

Barnett, R. E., Robinson, M. E., Tanner, J. E., Varela, J. L., and Klingner, R. E., "Design Examples for AAC Masonry Structures using U.S. Provisions," *Proceedings, 4th International Conference on Autoclaved Aerated Concrete*, Kingston University, London, September 8–9, 2005.

Klingner, R. E., Tanner, J. E., and Varela, J. L., "Development of Seismic Design Provisions for AAC Structures: An Overall Strategy for the U.S.," *Proceedings, 9th North American Masonry Conference*, Clemson, South Carolina, June 1–4, 2003.

Klingner, R. E., Tanner, J. E., Varela, J. L., and Barnett, R. E., "Development of Seismic Design Provisions for Autoclaved Aerated Concrete: An Overall Strategy for the United States of America," *Proceedings, 4th International Conference on Autoclaved Aerated Concrete*, Kingston University, London, September 8–9, 2005.

Tanner, J. E., Varela, J. L., and Klingner, R. E., "Seismic Performance of a Two-Story AAC Assemblage," *Proceedings, 9th North American Masonry Conference*, Clemson, South Carolina, June 1–4, 2003.

Tanner, J. E., Varela, J. L., and Klingner, R. E., "Seismic Testing of AAC Shear Walls: Technical Basis for Proposed Design Provisions," *Proceedings, 9th North American Masonry Conference*, Clemson, South Carolina, June 1–4, 2003.

Tanner, J. E., Varela, J. L., Brightman, M. T., Cancino, U., and Klingner, R. E., "Seismic Performance and Design of Autoclaved Aerated Concrete Structural Systems," *Proceedings, 13th World Conference in Earthquake Engineering*, Vancouver, Canada, August 1–6, 2004.

Tanner, J. E., Varela, J. L., Klingner, R. E., Fouad, F. H., and Barnett, R. E., "Technical Basis for U.S. Design Provisions for Autoclaved Aerated Concrete Masonry,": An Overall Strategy for the United States of America," *Proceedings, 4th International Conference on Autoclaved Aerated Concrete*, Kingston University, London, September 8–9, 2005.

Varela, J. L., Tanner, J. E., and Klingner, R. E., "Development of Seismic Force and Displacement Modification Factors for Design of AAC Structures," *Proceedings, 13th World Conference in Earthquake Engineering*, Vancouver, Canada, August 1–6, 2004.

Varela, J. L., Tanner, J. E., and Klingner, R. E., "Development of R and C_d Factors for Seismic Design of AAC Structures," *Proceedings, 9th North American Masonry Conference*, Clemson, South Carolina, June 1–4, 2003.

References

General References

ACI Committee 318, *Building Code Requirements for Reinforced Concrete (ACI 318-08)*, American Concrete Institute, Farmington Hills, Michigan, 2008.

AC1523.4R-09, *Guide for Design and Construction with Autoclaved Aerated Concrete Panels*. American Concrete Institute, Farmington Hills, Michigan, June 2009.

AISC LRFD, *Manual of Steel Construction, Load and Resistance Factor Design*, 2d ed., American Institute of Steel Construction, Chicago, Illinois, 1992.

ASCE 7-05, *Minimum Design Loads for Buildings and Other Structures (ASCE 7-05)*, American Society of Civil Engineers, Reston, Virginia, 2005 (with Supplement).

ATC 3-06, *Tentative Provisions for the Development of Seismic Regulations for Buildings (ATC 3-06)*, Applied Technology Council, National Bureau of Standards, Washington, D.C., 1978.

Beall, C., *Masonry Design and Detailing*, 5th ed., McGraw-Hill, New York, NY, 2003.

Brandow, G. E., Ekwueme, C., and Hart, G. C., *Design of Reinforced Masonry Structures*, 5th ed., Concrete Masonry Association of California and Nevada, 2003.

Drysdale, R. and Hamid, A., *Masonry Structures: Behavior and Design*, 3d ed., The Masonry Society, Boulder, Colorado, 2008.

Hillerborg, A., *Strip Method Design Handbook*, Taylor & Francis, London, UK, 1996.

IBC, *International Building Code*, 2009 ed., International Code Council, Washington, D.C., 2009.

MSJC, *Building Code Requirements for Masonry Structures (TMS 402-08/ACI 530-08/ASCE 5-08)*, The Masonry Society, Boulder, Colorado, the American Concrete Institute, Farmington Hills, Michigan, and the American Society of Civil Engineers, Reston, Virginia, 2008a.

MSJC, *Specification for Masonry Structures (TMS 602-08/ACI 530.1-08/ASCE 6-08)*, The Masonry Society, Boulder, Colorado, the American Concrete Institute, Farmington Hills, Michigan, and the American Society of Civil Engineers, Reston, Virginia, 2008.

NBC, *National Building Code*, Building Officials and Code Administrators International, Country Club, Illinois.

NEHRP, *NEHRP (National Earthquake Hazards Reduction Program) Recommended Provisions for the Development of Seismic Regulations for New Buildings (FEMA 450)*, Building Seismic Safety Council, Federal Emergency Management Agency, Washington, D.C., 2003.

SBC, *Standard Building Code*, Southern Building Code Congress International, Birmingham, Alabama.

Taly, N., *Design of Reinforced Masonry Structures*, McGraw-Hill, New York, NY, 2001.

UBC, *Uniform Building Code*, 1997 ed., International Conference of Building Officials, Whittier, California, 1997.

ASTM Standards

All ASTM standards are published by the American Society for Testing and Materials, West Conshohocken, Pennsylvania.

ASTM A82 (2007): Steel Wire, Plain, for Concrete Reinforcement

ASTM A153 (2009): Zinc Coating (Hot-Dip) on Iron and Steel Hardware

ASTM A167 (2009): Stainless and Heat-Resisting Chromium-Nickel Steel Plate, Sheet, and Strip

ASTM A193 (2009): Alloy-Steel and Stainless Steel Bolting Materials for High Temperature or High Pressure Service and Other Special Purpose Applications

ASTM A325 (2009): Structural Bolts, Steel, Heat Treated, 120/105 ksi Minimum Tensile Strength

ASTM A416 (2006): Steel Strand, Uncoated Seven-Wire for Prestressed Concrete

ASTM A496 (2007): Steel Wire, Deformed, for Concrete Reinforcement

ASTM A497 (2007): Steel Welded Wire Reinforcement, Deformed, for Concrete
ASTM A615 (2009): Deformed and Plain Carbon-Steel Bars for Concrete Reinforcement
ASTM A641 (2009): Zinc–Coated (Galvanized) Carbon Steel Wire
ASTM A653 (2009): Steel Sheet, Zinc-Coated (Galvanized) or Zinc-Iron Alloy-Coated (Galvannealed) by the Hot-Dip Process
ASTM A706 (2009): Low-Alloy Steel Deformed and Plain Bars for Concrete Reinforcement
ASTM A951 (2006): Steel Wire for Masonry Joint Reinforcement
ASTM A996 (2009): Rail-Steel and Axle-Steel Deformed Bars for Concrete Reinforcement
ASTM A1008 (2009): Steel, Sheet, Cold-Rolled, Carbon, Structural, High-Strength Low-Alloy, High-Strength Low-Alloy with Improved Formability, Solution Hardened, and Bake Hardenable
ASTM C55 (2006): Concrete Building Brick
ASTM C62 (2004): Building Brick (Solid Masonry Units Made from Clay or Shale)
ASTM C67 (2009): Sampling and Testing Brick and Structural Clay Tile
ASTM C90 (2009): Loadbearing Concrete Masonry Units
ASTM C91 (2005): Masonry Cement
ASTM C129 (2006): Nonloadbearing Concrete Masonry Units
ASTM C139 (2005): Concrete Masonry Units for Construction of Catch Basins and Manholes
ASTM C140 (2008): Sampling and Testing Concrete Masonry Units and Related Units
ASTM C144 (2004): Aggregate for Masonry Mortar
ASTM C207 (2006): Hydrated Lime for Masonry Purposes
ASTM C216 (2007): Facing Brick (Solid Masonry Units Made from Clay or Shale)
ASTM C270 (2008): Standard Specification for Mortar for Unit Masonry
ASTM C410 (2008): Industrial Floor Brick
ASTM C426 (2007): Linear Drying Shrinkage of Concrete Masonry Units
ASTM C476 (2009): Grout for Masonry
ASTM C652 (2004): Hollow Brick (Hollow Masonry Units Made from Clay or Shale)
ASTM C744 (2008): Prefaced Concrete and Calcium Silicate Masonry Units
ASTM C902 (2009): Pedestrian and Light Traffic Paving Brick
ASTM C936 (2009): Solid Concrete Interlocking Paving Units
ASTM C1006 (2007): Splitting Tensile Strength of Masonry Units
ASTM C1019 (2009): Sampling and Testing Grout
ASTM C1072 (2006): Measurement of Masonry Flexural Bond Strength

ASTM C1180 (2007): Standard Terminology of Mortar and Grout for Unit Masonry
ASTM C1232 (2009): Standard Terminology of Masonry
ASTM C1272 (2007): Heavy Vehicular Paving Brick
ASTM C1314 (2007): Compressive Strength of Masonry Prisms
ASTM C1319 (2006): Concrete Grid Paving Units
ASTM C1357 (2009): Evaluating Masonry Bond Strength
ASTM C1372 (2004): Dry-Cast Segmental Retaining Wall Units
ASTM C1386 (2007): Precast Autoclaved Aerated Concrete (AAC) Wall Construction Units
ASTM C1555 (2003): Autoclaved Aerated Concrete Masonry
ASTM C1611 (2009): Slump Flow of Self-Consolidating Concrete
ASTM C1717 (2009): Conducting Strength Tests of Masonry Wall Panels
ASTM E514 (2008): Water Penetration and Leakage Through Masonry
ASTM E518 (2009): Flexural Bond Strength of Masonry
ASTM E519 (2007): Diagonal Tension (Shear) in Masonry Assemblages

Index

Note: Page numbers referencing figures are followed by an "*f*," page numbers referencing tables are followed by a "*t*."

A

AAC. *See* autoclaved aerated concrete
AACPA. *See* Autoclaved Aerated Concrete Products Association
absorption. *See also* initial rate of absorption
  boiling-water, 31
  cold-water, 31
  of concrete masonry units, 36
acceleration response spectrum, 89–90, 89*f*, 99–100, 101*t*
acceleration-dependent site coefficient, 423
accessory materials
  coatings, 11, 39, 45
  connectors, 11, 39–41, 43*f*–44*f*
  flashing, 11, 14, 39, 42–45, 45*f*, 55
  moisture barriers, 11, 39, 46
  movement joints, 41, 46–47, 46*f*–47*f*
  preliminary discussion of, 11
  reinforcement, 38–39, 40*f*–42*f*
  sealants, 11, 39, 41
  specification of, 48–49
  types of, 38
  vapor barriers, 11, 39, 45–46, 55
accessory tools, 11
accidental torsion, 514
ACI. *See* American Concrete Institute
actual dimensions, 12
adhesion, 520
adjustable ties, 41, 43*f*–44*f*
adobe, 10*t*
air content, 23
airspace, 14
allowable flexural capacity, of cross section, 291–293, 291*f*–293*f*

allowable-stress checks
  of one-way shear, 237–238, 305
  for unreinforced panel walls, 237–238
allowable-stress design
  of anchor bolts, 265–274, 266*f*–267*f*, 269*f*–271*f*, 333–334, 334*t*
  approach of, 121, 128
  balanced conditions and, 308*f*, 312*f*
  flexural capacity of cross section and, 291–293, 291*f*–293*f*
  flexure and, 231–232, 231*t*, 295, 305, 336
  moment-axial force interaction diagrams and, 306–315, 306*f*, 308*f*, 311*f*–315*f*, 316*t*
  MSJC provisions, 109–113, 110*t*–113*t*, 231–232, 231*t*, 296–299, 331
  out-of-plane stress and, 245
  of reinforced beams, 295–300, 295*f*–298*f*, 334–335, 335*t*
  of reinforced bearing walls, 306–318, 306*f*, 308*f*, 311*f*–315*f*, 316*t*, 317*f*–318*f*, 336, 336*f*–337*f*
  of reinforced curtain walls, 300–306, 300*f*, 302*f*–303*f*, 305*f*, 335
  of reinforced lintels, 295–300, 295*f*–298*f*, 334–335, 335*t*
  of reinforced shear walls, 319–324, 319*f*, 321*f*, 323*f*, 336–337, 337*f*
  shear capacity in, 332–333
  shear reinforcement and, 223, 295
  simplification of, 305
  strength design *v.*, 195, 331–337, 332*t*–337*t*
  of unreinforced bearing walls, 243–259, 243*f*–244*f*, 246*f*, 246*t*–247*t*, 249*f*, 252*f*–253*f*, 257*f*–258*f*, 332

535

allowable-stress design (*Cont.*):
  of unreinforced panel walls, 229–243, 230*f*, 231*t*, 232*f*, 234*f*–235*f*, 237*f*, 239*t*, 240*f*–241*f*, 331–332, 332*t*
  of unreinforced shear walls, 259–265, 259*f*–262*f*, 265*f*, 332–333, 333*t*
allowable-stress interaction diagrams
  by hand, 307–311, 308*f*, 311*f*
  neutral axis and, 308*f*, 311
  reinforced bearing walls and, 306–315, 306*f*, 308*f*, 311*f*–315*f*, 316*t*
  by spreadsheet, 311–315, 312*f*–315*f*, 316*t*
allowable-stress loading combinations, 104–105, 109–110
allowable-stress reinforcement, 291–294, 291*f*, 293*f*
American Concrete Institute (ACI), 61, 113, 448
American Institute of Steel Construction, 113
American National Standards Institute (ANSI), 59, 115, 447
American Society for Testing and Materials (ASTM)
  A82, 39
  A153, 40
  A167, 39
  A185, 39
  A193-B7, 40
  A325, 39
  A416, 39
  A496, 39
  A497, 39
  A615, 39, 48, 421
  A641, 40
  A653, 40
  A706, 39
  A951, 39
  A996, 39
  A1008, 39
  AAC specifications of, 447
  C55 (Concrete Building Brick), 28
  C62 (Building Brick), 27, 30–31, 33–34, 34*t*
  C67 (Sampling and Testing Brick and Structural Clay Tile), 27
  C90 (Hollow Load-Building Concrete Masonry Units), 28, 36–37
  C91, 23–24
  C129 (Hollow Non-Load-Bearing Concrete Masonry Units), 28
  C139 (Concrete Masonry Units for Construction of Catch Basins and Manholes), 28
  C140, 28, 36
  C207 (Hydrated Lime for Masonry Purposes), 20

American Society for Testing and Materials (ASTM) (*Cont.*):
  C216 (Facing Brick), 27, 30–31, 33–34, 34*t*
  C270, 15, 19–24, 48, 421
  C410 (Industrial Floor Brick), 27
  C426 (Drying Shrinkage), 28, 36
  C476 (Grout for Masonry), 25, 48, 516
  C652 (Hollow Brick), 27, 421
  C744 (Prefaced Concrete and Calcium Silicate Masonry Units), 28
  C780, 56
  C902 (Pedestrian and Light Traffic Paving Brick), 27
  C936 (Solid Concrete Interlocking Paving Units), 28
  C1006 (Splitting Tensile Strength of Masonry Units), 27
  C1019 (Sampling and Testing Grout), 26
  C1072 (Measurement of Masonry Flexural Bond Strength), 28, 37–38
  C1180 (Standard Terminology of Mortar and Grout for Unit Masonry), 27
  C1232 (Standard Terminology of Masonry), 27
  C1272 (Heavy Vehicular Paving Brick), 27
  C1314 (Measurement of Compressive Strength of Masonry Prisms to Determine Compliance), 28, 37
  C1319 (Concrete Grid Paving Units), 28
  C1357 (Evaluating Masonry Bond Strength), 28, 37–38
  C1372 (Dry-Cast Segmental Retaining Wall Units), 28
  C1386, 439, 441
  C1388, 37
  C1389, 37
  C1390, 37
  C1391, 37
  C1611, 27
  C1717 (Conducting Strength Tests of Masonry Wall Panels), 28
  clay masonry units and, 27, 30–34, 33*f*, 34*t*, 48
  for concrete masonry units, 28, 35–38, 48
  contact information for, 115
  E72 (Strength Tests of Panels for Building Construction), 28, 37–38
  E514 (Water Permeance of Masonry), 28, 38
  E518 (Flexural Bond Strength of Masonry), 28, 38
  E519 (Diagonal Tension in Masonry Assemblages), 28, 38
  F1154, 39
  grout specifications of, 25–27, 48, 516

American Society for Testing and Materials (ASTM) (*Cont.*):
  hollow unit specifications of, 27–28, 36–37, 421
  masonry assemblages and, 28
  masonry unit specifications of, 27–29, 48
  for shale masonry units, 27
  specification development by, 61
  subcommittees, 447
American Society of Civil Engineers (ASCE), 369
  ASCE 7 of, 61, 63–64, 67, 79, 87, 105, 343, 369, 381, 422, 497
  contact information for, 113
  role of, 61
  seismic design provisions of, 422
anchor bolts
  allowable-stress design of, 265–274, 266*f*–267*f*, 269*f*–271*f*, 333–334, 334*t*
  bent-bar, 169–170, 172, 266–268, 272
  breakout areas of, 170, 170*f*, 267–268, 267*f*
  common uses of, 266*f*
  conical breakout cones for, 169, 169*f*, 266, 266*f*
  design, for curtain walls, 195–196, 196*f*, 305–306, 305*f*
  effective embedment of, 171, 272
  horizontally oriented, 168, 168*f*
  load factor for, 333
  loaded in combined tension and shear, 177, 273–274
  loaded in shear, 173–177, 173*f*–174*f*, 270–274, 270*f*–271*f*
  loaded in tension, 169–170, 169*f*–170*f*, 177, 266–270, 266*f*–267*f*, 269*f*, 273–274
  with masonry controls, 334*t*
  orientation of, 265
  safety factors for, 333, 334*t*
  shanks of, 170, 273
  single, 175–177, 269*f*, 272–273
  with steel controls, 334*t*
  strength design of, 168–177, 168*f*–171*f*, 173*f*–174*f*, 333–334, 334*t*
  tensile capacity of, 267–270
  vertically oriented, 168, 168*f*
ANSI. *See* American National Standards Institute
appearance, clay masonry unit, 30, 32–33
approximate approach, to flexible floor diaphragms, 362
arbitrary point, lateral load applied through, 351–352, 351*f*
arching action, 190–191, 300
architectural constraints
  for four-story building with clay masonry, 420
  of three-story AAC shear-wall hotel, 495

architectural details, 48
ASCE. *See* American Society of Civil Engineers
ASCE 7, 61, 63–64, 67, 79, 87, 105, 343, 369, 381, 422, 497
aspect ratio, 450
assemblages, 28, 37–38
ASTM. *See* American Society for Testing and Materials
autoclaved aerated concrete (AAC). *See also* three-story AAC shear-wall hotel design example
  applications of, 444, 445*f*–446*f*
  ASTM specifications for, 447
  axial load and, 452–456, 453*f*, 458–459, 475–476, 475*f*–476*f*, 485, 489, 515
  basic earthquake resistance mechanism of, 493
  characteristics of, 443*t*
  cladding, 446*f*
  compressive strength of, 448, 450, 452, 457, 460, 467
  construction, 440, 448–449
  critical strain condition for, 479, 479*f*, 490, 490*f*
  design and construction provisions for, 440, 447–448
  design background for, 444, 446*f*
  design provisions for, 440
  ductile shear-wall structures, 494–495, 494*t*
  elements, 440, 441*f*
  erection, 448–449
  field-reinforced, 473
  flexural capacity of, 481, 487
  flexural resistance of, 448
  floor diaphragms, 519–523, 519*f*–523*f*
  floor systems, 524
  handling, 448–449
  historical background of, 439–440
  in-plane shear resistance of, 448
  introduction to, 439, 440*f*
  loads for, 447, 464
  manufacturing of, 441–443, 442*f*
  masonry design, 122, 129
  materials used in, 441
  MSJC provisions for, 440, 447–448, 479–481, 479*f*, 489–493, 490*f*
  openings and, 462, 467
  PhD dissertations on, 525–526
  references on, 524–527
  reinforced beam design, 467–472, 468*f*–470*f*, 469*t*
  reinforced bearing wall design, 473–481, 473*f*, 474*t*, 475*f*–476*f*, 479*f*
  reinforced curtain wall design, 473

autoclaved aerated concrete (AAC) (*Cont.*):
  reinforced lintel design, 467–472, 468f–470f, 469t
  reinforced shear wall design, 481–493, 481f–484f, 487f, 490f
  reinforced units, 443
  reinforcement ratios for, 479–481, 479f, 489–493, 490f
  research on, 493
  roof systems, 524
  seismic design of, 493–495, 494t
  shear capacity of, 451, 462, 466, 493, 518–519
  shrinkage of, 495
  strength classes, 443, 443t
  strength design for, 447–448
  strength-reduction factors for, 449, 451–452, 463
  strip method and, 452
  unit dimensions, 444, 444t
  unreinforced bearing walls design, 452–462, 453f, 456f, 459f–460f
  unreinforced panel walls design, 449–452, 450f–451f
  unreinforced shear wall design, 462–467, 463f–465f
  in U.S., 440, 444, 446f, 447–448, 494–495, 494t
  walls loaded in-plane, 490, 490f
  walls loaded out-of-plane, 475–481, 475f–476f, 479f
Autoclaved Aerated Concrete Products Association (AACPA), 62, 114, 440
autoclaving, 443
axial capacity
  of columns, 244f
  slenderness influencing, 244, 244f
  of walls, 244f
axial forces. *See also* moment-axial force interaction diagrams
  calculation of, 263
  equilibrium of, 211–212, 221, 282, 286, 480, 491
  factored, 209, 454
axial load. *See also* eccentric axial load
  AAC and, 452–456, 453f, 458–459, 475–476, 475f–476f, 485, 489, 515
  compressive, 489–490
  concentric, 149–152, 150f, 150t, 246–248, 246f, 246t–247t, 452–455, 453f
  neutral axis within compression flange and, 431
  on pilasters, 402
  on reinforced clay shear wall, 216, 218
  in shear walls, 216, 218, 464, 472
  unfactored, 485, 515

B

balance point, 198–201, 307, 310
balanced reinforcement
  allowable-stress, 293–294, 293f
  percentage, 185f
bar joist, 389, 390f, 400, 400f, 407f
barrier walls
  composite, with filled collar joint, 14f
  multiwythe, 13–14
  single-wythe, 13–14, 14f, 378
barriers
  moisture, 11, 39, 46
  vapor, 11, 39, 45–46, 55
bars
  #5, 430, 434
  #4, 190, 409, 409f, 430, 434, 472, 486, 518
  maximum area of, 477
  reinforcing, 39, 40f, 187, 187t, 288, 289t, 468, 469t
  #7, 409, 409f, 414
  steel, 187, 187t, 288, 289t, 468, 469t
base shear. *See also* seismic base shear
  of four-story building with clay masonry, 425, 427–428
  wind load and, 380–384, 381t, 383t
basic allowable-stress loading combinations, 104–105
basic structural behavior
  of low-rise, bearing wall buildings, 3–4, 4f
  of unreinforced bearing walls, 147–148, 148f, 243–244, 243f–244f
  of unreinforced shear walls, 161–162, 161f, 259–260, 259f
basic structural design, of low-rise masonry buildings, 4–5, 5f
basic wind speeds, 68, 69f, 70t, 79–80, 83, 381, 384
beam columns
  cantilever, 5
  with centrally located reinforcement, 309
  design of, 306
  form of, 196
  vertical strips acting as, 4–5
beams. *See also* reinforced beams
  bond, 4, 190, 373, 486, 515
  critical strain condition for, 490f
  depth of, 187, 409, 468, 472
  flexural capacity of, 472
  shear for, 337
bearing plate, 249, 249f, 389, 389f, 455f
  under long-span joists, 406, 406f
  of pilasters, 402f
bearing stresses, 249, 249f, 389, 389f, 402, 455f

bearing walls. *See also* reinforced bearing walls; unreinforced bearing walls
  of low-rise, bearing wall buildings, 3–4, 4f, 399, 399f
  wall-to-floor connections and, 224f, 226
bed, 12
bed joints
  cracks at, 477
  reinforcement of, 40f, 122, 124
bedding, face-shell, 139–141, 139f, 141f, 234–237, 235f, 237f
bending moment, 450–451
bending stress, 155, 251, 254–255, 457
bent-bar anchor bolts, 169–170, 172, 266–268, 272
BIA. *See* Brick Industry Association
BOCA. *See* Building Officials and Code Administrators International
boiling-water absorption, 31
bolts. *See* anchor bolts
bond beams
  floor-level, 515
  #4 bars in, 190, 486
  horizontal reinforcement from, 4
  panel-to-bond beam joints and, 521f
  vertical reinforcement anchored to, 373
bond behavior of cracked, transformed sections, 286–288, 287f–288f
bond breaker, 56, 177, 226
bond patterns, 12–13, 13f
bottom reinforcement, 189f, 297, 297f, 470f
breakout areas
  of anchor bolts, 170, 170f, 267–268, 267f
  projected, 170, 170f, 267–268, 267f, 270
  tensile, 268, 270
breakout cones, 169, 169f, 266, 266f
breakout failure, 173–174, 173f–174f, 270–271, 270f–271f
Brick Industry Association (BIA), 62, 114
BSSC. *See* Building Seismic Safety Council
buckling, 244, 247, 255
Building Brick. *See* ASTM C62
building codes, U.S. *See also* International Building Code; Masonry Standards Joint Committee
  additional information on, 113–115
  development of, 59–60, 60f
  governmental organizations and, 62–63, 115
  industry organizations and, 61–62, 114
  model-code organizations and, 15, 63–64, 114
  NBC, 63
  SBC, 63
  technical specialty organizations and, 61, 113
  UBC, 63, 126

Building Officials and Code Administrators International (BOCA), 63
Building Seismic Safety Council (BSSC), 62, 115, 495

C

cambered planks, 327
cantilever beam columns, 5
cavity walls, 14, 55
cells, 4
cement. *See also* cementitious systems; portland cement
  gypsum, 17
  pozzolanic, 11, 16–18
  proportion requirements for, 19t, 21t–22t
  slag, 11, 18
cementitious systems, 18. *See also* cement-lime mortar; masonry-cement mortar; mortar-cement mortar
cement-lime mortar
  air content of, 23
  compressive strength of, 24
  definition of, 18
  masonry-cement mortar v., 24–25
  preliminary discussion of, 10
  property requirements for, 20t
  proportion requirements for, 19–29, 19t
  specification of, 15
  uses of, 48
center of rigidity
  concept of, 349–350, 349f–350f
  lateral load applied through, 352, 352f
  location of, 349f–350f, 350–351, 356f
  method 2b and, 349–355, 349f–353f, 356f, 357, 358f
  in one direction, 350f
  plan torsion and, 349–350, 350f
  torsional moment applied at, 352–354, 352f–353f
centrally located reinforcement, 309
centroidal moment of inertia, 283
chemistry, mortar, 15–18
chippage
  of clay masonry units, 32, 34t
  of concrete masonry units, 36
chord(s)
  checking of, 370
  compression, 371, 371f, 522
  force, 371, 371f, 410–411
  roof diaphragms and, 410–411
cladding
  AAC, 446f
  code basis of, 70t–71t, 75, 77f, 78–79, 83–87, 85t–87t

cladding (*Cont.*):
  design pressure on, 384–389, 384*t*, 386*t*–388*t*
  effective area of, 387
  reinforced panels as, 440
  wind load on, 83–87, 85*t*–87*t*
clay masonry units. *See also* four-story building with clay masonry, strength design example of
  appearance of, 30, 32–33
  ASTM specifications for, 27, 30–34, 33*f*, 34*t*, 48
  characteristics of, 30–32, 35
  chemistry associated with, 29
  chippage of, 32, 34*t*
  coefficient of thermal expansion of, 35
  color of, 35
  compressive strength of, 32, 34*t*, 247
  connection details for, 372–373, 372*f*
  curtain walls, 192, 193*f*, 301, 302*f*
  dimensional tolerances of, 30–31, 34*t*
  durability of, 30–31, 33, 34*t*
  expansion joints used in, 46, 55–56
  expansion of, 35, 420
  fire wall, 303*f*
  fired, 10*t*
  flashing placement and, 45*f*
  geology associated with, 29
  hollow, 4, 49, 150*t*, 246, 247*t*
  IRA of, 35, 56
  lintels, 123*f*
  manufacturing of, 29–30
  materials, 35–36
  modular, 38
  modulus of elasticity of, 35
  pilasters, 405*t*
  reinforced shear walls, 215–219, 217*f*, 320–323, 321*f*, 323*f*
  reinforcement in, 123*f*
  tensile strength of, 35
  through-wall, 420
  visual and serviceability characteristics of, 30–32
  wall-to-foundation connections and, 177, 178*f*, 223*f*, 226, 274, 274*f*, 325*f*, 327
  wet, 49
coatings, 11, 39, 45
coefficient of friction, 465, 485, 515
coefficient of thermal expansion
  of clay masonry units, 35
  of concrete masonry units, 37
cold-water absorption, 31
collar joint, 14*f*
collectors, diaphragm, 370
color
  of clay masonry units, 35
  of concrete masonry units, 36
columns. *See also* beam columns
  axial capacity of, 244*f*
  concrete, 302*f*
  critical strain condition for, 490*f*
  definition of, 196, 306
  panel wall connected to, 134*f*, 230*f*
  reinforced, 124, 302*f*
  steel, 302*f*
components
  basic, 9–15
  design pressure on, 384–389, 384*t*, 386*t*–388*t*
  wind load on, 83–87, 85*t*–87*t*
composite barrier walls, 14*f*
composite brick-block walls, 56
compression
  chord, 371, 371*f*, 522
  flange, 431
  pure, 197, 199, 307, 309, 314
  strut, 483, 523
compressive axial load, 489–490
compressive force, 211, 221, 287*f*, 308
compressive reinforcement, 220, 284, 290, 404, 490
compressive strength
  AAC, 448, 450, 452, 457, 460, 467
  of cement-lime mortar, 24
  of clay masonry units, 32, 34*t*, 247
  of concrete masonry units, 36
  crushing strength and, 119–120
  of masonry assemblages, 38
  of masonry-cement mortar, 24
  of reinforced lintels, 187–188
compressive stress
  blocks, 183, 285*f*, 286, 516
  flexural, 250
  maximum, 151–152, 154–157, 299, 391, 400, 453, 456, 459
  net, 455, 457
concentric axial load, 149–152, 150*f*, 150*t*, 246–248, 246*f*, 246*t*–247*t*, 452–455, 453*f*
concrete. *See also* autoclaved aerated concrete; concrete masonry units; one-story building with reinforced concrete masonry, strength design example of
  frame elements, 4
  ordering of, 9
  reinforced, 4–5, 477
  shrinkage, 55–56
  topping, 412
Concrete Building Brick. *See* ASTM C55

Concrete Grid Paving Units. *See* ASTM C1319
concrete masonry units
    absorption of, 36
    ASTM specifications for, 28, 35–38, 48
    characteristics of, 36–37
    chippage of, 36
    classification of, 10t
    coefficient of thermal expansion of, 37
    color of, 36
    columns, 302f
    compressive strength of, 36
    control joints used in, 46
    curtain walls, 192, 194f, 301, 303f
    dimensional tolerances of, 36
    hollow, 4, 38, 49
    IRA of, 37
    manufacturing of, 35–36
    materials, 35–36
    modulus of elasticity of, 37
    tensile strength of, 37
Concrete Masonry Units for Construction of Catch Basins and Manholes. *See* ASTM C139
Conducting Strength Tests of Masonry Wall Panels. *See* ASTM C1717
conference proceedings, on AAC, 526–527
confined masonry, 124
conical breakout cones, 169, 169f, 266, 266f
connection details
    for clay masonry units, 372–373, 372f
    for roof and floor diaphragms, 372–373, 372f
connectors, 11, 39–41, 43f–44f
consensus rules, 59–60
construction, AAC, 440, 448–449
construction details, for structures requiring little structural calculation, 49–54, 51f–54f
construction joints, 41
construction process, 49
contract documents, 119
control joints
    in concrete masonry units, 46
    location of, 378, 379f
    materials, 41, 46, 47f, 50, 53, 53f, 56
    vertical, 401, 495
convergence, 478
cooling, 30
cracked, transformed sections
    allowable flexural capacity of cross section and, 291–293, 291f–293f
    allowable-stress balanced reinforcement and, 293–294, 293f
    bond behavior of, 286–288, 287f–288f
    flexural behavior of, 281–283, 282f

cracked, transformed sections (*Cont.*):
    neutral axis of, 282–284, 283f, 288–290, 289f, 294
    physical properties of steel reinforcing wire and bars and, 288, 289t
    shear behavior of, 284–286, 285f–286f
cracked inertia, 477
cracking
    at bed joints, 477
    moment, 209, 477
    web-shear, 463, 465, 470, 482
critical strain condition
    for AAC walls, 479, 479f, 490, 490f
    for beams, 490f
    for columns, 490f
    out-of-plane loads and, 479, 479f
    strength design of reinforced masonry and, 210–211, 211f, 219–220, 219f
cross section
    equilibrium of forces on, 298, 298f
    flexural capacity of, 291–293, 291f–293f
    pilaster, 401f
crushing
    of diagonal strut, 463, 470, 482
    masonry, 272–273
    strength, 119–120
cryptoflorescence, 32
curing, 49
curtain walls. *See also* reinforced curtain walls
    anchor design for, 195–196, 196f, 305–306, 305f
    background on, 191–192, 191f, 300–301, 300f
    clay masonry, 192, 193f, 301, 302f
    concrete masonry, 192, 194f, 301, 303f
    flexural tensile stress and, 191–192, 301
    plan view of, 300f
    structural action of, 192, 301

D

dead load
    eccentric, 318, 318f, 392, 392f, 403, 403f
    floor, 422f, 497
    according to IBC, 64
    1/3 stress increase and, 104–105
    roof, 422f, 497
deflection
    mid-height, 210
    out-of-plane loads and, 210, 239, 478
deformed reinforcement, 39, 40f, 421
deformed reinforcing bars, 39, 40f
degrees of freedom, of rigid floor diaphragms, 343
dehydration, 30

design acceleration response spectrum, 99–100, 101t
design criteria
   for four-story building with clay masonry, 420–428, 420f–421f, 424t, 427f, 427t
   for one-story building with reinforced concrete masonry, 378, 379f, 380
   for three-story AAC shear-wall hotel, 495–514, 496f, 500f–504f, 505t–506t, 509f, 509t–510t
design horizontal seismic load effect, 511
design intent, classification by, 120–121
design moment, 370–371, 370f–371f, 400, 458, 468, 513f, 513t
design pressure, on components, 384–389, 384t, 386t–388t
design response acceleration parameter, 423, 498, 506–507
design response spectrum, 99–100, 101t, 423, 424f, 426, 498, 508, 509f
design shear
   capacity, 165, 434, 466, 471, 517
   flexible diaphragms and, 370–371, 370f–371f
   of four-story building with clay masonry, 427f
   on walls, 412, 413t
diagonal strut
   in compression, 483
   crushing of, 463, 470, 482
Diagonal Tension in Masonry Assemblages. *See* ASTM E519
diaphragms. *See also* chord(s); flexible diaphragms; floor diaphragms; horizontal diaphragms; rigid diaphragms; roof diaphragms
   anchorage of, 325
   collectors, 370
   design of, 369–371, 370f–371f
   integral, 371
   loaded nodes for, 522f
   nonintegral, 371
   reactions of, 362
   seismic design and, 365
   semirigid, 342
   shear and, 364, 364f, 369, 370–371, 370f–371f
   unloaded nodes for, 523f
dimensional tolerances
   of clay masonry units, 30–31, 34t
   of concrete masonry units, 36
dimensions, preliminary discussion of, 11–12
direction of span, 5f
drag strut, 370
drainage walls, 13–14, 14f, 48, 55
dry press process, 29

Dry-Cast Segmental Retaining Wall Units. *See* ASTM C1372
Drying Shrinkage. *See* ASTM C426
ductile AAC shear-wall structures, 494–495, 494t
ductile behavior, 489
durability, of clay masonry units, 30–31, 33, 34t
dynamic loads, 344

E

earthquake loading
   background on, 88–89, 88f–89f
   according to IBC, 87–103, 88f–89f, 92f–96f, 97t–98t, 101f, 101t–102t, 419
   seismic base shear determined and distributed in, 90–103, 92f–96f, 97t–98t, 101f, 101t–102t
   seismic ground motion values in, 89–90, 498, 504
   seismic resistance and, 126–127
   special reinforced shear wall for, 431, 434
   transverse shear walls designed for, 428–430, 429f, 514–517, 514f, 516f
earthquakes
   basic AAC resistance mechanism to, 493
   connection details for roof and floor diaphragms and, 372–373, 372f
   500-year, 98–99, 507
   lateral load from, 422–428, 424t, 427f, 427t, 484, 484f, 497–514, 500f–504f, 505t–506t, 509f, 509t–510t
   maximum considered, 423, 498–499, 500f–503f
   NEHRP for, 62–63
   2500-year, 91, 499
   unfactored in-plane lateral loads and, 484, 484f
eccentric axial load
   on AAC masonry walls loaded out-of-plane, 476, 476f
   tension and, 391
   unfactored moment diagrams due to, 253, 253f, 318, 318f, 460, 460f, 476, 476f
   unreinforced bearing walls with, 153–159, 153f, 156f, 158f, 248–252, 249f, 253, 253f, 318, 318f, 455f
   unreinforced bearing walls with, plus wind, 156–159, 156f, 158f, 318, 318f, 459–461, 459f–460f, 476, 476f
eccentric dead load, 318, 318f, 392, 392f, 403, 403f, 422f, 497
eccentric gravity load, 243, 248–249, 254, 401, 455, 459
eccentricity, minimum, 248, 455
effective embedment, of anchor bolts, 171, 272
efficiency factor, 483
efflorescence, 32, 34t

elastic buckling, 244
embedded steel elements, 4
empirical design, 122
enclosure classification, 74, 81, 85, 381, 384
equilibrium
    axial, 211–212, 221, 282, 286, 480, 491
    of forces on cross section, 298, 298f
    moment, 282
    rotational, 353
equivalent lateral force procedure, 422–423, 498
ESCSI. *See* Expanded Shale Clay and Slate Institute
Evaluating Masonry Bond Strength. *See* ASTM C1357
exact approach, to flexible floor diaphragms, 362
Expanded Shale Clay and Slate Institute (ESCSI), 62
expansion
    of clay masonry units, 35, 420
    coefficient of thermal, 35, 37
    freeze-thaw, 35
    moisture, 35
expansion joints, 420
    in clay masonry units, 46, 53f, 55–56
    horizontally oriented, 46f, 55
    sealants and, 41
    vertical strips and, 242
    vertically oriented, 46f, 56
exposure category, 69, 80, 83, 381, 384
exterior walls
    of four-story building with clay masonry, 435
    for gravity load, 435, 495, 517–519, 518f
    of three-story AAC shear-wall hotel, 495, 517–519, 518f
external nominal moment, 184f
external pressure coefficients, 72f–73f, 75, 76f–77f, 81, 85, 381–382, 385, 387
extreme-fiber tensile stress, 245

F

face, 12
face-shell bedding, 139–141, 139f, 141f, 234–237, 235f, 237f
Facing Brick. *See* ASTM C216
factored axial force, 209, 454
factored design lateral forces
    for four-story building with clay masonry, 427f
    for three-story AAC shear-wall hotel, 512–513
factored design shears, 513f, 513t
factored moment
    design, 400, 427f, 450–451, 513f, 513t
    diagrams, 403, 403f

factory-reinforced panels, 473
failure
    pryout, 173–174, 173f, 270, 270f, 273
    shear breakout, 173–174, 173f–174f, 270–271, 270f–271f
    steel, 333
    surface, 214–215, 482
Federal Emergency Management Agency (FEMA), 62
fiber tension, 452
field-reinforced AAC masonry, 473
filled collar joint, 14f
finite element method. *See* method 1
finite element programs, 362
fire
    IBC and, 378, 380
    rating, 439
    wall, 303f
fire design
    for four-story building with clay masonry, 420–421
    for one-story building with reinforced concrete masonry, 378, 380
    of three-story AAC shear-wall hotel, 496
fired clay masonry units, 10t
#5 bars, 430, 434
500-year earthquake, 98–99, 507
fixed lintels, 54, 54f
flange
    compression, 431
    compressive stress block and, 516
    width, 429f
flashing, 11, 14, 39, 42–45, 45f, 55
Flemish bond, 13, 13f
flexible diaphragms
    design for moment in, 370–371, 370f–371f
    design shear and, 370–371, 370f–371f
    introduction to design of, 370
    lateral load analysis of shear-wall structures with, 362–364, 363f–364f
    roof, 363f
    shears and, 364, 364f, 370–371, 370f–371f
    shear-wall building with, 362–364, 363f–364f
flexible floor diaphragms
    approximate approach to, 362
    classification of, 342–343, 364–365
    exact approach to, 362
    general characteristics of, 343
    lateral load analysis of shear-wall structures with, 362–364, 363f–364f
Flexural Bond Strength of Masonry. *See* ASTM E518

flexural capacity, 207, 212, 217
  of beams, 472
  of cross section, 291–293, 291f–293f
  of reinforced AAC masonry shear walls, 481, 487
flexural compressive stress, 250
flexural deformation, 361
flexural design, 468
  in-plane, of transverse shear walls, 516–517, 516f
  out-of-plane loads and, 210–212, 211f
  of panel walls, 135, 136t
  of reinforced shear walls, 219–222, 220f
flexural reinforcement, 210–212, 211f, 219–222, 221f, 298, 323–324, 409, 431, 471–472, 489
flexural resistance, AAC, 448
flexural tensile stress, 135–142, 165–166
  curtain walls and, 191–192, 301
  net, 257, 263, 317, 466
  panel walls and, 233–236, 301, 450–451
  shear walls and, 260
flexure
  allowable-stress design and, 231–232, 231t, 295, 305, 336
  cracked, transformed sections and, 281–283, 282f
  pure, 198, 200, 307, 309–310
  reinforced beams and, 183–186, 184f–185f, 189–192
  reinforced curtain walls and, 192–195
  reinforced lintels and, 186, 189–192
  reinforced shear walls and, 219–222, 220f
  unreinforced panel walls and, 135, 136t, 231–232, 231t, 449
  Wall Segment A and, 414
floor diaphragms. *See also* flexible floor diaphragms; rigid floor diaphragms
  AAC, 519–523, 519f–523f
  connection details for, 372–373, 372f
  horizontal, 3, 162, 259, 342–343, 362
  for in-plane actions, 519–523, 519f–523f
  lateral load transferred by, 421
  plan torsion and, 344–345, 347
  reinforced panels as, 440
floor load
  dead, 422f, 497
  due to gravity, 422, 497
  live, 64–66, 65t–66t, 497
floor-level bond beams, 515
floors. *See also* wall-to-floor connections
  AAC systems, 524
  actions of, 362
  slabs of, 50, 51f
  weight of, 426, 512

flow, 23
force coefficients, 381t, 385, 387
foundation. *See also* wall-to-foundation connections
  dowels, 415
  walls, 51f
#4 bars, 190, 409, 409f, 430, 434, 472, 486, 518
#4 foundation dowels, 415
four-story building with clay masonry, strength design example of
  architectural constraints for, 420
  base shear of, 425, 427–428
  design criteria chosen for, 420–428, 420f–421f, 424t, 427f, 427t
  design shear of, 427f
  design steps for, 419–420
  exterior walls of, 435
  factored design lateral forces and moment for, 427f
  fire design for, 420–421
  lintels of, 435
  materials specified for, 421
  overall comments on, 435
  plan view of, 420f–421f
  seismic loads on, 419
  structural irregularities of, 428
  structural systems of, 420
  transverse shear walls of, 419, 421, 426, 428–435, 429f–430f, 432t–433t
  water-penetration resistance in, 420
frame buildings, wall buildings *v.*, 264
frame elements, concrete, 4
free body, 264, 265f, 351f
freeze-thaw expansion, 35
freeze-thaw resistance, 30–33
fresh grout, 26
fresh mortar, 23
fully reinforced masonry, 127

G

geometric centroid, 350
glass-block masonry design, 122
governmental organizations, 62–63, 115
gravity load
  arching action and, 300
  eccentric, 243, 248–249, 254, 401, 455, 459
  exterior walls designed for, 435, 495, 517–519, 518f
  factored, 407, 407f
  floor, 422, 497
  according to IBC, 64–67, 65t–66t, 67f
  on lintels, 399, 399f, 406–409, 406f–408f, 407t

gravity load (*Cont.*):
   low-rise, bearing wall buildings designed for, 4
   plus out-of-plane loads of walls, 389–406, 389f–390f, 392f, 394f, 395t, 396f–397f, 398t, 399f–404f, 405t, 406f
   reinforced panels supporting, 440
   on reinforced shear walls, 212, 319, 481
   roof, 380, 422, 464, 466, 497
   structural systems for, 389, 421, 497
   transverse shear walls designed for, 514–517, 514f, 516f
   on unreinforced bearing walls, 243
   on unreinforced shear walls, 162, 260, 462
gross inertia, 477
grout
   ASTM specifications for, 25–27, 48, 516
   fresh, 26
   hardened, 26–27
   hollow units and, 4, 140, 235–236
   portland cement in, 25
   preliminary discussion of, 11
   proportion requirements for, 26, 26t
   self-consolidating, 27
   step of, 49
   strength, 516
Grout for Masonry. *See* ASTM C476
gust effect factor, 74, 81, 83, 381, 384
gypsum cement, 17

H

hardened grout, 26–27
hardened mortar, 23–24
head, 12
height, 11–12
high-retentivity mortar, 56
Hollow Brick. *See* ASTM C652
Hollow Load-Building Concrete Masonry Units. *See* ASTM C90
Hollow Non-Load-Bearing Concrete Masonry Units. *See* ASTM C129
hollow units
   ASTM specifications for, 27–28, 36–37, 421
   clay masonry, 4, 49, 150t, 246, 247t
   concrete masonry, 4, 38, 49
   grout and, 4, 140, 235–236
   single-wythe unreinforced panel walls using, 137–138, 138f, 233–234, 234f
   single-wythe unreinforced panel walls using, face-shell bedding only, 139–140, 139f, 234–235, 235f

hollow units (*Cont.*):
   single-wythe unreinforced panel walls using, fully grouted, 140, 235–236
   two-wythe unreinforced panel walls using, face-shell bedding only, 140–141, 141f, 236–237, 237f
hollow-core planks, 421
horizontal diaphragms
   classification of, 342–343, 364
   floor, 3, 162, 259, 342–343, 362
   in-plane flexibility of, 342
   roof, 3, 162, 164–166, 164f–165f, 168, 259–263, 262f
horizontal reinforcement, 4, 5f, 49, 124, 127, 301, 483
horizontal seismic load effect, 511
horizontal tributary areas, 362
horizontally distributed seismic base shear, 498–514, 500f–504f, 505t–506t, 509f, 509t–510t
horizontally oriented anchor bolts, 168, 168f
horizontally oriented expansion joints, 46f, 55
Hydrated Lime for Masonry Purposes. *See* ASTM C207
hydrated masons' lime, 10, 19t, 25
hydraulic mortars, 15–17
hydraulic-cement mortars, 16–17

I

IBC. *See* International Building Code
ICBO. *See* International Conference of Building Officials
ICC. *See* International Code Council
ICC AC 15, 446
ICC ES, 494
ICC ESR-1371, 446
ICC Structural Committee, 494–495
idealized single-degree-of-freedom system, 88, 88f
IMI. *See* International Masonry Institute
impact resistance, 383
importance factor, 68–69, 80, 83, 101t, 381, 384, 424, 424t, 498, 508, 509t
Industrial Floor Brick. *See* ASTM C410
industry organizations, 61–62, 114
inelastic buckling, 244, 247, 255
inertia, 241, 283, 477
inherent torsion, 514
initial rate of absorption (IRA)
   of clay masonry units, 35, 56
   of concrete masonry units, 37
   over 30, 49
in-plane bending, 435

in-plane discontinuity, 514
in-plane flexibility, 342, 362
in-plane lateral loads, 216f, 321f, 484, 484f
in-plane shear loads, 212, 216, 260, 319, 448, 462
in-plane strength, 369
in-plane walls
   AAC, 490, 490f
   east, 413–414, 414f
   flexural design of, 516–517, 516f
   west, 413
integral diaphragms, 371
interface, unbonded, 466
intermediate walls, 490
internal lever arm, 185, 195
internal porosity, 439
internal pressure coefficients, 75, 75f, 81, 85, 381, 384
internal pressure evaluation, 383
internal shears, roof diaphragms resisting, 371
International Building Code (IBC)
   allowable-stress loading combinations of, 104–105, 109–110
   dead load according to, 64
   development of, 63
   earthquake loading according to, 87–103, 88f–89f, 92f–96f, 97t–98t, 101f, 101t–102t, 419
   fire and, 378, 380
   gravity loads according to, 64–67, 65t–66t, 67f
   introduction to, 64
   strength loading combinations of, 103, 105
   wind loading according to, 67–87, 69f, 70t–71t, 72f–73f, 75f–77f, 80f, 80t, 84t–87t
International Code Council (ICC), 63–64, 114, 446, 494–495
International Conference of Building Officials (ICBO), 63
International Masonry Institute (IMI), 62, 114
interstitial condensation, 55
inward pressure, maximum, 386
IRA. *See* initial rate of absorption

J

joint. *See also* control joints; expansion joints; movement joints
   bed, 40f, 122, 124, 477
   construction, 41
   filled collar, 14f
   panel-to-bond beam, 521f
   panel-to-panel, 521f
   reinforcement, 39, 40f, 122, 124, 187, 192, 303f, 483

joists. *See also* long-span joists
   bar, 389, 390f, 400, 400f, 407f
   steel, 372, 372f
   wooden, 373, 373f
journal publications, on AAC, 526

K

kinematics, 281

L

lateral forces. *See* factored design lateral forces
lateral load analysis, of shear-wall structures
   comments on, 361
   conclusions regarding, 361–362
   with flexible diaphragms, 362–364, 363f–364f
   for one-story building with reinforced concrete masonry, 410–412, 410f–411f, 413t
   method 1 (finite element method) for, 343–345, 345f, 361
   method 2a (simplest hand method) for, 344–348, 346f–348f, 360t, 361
   method 2b (more complex hand method) for, 344, 349–359, 349f–353f, 355f–356f, 358f, 361
   method 2c (most complex hand model) for, 344, 359, 359f, 360t, 361
   with rigid floor diaphragms, 343–362, 345f–353f, 355f–356f, 358f–359f, 360t
lateral loads
   AAC and, 464
   through arbitrary point, 351–352, 351f
   on box-type structures, 161–162, 161f, 259–260, 260f
   through center of rigidity, 352, 352f
   decomposition of, 350f–351f
   from earthquakes, 422–428, 424t, 427f, 427t, 484, 484f, 497–514, 500f–504f, 505t–506t, 509f, 509t–510t
   in-plane, 216f, 321f, 484, 484f
   low-rise, bearing wall buildings designed for, 4
   on low-rise building with reinforced concrete masonry, 422–428, 424t, 427f, 427t
   out-of-plane, 324
   resistance of, 3–4, 4f, 161–162, 161f
   structural systems for, 389, 421, 497
   transfer of, 421
   transmission of, 262, 262f, 324, 421
   transverse, 519
leeward side
   code basis for structural design of, 78–79, 81–83, 84t, 87, 87t
   of one-story building with reinforced concrete masonry, 382–384, 386–387, 387t

length, 11–12
leveling bed mortar, 465, 470
lime. *See* cement-lime mortar; hydrated masons' lime; sand-lime mortar
linear interaction equation, 274
linear stress-strain behavior, 119
lintels. *See also* reinforced lintels
   clay masonry, 123*f*
   definition of, 186
   depth of, 186, 188, 295–297, 334, 467–470
   fixed, 54, 54*f*
   in four-story building with clay masonry, 435
   gravity load on, 399, 399*f*, 406–409, 406*f*–408*f*, 407*t*
   loose, 54, 54*f*
   of one-story, bearing wall buildings, 399, 399*f*
   of one-story building with reinforced concrete masonry, 399, 399*f*, 406–409, 406*f*–408*f*, 407*t*
   span of, 469
   wall sections at, 54, 54*f*
live load
   floor, 64–66, 65*t*–66*t*, 497
   reduced, 401
   roof, 67, 67*f*, 497
   wall, 66–67
load. *See also* dead load; floor load; gravity load; live load; roof load; wind load
   AAC, 447, 464
   dynamic, 344
   effects, 331–332, 332*t*, 335
   factors, 135, 333, 449
   in-plane shear, 212, 216, 260, 319, 448, 462
   at roof diaphragm level, 344–345
   seismic, 419, 511
load-bearing masonry, 120
long-span joists
   bearing plate under, 406, 406*f*
   effective area of, 387
   on pilasters, 399–400, 402
   reactions of, due to wind, 380–384, 381*t*, 383*t*
loose lintels, 54, 54*f*
low-modulus acrylic stucco, 495
low-rise, bearing wall buildings
   basic structural behavior of, 3–4, 4*f*
   bearing walls of, 3–4, 4*f*
   gravity loads and, 4
   lateral loads and, 4
low-rise building with reinforced concrete masonry, strength design example of
   connections of, 415
   control joints in, 378, 379*f*
   design criteria chosen for, 378, 379*f*, 380
   design floor load due to gravity in, 422

low-rise building with reinforced concrete masonry, strength design example of (*Cont.*):
   design lateral load from earthquake for, 422–428, 424*t*, 427*f*, 427*t*
   design steps of, 378
   east wall of, 393–399, 394*f*, 396*f*–397*f*, 398*t*, 399*f*, 406*f*, 409, 409*f*, 412–415, 413*t*, 414*f*
   elevation of, 379*f*
   fire design for, 378, 380
   introduction to, 377
   lateral force analysis for, 410–412, 410*f*–411*f*, 413*t*
   lateral loads on, 422–428, 424*t*, 427*f*, 427*t*
   leeward side of, 382–384, 386–387, 387*t*
   lintels of, 399, 399*f*, 406–409, 406*f*–408*f*, 407*t*
   materials specified in, 380
   north wall of, 399–404, 400*f*–404*f*, 405*t*, 409, 415
   pilasters of, 399–404, 401*f*–404*f*, 405*t*
   plan of, 379*f*
   roof diaphragm of, 410–412, 410*f*–411*f*, 413*t*
   roof load due to gravity and, 380, 422
   south wall of, 399–404, 400*f*–404*f*, 405*t*, 409, 415
   structural systems of, 389
   Wall Segment A of, 414–415
   Wall Segment B of, 396–397, 397*f*, 398*t*
   walls designed in, 413–415, 414*f*
   walls for gravity plus out-of-plane loads of, 389–406, 389*f*–390*f*, 392*f*, 394*f*, 395*t*, 396*f*–397*f*, 398*t*, 399*f*–404*f*, 405*t*, 406*f*
   wall-to-roof details of, 415
   water-penetration resistance in, 378
   west wall of, 389–394, 389*f*–390*f*, 392*f*, 394*f*, 395*t*, 409, 412–413, 415
   wind load for, 380–389, 380*f*, 381*t*, 383*t*–384*t*, 386*t*–388*t*
   windward side of, 382–383, 386, 386*t*
low-rise buildings. *See also* low-rise, bearing wall buildings; low-rise building with reinforced concrete masonry, strength design example of
   basic structural design of, 4–5, 5*f*
   with rigid floor diaphragms, 361–362

M

magnifier, 156, 245
main wind force-resisting system (MWFRS), 75, 76*f*, 79–83, 80*f*, 80*t*, 84*t*, 380–384, 381*t*, 383*t*
manufacturing
   AAC, 441–443, 442*f*
   of clay masonry units, 29–30
   of concrete masonry units, 35–36
Mason Contractors' Association of America (MCAA), 62, 114

masonry. *See also* masonry elements; masonry units; reinforced masonry
   accessory tools, 11
   applications of, 9
   assemblages, 28, 37–38
   basic components of, 9–15
   confined, 124
   controls, anchor bolts with, 334t
   crushing, 272–273
   curing of, 49
   load-bearing, 120
   mechanical behavior of, 119–120
   modulus, 344
   nonload-bearing, 120, 142, 238, 451
   partially reinforced, 124–127
   sand, 10
   specified, 15
   tensile breakout of, 268
   unreinforced, 121, 124
   walls, 13–14, 14f, 122–124, 125f, 143f
Masonry Designers' Guide, 493
masonry elements
   classification of, 120–121, 127–129
   design approaches for, 121–122
   reinforcement used in, 122–127, 123f, 125f–126f
   structural, behavior of, 3
Masonry Standards Joint Committee (MSJC), 5, 61, 104–105
   AAC design provisions of, 440, 447–448, 479–481, 479f, 489–493, 490f
   allowable-stress design provisions of, 109–113, 110t–113t, 231–232, 231t, 296–299, 331
   design approaches and, 122
   flexural design of panel walls and, 135, 136t
   flexural reinforcement and, 210–212, 211f, 219–222, 221f
   1/3 stress increase allowed by, 104–105
   strength design provisions of, 105–109, 106t–109t, 331
   strength-reduction factors of, 106, 106t
   unit strength method of, 120, 187
masonry units. *See also* clay masonry units; concrete masonry units; hollow units
   ASTM specifications for, 27–29, 48
   bond patterns of, 12–13, 13f
   orientation of, 12, 12f
   preliminary discussion of, 9, 10t
   section properties for, 150t, 246t
   shale, 27
   solid, 135–137, 137t, 232–233, 232f
   stretcher, 122
   through-wall, 420

masonry units (*Cont.*):
   trough, 122
   unfired, 10t
masonry-cement mortar
   air content of, 23
   cement-lime mortar *v.*, 24–25
   compressive strength of, 24
   preliminary discussion of, 10–11
   proportion requirements for, 20–21, 21t
   specification of, 15
mason's sand, proportion requirements for, 19t, 21t–22t
mass irregularity, 428, 513
material tests, 120
materials. *See also* accessory materials
   in AAC, 441
   clay masonry, 35–36
   concrete masonry units, 35–36
   control joint, 41, 46, 47f, 50, 53, 53f, 56
   for four-story building with clay masonry, 421
   for low-rise building with reinforced concrete masonry, 380
   for three-story AAC shear-wall hotel, 497
maximum bar area, 477
maximum compressive stress, 151–152, 154–157, 299, 391, 400, 453, 456, 459
maximum considered earthquake (MCE), 423, 498–499, 500f–503f
maximum flexural reinforcement, 210–212, 211f, 219–222, 221f, 323–324, 472
maximum moment, 475, 477
maximum reinforcement ratios, 323–324, 479–481, 479f, 489–493, 490f
maximum tensile stress, 151, 154, 157–158, 391, 393, 400, 453, 456, 458, 461
MCAA. *See* Mason Contractors' Association of America
MCE. *See* maximum considered earthquake
Measurement of Compressive Strength of Masonry Prisms to Determine Compliance. *See* ASTM C1314
Measurement of Masonry Flexural Bond Strength. *See* ASTM C1072
method 1 (finite element method)
   comments on, 361
   definition of, 343–344
   example of, 344–345, 345f
   for lateral load analysis of shear-wall structures, 343–345, 345f, 361
   results of, 359, 360t

method 2a (simplest hand method)
  comments on, 361
  definition of, 344
  distribution of shears among wall segments and, 346
  example of, 346–348, 347f–348f
  for lateral load analysis of shear-wall structures, 344–348, 346f–348f, 360t, 361
  results of, 360t
  shearing stiffness of wall segments and, 344–346, 346f
method 2b (more complex hand method)
  center of rigidity and, 349–355, 349f–353f, 356f, 357, 358f
  comments on, 361
  definition of, 344
  example of, 354–359, 355f–356f, 358f
  for lateral load analysis of shear-wall structures, 344, 349–359, 349f–353f, 355f–356f, 358f, 361
  response due to direct shear plus plan torsion and, 354, 355f, 358f
  response due to lateral load applied through center of rigidity and, 352, 352f
  response due to torsional moment applied at center of rigidity and, 352–354, 352f–353f
  response to lateral load applied through arbitrary point and, 351–352, 351f
  results of, 359, 360t
method 2c (most complex hand model)
  comments on, 361
  definition of, 344
  for lateral load analysis of shear-wall structures, 344, 359, 359f, 360t, 361
  results of, 359, 360t
mid-height deflection, 210
midspan moment, 400
minimum eccentricity requirements, 248, 455
minimum flexural reinforcement, 201, 219, 323–324
minimum reinforcement ratios, 323–324, 479–481, 479f, 489
modal response-spectrum analysis, 422
model code(s)
  adoption of, 60
  organizations, 15, 63–64, 114
Moderate weathering regions, 33
modular clay units, 38
modularity, 47
modulus, masonry, 344
modulus of elasticity
  of clay masonry units, 35
  of concrete masonry units, 37
modulus of rupture
  panel walls and, 135, 136t, 138, 140, 331–332, 449, 451
  for unreinforced bearing walls of AAC masonry, 453, 461
moisture barriers, 11, 39, 46
moisture expansion, 35
moment. *See also* unfactored moment diagrams
  bending, 450–451
  cracking, 209, 477
  design, 370–371, 370f–371f, 400, 458, 468, 513f, 513t
  diagram, 484f
  equilibrium, 282
  external nominal, 184f
  factored, 400, 403f, 427f, 450–451, 513f, 513t
  in flexible diaphragms, 370–371, 370f–371f
  of inertia, 241, 283
  maximum, 475, 477
  midspan, 400
  secondary, 477
  torsional, 352–354, 352f–353f
moment-axial force interaction diagrams
  by allowable-stress approach, 306–315, 306f, 308f, 311f–315f, 316t
  for east wall of example low-rise building, 396, 396f
  by hand, 199–201, 201f, 307–314, 308f, 311f–315f
  for pilasters, 403–404, 404f, 405f
  for reinforced bearing walls, 197, 197f, 199–201, 201f, 204–205, 205f, 206t, 473–475, 473f, 474t
  for reinforced shear wall, 217, 217f, 487, 487f
  by spreadsheet, 204–205, 205f, 206t, 311–315, 312f–315f, 316t
  by strength approach, 197–205, 197f–198f, 201f–203f, 205f, 206t
  for transverse shear wall, 430, 430f, 432t–433t
  for Wall Segment B of example low-rise building, 396–397, 397f, 398t
  for west wall of example low-rise building, 394, 394f
more complex hand method. *See* method 2b
mortar. *See also* cementitious systems
  chemistry of, 15–18
  fresh, 23
  hardened, 23–24
  high-retentivity, 56
  hydraulic, 15–17
  leveling bed, 465, 470
  mixing and batching, 56
  portland cement in, 10–11, 16–18, 19t, 20, 22, 21t–22t, 24–25

mortar (*Cont.*):
  pozzolanic cement in, 11, 16–18
  preliminary discussion of, 10–11
  sand-lime, 15–16
  specification of, 19–23, 19*t*–22*t*, 27, 48
  standards for, 15
  thin-bed, 457, 463, 468, 497
  Type K, 19
  Type M, 19–20, 19*t*–22*t*, 22–23, 187, 231*t*, 290
  Type N, 19–20, 19*t*–22*t*, 22–23, 135, 137–138, 140–141, 231*t*, 232–236
  Type O, 19, 19*t*–22*t*, 23
  Type S, 19–20, 19*t*–22*t*, 22, 138, 140–141, 154, 158, 166, 187–188, 192, 215, 231*t*, 234–235, 237, 250, 254–255, 290, 380, 421
mortar-cement mortar
  air content of, 23
  preliminary discussion of, 11
  property requirements for, 22, 22*t*
  proportion requirements for, 21–22, 22*t*
  specification of, 15
  tensile bond strength of, 24–25
  uses of, 48
most complex hand model. *See* method 2c
movement joints
  as accessory materials, 41, 46–47, 46*f*–47*f*
  in four-story building with clay masonry, 420
  in three-story AAC shear-wall hotel, 495
MS theses, on AAC, 525
MSJC. *See* Masonry Standards Joint Committee
multiwythe, noncomposite wall. *See* panel walls
multiwythe barrier walls, 13–14
MWFRS. *See* main wind force-resisting system

N

National Building Code (NBC), 63
National Concrete Masonry Association (NCMA), 61–62, 114
National Earthquake Hazard Reduction Program (NEHRP), 62–63
National Fire Protection Association (NFPA), 64, 114
National Institute of Building Sciences (NIBS), 62
National Lime Association (NLA), 62
NBC. *See* National Building Code
NCMA. *See* National Concrete Masonry Association
negative votes, 60
Negligible weathering regions, 33
NEHRP. *See* National Earthquake Hazard Reduction Program
net compressive stress, 455, 457
net flexural tensile stress, 257, 263, 466
net tensile stress, 246, 250–251, 254–256, 317, 392, 400
net uplift, 390
neutral axis
  allowable-stress interaction diagrams and, 308*f*, 311
  axial load and, 431
  balance point and, 310
  within compression flange, 431
  of cracked, transformed section, 282–284, 283*f*, 288–290, 289*f*, 294
  reinforced masonry and, 198, 198*f*, 201–202, 202*f*–203*f*, 210, 220
  shear greatest at, 286
NFPA. *See* National Fire Protection Association
NIBS. *See* National Institute of Building Sciences
NLA. *See* National Lime Association
nominal dimensions, 12
nominal pryout capacity, 176
nominal shear capacity, 165, 174, 214*f*, 434, 470, 483*f*
nominal shear strength, 481–482
nominal tensile capacity, 169–172
noncalculated reinforcement, 238, 451
nonintegral diaphragms, 371
nonlinear stress-strain behavior, 119
nonload-bearing masonry, 120, 142, 238, 451
nonparallel systems, 428
north-south shear, 341

O

occupancy category, 424*t*, 498
1-s period response acceleration parameter, 423, 424*t*, 506–507, 510*t*
one-story building. *See* one-story building with reinforced concrete masonry, strength design example of
1/3 running bond, 13, 13*f*
1/3 stress increase, 104–105
one-way shear
  allowable-stress checks of, 237–238, 305
  strength checks for, 141–142
openings
  AAC and, 462, 467
  building example with, 341–342, 342*f*
  unreinforced bearing walls with, 159–161, 160*f*, 257–258, 257*f*–258*f*, 462
  unreinforced shear walls with, 167–168, 167*f*–168*f*, 264–265, 265*f*
out-of-plane bending, 435
out-of-plane deflection, 210, 239, 478

Index

out-of-plane loads
  deflection and, 210, 239, 478
  distribution of, to vertical and horizontal strips of unreinforced single-wythe panel wall, 144–145, 144f, 239–241, 240f
  distribution of, to vertical and horizontal strips of unreinforced two-wythe panel wall, 145–147, 146f, 241–243, 241f
  flexural design and, 210–212, 211f
  gravity load plus, on one-story building with reinforced concrete masonry, 389–406, 389f–390f, 392f, 394f, 395t, 396f–397f, 398t, 399f–404f, 405t, 406f
  lateral, 324
  on reinforced bearing walls, 205–212, 207f–208f, 211f, 315–318, 316t, 317f–318f
  on unreinforced bearing walls, 332
  on unreinforced panel walls, 144–147, 144f, 146f, 239–243, 240f–241f, 331–332, 332t
  vertical strips resisting, 4
  wind, 399–401, 400f, 435, 495, 517–519, 518f
out-of-plane shear capacity, 451
out-of-plane stress, 245
out-of-plane walls
  AAC, 475–481, 475f–476f, 479f
  critical strain condition of, 479, 479f
  reinforcement ratios for, 479–481, 479f
  single-wythe unreinforced panel, 144–145, 144f, 239–241, 240f
  two-wythe unreinforced panel, 145–147, 146f, 241–243, 241f
  unreinforced bearing, 332
out-of-plane wind, 399–401, 400f, 435, 495, 517–519, 518f
overall modularity, 50, 51f
overall structural configuration layout, 47–48
oxidation, 30
oxygen control, 30

P

paint, 45
panel walls. *See also* unreinforced panel walls
  columns connected to, 134f, 230f
  design of, 135, 136t, 191
  flexural tensile stress and, 233–236, 301, 450–451
  modulus of rupture and, 135, 136t, 138, 140, 331–332, 449, 451
panels, 435, 440. *See also* reinforced panels
panel-to-bond beam joints, 521f
panel-to-panel joint, 521f
parapet, 403, 460
partially reinforced masonry, 124–127

PCA. *See* Portland Cement Association
Pedestrian and Light Traffic Paving Brick. *See* ASTM C902
perforated wall, 341, 365
perimeter wall, 426, 512
PhD dissertations, on AAC, 525–526
pilasters
  axial load on, 402
  bearing plates of, 402f
  clay masonry, 405t
  cross section of, 401f
  effective depth of, 404f
  long-span joists on, 399–400, 402
  moment-axial force interaction diagrams for, 403–404, 404f, 405t
  of one-story building with reinforced concrete masonry, 399–404, 401f–404f, 405t
  reinforced, 124, 126f, 302f
  spacing of, 303f
  strength interaction diagram for, 403–404, 404f, 405t
pintle ties, 43f
plan torsion
  accuracy and, 361
  center of rigidity and, 349–350, 350f
  effects of, 344–345, 347
  floor diaphragms and, 344–345, 347
  response due to direct shear plus, 354, 355f, 358f
planks
  cambered, 327
  hollow-core, 421
Plaster of Paris, 17
plastic hinge zone, 477
portland cement
  chemistry of, 17–18
  in grout, 25
  in mortar, 10–11, 16–18, 19t, 20, 21t–22t, 22, 24–25
  pozzolanic cement used with, 18
  proportion requirements for, 19t, 21t–22t
Portland Cement Association (PCA), 61, 114
posttensioning tendons, 39, 42f
pozzolanic cement
  chemistry of, 16–17
  in mortar, 11, 16–18
  portland cement used with, 18
Prefaced Concrete and Calcium Silicate Masonry Units. *See* ASTM C744
preheating, 30
Prestressed Concrete Institute, 114
projected breakout areas, 170, 170f, 267–268, 267f, 270

property requirements
  for cement-lime mortar, 20t
  for mortar-cement mortar, 22, 22t
proportion requirements
  for cement, 19t, 21t–22t
  for cement-lime mortar, 19–29, 19t
  for grout, 26, 26t
  for hydrated masons' lime, 19t
  for masonry-cement mortar, 20–21, 21t
  for mason's sand, 19t, 21t–22t
  for mortar-cement mortar, 21–22, 22t
  for portland cement, 19t, 21t–22t
pryout failure, 173–174, 173f, 270, 270f, 273
public comment, 60
pullout, 333
pure compression, 197, 199, 307, 309, 314
pure flexure, 198, 200, 307, 309–310

R

Recommended Provisions (NEHRP), 62–63
rectangular stress block, 479–480, 490
redundancy
  factor, 426
  low, 428
re-entrant corners, 428
reinforced beams
  AAC, 467–472, 468f–470f, 469t
  allowable-stress design of, 295–300, 295f–298f, 334–335, 335t
  code basis for, 107, 108t, 110, 111t
  flexure and, 183–186, 184f–185f, 189–192
  strength design of, 107, 108t, 183–191, 184f–186f, 187t, 188f–189f, 334–335, 335t
  structural design of, 122, 128
reinforced bearing walls
  AAC, 473–481, 473f, 474t, 475f–476f, 479f
  allowable-stress design of, 306–318, 306f, 308f, 311f–315f, 316t, 317f–318f, 336, 336f–337f
  allowable-stress interaction diagrams and, 306–315, 306f, 308f, 311f–315f, 316t
  code basis for, 110, 111t, 112, 112t
  moment-axial force interaction diagrams for, 197, 197f, 199–201, 201f, 204–205, 205f, 206t, 473–475, 473f, 474t
  out-of-plane loads on, 205–212, 207f–208f, 211f, 315–318, 316t, 317f–318f
  required details for, 223–226, 223f–225f, 324–327, 325f–327f
  strength design of, 108, 109t, 196–212, 197f, 201f–203f, 205f, 206t, 207f–208f, 211f, 336, 336f–337f

reinforced bearing walls (*Cont.*):
  strength interaction diagrams and, 197–205, 197f–198f, 201f–203f, 205f, 206t
  structural design of, 128
  unity equation and, 317
  wall-to-floor connections and, 224f, 226, 325f–326f, 327
  wall-to-foundation connections and, 223f, 226, 325f, 327
  wall-to-roof details and, 225f, 226, 326f, 327
  wall-to-wall connections and, 225f, 226, 327, 327f
reinforced columns, 124, 302f
reinforced concrete, 4–5, 477. *See also* one-story building with reinforced concrete masonry, strength design example of
reinforced curtain walls
  AAC, 473
  allowable-stress design of, 300–306, 300f, 302f–303f, 305f, 335
  code basis of, 112, 112t
  flexure and, 192–195
  strength design of, 107, 108t, 191–196, 191f, 193f–194f, 196f, 335
  structural design of, 122, 128
reinforced lintels
  AAC, 467–472, 468f–470f, 469t
  allowable-stress design of, 295–300, 295f–298f, 334–335, 335t
  bottom reinforcement of, 297, 297f
  code basis of, 110, 111t
  compressive strength of, 187–188
  flexure and, 186, 189–192
  strength design of, 107, 108t, 183–191, 184f–186f, 187t, 188f–189f, 334–335, 335t
  structural design of, 122, 123f, 128
reinforced masonry. *See also* one-story building with reinforced concrete masonry, strength design example of
  basic structural configuration of, 4–5, 5f
  classification of, 121
  fully, 127
  neutral axis and, 198, 198f, 201–202, 202f–203f, 210, 220
  strength design of, 210–211, 211f, 219–220, 219f
  structural design of, 126–128
reinforced panels
  as cladding, 440
  factory-reinforced, 473
  as floor and roof diaphragms, 440
  gravity load and, 440
reinforced pilasters, 124, 126f, 302f

reinforced shear walls
  AAC, 481–493, 481f–484f, 487, 487f, 490f
  allowable-stress design of, 319–324, 319f, 321f, 323f, 336–337, 337f
  axial load on, 216, 218
  clay, 215–219, 217f, 320–323, 321f, 323f
  comments on design of, 223
  design actions for, 319f
  flexural design of, 219–222, 220f
  gravity loads on, 212, 319, 481
  moment-axial force interaction diagrams for, 217, 217f, 487, 487f
  required details for, 223–226, 223f–225f, 324–327, 325f–327f
  special, 431, 434
  strength design of, 107, 107t, 212–223, 213f–214f, 217f, 220f, 336–337, 337f
  wall-to-floor connections and, 224f, 226, 325f–326f, 327
  wall-to-foundation connections and, 223f, 226, 325f, 327
  wall-to-roof details and, 225f, 226, 326f, 327
  wall-to-wall connections and, 225f, 226, 327, 327f
reinforcement
  accessory materials, 38–39, 40f–42f
  allowable stress in, 291–294, 291f, 293f
  balanced, 185f, 293–294, 293f
  from bars, 39, 40f, 187, 187t, 288, 289t, 468, 469t
  bottom, 189f, 297, 297f, 470f
  centrally located, beam columns with, 309
  in clay masonry units, 123f
  compressive, 220, 284, 290, 404, 490
  deformed, 39, 40f, 421
  in design of structures requiring little structural calculation, 49, 50f
  flexural, 210–212, 211f, 219–222, 221f, 298, 323–324, 409, 431, 471–472, 489
  horizontal, 4, 5f, 49, 124, 127, 301, 483
  joint, 39, 40f, 122, 124, 187, 192, 303f, 483
  masonry elements using, 122–127, 123f, 125f–126f
  nomenclature of, 124–127
  noncalculated, 238, 451
  overall starting point for, 5, 5f, 49, 50f
  seismic, 218
  shear, 213–214, 214f, 217–218, 223, 295–296, 320, 324, 482, 482f
  from steel wire, 187, 187t, 288, 289t
  tension, 184–185, 210, 299
  vertical, 4, 5f, 49, 124, 127, 303f, 373
  welded wire, 39, 41f
reinforcement ratios
  for AAC masonry, 479–481, 479f, 489–493, 490f
  maximum, 323–324, 479–481, 479f, 489–493, 490f
  minimum, 323–324, 479–481, 479f, 489
  for out-of-plane walls, 479–481, 479f
response acceleration parameters
  design, 423, 498, 506–507
  1-s period, 423, 424t, 506–507, 510t
  short-period, 506–507, 510t
  spectral, 499, 500f–503f
response spectrum
  acceleration, 89–90, 89f, 99–100, 101t
  design, 99–100, 101t, 423, 424f, 426, 498, 508, 509f
  design acceleration, 99–100, 101t
  in modal response-spectrum analysis, 422
retentivity, 23
rigid diaphragms
  classification of, 342–343, 364–365
  design of, 369–370
  in-plane strength of, 369
rigid floor diaphragms
  classification of, 342–343, 364–365
  degrees of freedom of, 343
  free-body diagram of, 351f
  lateral load analysis of shear-wall structures with, 343–362, 345f–353f, 355f–356f, 358f–359f, 360t
  low-rise buildings with, 361–362
  prescriptive characteristics of, 343
  shearing stiffness and, 345–346, 346f
rigidity. *See* center of rigidity
roof(s). *See also* wall-to-roof details
  AAC systems, 524
  reaction, 250, 253, 459, 464
  rotational equilibrium of, 353
  weight of, 426
  wind pressures on, 387–389, 388t
roof diaphragms
  checking of, 410–412, 410f–411f
  chords and, 410–411
  connection details for, 372–373, 372f
  flexible, 363f
  horizontal, 3, 162, 164–166, 164f–165f, 168, 259–263, 262f
  internal shears resisted by, 371
  lateral load transferred by, 421
  load at level of, 344–345
  of one-story building with reinforced concrete masonry, 410–412, 410f–411f, 413t
  reaction on, 464f
  reinforced panels as, 440
  shear capacity of, 412

roof diaphragms (*Cont.*):
   shear walls and, 465f
   in wall-to-roof details, 415
   wind load transferred to, 410, 410f–411f
roof load
   dead, 422f, 497
   due to gravity, 380, 422, 464, 466, 497
   live, 67, 67f, 497
rotational equilibrium, 353
running bond, 13, 13f, 49

S

safety factors, for anchor bolts, 333, 334t
Sampling and Testing Brick and Structural Clay Tile. *See* ASTM C67
Sampling and Testing Grout. *See* ASTM C1019
sand masonry, 10
sand-lime mortar, 15–16
saturation coefficient, 31–32
SBC. *See* Standard Building Code
SBCCI. *See* Southern Building Code Congress International
sealants, 11, 39, 41
secondary moments, 477
second-order effects, 452
section properties
   for masonry units, 150t, 246t
   per foot of plan length, 478
   for walls, 143t, 238, 239t
seismic base shear
   earthquake loading and, 90–103, 92f–96f, 97t–98t, 101f, 101t–102t
   for four-story building with clay masonry, 425, 428
   horizontally distributed, 498–514, 500f–504f, 505t–506t, 509f, 509t–510t
   for three-story AAC shear-wall hotel, 498–514, 500f–504f, 505t–506t, 509f, 509t–510t
   vertically distributed, 498–514, 500f–504f, 505t–506t, 509f, 509t–510t
seismic design. *See also* four-story building with clay masonry, strength design example of
   of AAC structures, 493–495, 494t
   ASCE provisions for, 422
   categories, 101t, 325, 369, 424–425, 424t, 509, 510t
   diaphragms and, 365
   factors, 494–495, 494t
seismic force-reduction factor, 494–495, 494t
seismic ground motion values
   in earthquake loading, 89–90, 498, 504
   for three-story AAC shear-wall hotel, 498, 504

seismic load, 419, 511
seismic reinforcement, 218
seismic resistance, 126–127
seismic response
   history procedure, 422
   increased, 428, 513–514
self-consolidating grout, 27
semirigid diaphragms, 342
#7 bars, 409, 409f, 414
#7 foundation dowels, 415
Severe weathering regions, 33–34
shale masonry units, 27
shanks, anchor bolt, 170, 273
shear. *See also* base shear; design shear
   anchor bolts loaded in, 173–177, 173f–174f, 270–274, 270f–271f
   for beams, 337
   cracked, transformed sections and, 284–286, 285f–286f
   diaphragms and, 364, 364f, 369, 370–371, 370f–371f
   direct, response to, 354, 355f, 358f
   distribution, 345–346, 358f, 361–362, 364, 364f
   in-plane loads, 212, 216, 260, 319, 448, 462
   internal, roof diaphragms resisting, 371
   at neutral axis, 286
   nominal strength of, 481–482
   north-south, 341
   one-way, 141–142, 237–238, 305
   reinforcement, 213–214, 214f, 217–218, 223, 295–296, 320, 324, 482, 482f
   strength, of masonry assemblages, 38
   strength checks of, 141–142, 163
   strength-reduction factor for, 463, 482
   transfer, 370
   Wall Segment A and, 414–415
   in walls, 347, 356–357, 358f
   web-shear cracking and, 463, 465, 470, 482
shear breakout failure, 173–174, 173f–174f, 270–271, 270f–271f
shear capacity
   AAC, 451, 462, 466, 493, 518–519
   allowable-stress, 332–333
   in allowable-stress design, 332–333
   calculation of, 188–189, 260–261, 297, 408, 411
   design, 165, 434, 466, 471, 517
   masonry crushing governing, 272–273
   nominal, 165, 174, 214f, 434, 470, 483f
   out-of-plane, 451
   pryout governing, 270, 273
   of roof diaphragm, 412

shear capacity (*Cont.*):
   strength check for, 142, 163
   of unreinforced panel walls, 142, 451
   of unreinforced shear walls, 260–261, 264, 467
shear walls. *See also* lateral load analysis, of shear-wall structures; reinforced shear walls; three-story AAC shear-wall hotel; transverse shear walls; unreinforced shear walls
   axial load in, 216, 218, 464, 472
   as cantilever beam columns, 5
   ductile AAC, 494–495, 494*t*
   with flexible diaphragms, 362–364, 363*f*–364*f*
   flexural tensile stress and, 260
   roof diaphragms and, 465*f*
   tall, 472
shearing deformation, 345, 346*f*, 361
shearing stiffness, 345–346, 346*f*
short-period response acceleration, 506–507, 510*t*
shrinkage, 37, 46, 55
   AAC, 495
   ASTM C426 for, 28, 36
   concrete, 55–56
simplest hand method. *See* method 2a
simplified alternative procedure, 422
single anchor bolts, 175–177, 269*f*, 272–273
single-wythe barrier walls
   classification of, 13–14, 14*f*
   for water-penetration resistance, 13, 378
single-wythe unreinforced panel walls
   AAC, 450–451, 451*f*
   allowable-stress design of, 232–233, 232*f*
   using hollow units, 137–138, 138*f*, 233–234, 234*f*
   using hollow units, face-shell bedding only, 139–140, 139*f*, 234–235, 235*f*
   using hollow units, fully grouted, 140, 235–236
   out-of-plane loads distributed to vertical and horizontal strips of, 144–145, 144*f*, 239–241, 240*f*
   using solid units, 135–137, 137*t*, 232–233, 232*f*
site class
   assignment of, 505*t*
   B, 498–499, 500*f*–503*f*, 505*t*
   D, 505*t*
site coefficient, 423, 505, 506*t*
#6 foundation dowels, 415
slag cement, 11, 18
slenderness, 148, 148*f*
   axial capacity influenced by, 244, 244*f*
   maximum compressive stress influenced by, 459
   transition, 247, 250, 254–255, 454, 457
slenderness-dependant reduction factor, 208, 477

sliding, 463, 482
soft mud process, 29
Solid Concrete Interlocking Paving Units. *See* ASTM C936
solid masonry units, 135–137, 137*t*, 232–233, 232*f*
Southern Building Code Congress International (SBCCI), 63
special reinforced walls
   shear, 431, 434
   yield strain multiple for, 490
specified dimensions, 12
spectral response acceleration parameter, 499, 500*f*–503*f*
spine wall, 426, 512
splitting tensile strength, 27, 460, 467, 472
Splitting Tensile Strength of Masonry Units. *See* ASTM C1006
spreadsheet
   allowable-stress interaction diagrams by, 311–315, 312*f*–315*f*, 316*t*
   for calculation of wind pressure, 386*t*–388*t*
   moment-axial force interaction diagrams by, 204–205, 205*f*, 206*t*, 311–315, 312*f*–315*f*, 316*t*
   strength interaction diagrams by, 201–205, 202*f*–205*f*, 206*t*, 217, 217*f*, 395*t*, 473–475, 473*f*, 474*t*, 516*f*
stability check, 250–251, 255, 257, 317
stack bond, 13, 13*f*, 49
Standard Building Code (SBC), 63
Standard Terminology of Masonry. *See* ASTM C1232
Standard Terminology of Mortar and Grout for Unit Masonry. *See* ASTM C1180
statically determinate cantilevers, 421, 435
statics, 281
steel
   in AAC manufacturing, 443
   area, 411
   columns, 302*f*
   controls, anchor bolts with, 334*t*
   failure, 333
   joists, 372, 372*f*
   percentage, 185
   reinforcing bars, 187, 187*t*, 288, 289*t*, 468, 469*t*
   reinforcing wire, 187, 187*t*, 288, 289*t*
   yield, 268–269
stiff mud process, 29
stiffness
   irregularity, 428, 513
   parameters related to, 247, 250, 254–255, 454, 457
   shearing, 344–346, 346*f*
   torsional, 370
   wall, 344–346, 346*f*, 349–350

strain. *See also* critical strain condition
  allowable-stress balanced conditions and, 312f–314f
  maximum useful value of, 183
  yield, 479, 490
strength checks, 141–142, 163
strength classes, AAC, 443, 443t
strength design. *See also* four-story building with clay masonry, strength design example of; one-story building with reinforced concrete masonry, strength design example of
  for AAC, 447–448
  allowable-stress design v., 195, 331–337, 332t–337t
  of anchor bolts, 168–177, 168f–171f, 173f–174f, 333–334, 334t
  approach of, 121, 127–129
  critical strain condition and, 210–211, 211f, 219–220, 219f
  moment-axial force interaction diagrams and, 197–205, 197f–198f, 201f–203f, 205f, 206t
  MSJC *Code* provisions for, 105–109, 106t–109t, 331
  of reinforced beams, 107, 108t, 183–191, 184f–186f, 187t, 188f–189f, 334–335, 335t
  of reinforced bearing walls, 108, 109t, 196–212, 197f, 201f–203f, 205f, 206t, 207f–208f, 211f, 336, 336f–337f
  of reinforced curtain walls, 107, 108t, 191–196, 191f, 193f–194f, 196f, 335
  of reinforced lintels, 107, 108t, 183–191, 184f–186f, 187t, 188f–189f, 334–335, 335t
  of reinforced masonry, 210–211, 211f, 219–220, 219f
  of reinforced shear walls, 107, 107t, 212–223, 213f–214f, 217f, 220f, 336–337, 337f
  of unreinforced bearing walls, 107, 107t, 127–128, 147–161, 148f, 150f, 150t, 153f, 156f, 158f, 160f, 332
  of unreinforced panel walls, 106, 106t, 133–147, 134f, 136t, 137f–139f, 141f, 143f–144f, 146f, 331–332, 332t
  of unreinforced shear walls, 108, 109t, 161–168, 161f–165f, 167f, 332–333, 333t, 463f–464f
strength interaction diagrams
  for AAC reinforced bearing wall design, 473–475, 473f, 474t
  background on, 197–199, 197f–198f
  for east wall of example low-rise building, 396, 396f
  by hand, 197–201, 201f
  for pilasters, 403–404, 404f, 405t
  plot of, 201, 201f, 205, 205f, 206t

strength interaction diagrams (*Cont.*):
  reinforced bearing walls and, 197–205, 197f–198f, 201f–203f, 205f, 206t
  by spreadsheet, 201–205, 202f–205f, 206t, 217, 217f, 395f, 473–475, 473f, 474t, 516f
  for transverse shear walls, 430, 430f, 432t–433t, 516, 516f
  for west wall of example low-rise building, 394, 394f
strength loading combinations, IBC, 103, 105
Strength Tests of Panels for Building Construction. *See* ASTM E72
strength-reduction factors, 135, 142, 176–177, 194
  AAC, 449, 451–452, 463
  MSJC, 106, 106t
  for shear, 463, 482
stress. *See also* allowable-stress design; compressive stress; tensile stress
  allowable-stress balanced conditions and, 312f–314f
  bearing, 249, 249f, 389, 389f, 402, 455f
  bending, 155, 251, 254–255, 457
  maximum tensile, 151, 154, 157–158, 391, 393, 400, 453, 456, 458, 461
  1/3 stress increase and, 104–105
  out-of-plane, 245
  yield, 336
stress block
  compressive, 183, 285f, 286, 516
  rectangular, 479–480, 490
stress-strain relation, 119, 281, 282f
stretcher units, 122
strip method
  AAC and, 452
  theoretical derivation of, 143–144, 238–239
structural function, classification by, 120
structural irregularities
  of four-story building with clay masonry, 428
  of three-story AAC shear-wall hotel, 513
  vertical, 428, 513
structural systems
  of four-story building with clay masonry, 420
  for gravity load, 389, 421, 497
  for lateral load, 389, 421, 497
  of one-story building with reinforced concrete masonry, 389
  of three-story AAC shear-wall hotel, 497
structures requiring little structural calculation, design of
  construction details for, 49–54, 51f–54f
  design steps for, 47–49
  reinforcement in, 49, 50f

strut
  compression, 483, 523
  diagonal, 463, 470, 482–483
  drag, 370
suction, 391, 399
surface texture, 37
symmetrical buildings, 349f

T

technical specialty organizations, 61, 113
tensile bond strength, 48, 120
  of clay masonry units, 35
  of concrete masonry units, 37
  of masonry assemblages, 38
  of mortar-cement mortar, 24–25
tensile breakout, 268, 270
tensile capacity
  of anchor bolts, 267–270
  nominal, 169–172
tensile force, 211, 221, 287f–288f, 308, 491
tensile strength. *See also* tensile bond strength
  of clay masonry units, 35
  of concrete masonry units, 37
  splitting, 27, 460, 467, 472
tensile stress. *See also* flexural tensile stress
  area, 269, 271
  extreme-fiber, 245
  maximum, 151, 154, 157–158, 391, 393, 400, 453, 456, 458, 461
  net, 246, 250–251, 254–256, 317, 392, 400
  in tensile reinforcement, 299
tension
  anchor bolts loaded in, 169–170, 169f–170f, 177, 266–270, 266f–267f, 269f, 273–274
  ASTM E519 for, 28, 38
  eccentric axial load and, 391
  fiber, 452
  reinforcement, 184–185, 210, 299
The Masonry Society (TMS), 61, 113, 447
thermal conductivity, 439
thickness, 11–12
thin-bed mortar, 457, 463, 468, 497
three-story AAC shear-wall hotel design example
  architectural constraints of, 495
  design criteria chosen for, 495–514, 496f, 500f–504f, 505t–506t, 509f, 509t–510t
  elevation of, 496f
  exterior walls of, 495, 517–519, 518f
  factored design lateral forces for, 512–513
  fire design of, 496
  floor diaphragm of, 519–523, 519f–523f
  materials specified for, 497

three-story AAC shear-wall hotel design example (*Cont.*):
  MCE for, 498–499, 500f–503f
  plan of, 496f
  seismic base shear for, 498–514, 500f–504f, 505t–506t, 509f, 509t–510t
  seismic ground motion values for, 498, 504
  structural irregularities of, 513
  structural systems of, 497
  transverse shear walls of, 495, 514–519, 514f, 516f, 518f
  water-penetration resistance in, 495
through-wall units, 420
ties
  adjustable, 41, 43f–44f
  pintle, 43f
  specification of, 55
  veneer, 43f
TMS. *See* The Masonry Society
topographic factor, 74, 81, 83, 381, 384
torsion, 514. *See also* plan torsion
torsional moment, 352–354, 352f–353f
torsional stiffness, 370
transition slenderness, 247, 250, 254–255, 454, 457
transverse lateral load, 519
transverse shear walls
  for earthquake loads, 428–430, 429f, 514–517, 514f, 516f
  of four-story building with clay masonry, 419, 421, 426, 428–435, 429f–430f, 432t–433t
  for gravity loads, 514–517, 514f, 516f
  in-plane flexural design of, 516–517, 516f
  moment-axial force interaction diagram for, 430, 430f, 432t–433t
  as statically determinate cantilevers, 421, 435
  strength interaction diagrams for, 430, 430f, 432t–433t, 516, 516f
  of three-story AAC hotel, 495, 514–519, 514f, 516f, 518f
tributary area, 389, 390f, 401, 402f, 407, 407f
tributary length, 370
tributary width, 161, 258, 258f, 394, 429
trough units, 122
truss mechanism, 483
truss model, 521, 522f
2500-year earthquake, 91, 499
2009 IBC. *See* International Building Code (IBC)
two-wythe unreinforced panel walls
  out-of-plane loads distributed to vertical and horizontal strips of, 145–147, 146f, 241–243, 241f

two-wythe unreinforced panel walls (*Cont.*):
    as two sets of horizontal and vertical crossing strips, 230*f*
    using hollow units, face-shell bedding only, 140–141, 141*f*, 236–237, 237*f*
Type K mortar, 19
Type M mortar, 19–20, 19*t*–22*t*, 22–23, 187, 231*t*, 290
Type N mortar, 19–20, 19*t*–22*t*, 22–23, 135, 137–138, 140–141, 231*t*, 232–236
Type O mortar, 19, 19*t*–22*t*, 23
Type S mortar, 19–20, 19*t*–22*t*, 22, 138, 140–141, 154, 158, 166, 187–188, 192, 215, 231*t*, 234–235, 237, 250, 254–255, 290, 380, 421

U

UBC. *See* Uniform Building Code
unbonded interface, 466
unfactored axial load, 485, 515
unfactored in-plane lateral loads, 484, 484*f*
unfactored moment diagrams, 157, 158*f*, 207, 208*f*
    due to eccentric axial load, 253, 253*f*, 318, 318*f*, 460, 460*f*, 476, 476*f*
    due to eccentric dead load, 392, 392*f*
    due to wind, 253, 253*f*, 318, 318*f*, 392, 392*f*, 460, 460*f*, 476, 476*f*
unfired masonry units, 10*t*
Uniform Building Code (UBC), 63, 126
unit strength method, 120, 187
United States. *See also* building codes, U.S.
    AAC in, 440, 444, 446*f*, 447–448, 494–495, 494*t*
    MCE for, 498–499, 500*f*–503*f*
    seismic design factors in, 494–495, 494*t*
unity equation
    reinforced bearing walls and, 317
    unreinforced bearing walls and, 245–246, 248, 250–251, 254, 256–257
unreinforced bearing walls
    AAC, 452–462, 453*f*, 456*f*, 459*f*–460*f*
    allowable-stress design of, 243–259, 243*f*–244*f*, 246*f*, 246*t*–247*t*, 249*f*, 252*f*–253*f*, 257*f*–258*f*, 332
    basic structural behavior of, 147–148, 148*f*, 243–244, 243*f*–244*f*
    with concentric axial load, 149–152, 150*f*, 150*t*, 246–248, 246*f*, 246*t*–247*t*, 452–455, 453*f*
    with eccentric axial load, 153–159, 153*f*, 156*f*, 158*f*, 248–252, 249*f*, 253, 253*f*, 318, 318*f*, 455*f*
    with eccentric axial load plus wind, 156–159, 156*f*, 158*f*, 318, 318*f*, 459–461, 459*f*–460*f*, 476, 476*f*
    gravity loads on, 243

unreinforced bearing walls (*Cont.*):
    with openings, 159–161, 160*f*, 257–258, 257*f*–258*f*, 462
    out-of-plane loads on, 332
    required details for, 177–178, 178*f*–179*f*, 274–275, 274*f*–277*f*
    strength design of, 107, 107*t*, 127–128, 147–161, 148*f*, 150*f*, 150*t*, 153*f*, 156*f*, 158*f*, 160*f*, 332
    unity equation and, 245–246, 248, 250–251, 254, 256–257
    wall-to-floor details and, 178, 179*f*, 275, 275*f*–276*f*
    wall-to-foundation connections and, 177, 178*f*, 274, 274*f*
    wall-to-roof details and, 178, 180*f*, 275, 276*f*
    wall-to-wall connections and, 178, 180*f*, 275, 277*f*
unreinforced masonry, classification of, 121, 124
unreinforced panel walls, 106, 106*t*, 110, 110*t*, 127–128. *See also* single-wythe unreinforced panel walls; two-wythe unreinforced panel walls
    AAC, 449–452, 450*f*–451*f*
    allowable-stress checks for, 237–238
    allowable-stress design of, 229–243, 230*f*, 231*t*, 232*f*, 234*f*–235*f*, 237*f*, 239*t*, 240*f*–241*f*, 331–332, 332*t*
    examples of use of, 133–135, 134*f*, 229–231, 230*f*
    flexure and, 135, 136*t*, 231–232, 231*t*, 449
    out-of-plane loads on, 144–147, 144*f*, 146*f*, 239–243, 240*f*–241*f*, 331–332, 332*t*
    shear capacity of, 142, 451
    strength checks for, 141–142
    strength design of, 106, 106*t*, 133–147, 134*f*, 136*t*, 137*f*–139*f*, 141*f*, 143*f*–144*f*, 146*f*, 331–332, 332*t*
unreinforced shear walls, 107, 107*t*, 110, 111*t*, 128
    AAC, 462–467, 463*f*–465*f*
    allowable-stress design of, 259–265, 259*f*–262*f*, 265*f*, 332–333, 333*t*
    basic structural behavior of, 161–162, 161*f*, 259–260, 259*f*
    design actions for, 463*f*
    gravity loads on, 162, 260, 462
    with openings, 167–168, 167*f*–168*f*, 264–265, 265*f*
    required details for, 177–178, 178*f*–179*f*, 274–275, 274*f*–277*f*
    shear capacity of, 260–261, 264, 467
    strength design of, 108, 109*t*, 161–168, 161*f*–165*f*, 167*f*, 332–333, 333*t*, 463*f*–464*f*
    wall-to-floor details and, 178, 179*f*, 275, 275*f*–276*f*
    wall-to-foundation connections and, 177, 178*f*, 274, 274*f*

unreinforced shear walls (*Cont.*):
    wall-to-roof details and, 178, 180*f*, 275, 276*f*
    wall-to-wall connections and, 178, 180*f*, 275, 277*f*
unsymmetrical buildings, 349*f*
uplift, 390, 404
U.S. *See* United States

V

vapor barriers, 11, 39, 45–46, 55
velocity pressure, 69, 71*t*, 78, 80–81, 80*t*, 85–86, 85*t*
    exposure coefficients, 381*t*, 384, 384*t*
    wind load and, 382, 385, 387–388
velocity-dependent site coefficient, 423, 505, 506*t*
veneer design, 122
veneer ties, 43*f*
vertical control joints, 401, 495
vertical geometric irregularity, 428, 514
vertical reinforcement, 4, 5*f*, 49, 124, 127, 303*f*, 373
vertical strips, 3–5, 4*f*
vertical structural irregularities, 428, 513
vertically distributed seismic base shear, 498–514, 500*f*–504*f*, 505*t*–506*t*, 509*f*, 509*t*–510*t*
vertically oriented anchor bolts, 168, 168*f*
vertically oriented expansion joints, 46*f*, 56
vitrification, 30

W

wall(s). *See also* bearing walls; curtain walls; in-plane walls; out-of-plane walls; panel walls; shear walls; wall-to-floor connections; wall-to-foundation connections; wall-to-roof details; wall-to-wall connections
    AAC, critical strain condition for, 479, 479*f*, 490, 490*f*
    axial capacity of, 244*f*
    barrier, 13–14, 14*f*, 378
    cavity, 14, 55
    composite brick-block, 56
    configuration of, 4–5, 5*f*
    design of, in one-story building with reinforced concrete masonry, 413–415, 414*f*
    design pressure on, 384–389, 384*t*, 386*t*–388*t*
    design shear on, 412, 413*t*
    drainage, 13–14, 14*f*, 48, 55
    east, of one-story building with reinforced concrete masonry, 393–399, 394*f*, 396*f*–397*f*, 398*t*, 399*f*, 406*f*, 409, 409*f*, 412–415, 413*t*, 414*f*
    elements, design pressure on, 384–389, 384*t*, 386*t*–388*t*
    exterior, 435, 495, 517–519, 518*f*
    fire, 303*f*
    floor slab connected with, 50, 51*f*

wall(s) (*Cont.*):
    foundation, 51*f*
    gravity plus out-of-plane loads of, 389–406, 389*f*–390*f*, 392*f*, 394*f*, 395*t*, 396*f*–397*f*, 398*t*, 399*f*–404*f*, 405*t*, 406*f*
    intermediate, 490
    live load on, 66–67
    masonry, 13–14, 14*f*, 122–124, 125*f*, 143*f*
    north, of one-story building with reinforced concrete masonry, 399–404, 400*f*–404*f*, 405*t*, 409, 415
    perforated, 341, 365
    perimeter, 426, 512
    roof connected with, 50, 51*f*
    section properties for, 143*t*, 238, 239*t*
    sections at lintels, 54, 54*f*
    Segment A, of one-story building with reinforced concrete masonry, 414–415
    Segment B, of one-story building with reinforced concrete masonry, 396–397, 397*f*, 398*t*
    shears, 347, 356–357, 358*f*
    south, of one-story building with reinforced concrete masonry, 399–404, 400*f*–404*f*, 405*t*, 409, 415
    spine, 426, 512
    stiffness of, 344–346, 346*f*, 349–350
    systems of, 48
    types of, 13–14, 14*f*, 55
    as vertically oriented strips, 3, 4*f*
    weight of, 512
    west, of one-story building with reinforced concrete masonry, 389–394, 389*f*–390*f*, 392*f*, 394*f*, 395*t*, 409, 412–413, 415
wall buildings
    behavior and design of, 166–167, 264
    frame buildings *v.*, 264
wall-to-floor connections
    bearing walls and, 224*f*, 226
    cambered planks in, 327
    reinforced bearing walls and, 224*f*, 226, 325*f*–326*f*, 327
    reinforced shear walls and, 224*f*, 226, 325*f*–326*f*, 327
    slab in, 50, 51*f*
    unreinforced bearing walls and, 178, 179*f*, 275, 275*f*–276*f*
    unreinforced shear walls and, 178, 179*f*, 275, 275*f*–276*f*
wall-to-foundation connections
    clay masonry and, 177, 178*f*, 223*f*, 226, 274, 274*f*, 325*f*, 327

wall-to-foundation connections (*Cont.*):
    reinforced bearing walls and, 223f, 226, 325f, 327
    reinforced shear walls and, 223f, 226, 325f, 327
    unreinforced bearing walls and, 177, 178f, 274, 274f
    unreinforced shear walls and, 177, 178f, 274, 274f
wall-to-roof details, 50, 51f
    in one-story building with reinforced concrete masonry, 415
    reinforced bearing walls and, 225f, 226, 326f, 327
    reinforced shear walls and, 225f, 226, 326f, 327
    roof diaphragms and, 415
    unreinforced bearing walls and, 178, 180f, 275, 276f
    unreinforced shear walls and, 178, 180f, 275, 276f
wall-to-wall connections
    reinforced bearing walls and, 225f, 226, 327, 327f
    reinforced shear walls and, 225f, 226, 327, 327f
    unreinforced bearing walls and, 178, 180f, 275, 277f
    unreinforced shear walls and, 178, 180f, 275, 277f
water permeability, 28, 38
Water Permeance of Masonry. *See* ASTM E514
water-penetration resistance
    in four-story building with clay masonry, 420
    increased, 54–56
    level of, 48
    in one-story building with reinforced concrete masonry, 378
    single-wythe barrier walls for, 13, 378
    in three-story AAC shear-wall hotel, 495
water-repellent coatings, 45
weak story, 428
weathering index, 33, 33f
weathering regions, 33–34
web-shear cracking, 463, 465, 470, 482
weepholes, 14, 55
welded wire reinforcement, 39, 41f
wet clay units, 49
wind
    basic speeds, 68, 69f, 70t, 79–80, 83, 381, 384
    design pressure on wall elements due to, 384–389, 384t, 386t–388t

wind (*Cont.*):
    directionality factor, 68, 70t, 79–80, 83, 381
    long-span joist reactions due to, 380–384, 381t, 383t
    MWFRS and, 75, 76f, 79–83, 80f, 80t, 84t, 380–384, 381t, 383t
    out-of-plane, 399–401, 400f, 435, 517–519, 518f
    unfactored moment diagrams due to, 253, 253f, 318, 318f, 392, 392f, 460, 460f, 476, 476f
    unreinforced bearing walls with eccentric axial load and, 156–159, 156f, 158f, 318, 318f, 459–461, 459f–460f, 476, 476f
    uplift, 390, 404
wind load
    base shear and, 380–384, 381t, 383t
    on components and cladding, 83–87, 85t–87t
    critical locations and, 159
    design procedure for, 68–75, 69f, 70t–71t, 72f–73f, 75f
    IBC and, 67–87, 69f, 70t–71t, 72f–73f, 75f–77f, 80f, 80t, 84t–87t
    load factor for, 331
    on main wind force-resisting system, 79–83, 80f, 80t, 84t
    MWFRS and, 380–384, 381t, 383t
    nonload-bearing masonry and, 238
    for one-story building with reinforced concrete masonry, 380–389, 380f, 381t, 383t–384t, 386t–388t
    out-of-plane, 399–401, 400f, 435, 495, 517–519, 518f
    transferred to roof diaphragm, 410, 410f–411f
    velocity pressure and, 382, 385, 387–388
wind pressure
    on roof, 387–389, 388t
    spreadsheet for calculation of, 386t–388t
    suction and, 391, 399
    uniformly distributed, 341, 363
windward side, 78–79, 81, 84t, 86, 86t, 382–383, 386t
wooden joists, 373, 373f
workability, 23–25

Y

yield
    steel, 268–269
    strain, 479, 490
    stress, 336